Highways

Volume
Traffic Planning and Engineering

This textbook is dedicated once more to

Nuala

who helped, as always,
with 'heart an' hand'

Highways

Volume 1
Traffic Planning and Engineering

Third edition

C A O'Flaherty

BE (NUI), MS, PhD (Iowa State), C Eng, FICE, FCIT, FIE(Aust), FIHT, MIEI
Director and Principal, Tasmanian State Institute of Technology
Formerly Professor of Transport Engineering and Director of the Institute for Transport Studies, University of Leeds

Edward Arnold

© CA O'Flaherty 1986

First published in Great Britain 1967 by
Edward Arnold (Publishers) Ltd, 41 Bedford Square, London WC1B 3DQ

Edward Arnold, 3 East Read Street, Baltimore, Maryland 21201, USA

Edward Arnold (Australia) Pty Ltd, 80 Waverley Road, Caulfield East, Victoria
3145, Australia

Second edition 1974
Third edition 1986

British Library Cataloguing in Publication Data

O'Flaherty, C.A.
 Highways.——3rd ed.
 Vol. 1: Traffic planning and engineering
 1. Highway engineering
 I. Title
 625.7   TE145

ISBN 0-7131-3526-3

Text set in 10/11pt Times Lasercomp
Printed and bound in Great Britain by Butler & Tanner Ltd, Frome and London

Preface

The decision taken during the preparation of the second edition of *Highways*, to divide the book into two volumes, appears to have been well accepted and this has been carried forward into the third edition. Thus Volume 1, now more appropriately entitled *Traffic Planning and Engineering*, covers the historical and administrative aspects of highways together with those other matters which are of particular interest to the traffic planner and engineer. Volume 2: *Highway Engineering* continues to deal with the 'hard' side of highways, i.e. their physical location and structural design as well as the materials used in their construction and maintenance.

The intention underlying the preparation of this third edition has not changed from that described in the previous editions—to provide a basic textbook for the young engineer about to centre his or her career on highways. It is hoped that this new edition may also be of value to those about to enter the town planning profession because of the emphasis placed upon traffic aspects in Volume 1.

There are a number of significant changes in this third edition of Volume 1.

First and most important, the text has been completely revised and updated, and every chapter has been extensively modified. The published information and research on this very dynamic area of study has expanded tremendously over the past decade, and hence Volume 1 has been lengthened in order to attempt to cover the various technical topics in a reasonable way.

Chapter 1: *The road in perspective* has been expanded to give an overview of major road transport happenings and policies of recent years, with the aim of pointing towards developments for the future. Special attention has been paid to the 'oil crisis' in this overview with the objective of placing it in its proper perspective. Chapter 2: *Highway finance and administration* has been updated to the extent practicable, i.e. public administration is an ever-evolving process and changes are continually taking place.

As well as having a major updating, Chapter 3: *Highway and traffic planning* has been expanded to introduce discussion regarding basic traffic strategies and options in urban areas. Particular attention has been paid in this chapter also to the public participation process whereby the public is encouraged to take a more active role in highway planning decisions.

Greater attention has been paid in this edition to the design of parking facilities in Chapter 4: *Parking*. The next chapter was entitled *The economics of road improvements* in the second edition; in this third edition, it is more aptly called Chapter 5: *The assessment of road improvements*. As well as presenting data regarding vehicle operating and accident cost analyses, the tenor of this chapter has been varied to allow for the current trend towards the inclusion of non-economic considerations in the appraisal of major highway proposals.

Whilst the outline of Chapter 6: *Geometric design of highways* has not changed, this chapter has had a major rewrite and expansion to allow for the major changes in highway capacity and design criteria which have been introduced in Britain over the past decade. Also included in this chapter (following on from a discussion in Chapter 3) are details of traffic noise design in relation to new highways.

Chapter 7: *Road safety and traffic management* has also been completely rewritten and expanded to allow for the very great advances in knowledge that have taken place during the 1970s. The road safety component of this chapter has received greater attention than previously, and new sections have been included on cycle paths and priority for buses. The last chapter Chapter 8: *Highway lighting* has been updated.

Particular attention has also been paid in this third edition to the referencing system. As in the two previous editions, authoritative references have been quoted where it has been appropriate to authenticate certain statements and to provide the source of facts, tables and figures. In this edition, however, an attempt has been made to point the reader towards significant references as an aid to further study; on the whole, these references are fairly recent, although 'classics' are quoted in many instances. It is important for the young engineer to appreciate that a tremendous amount of research is currently being carried out in traffic planning and engineering with the result that many previously held tenets are now being questioned and discarded; even at the undergraduate level, it is necessary for the student to read the professional journals and governmental publications in order to keep abreast of these changes and to find out 'the reason why'.

Whilst this book is primarily aimed at students in Britain, the writer is conscious that this edition is being written from afar. Every attempt has been made to minimize the likelihood of error, but mistakes may inadvertently have crept in, and so the writer will welcome any advice on these. Meanwhile the reader is reminded that the factual material in the text is primarily given for demonstration purposes, and is urged to seek out the original material and to consult the most up-to-date 'official' versions of recommended practices, standards, etc., when engaged in actual planning, design and fieldwork.

Coleman A O'Flaherty

Acknowledgements

I am indebted to the many organizations and journals which allowed me to reproduce diagrams and tables from their publications. The number in the title of each table and figure indicates the reference at the end of each chapter where the source is to be found. It should be noted that material quoted from British government publications is Crown copyright and reproduced by permission of the Controller of Her Majesty's Stationery Office.

I should also like to thank again my former colleagues at the University of Leeds, most particularly G Leake and J Cabrerra, for their encouragement during the preparation of the first and second editions, when the tone for this third edition was set.

Not only I but the publishers are grateful to Mrs G McCulloch for her excellent typing and interpretation of the handwritten word. This text has been written from afar, and this would not have been possible without the help of the many library staff at my Institute who sought and obtained numerous references on my behalf. Above all, however, I am indebted to Miss J Jensen, Librarian-in-Charge, at the State Offices Library, Hobart, for her very considerable, unassuming, professional help in obtaining the reference material on which this third edition is mainly based.

Last, but far from being least, I thank my wife, Nuala, who has helped me immeasurably in this work. Not only has she participated actively in the preparation of this text, but her patience and forebearance have made it possible.

Coleman A O'Flaherty

Contents

x *Contents*

1
The road in perspective

Roads can be considered as much a cause as a consequence of civilization; they both precede and follow it. History testifies that the provision of roadways is necessary to draw a country out of a state of barbarism, but that civilization is not attained until communication between neighbours is made so easy that the local differences which breed narrowness and bigotry are minimized. As civilization is advanced and prosperity increases, there is an inevitable demand for better and speedier communication facilities, especially for roads. Indeed, it can truly be said that the prosperity of a nation is bound up with the state of its roads, and that the roads act as a palimpsest of a nation's history.

It used to be fashionable to heap abuse upon the roads of Britain because of their relative inability to absorb the avalanche of motor vehicles descending upon them. That they are not capable of handling at all times the demands placed upon them cannot be denied; what is not always appreciated, however, is that the roads of Britain were never purpose-built to meet modern communication needs but that, instead, the motor vehicle was thrust upon a road system already in existence and possessed of a deep association with the history of the country. For this reason, therefore, it is desirable to present a picture of how the existing roads evolved; by so doing it is possible to explain the existence of certain obstacles in the way of adapting the inherited roads to meet modern requirements. Included in this historical perspective is some brief account of particular financial and administrative problems affecting the roads since they, as much as the technical limitations of the times, have greatly influenced the development of the roads in Britain. Finally, this chapter closes with an overview of highway and traffic development over the past two decades—the 'motorway age'—and discusses some current policy trends, and their likely implications for the late 1980s and the 1990s.

Ancient roads

The birth of the road as a formal entity is lost in the mists of antiquity. Most certainly, however, the trails deliberately chosen and travelled by ancient man and his pack animals were the forerunners of the road as we know it today. As man developed and his desire for communication increased, so inevitably trails became pathways and pathways developed to become recognized travelways.

The invention of the wheel over 5000 years ago made necessary the construction of special hard surfacings capable of carrying concentrated and greater loads than hitherfore. That this was realized by the ancients is evidenced by the multiplicity of sometimes sophisticated but more often crude roadways that have been discovered by archaeologists.

Non-British roadways

Perhaps the most monumental of the earliest recorded 'roadways' was the stone-paved sloping causeway constructed some 5000 years ago at the direction of the Egyptian king Cheops to facilitate the conveyance of the huge limestone blocks used in the building of the Great Pyramid. This causeway consisted of an embankment having a bottom layer of sandy soil beneath a layer of broken limestone, on top of which were placed layers of tiles or bricks; the various constructional layers were contained within two walls of quarried stone which were strengthened by placing sloping embankments of sand against either side. This causeway took ten years to complete and the finished facility was 'five furlongs long, ten fathoms wide and of height, at the highest part, eight fathoms'[1]. Indeed, it can be said that this construction was in no way inferior to the pyramid itself.

What has been described as the oldest road in Europe, and which might also be called the first dual carriageway, was the roadway built in Crete about 2000 BC to connect Cortina on the south-east of the island to Knossos on the north. Archaeologists have described the method of construction as follows. First of all, the earth was excavated to a depth of about 200 mm over a width of about 4.5 m, and the subsoil levelled and compacted. Into this excavation was placed a watertight foundation course of sandstone embedded in a clay–gypsum mortar about 200 mm deep. On top of this, a 50 mm layer of clay was laid to form a bed for the pavement. A 300 mm wide pavement composed of 50 mm thick basalt slabs formed the crown of the road and this was flanked by pavements of rough limestone blocks grouted with gypsum mortar. It is believed that the central pavement was for the use of pedestrians while light-wheeled traffic ran on the sides.

A third road of constructional interest is the royal processional route in Babylon which was probably built about 620 BC. This road was only about 1220 m long but it is notable for the materials utilized in its making. The road was paved to a width of 10–20 m with stone slabs placed on a foundation of three or more layers of brick jointed with bitumen. The middle part of the surface was paved with limestone slabs, while the sides were deemed to be adequately surfaced with red and white breccia slabs set in bitumen.

These few examples illustrate that road construction had achieved a certain sophistication many thousands of years ago.

That the *purpose* of a road was also appreciated by at least some of the ancients is illustrated by the character of the route known as the Persian Royal Road. This trade-cum-military road is believed to have run from Smyrna eastward across Turkey and then southward through Persia to the Persian Gulf, for a total length of 2400 km. The territory through which it passed was part of the Persian empire which was then (520–485 BC) ruled by King Darius I. This monarch was determined to keep his empire under control and decided to have the route laid out so that he could move his troops quickly. To facilitate movement, he provided resting places at intervals of one day's walking along the way. He further arranged for it to follow the shortest practical route, even to the extent of bypassing some of the largest towns in the empire; he did, however, provide service roads into these towns. His purpose in bypassing the towns was to ensure swift movement unhindered by congestion and other obstacles which were even then prevalent in the built-up areas.

Another who appreciated the importance of the road was the Indian ruler Chandragupta who, in the period 322–298 BC, constructed a 2400 km road across the subcontinent. This monarch set up a special ministry to organize and carry out maintenance on the route, provide milestones and resthouses, and operate the ferry system at river crossings.

British trackways

As far as is known, there were no formal roads constructed in Britain prior to AD 43. Certainly the driftways used by the early inhabitants of this country cannot be dignified as roadways, being nothing more really than the customary paths marked out by the frequent passing of man.

The oldest of the British pathways are located on the downs of south and south-east England. These were not haphazard tracks wandering aimlessly about the country, but had instead a very definite sense and purpose. The longest and most definite of these are known as Ridgeways because of their location just below the crests of hills. These tracks, which came into existence about 2800 BC in order to join centres of occupation, only descended into the valleys when it became necessary to cross rivers at fordable locations; other than this, they stuck to the hill crests where the underbrush was less dense and there was more freedom of movement.

As time progressed, the Ridgeways developed almost imperceptibly into what are now known as Trackways. These included pathways which ran on the lower slopes; they were sufficiently above the bottoms of valleys to ensure good drainage but yet were low enough to obviate unnecessary climbing. The Trackways were primarily trade routes used in the export of minerals, and along them settlements and shrines developed; perhaps the most famous of these is the great sun temple at Stonehenge which was constructed about the beginning of the Bronze Age.

Roman roads

The Romans are generally given credit for being the first real highway designers. Roads played a vital role in the maintenance of the Roman empire; in fact it can be said that the empire was held together by a well-designed system of roadways extending from Rome into Italy, Gaul, Britain, Illyria, Thrace, Asia Minor, Pontus, the East, Egypt and Africa.

These regions were divided into 113 provinces traversed by 372 great roads totalling 52 964 Roman miles[2]. This mileage is equivalent to about 78 000 km.

There is much discussion as to whether or not the Roman roads should be classified as military facilities. They are usually regarded in this way as their construction was motivated by military expediency, it being necessary for occupying military forces to be able to travel quickly to any point in the province threatened by attack. They were non-military roads, however, in that their main purpose was not for conquest but to aid administration and quell rebellion *after* a country had been occupied.

The great Roman road system was based on 29 major roads radiating from Rome to the outermost fringes of the empire. The first and perhaps the greatest of these roads was the Via Appia which linked Rome with Brundisium in the heel of Italy. This 660 km road, which was started in 312 BC and reached Brundisium in 244 BC, provided improved communication between Rome and North Africa and the East. This roadway—sections of which can still be seen today—was usually constructed at least 4.25 m wide; this enabled two chariots to pass each other with ease and legions to march six abreast. It was common practice to reduce the gradient by cutting tunnels and one such tunnel on the Via Appia was about 0.75 km long.

Two geometrical features of the Roman roads are of particular interest. The first of these refers to the habit of constructing roads well above ground level, i.e. nearly all of the great Roman roads were built on embankments, the major ones averaging some 2 m in height[3]. While the raising of the way had the engineering byproduct of serving to keep the surface dry, its main purpose was to give a commanding view of the surrounding countryside so that troops passing along were less liable to surprise attack. From this safety measure has arisen the modern term for a road, that of *Highway*. The second feature of the Roman road was its directness. Thus, for instance, Britain's Watling Street appears on the map as a gradual curve, whereas in reality it was composed of nine separate straight lines connecting points along the way. The adoption of the straight line by the Romans was primarily motivated by the fact that a straight road obviously provided the most direct communication in hostile districts. Notwithstanding this, deviations from the straight line connecting two adjacent stations were common. For instance, if an established trackway was available, the Romans tolerated a great many curves and only straightened those which were too winding to fit their scheme. In hilly regions also, the Romans, of necessity, gave up the straight line.

Roman roads in Britain

Following the occupation of England in 55 BC, the Romans set about constructing a complete road system radiating out from London—chosen as their capital—and linking the main centres of population. This system is undoubtedly the greatest initiative in the story of early British road

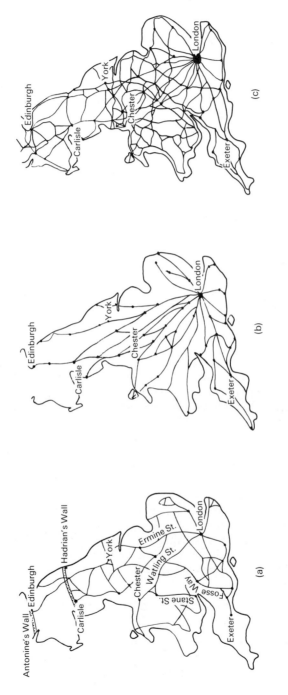

Fig. 1.1 Historic road systems in Britain: (a) some of the more important Roman roads, (b) chief directional roads, 1771, and (c) pre-motorway trunk road system

communications. As is illustrated in Fig. 1.1, it originated all that followed; indeed, it has been said that there was no real development, no essentially new departure, between the planning of the Roman military scheme and the coming of the railway[4].

In the space of 150 years, the Romans constructed some 5000 km of major road in Britain, extending well into Wales and far north to beyond Hadrian's Wall. These roads can be classified according to their method of construction as follows: (a) viae terrenae, which were made of levelled earth, (b) viae glareatae, which had a gravelled surface, and (c) viae munitae—the highest type of Roman road—which were paved with rectangular or polygonal stone blocks. The type of construction was also varied to suit the needs of a particular location, e.g. construction thicknesses of 1 m were common on poor ground and, when swampy soils were met, it was not uncommon to drive wooden piles and then to construct the road on these piles (see Fig. 1.2).

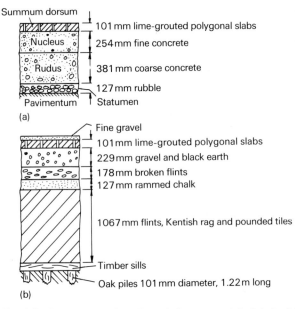

Summmum dorsum
101 mm lime-grouted polygonal slabs
Nucleus — 254 mm fine concrete
Rudus — 381 mm coarse concrete
127 mm rubble
Pavimentum   Statumen
(a)

Fine gravel
101 mm lime-grouted polygonal slabs
229 mm gravel and black earth
178 mm broken flints
127 mm rammed chalk

1067 mm flints, Kentish rag and pounded tiles

Timber sills
Oak piles 101 mm diameter, 1.22 m long
(b)

Fig. 1.2   Pavement sections through Roman roads in Britain: (a) Fosse Way, near Radstock, and (b) through Medway Valley, near Rochester

Post-Roman period to the 18th century

The withdrawal of the Roman legions from Britain in AD 407 not only anticipated the death of the Roman empire but it also preceded the complete breakdown of the only organized road system in Europe. Only once or twice were improvements attempted, notably by Charlemagne who made strenuous efforts to build and maintain good local roads. After his death in AD 812, the European Roman road system deteriorated still further.

Roads prior to the Middle Ages

It does not appear that following the departure of the legions a single new road was deliberately planned and constructed in Britain until the advent of the 17th century. Although the Roman roads were the main highways of internal communication for a very long period, with time, however, they began to decay and disintegrate as they were exposed to the wear and tear of natural and artificial agencies. Their decay was not due simply to carelessness and ignorance of maintenance practices, but in reality it had its roots in deep-seated political causes[5]. During the Dark Ages, the country was split into a number of small kingdoms whose rulers preferred the political ideal of isolation rather than coordination. Their needs and desires were parochial rather than national and thus they exerted little or no effort to preserve the through roads—except as it suited their internal needs—and so these soon fell into disrepair.

Inevitably, the conditions of the roads became deplorable. New roads simply consisted of tracks worn haphazardly according to need. New tracks were made about every form of obstruction, care being usually taken to avoid private property and cultivated land. When sections of an existing road became untraversable, they were simply abandoned and new trackways created around them. These practices largely account for the winding and indeed tortuous character of many present-day British roads and lanes.

Roads during the Middle Ages

Throughout the Middle Ages, the main through roads all over Britain were nothing more than miry tracks full of ruts and holes, and the rivers and the seas were relied upon as the main arteries of trade. As a result, it was practically impossible to travel long distances in the winter, no matter how urgent the reason. The difficulties encountered in transporting commodities during inclement weather were such that a town's preparations for the winter often resembled preparations for a long siege.

Fairs were a most important feature of medieval life and the biggest fairs were always held in the autumn and spring so that towns could reprovision. The potholes and mud on the roads were not the only obstruction to travel to the fairs during these times. While they lasted it was customary to post specially mounted patrols along the main roads leading to the fairs so that the traveller could be provided with some protection against the terrors of the robber and the highwayman. Historic evidence of the need for these guards is given in the Statute of Winchester which, in 1285, was the first statute to attempt to deal with roads. This statute ordained 'That highways leading from one market town to another shall be enlarged where as bushes, woods, or dykes be, so that there be neither dyke nor bush whereby a man may lurk to do hurt within

200 feet [61 m] of the one side and 200 feet [61 m] of the other side of the way'. Unfortunately, there is ample evidence to testify that this statute was more noted in its breach than in its observance.

In general, the care and maintenance applied to roads during medieval times were merely a matter of chance. The only systematic approach to road maintenance came about at the instigation of the religious authorities. In keeping with the times, road repairs were considered as pious and meritorious work of the same sort as visiting the sick or caring for the poor; road repairing was regarded as a true charity for the unfortunate travellers[6]. To encourage the faithful to take part in constructing or repairing roads and bridges, the religious authorities often granted indulgences to those who performed such public services.

British roads were in a state of continual decline as the medieval times came to a close. The suppression of the monasteries by Henry VIII completed the road decay by removing the most important class of road-maintainers; unfortunately, the new owners of the ecclesiastical estates were not inclined to accept the road obligations imposed by custom on the original landowner.

Coincident with the decline of the religious orders, an agricultural revolution took place which substituted pasture for arable land all over Britain. This had the additional effect of lessening the need to move produce by cart and resulted in the generally held consideration that there was less need to take care of the roads.

These reasons, taken in conjunction with the long and expensive wars to which the country was subjected, resulted in the roads of Britain rapidly reaching a very low state indeed after the turn of the 16th century.

Roads between the 16th and 18th centuries

By the middle of the 16th century, the road conditions were so bad that in 1555 Parliament was forced to pass an Act 'For amending the Highways being now very noisom and tedious to travel and dangerous to all Passengers and Carriages'. This was the first Act which provided by statute for the maintenance of highways. It required the individual parishes to be responsible for the maintenance of their own roads, and laid down rules by which they had to supply all the materials and labour needed for this purpose. In addition, each parish had to provide from amongst its own inhabitants an official to superintend, gratuitously, this road maintenance; these officials were usually known as Surveyors of Highways. The lack of esteem in which the office of Surveyor was held is reflected in the fact that the official was rarely either a person of high social position or a person with technical skills. Invariably, the post was held by a local farmer or shopkeeper who was both incompetent and unpopular.

At the same time as road conditions were deteriorating, the opportunities for travel were increasing. About the middle of the 16th century, the first wheeled coaches appeared in Europe. Until that time, most travellers, even the Courts on their royal progresses, rode on horseback or travelled on foot. The first coach to be seen in Britain appeared in London in 1555, having been specially constructed for the Earl of Rutland. Long-distance passenger coach services were commenced from London in 1605, and hackney coaches were plying the streets of the capital city in 1634.

It is ironic to note that even during these times the traffic problem in London was considered to be chaotic. In 1635, the situation was considered to be so undesirable that a Royal Proclamation was published which stated 'That the great number of hackney coaches of late time seen and kept in London, Westminster and their suburbs, and the general and promiscuous use of coaches there, were not only a great disturbance to his Majesty, his dearest Consort the Queen, the Nobility, and others of Place and Degree, in their passage through the streets; but the streets themselves were so pestered, and the pavements so broken up, that the common passage is thereby hindered and more dangerous; and the prices of hay and provender, and other provisions of stable, thereby made exceedingly dear. Wherefore We expressly command and forbid, that no hackney or hired coach be used or suffered in London or Westminster or the suburbs thereof. And also, that no person shall go in a coach in the said streets, except the owner of the coach shall constantly keep up four able horses for our service, when required.'.

Two features of this 'environmental' proclamation make it notable. The first of these is the fact that little attention was paid to it by the populace, illustrating that even then the people accepted as their right that they should have usage of the roads. More significant, perhaps, is the fact that it marked the onset of the concept that if the roads were not capable of handling the traffic, then the traffic should be reformed to suit the roads.

As time wore on, the official attitude became that traffic of all types was a nuisance that had to be suppressed if possible and, if not, to be allowed to exist only in accordance with the most rigorous regulations. Thus a 1621 Act forbade the use of four-wheeled wagons and the carriage of more than one tonne of goods because vehicles with 'excessive burdens so galled the highways and the very foundations of bridges, that they were public nuisances'[3]. This type of legislation continued during the 17th and 18th centuries, becoming increasingly detailed and increasingly incomprehensible.

Roads and road transport into the 20th century

In spite of the legislative attempts to curb its growth, the 17th century saw a steady increase in the amount of wheeled vehicular traffic. With the onset of the 18th century, foreign trade became more important to the steadily developing manufacturing industries in Britain, and soon long trains of carts were laboriously dragging coal from the mines to the ironworks, glassworks and potteries, and manufactured goods to the ports and harbours. The state of at least part of the road system at this time is probably best described by this rather graphic description of London's

Mile End Road in 1756 which 'resembled a stagnant lake of deep mud from Whitechapel to Stratford, with some deep and dangerous sloughs; in many places it was hard work for the horses to go faster than a foot pace, on level ground, with a light four-wheeled post chaise'[7].

Acceptance of the need for good roads

The first half of the 18th century saw the occurrence of two events which signified the growing acceptance of the need for competent road-makers and a viable road system. These events were the formal construction of 400 km of roadway in Scotland by General Wade, and the formation of the Turnpike Trusts in England.

Military roads

The most famous of the British roads of the early part of the 18th century were those constructed by General George Wade in the southern highlands of Scotland. These roads, construction of which began in 1726, were military roads in the same sense as the Roman roads were military roads; they were constructed to ensure that the near success of the Insurrection of 1715 was not repeated. Thus, in the course of twelve years, General Wade built no less than 400 km of road and forty bridges[8].

General Wade's roads were planned to make the homes of the clans easily accessible to the forts and garrison stations located at strategic positions throughout the highlands. In addition, they were intended to have the pacific purpose of improving the general conditions in the highlands, which were at that time deplorably backward. Unlike the Roman military roads, however, General Wade's roads were mainly excavated tracks between 3.75 and 5.00 m wide, bedded with stones and surfaced with gravel; they often followed direct lines over the hills with little regard for the steepness of grade and were poorly drained.

Following Wade's departure from Scotland, his work was taken over by Major William Caulfeild who built roads and bridges which were often of better quality than those of his predecessor. By the time of Caulfeild's death in 1767, 1380 km of road had been completed with another 225 km under construction. Unfortunately, little maintenance was applied to the roads and, following the withdrawal of the military labour used for this purpose in 1790, these roads fell rapidly into disrepair. As a result, communications in most of the highlands at the end of the 18th century were little improved over those at its beginning, but at least the foundations of a transport system had been laid in Scotland.

The Turnpike Trusts

At the turn of the 18th century, the methods employed in making and maintaining roads in England were very primitive. The usual procedure was to dig loose earth from the ditches and throw it on the roadway in the hope that the traffic would beat it into a hard compact surface. However, these surfaces were not capable of withstanding the stresses imposed on them by the wheeled traffic, and they were soon churned into bands of mud with the onset of wet weather.

Confronted by the abominable state of the highways, a new principle of road maintenance became accepted; this was that the road users should pay for the upkeep of the highways. In 1706, there was passed in Parliament the first of a succession of statutes which created special bodies known as Turnpike Trustees. The number of these bodies came eventually to over 1100, administering some 36 800 km of road, constructed at the cost of an accumulated debt of £7 million, and raising and spending an annual revenue of more than £1.5 million[7]. These Trusts, established by literally thousands of separate (albeit similar) Acts of Parliament from 1706 onwards, were each empowered to construct and maintain a specified length of an important roadway and to levy tolls upon certain kinds of traffic. The powers were given only for a limited term of years, usually twenty-one, since it was the intention to remove the toll-gates as soon as the roads were in sufficiently good shape to do this. However, since this happy state never came to pass, every Trust eventually applied for a new Act extending its existence and so the Trustees became virtually permanent Local Authorities for roads.

Whatever the faults of the Turnpike Trusts—and they were indeed very many—there can be no doubt that they marked a most important stage in the development of the road system in Britain. First of all, the system made it possible for people to choose to travel, whereas before they only travelled when absolutely necessary. More important, they established the principle that the road user should pay a share of the road costs. Most important of all, however, is the fact that the advent of the turnpikes brought about and made definite what is in effect the present-day major road network.

If during the 18th century anybody had bothered to make a turnpike map of England, it would have shown, not a system of radiating arteries of communication, but scattered instances of turnpike administration that were unconnected with each other. At first, these sections would have appeared as mere dots on the map, but with time it would be seen that the dots were gradually increasing in number and size so that, eventually, they formed continuous lines. In 1794, it was stated that they 'so multiplied and extended as to form almost a universal plan of communication throughout the kingdom'. It is this network that finally determined the 20th century's network of roads. That this is so can be seen from a comparison of the road system described by Patterson in 1771 and the present system; these are illustrated in Fig. 1.1. The influence of the Roman system is also apparent in this figure.

The Turnpike Trusts were important for another reason—they emphasized the need for skilled road-makers in the country.

Developments in road construction and administration

Although towards the end of the 18th century considerable emphasis was placed upon the need for improved communications in the country, little engineering skill was applied to the construction of roads. The idea that road-making needed any special knowledge or education was incompatible with the professional thinking of that time, with the unhappy result that the application of engineering principles to road-making was entirely neglected. The 'science' of road construction as practised in Britain in the 18th century was simply to heap more soil on top of the existing mud and hope that traffic would compact it into a hard surface. Indeed, so little was known about the repairing of roads that sometimes when ruts were being filled the sides of the travelled way were made higher than the centre. The concept of providing drainage seems never to have been considered in practice and, as a result, the ruts made by wheeled vehicles were continually filled with water. When it is considered that the ruts on certain turnpikes were over 1 m deep, the need for stating the importance of good drainage seems rather superfluous.

By the turn of the 19th century, however, the value of firm roads with solid surfaces had become better accepted and, as a result, it became standard practice throughout British cities to pave main streets with stone setts, hand-laid and supported on a granular or cemented bed. The noise problem resulting from horse-drawn traffic moving over the cobbled streets was such that a significant number of quiet streets were also constructed of wooden blocks, from the 1830s onwards. Asphalt, which was first utilized as a road surface in Paris in the 1850s, was employed in Britain about 1870; this was considered a major advance in highway engineering because of the ease with which the surfacing could be cleansed (a most important environmental consideration in the era of the horse-drawn vehicle).

This nineteenth century acceptance of the need for much improved roads can be attributed at least indirectly, if not directly, to a number of 18th century personalities who contributed significantly to the development of the road and the science of road-making.

Pierre Trésaguet

The father of modern highway engineering can truly be said to be Pierre Trésaguet, the Inspector General of Roads in France between 1775 and 1785. He was the first engineer to appreciate fully the importance of the moisture content of the subsoil and its effect upon the stability of the road foundation. Hence he made provision for drainage and, in contrast to the Romans, designed a relatively light cross-section on the principle that a good subsoil should be able to support the load laid upon it.

Trésaguet not only emphasized the importance of drainage but he also advocated a diminished road thickness to meet the varied requirements of traffic (see Fig. 1.3(a)). He enunciated the necessity for continuous organized maintenance instead of intermittent repair if the roads were to be kept usable at all times, and so he divided the roads between villages into sections so that an entire road could be covered by maintenance men living nearby.

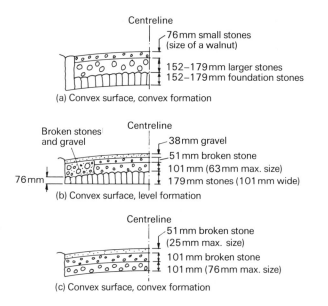

(a) Convex surface, convex formation

- 76 mm small stones (size of a walnut)
- 152–179 mm larger stones
- 152–179 mm foundation stones

Broken stones and gravel

76 mm

(b) Convex surface, level formation

- 38 mm gravel
- 51 mm broken stone
- 101 mm (63 mm max. size)
- 179 mm stones (101 mm wide)

(c) Convex surface, convex formation

- 51 mm broken stone (25 mm max. size)
- 101 mm broken stone
- 101 mm (76 mm max. size)

Fig. 1.3 Road cross-sections of historical importance: (a) Trésaguet's construction, (b) Telford's construction, and (c) Macadam's construction

Robert Phillips

In Britain, the man who perhaps might be termed the pioneer of scientific road design is a certain Robert Phillips. As early as 1736, he presented a paper to the Royal Society entitled *A Dissertation concerning the Present State of the High Roads of England, Especially of those near London, wherein is proposed a New Method of Repairing and Maintaining Them.* In this paper, Phillips gave a very comprehensive account of the slovenly practices which were then employed in road-making and maintenance. He was concerned with the clay and gravel roads then in use and emphasized that a layer of gravel, if resting upon a well-drained 'sole', would be beaten by the traffic into a solid road surface.

John Metcalf

Unfortunately, Phillips' lead was not accepted at the time and road-making remained a lost art in Britain until about 1765 when John Metcalf came to prominence. Known as 'Blind Jack of Knaresborough', this master road-builder was a most extraordinary character. Born of humble parentage in 1717, he contracted smallpox and became blind at the age of six. He refused, however, to let his affliction become a disability and in the course of a full and varied life he was in turn a musician, soldier, waggoner, horse-trader and eventually, when he was well over forty years old, a road-maker.

Two aspects of Blind Jack's work make him particularly notable. The first of these was his insistence on good drainage. In constructing the road, he emphasized that it should have a good foundation and used large stones to obtain this quality. On top of this he placed a layer of excavated road material in order to raise the roadbed, and on top of this he placed a surface layer of gravel. The carriageway was arched to throw off the rainwater, which was then drained away by capacious ditches on either side of the road.

Perhaps the most individualistic of his road achievements, however, was his approach to building roads over soft ground. Instead of avoiding boggy terrain as had been the custom until that time, he devised a special type of road to go across it. This consisted of constructing his normal roadway on top of a raft subbase composed of bundles of heather carefully prepared and placed in position on the soft ground.

Thomas Telford

At the end of the 18th century, the condition of the populace in Scotland was so bad that the Government had fears that the countryside might become depopulated through emigration. In 1802, Thomas Telford was sent to the highlands to determine methods for their development and improvement. This was a particularly well-informed choice since Telford, in retrospect, can beyond doubt be ranked as one of the very great civil engineers of all time. Born in 1757, this Scotsman was the son of a farmherd and as such he had the very minimum of formal education. Early in life he was apprenticed to a stonemason and afterwards worked at his trade in Edinburgh, London and Portsmouth. In 1786, he was appointed County Surveyor of Shropshire and soon distinguished himself by the stature of his building projects, particularly those in roads and bridges. In 1793, he became engineer to the Ellesmere Canal Company and constructed some remarkable aqueducts at Chirk and Pont Cysylltan. In 1801, his proposal for a new London Bridge of a single span of iron won him considerable acclaim as a designer of vision.

This (then) was the man who when he went on tour in Scotland found neither roads nor carts north of Inverness, and was able to state in his report that all goods were carried by ponies or women. In the southern highlands, he found the stone-covered roads of General Wade to be in disuse, primarily because they had been built to lead to military strategic positions and not to the ports or market centres. He wrote a report containing five proposals, the most important being that there was a great need for improving communications by means of roads and bridges; he stated that 'they will not only furnish present Employment, but promise to accomplish all the leading Objects which can reasonably be looked forward to for the Improvement and future Welfare of the country, whether as regards its Agriculture, Fisheries or Manufactures'.

Thomas Telford's proposals were accepted and, in 1803, he was given governmental authority to put them into effect. His industry in so doing is testified by the 1470 km of new roads and 1117 bridges constructed under his personal supervision during the following 18 years. He perfected the broken stone method of construction for which he is famous (see Fig. 1.3(b)), while recognizing that the road should be designed for the conditions prevailing. Thus he appreciated that the wear of the highland roads by traffic was insignificant and foresaw that the main enemies were the natural ones of storm, flood and frost. Hence he provided for maintenance by appointing road superintendents to ensure that the roads were kept well drained and damage caused by winter frost was promptly repaired.

Thomas Telford's most renowned achievement as a road-maker was the reconstruction of the London–Holyhead road. Following the union with Ireland in 1805, the need for improving the means of travel between London and Holyhead was raised by both the Irish Members of Parliament, who had to use the route perpetually, and the Postmaster General, who considered the way to be in such poor condition as to be unsafe for the mail coaches. In 1810, Telford was asked to survey the road and, in 1815, he began its reconstruction.

Not only did Telford develop a structurally sound method of road construction on this project, but he also paid particular attention to the alignment of the road. It is recorded that the old road near Anglesey rose and fell a total vertical distance of 1079 m between its extremities 38.5 km apart; its replacement, built by Telford, was 3.22 km shorter and rose and fell 391 m less than the old road. He is said to have taken 32.2 km off the total journey from London to Holyhead, while at the same time easing the gradients so that there was no slope greater than 5 per cent. When completed in 1830, this road was considered to provide by far the best facility for land travelling of its time. Its quality is attested to by the fact that much of this road is incorporated in today's A5 trunk road.

John Loudon Macadam (McAdam)

Whereas Thomas Telford was first and foremost a civil engineer with a remarkable aptitude for road-building, it can be justifiably stated that his contemporary, John Loudon Macadam, was Britain's first true highway engineering specialist. Telford was an architect, builder of canals and ports, and bridge designer and constructor who was also a perfectionist in road-making; it was he who made highway-making a respectable art. Macadam made it an economical one.

John Loudon Macadam was born in 1756 at Carsphairn, but early in his life he was moved to Ayrshire where his father owned an estate and founded a bank at Ayr. In 1770, after the death of his father, he went to America where he was able to amass a considerable fortune during the American War of Independence. In 1783, he returned to his native Scotland where he soon began to take an interest in public affairs. For some reason roads fascinated him, and when he was made a Trustee of his local turnpike Macadam began to take a great interest in all matters relating to roads and their administration. In 1798, he removed to Falmouth where he went into business victualling the navy. His pastime during this period was to travel about the south-west studying the roads and devising means for their betterment. The knowledge and concepts he then developed were later put to good use when, in 1816, he was appointed Surveyor for the 238 km long Bristol Turnpike Trust. He succeeded so well in this post that in a few years he was in demand all over England as an adviser to other Turnpike Trusts and eventually, in 1826, to the Government itself.

Macadam is best known for the economical method of road construction he advocated and which, in modified form, bears his name today (see Fig. 1.3(c)). The two fundamental principles which he emphasized were 'that it is the native soil which really supports the weight of the traffic; that while it is preserved in a dry state it will carry any weight without sinking' and 'to put broken stone upon a road, which shall unite by its own angles so as to form a solid hard surface'. These concepts are as valid today as they were over a century and a half ago.

Under Macadam's system of construction, the foundation was shaped to the intended surface camber, thereby giving good side drainage to the foundation as well as a uniform construction thickness throughout the entire width of the road. The amount of crossfall which he called for was about half of that required by Telford, being at the very most 76 mm on a 9.14 m wide road.

While Macadam is renowned for his method of construction, his roads were actually in some ways inferior to those of Telford. It is no coincidence that many modern roads still exist on their original Telford foundations, whereas it is doubtful indeed whether any of the original Macadam roads have survived. The greatness of his construction lay in the fact that it was a technique which for efficiency and cheapness was an enormous improvement over any method used by his contemporaries.

Macadam's real claim to fame should perhaps be based more legitimately on his advocation of the importance of effective highway administration. He persistently urged that the essential feature of any efficient service must necessarily be the trained professional official, paid a salary which enabled him to be above corruption—Macadam said that he must be of good social status—and who would give his entire time to his duties and be held responsible for the success of any undertaking. Macadam also advocated that roads be constructed by skilled labour using the best instruments available. He urged the creation of a central highway authority which would act in an advisory and monitory capacity on all matters relating to the roads, although at the same time he depreciated any attempt to nationalize the roads on the prophetic grounds that the Government would utilize them as a source of revenue, instead of looking upon their upkeep as a public service of the first magnitude[5].

Macadam's proposals were considerably ahead of their time, and it cannot be said that they were accepted in bulk during his lifetime. In general, however, they were regarded with favour and there is no doubt that they initiated concepts and shaped events influencing the future of the whole transportation industry. Whereas to Trésaguet can be given the credit of bringing science to road-making, there is no doubt that John Loudon Macadam was the first Highway Engineer.

The roads and the railways

What all the travel of the coaching days and all the urgings of Macadam and his contemporaries failed to do, the railways did quite incidentally and without effort; they forced the Government to reform the administration of the national roads. The railways were invented to satisfy a need that the roads proved incapable of meeting, that of keeping pace with the progress of industry. From the time that the first railway—it was known as the Surrey Iron Railway, used horse traction, and ran from Wandsworth to Croydon—was opened for traffic in 1805, the railways never looked back. When the Stockton-Darlington railway was opened in 1825, it marked a complete supersession of through passenger and goods traffic by road. The transfer of long-distance traffic from road to rail was instantaneous as soon as a railway became open to the towns along its route, e.g. the last stage-coach between London and Birmingham ran in 1839, whereas the railway line was opened only a year previously in 1838.

'Democracy in rail transport' was initiated in 1844, when an Act was passed which compelled the railways to operate a certain number of cheap trains, and effectively implemented in 1872, when the Midland Railway announced that third-class passengers would be allowed on their best and fastest trains. In 1850, the railways carried some 67m passengers, 300m in 1860, 600m in 1880, well over 1000m in 1900, and nearly 1300m in 1910.

What Macadam came to describe as the 'calamity of railways' fell upon the Turnpike Trusts in particular between the years 1830 and 1850. As a result of the ease of train travel, long-distance traffic by road was so heavily curtailed that very many of the Trusts were soon brought to a state of chronic insolvency and they began to disappear. All projected highway improvements were either abandoned or shelved and during the second half of the 19th century the roads went back to practically the same state as they had been in before the establishment of the toll system. In 1864, conditions were so desperate that a Commission of the House of Commons was appointed to inquire into the Turnpike Trusts; its final report strongly recommended the abolition of the entire toll system. This advice was accepted and acted upon by refusing to grant a renewal when the authorized term of a Turnpike Trust's existence was at an end. While this means of terminating the toll system was a slow one, it was also a thorough as well as an economic method. By 1890, there were only two Turnpike Trusts as against 854 in the year 1870. The final Trust, that on the Anglesey portion of the London-Holyhead road, collected its last toll on 1 November 1895.

From stage-coach to motor vehicle

The abolition of the Turnpike Trusts meant that all roads reverted to the old system of parish maintenance. A measure of the chaos into which these turnpike roads were thrust is reflected in the statistic that at the turn of the latter half of the 19th century there were some 15 000 separate and mutually independent highway boards in England and Wales.

The transfer of some 37 000 km of turnpike roads naturally caused resentment in those parishes that had to support them out of local funds, especially as these were the 'main' roads and had to receive considerable upkeep. Eventually, however, in 1882, the State accepted financial responsibility for aiding the highways and made a grant of one-quarter of the cost of the main roads to the highway authorities; in 1887, this grant was increased to one-half. The Local Government Acts of 1888 and 1894 resulted in changes which essentially stabilized responsibility for the roads by assigning the rural main roads to county councils, county boroughs became responsible for all their own roads, and the rural, urban and borough councils had to look after the remaining roads.

It must be mentioned here that all of these reforms in road administration were undertaken at a time when the roads were practically deserted, unrepaired and almost universally considered useless. The reforms were primarily the outcome of despair over what to do with these derelict roads, rather than in any constructive and pioneering spirit. Fortunately, however, these reforms came at a time, still unsuspected, when they could be of great use, for by the turn of the 20th century the mixed blessing of the motor vehicle had arrived.

Urban public transport

Prior to the Industrial Revolution, travel was, on the whole, by horse or foot—indeed, as noted previously, it was only about the middle of the 16th century that the first wheeled coach appeared in Britain. With the advent of the Industrial Revolution, the situation radically changed; civic life and form was altered almost beyond recognition as a great wave of migration from the countryside was initiated. The result was that villages became towns and towns became cities—and this was accompanied by the beginning of what is now known as the population explosion (e.g. in 1800, the population of England and Wales was only about nine million, two-thirds of whom could be classified as rural dwellers).

As the population increased, the need for efficient public transport facilities became evident, particularly in towns, and many far-sighted and entrepreneurial businessmen saw profits in providing these transport services (for it must be remembered that public transport was a profitable enterprise until well into the 20th century).

4 July 1829 saw the initiation of mass public transport in towns when Mr George Shillibeer started his 20-passenger horse 'omnibus' service in London between the village of Paddington and the Bank.

22 April 1833 saw Mr Walter Hancock testing his 14-passenger steam omnibus service over the same route.

30 August 1860 saw the beginning of the end of the horse-bus when Mr George Train initiated a horse-tram service in Birkenhead. It could carry more people more comfortably and more quickly than could the horse-bus.

10 January 1863 saw the opening of the world's first underground railway, the first part of the Metropolitan (steam) Railway between Paddington and Farringdon.

29 September 1885 saw the first electric street tramway at Blackpool—and this initiated the death of the horse-tram. The electric tram had a really tremendous effect upon town development, for, within twenty-five years, nearly every town in Britain had its own network of electric tramways. Typically, these radiated outwards from the central area and serviced settlements clustered along these routes. Thus transport development began to be clearly associated with the now-familiar pattern whereby residential densities decline from zone to zone from the town centre outwards.

12 April 1903 saw the start of the first municipally-operated motor bus service at Eastbourne. This was important for two reasons. Firstly, it marked the first occasion that a municipality assumed responsibility for providing a *public* transport system in a town. Secondly, the motor bus service brought the suburbs within easier reach of the town centre and, since it was not forced to stay on fixed routes, it could service housing not astride the radial arteries—and thus it helped to consolidate urban sprawl.

The bicycle

The first report in recorded history of man being able to balance on a two-wheeled vehicle was in 1791 when a Frenchman, the Compte de Sivrac, rode his 'hobby horse' in the garden of the Royal Palace[9]. This hobby horse was not a toy: rather its rider had established the exact height of the seat so that his feet could touch the ground in a manner to obtain the most propelling power. By 1820, trips of nearly 40 km in 2.5 h had been accomplished on these vehicles, which were of about 27 kg mass, and about 3000 were estimated to be in use in Europe.

The Macmillan bicycle of 1839 is generally given the credit as being the first true bicycle, i.e. forward motion was obtained with the aid of pedals, without the rider's feet touching the ground. As time went by, the mass of the bicycle became less, and it was noted that the rider could travel a much greater distance with one revolution of his feet by making the front wheel larger—hence the development of the 'penny farthing' bicycle. Thus, in 1886, a Mr G P Mills rode his penny farthing bicycle with solid tyres from Lands End to John O'Groats, a distance of about 1392 km, in a fraction over five days.

The invention (in 1873) and development of the low bicycle was responsible for a tremendous cycling boom in Britain into the 1900s. Technological factors which encouraged this were the invention of the pneumatic tyre (by Dunlop in 1888), the improvement in the reliability of the bicycle chain, the advancement of frame design, the roller-bearing, the ball-bearing, drawn-steel tubing, expansion brakes and the use of light, non-ferrous metals—and, of course, improvements in the highways. The bicycle was also seen as a cheap and easy alternative to public transport and, with the growing emancipation of women, it became socially accepted for it to be used by both sexes.

By the turn of the century, the development of the bicycle had reached quite a sophisticated stage—it is probably true to say that no major improvements in its design have taken place since then, except for the deraileur (gear change) and some minor improvements in metals—and many highways were reported to carry more bicycles than horse-drawn vehicles. Indeed, many complaints were voiced in the newspapers of the day about the dangers that cyclists were causing to pedestrians and other users of the carriageway.

In the early years of the 20th century, it had been considered fashionable to be a cyclist; however, in the economically depressed years of the 1920s and 1930s, the bicycle became the vehicle of the mass of the populace. Its usage then grew to the extent that, in 1952, about one-quarter of all of Britain's vehicle-kilometres were attributed to pedal cycles.

The motor car
The history of the motor car really began in October 1885 when Karl Benz of Mannheim completed his petrol-driven tricar (patented January 1886); this had a specially designed chassis of steel tubing, with a rear-mounted engine of a four-stroke gas type, adapted to run on benzine through a surface carburettor. Very soon thereafter, in 1886, Gottlieb Daimler of Cannstatt, working independently and unbeknownst to Benz, completed building the world's first four-wheeled petrol-driven car; this was virtually a coach minus the shafts, with a watercooled 1.1 hp single-cylinder four-stroke engine fitted between the front and back seats. Little did these pioneer engineers realize that they were about to usher in a transport revolution. Their vehicles, while not very ornamental, were operational and heralded a return of traffic to roads, and the eventual need for new initiatives in respect of road development.

The beginning of the motor age in Britain can perhaps be precisely defined as the year 1896, for this year marked the passing of the Locomotive on Highways Act. Up to that date, all road locomotives were limited to speeds of 6.44 km/h in open country and 3.22 km/h in a populous neighbourhood, and each vehicle had to be preceded by a man carrying a red flag. The Act of 1896 gave to the light locomotive, in which was included the motor car, the right to travel upon the highways at a speed not exceeding 22.53 km/h. The enfranchisement of the motor vehicle in 1896 was celebrated on 14 November of that year (the day the Act became law) by an organized run from the Hotel Metropole in Whitehall, London, to the Hotel Metropole, Brighton.

The highway and motor vehicle today

In 1895, prior to the implementation of the Locomotive on Highways Act, a Scotsman named Joseph Wright sought permission to drive his horseless carriage on Glasgow's local roads. He was advised that this would be imprudent, and in reply he wrote[10]: 'Allow me to say that those in authority ... might as well try to beat back the waves of the sea with a broom as try to stem the tide of horseless carriages which are looming in the distance.'.

The advent of the motor vehicle had little effect upon mass movement or town development for some time, however, for the very practical reason that in those early days it could only be enjoyed by the wealthy. Of much more importance to the general populace of that time was the low bicycle and public transport. Thus the first concerted pressure on the Government for the improvement of rural roads came from the cyclists. Indeed, it was the cyclists who formed a Roads Improvement Association in 1886 which, as the century turned, was joined by the newly-formed motoring organizations.

In 1903, there were approximately 18 000 mechanically propelled vehicles registered as using Britain's roads, of which about 44 per cent were cars and vans. By 1909, the number of vehicles had risen to 144 000, of which 37 per cent were cars and vans.

Interwar years

Notwithstanding the fairly rapid growth in the numbers of motor vehicles between the two world wars (cars from 187 000 to over 2m; goods vehicles from 100 000 to 490 000), the first forty years of the 20th century were years of evolutionary development rather than revolutionary change for roads in Britain. Initially, the emphasis was on 'laying the dust' and eventually on reconstructing existing roads; thus, in the period 1911–35, the formal road system was only increased by less than 5000 km. Most passenger travel was still by public transport and the bicycle, with the horse-and-cart and the railway still being major movers of goods and commodities. Traffic congestion, very often associated with the mixed use of the street system in towns, was tackled by increasing capacity through local road widenings and improvements.

Prior to World War I, the objectives of the public transport operator were almost entirely dictated by the profit motive, and there were many examples of operators implementing new services and then selling off surplus land at enhanced prices for industrial and residential development[11]. However, the over-provision of services and fierce competition between rival operators led eventually, during the interwar period, to mergers and takeovers through the normal processes of a free market, which in turn led to rationalization of service levels.

The interwar period also saw the continuation of the flight of home-dwellers to the suburbs, and the movement of industry to sites adjacent to good road facilities. This development, which was accelerated after World War II, was mainly a product of inner city slum clearance and higher incomes, but it was aided by the mobility given by the motor vehicle. This move was also encouraged by town planning practice which generally favoured the development of the centripetal type of town—indeed city region—of the idealized form shown in Fig. 1.4(b), which has had a very constraining effect upon the planning of transport facilities to this day.

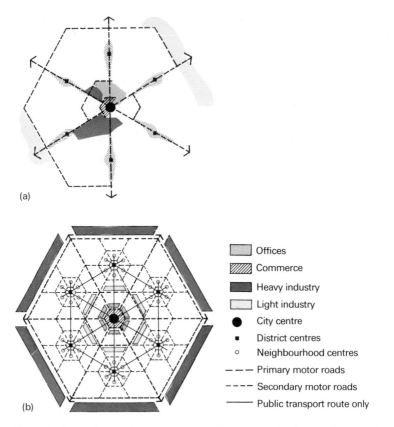

Fig. 1.4   General town planning practice with respect to land use and transport routes: (a) general existing situation and (b) idealized centripetal town

Post-World-War-II developments

With the end of World War II, it can be said that the trends in urban development, and highways, in Britain simply took up where they had left off in 1939. Immediate post-war priorities were for the repair of war damage and housing and the provision of social services, and the need for road or transport facilities was not given a high priority in the national effort.

In contrast to what was happening in Britain, road development in the USA took a revolutionary step forward with the passing of the Federal Aid Highway Act of 1944 which authorized the development of a highway system 'so located as to connect by routes, as direct as practicable, the principal metropolitan areas, cities, and industrial centers, to serve the national defense, and to connect at suitable border points with routes of continual importance in the Dominion of Canada and the Republic of Mexico'. This network, known as the Interstate and Defense Highway System, connects 90 per cent of the American cities with populations of over 50 000 by means of about 70 000 km of motorway. Nearly 13 000 km of the system are composed of extensions into urban areas and circumferential urban routes.

The Interstate system, although constituting only about 1.2 per cent of the USA's total road and street system, is expected to carry some 20 per cent of its traffic. Some 85 per cent of its length has had to be located on new alignments as the existing highway alignments could not be reconciled with the high standards required of the new system. Most of the routes are composed of four-lane dual carriageways growing to six and eight lanes in and near metropolitan areas; access to these highways is rigidly controlled over the entire system, entry only being permitted at carefully selected locations. Traffic interchanges, overpasses and underpasses have eliminated all at-grade crossings, whether they be highway or railway. When the Interstate system is essentially completed this decade, it will be possible to drive from coast to coast across the USA without encountering a single traffic light, roundabout or stop sign.

This American programme was well noted internationally, and for many years it set the standards against which highway developments elsewhere in the world were evaluated. Its beneficial impact on organized research and on the planning of highways was of particular importance; apart from anything else it provided a significance to highway and traffic studies in institutions of higher education which had been lacking for some time.

Britain's strategic road network

As Table 1.1 shows, the 1950s saw the beginning of significant real growth in personal incomes in Britain, and with it a major growth in the numbers of motor vehicles. The total number of cars increased by almost 80 per cent between 1954 and 1960, whilst freight traffic by roads

Table 1.1 Historical trends in car ownership and income[12]

Year	Cars per person	GDP per person* (£)
1903	0.0002	266
1905	0.0004	270
1910	0.0013	266
1915	0.0033	n.a.
1920	0.0044	n.a.
1925	0.0135	275
1930	0.0237	288
1935	0.0324	301
1940	0.0303	n.a.
1945	0.0311	n.a.
1950	0.0459	376
1955	0.0712	430
1960	0.1085	470
1965	0.1692	534
1970	0.2137	577
1975	0.2526	n.a.

*Gross domestic product per person, at constant 1963 prices

increased by over 40 per cent. Pressures developed for better highways, and local authorities developed major road programmes to meet what were then seen as traffic demands which had to be satisfied. The need for a better road system was accepted by the Government, and the highway programme was expanded quickly. Britain's first motorway, the Preston bypass, was opened to traffic in December 1958, followed by the first stage of the M1 motorway in November 1959. Money spent on building and improving trunk roads and motorways more than trebled in real terms in the decade 1960–61 to 1970–71. Over the same period, the number of cars again doubled (from 5.5m to over 11m), and freight traffic by road increased by two-thirds.

In 1970, when completion of its (then) first target of 1600 km of motorway was in sight, the Government announced plans[13] for the comprehensive improvement, to at least dual carriageway standard, of 6750 km of road to form a strategic inter-urban highway network for England. Previous proposals for strategic highway networks for Scotland and Wales, which tied in with the English proposals, had already been published[14, 15]. Since then, as is discussed later, national transport and highway policy has been modified. Figure 1.5 shows the strategic network of intercity highways as it existed in the early/mid-1980s. The network, when completed, was intended to be easily connected to all major centres of population, to promote economic growth and regional development, and to achieve environmental improvements by diverting traffic from a large number of towns and villages, especially historic towns.

In addition to the intercity routes shown in Fig. 1.5, many new high-standard highways with limited access control to ensure fast traffic movement were constructed in urban areas in the 1960s and 1970s. These included motorways in Birmingham, Bristol, Glasgow, Greater London, Greater Manchester, Leeds and Merseyside. However, since the early 1970s, progress on new urban highways has been significantly reduced as a result of changes in public attitudes towards the role of the highway in the city, the limits on public expenditure, and the shift towards favoured treatment for public transport and comprehensive traffic management schemes. Priority for roads in urban areas is now being given to those which will meet the needs of industry and commerce, serve new industrial or housing developments, provide links with the national trunk road network, and bypass severe congestion bottlenecks.

The development of the highway system has had a significant effect also upon the methods of moving goods. This is most easily illustrated by noting that, notwithstanding the growth in the national economy, British railways only moved 145×10^6 t of freight in 1983 versus 427×10^6 t in 1900, or that the British freight workload in 1920 was estimated at 31.35×10^9 t-km, whereas, in 1983, it had reduced to 17.1×10^9 t-km[16].

Over the years, there has also been a major shift towards using heavier vehicles to move goods by road. Thus goods vehicles over 38 t gross, which were non-existent prior to 1983, represented 3 per cent of all goods vehicles over 3.44 t gross by early 1984. Current gross mass and length limits on road freight vehicles are given in Table 1.2.

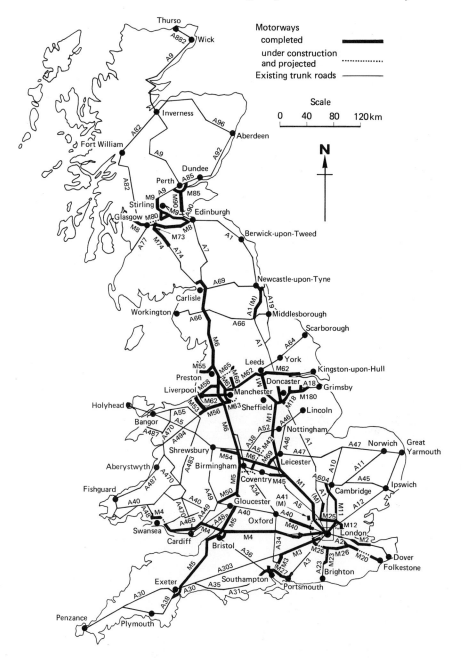

Fig. 1.5 Britain's intercity strategic trunk road network

Table 1.2   Gross mass and length limits on road freight vehicles

Vehicle type	Mass limit		Length limit	
	Year introduced	Limit (t)	Year introduced	Limit (m)
Rigid				
2-axle	1964	16.26 ⎫		
3-axle	1972	24.38 ⎬	1964	3.55
4-axle	1972	30.48 ⎭		
Articulated				
3-axle	1966	24.38 ⎫		
4-axle	1964	32.57* ⎭	1967	4.57

* Increased to 38 t on 1 May 1983

Table 1.3, which summarizes the changes in respect of commodities moved by road and rail in the years 1962 and 1982, shows that road haulage work is concentrated on food products, manufactured goods and building materials, as compared with coal, coke and iron and steel products by rail. For road transport, the average length of haul for most commodities has increased from 43 to 71 km over the same period. This increase is related to the shift to the use of heavier vehicles, i.e. haul lengths for vehicles of a given size changed relatively little, whereas maximum mass vehicles had by far the longest length of haul. Apart from the smallest vehicles, the average load factor by mass (ratio of payload to carrying capacity) for vehicles of a given size also declined[17].

Table 1.3   Billion-tonne-kilometres of selected commodities carried by the road and rail modes, in 1962 and 1982[16]

Mode	Commodity	1962	1982	% change
Road	All	55	90	+64
	of which			
	Food, drink and tobacco	14	23	+64
	Crude minerals	7	10	+43
	Building materials	6	7	+17
	Iron and steel products	3	6	+100
	Chemicals	3	5	+67
	Other manufactures	9	7	−22
Rail	All	26	16	−38
	of which			
	Coal and coke	12	6	−50
	Iron and steel products	4	2	−50

Public transport
While the 1960s and 1970s saw a massive investment in, and positive development of, the trunk road and motorway system, the same period saw an equally dramatic decline in public transport usage. Given a free

Table 1.4 Inland passenger transport in Britain, 1963 and 1983[16]

Transport mode	1963 Passenger-km ($\times 10^9$)	%	1983 Passenger-km ($\times 10^9$)	%
Rail	36	13	35	7
Road				
public transport	64	24	42	8
private motor transport	163	60	414	84
pedal cycles	7	3	5	1
Total	270	100	496	100

choice, people elected to use their motor cars in preference to public transport for other than long trips. This trend is demonstrated by the basic passenger statistics shown in Table 1.4. While these data reflect only national trends, there is ample evidence elsewhere that there has been a clear and continued swing from the use of public transport in towns, although the amount of the decline has varied from town to town, and the peak year for public transport passengers (prior to the decline) was much later for some towns than for others. At the present time, it would appear that this swing with respect to passengers is likely to continue, albeit at a slower rate; however, the number of public transport vehicles per head of population may well have reached its base level and not go much lower.

The mobility provided by the private car, as compared with public transport, is well illustrated in Table 1.5. These data from the seminal London Traffic Survey of the early 1960s show that the effect of increasing car availability is not only to reduce travel by public transport, but also to increase travel from each household, especially for non-work purposes. Numerous transport studies since then have reported results of a similar nature.

Since 1930, public transport (road) has been controlled by a system of licensing under the jurisdiction of Area Traffic Commissioners (see Chapter 2). This system was applied to an industry which (until the

Table 1.5 Trip generation per household per day for households in London with one employed resident per household in an urban area of average rail and bus accessibility[18]

Trip purpose	Household income* (£)	Private transport No car	1-car	Multicar	Public transport No car	1-car	Multicar
Work	0–1000	0.2	1.2	—	0.9	0.4	—
	1000–2000	0.3	1.2	1.6	1.1	0.6	0.5
	>2000	—	1.0	1.5	—	0.5	0.2
Non-work	0–1000	0.2	2.0	—	0.8	0.4	—
	1000–2000	0.3	2.7	5.2	1.4	0.7	0.3
	>2000	—	3.9	8.4	—	1.1	0.9

*1963 salaries

1960s) generally had a fairly healthy financial and operating basis. As a result, the commissioners were under no pressure to encourage bus companies to reorganize their services to obtain greater efficiency and/or coordination; rather they saw their function as fixing routes, timetables and fares, and applying control in the public interest. Governmental authorities generally took the view that public transport should pay its way, and the periodic injections of funds to wipe-out particular recurring debts and assist in capital expenditure were usually accompanied by admonitions to this effect.

With the downturn in public transport usage, however, there was a further rationalization of public transport operations, and a reassessment of its role took place. As soon as the profitability of the system disappeared, the governmental system had to accept a greater responsibility for public transport, albeit reluctantly in many instances. This responsibility now extends for all practical purposes to the complete planning, coordination, operational control and financing of the system as a whole, e.g. under the Transport Act of 1978, County Councils are required to prepare Public Transport Plans in conjunetion with the preparation of proposals under the Transport Policy and Programme System.

Improvements to bus services have been undertaken in many areas in attempts to make the public transport system more attractive to users; these have included the introduction of express and/or limited-stop services, park-and-ride services, and Post Office minibuses[19] carrying mail and passengers in rural areas. Bus operators have had to take steps to contain costs and improve productivity by introducing larger buses, increasing the number of buses capable of being operated by one person, and reducing or rationalizing services. Many uneconomic bus services have been withdrawn, particularly in rural areas, and experimental services introduced into both urban and rural areas (see, for example, references 20, 21 and 22). Comprehensive traffic management schemes involving priority for buses have now become fairly commonplace in and about town centres.

The environment
As noted previously, concerns have been expressed at various times in history at the detrimental environmental impacts associated with the development of highways. These concerns, in turn, led to various limited attempts at reform and change, e.g. following the expansion of towns and cities which resulted from the Industrial Revolution, the greater attention then being given to public health resulted in an outcry against the use of granite sett-paving in town streets; the paving stones not only gave an uncomfortable ride but were slippery and very noisy and, being difficult to wash, were generally foul as a result of the passage of the (then) ever-increasing number of animals. Wood-paving, which was substituted initially to help to correct the situation, was not very successful; not only was the carriageway still slippery, but the unsanitary aspects of the problem were little improved due to the porosity of the surfacing. With

the construction of modern bituminous surfacings in the latter half of the 19th century, this particular environmental problem began to be alleviated—at least cleansing became easier—but it was not until the horse was replaced by the motor vehicle that it was eradicated.

The eventual solving of the above environmental problem can be said to have given rise to many others. The growth in motor vehicle numbers and construction of new roadways soon gave rise to problems of noise, vibration, pollution, visual intrusion and community severance. However, it was not until the 1960s that the scale of these problems was properly recognized[23], and the Government clearly accepted the responsibility for their resolution. A measure of the concerns that people have regarding certain types of environmental impacts is reflected in the data in Fig. 1.6 which are from a nationwide interview survey[24].

In 1963, a thorough examination of the *noise* problem created by motor vehicles reported the representative existing noise climates listed in Table 1.6. To illustrate how these measured levels compare with desired levels, Table 1.7 gives the investigating committee's recommendations regarding the maximum noise levels inside living rooms and bedrooms which should not be exceeded for more than 10 per cent of the time.

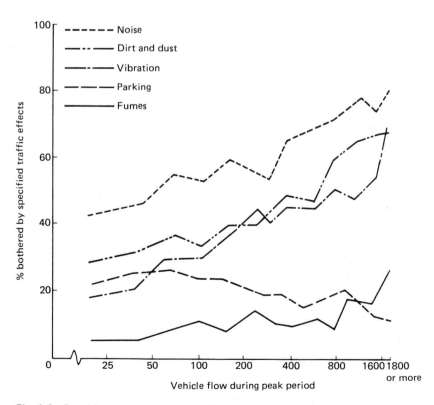

Fig. 1.6 People's response to type of traffic disturbance at home

Table 1.6   Representative existing noise climates[25]

| | | Noise climate* (dB(A)) | |
| | | Daytime (8.00 a.m.– 6.00 p.m.) | Night-time (1.00 a.m.– 6.00 a.m.) |
Group	Location		
A	Arterial roads with many heavy vehicles and buses (kerbside)	80–68	75–50
B1	Major roads with heavy traffic and buses	75–63	61–49
B2	Side roads within 13.75–18.25 m of roads in Groups A or B1 above	75–63	61–49
C1	Main residential roads	70–60	55–44
C2	Side roads within 18.25–45.75 m of heavy traffic routes	70–60	55–44
C3	Courtyards of blocks of flats, screened from direct view of heavy traffic	70–60	55–44
D	Residential roads with local traffic only	65–56	53–45

*Noise climate is the range of noise level recorded for 80 per cent of the time; the level exceeds the higher figure for 10 per cent of the time, and is less than the lower figure for 10 per cent of the time

Table 1.7   Recommended maximum noise levels in homes[25]

| | Maximum noise level (dB(A)) | |
Type of area	Daytime	Night-time
Country areas	40	30
Suburban areas	45	35
Busy urban areas	50	35

In 1970, it was estimated[26] that 19 to 46 per cent of Britain's urban population of 45m, and 13 to 32 per cent of the rural population of 9m, lived in roads with traffic flows producing noise which was undesirable for residential areas.

The Noise Regulations[27, 28] made under the Land Compensation Act of 1973 have now made it the duty of the highway authority constructing a new road to carry out noise insulation work on a dwelling if, due to the construction of the new road, the L_{10} (18 h) noise level (i.e. the average of the hourly values of L_{10} between 6 a.m. and midnight) at a point 1 m from the external facade of the dwelling is expected to exceed 68 dB(A). The L_{10} (18 h) standard for entitlement to remedial action is based on a social survey carried out in 1968 which showed that this noise level corresponded very roughly to 50 per cent dissatisfaction with free-flow traffic noise[29].

Pollution of the atmosphere by motor vehicles takes several forms, viz. colourless, odourless toxic gases; colourless, malodorous toxic gases; coloured, malodorous toxic gases; particulate material (containing undesirable materials such as lead and carcinogens) which soils buildings and affects the growth of vegetation. Measurements of pollutants at the kerbside in typical urban situations clearly show a direct relationship between pollutant levels and traffic flow volumes. Furthermore, the degree of pollution varies with traffic conditions, topography and meteorology; the levels also fall away rapidly as the distance from the road increases.

It is generally accepted[30] that there is no evidence that the present traffic pollution levels constitute a danger to health, although the view is held that they should, if possible, be reduced. One report[31] recommended, for example, that steps should be taken to reduce progressively emissions of lead to the air from traffic, and that the concentrations of lead in the air should not exceed two microgrammes per litre in areas where people are likely to be exposed for some time. (As a result, the Government reduced the permissible lead limit in petrol to 0.40 g/l in 1981.)

Overall, it must be said that there is a paucity of reliable information on the effects of long-term, low-concentration exposure to pollution from motor vehicles upon people. The black smoke from diesel engines is not a health hazard either; however, it is unpleasant and a road accident risk, particularly in traffic congestion or in narrower enclosed streets where it may obscure vision.

Another environmental 'disamenity' of significance is the aesthetic deterioration associated with *vehicle intrusion*[23], e.g. 'the dreary, formless car parks, often absorbing large areas of towns, whose construction has involved the sacrifice of the closely knit development which has contributed so much to the character of the inner areas of our towns; the severing effects of heavy traffic flows; and the modern highway works whose great widths are violently out of scale with the more modest dimensions of the towns through which they pass'.

Attempts have been made[32] to develop an objective index of visual attractiveness which can be used in conjunction with highway design. However, at this time all practical inputs are subjective ones, and as such they inevitably vary according to the outlook of the adviser. In the case of the motorway programme, it has been governmental policy from the onset to utilize the skills of the Department of Transport's expert Landscape Advisory Committee and the Fine Arts Commission in selecting highway alignments, and ensuring that roads are well fitted into the landscape. As part of its overall trunk road programme, the Department of Transport typically plants about two million trees and bushes per year for landscaping purposes.

Severance is the dividing effect that highways and traffic have on the human community through the breaking of connections with friends, schools, shops, etc. Since it forces changes to established habits, severance is often perceived by the general public as being a greater problem than it actually is. In practice, its detrimental effects are most directly felt by pedestrians in built-up areas. In rural areas, it has its greatest impact on farms with one or more of the following characteristics[33]: small size, existing close to the economic margin, intensive crop/livestock system.

As with visual intrusion, subjective judgement is still the only practical approach to the evaluation of the severance impact of a highway upon a community.

As with noise, *vibrations* from highway traffic are capable of scientific measurement. In practice, however, vibration (with associated infra-sound) is generally unlikely to be of significance.

One of the most significant developments of the 1960s and 1970s has been the considerable increase in public awareness of the concept 'quality of life'. This has manifested itself in many ways in civic life; not least of these has been in increased criticism of traditional criteria and thinking on matters affecting the environment as a whole, and most commonly in the questioning and scrutiny of the goals and performance of urban transport. As a result, the highway and transport planner has had to learn to live with uncertainty in relation to the implementation future of any proposals, and to adjust to what might appear to many as relatively irrational behaviour by the public. To caricature the latter kind of issue, the highway planner has had to accept that car users who want fast and safe roads will often strongly support house owners who object to having their homes demolished in order to provide space for these roads, even though their own personal transport interests may be aided by siding with the planner.

Very often, the planner (correctly) attributed the public's stance in relation to road provision to inadequate information—or, more fundamentally, to a lack of awareness of the trade-offs inherent in the decision-making process. The constructive reaction to this dilemma has been to seek to understand this irrational behaviour, involve the public in the planning process, and steer a professional middle course in order to obtain an acceptable answer. The result has been the evolution of a process of *public participation* in highway planning, the utilization of which is now governmental policy.

Initially, the approach to public participation was to rely upon mass publicity to inform the public regarding particular proposals, and then to utilize public opinion surveys, exhibitions and public meetings to get feedback re their acceptability. In small areas such as villages, with which local organizations and people could identify closely, the system worked reasonably well; in most places, however, the people who made spontaneous contributions to the wider policy issues over a large geographic area were not usually typical of the whole population. As a result, local authorities now spend considerable effort trying to improve their methods of publicity and participation by use of local radio and regional television, and by involving street groups at local information points during the preparation of plans. In addition, deeper responses are being sought through conferences, seminars and the follow-up of public meetings by discussion groups involving the public and the planners.

The core of the participation process in Britain is still undoubtedly the public inquiry[34] which generally follows the consultation process described above. As part of the preparation for the inquiry, the Department of Transport is now required to provide members of the public with the facts and assumptions on which the case for the scheme in question is based, with results of analyses and explanations of methodologies, and with any other requested information which, although not part of the Department's case, can be provided without costly special research. The aim is to make available as much relevant information as possible to the public within the constraints of cost. In addition, the onus is on the Department of Transport to provide library facilities before and during the inquiry, which are in proportion to the complexity and size of the scheme and local interest in it. The 'library' must contain all the material which the Government intends to present and on which it bases its case for the road. Objectors can copy any documents held in this library.

Energy
Prior to the 17th century, the rate of world population increase remained near zero, with high birth and death rates in balance with each other. With the advent of the Industrial Revolution, however, people's standards of living began to increase, people lived longer (in Europe), and the world population growth rate increased to 0.5 per cent by the 19th century. By 1950, the growth rate had reached 1.0 per cent and, by 1970—with increasing standards in the newly developing third world countries—the rate had accelerated to a figure in excess of 2.0 per cent. At 2.0 per cent growth rate, the world population doubles in 35 years; at 2.5 per cent, it doubles in about 20 years.

A corollary of growth in standards of living and life expectancy has been a greater utilization of the earth's resources, including its fossil fuels (see Table 1.8).

Over the years, various warnings were sounded to the effect that if the rate of growth of energy demand continued to grow exponentially, then impossible levels of demand must be reached eventually; that there were finite amounts of fossil fuels on earth; that it was only a matter of time before mankind must undergo the traumatic experience of changing from an ever-expanding economy to a steady-state non-growth one, from an oil-dominated economy (typically 50 per cent satisfied by crude oil, 20 per cent each by coal and natural gas, and 10 per cent other) back to a coal-dominated one, and/or onward to a solar energy, hydroelectric and nuclear power era (provided that environmental and safety concerns can be resolved). However, these warnings were not heeded; while the logic was irrefutable, it was simply assumed that 'technology will solve the problem'.

Table 1.8 Per capita world production of energy, 1900–1985[35]

Year	Energy (gigajoules per person per year)
1900	14
1925	23
1950	32
1960	40
1965	47
1970	55
1973	60*
1985	87*

*Estimated

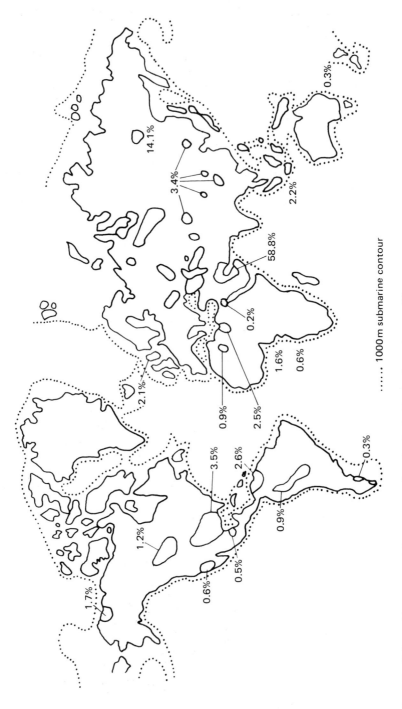

Fig. 1.7 Estimated known remaining reserves of crude oil: distribution by major areas[36]

Labels within figure:
14.1%
3.4%
58.8%
0.2%
2.2%
0.3%
1.6%
0.6%
0.9%
2.5%
2.1%
3.5%
2.6%
1.2%
0.6%
0.5%
0.9%
1.7%
0.3%

······ 1000 m submarine contour

Then, on 16 October 1973, the Arab Gulf component of the Organization of Petroleum Exporting Countries (OPEC), on political grounds (i.e. the Arab-Israeli conflict), decreed a 25 per cent cut of its oil production and raised the posted price of crude oil at the wellhead by 70 per cent. This signalled the start of a rapidly spiralling, entirely new level of energy pricing and the onset of a long period of international economic recession. The world concluded that it had an 'energy crisis'—meaning that there was a sudden realization of the dependence by whole national economies upon the supply of a vital fuel, oil.

There are a number of points which should be stressed at this stage. Firstly, there was (and is) no real worldwide shortage of energy resources; from a resource viewpoint, the world's total fossil fuel energy resources could last for many hundreds of years. Secondly, however, some 60 per cent of the world's *known* reserves of oil is located in the Middle East (see Fig. 1.7), a resource component that is subject to sudden disruption in times of regional upset. Thirdly, while the exact size of the *ultimate* reserves of any of the fossil fuels is unknown, in the case of oil the limits were foreseeable, e.g. see Table 1.9 which estimates the duration of recoverable world reserves of oil/oil equivalent, as predicted in 1976.

Throughout the 1970s and early 1980s, there were further significant real increases in the price of oil, as a result of which governments in most countries placed a very considerable emphasis upon policies which encouraged oil conservation and substitution, and the overall more efficient usage of energy. The beneficial effects of these policies were such that in 1984 the general expectation was that the growth in the demand for energy in third world and western economies during the period 1985–2010 was expected to average less than 2 per cent per year; this is roughly half the rate of energy growth experienced during the decade prior to 1973.

A point which might also be emphasized here, however, is that the majority of oil reserves so far found are those which are capable of being exploited most cheaply. Future finds will be more expensive to tap (e.g. deep offshore, in the polar zones, heavy oils, oil shales and tar sands), whilst enhanced recovery (i.e. the current recovery rate from oil deposits is only 25 per cent) will also be more expensive. Although oil prices are not

Table 1.9 Duration of recoverable world reserves of oil/oil equivalent[37]

| Resource | Total recoverable reserves* ($\times 10^9$ t) | 1974 consumption ($\times 10^9$ t) | Duration of reserves in years at exponential consumption growth rates of | | |
			0%	2%	4%
Oil	237	2.84	82	49	37
Gas	174	1.12	155	72	51
Coal and lignite					
10% recovery rate	655	1.73	379	108	71
50% recovery rate	3277	1.73	1895	185	110

*Includes probable and undiscovered

expected to rise in the mid/late 1980s, mainly due to surplus capacity in virtually all the OPEC producing countries, it is inevitable that the real cost of extracting oil will eventually rise. Just how large the price rises or their trajectory will be is difficult to project, although the Government has forecast[38] that the 1977 OPEC price of oil will at least double (in real terms) by the year 2000, and that the price of petrol (net of tax) will probably increase by at least 50 per cent over the same period.

It is likely therefore that the real cost of road transport will eventually also rise.

Britain, by comparison with most industrial countries, is well placed for the foreseeable future in respect of the availability of oil. Taking account of cumulative production (572m tonnes prior to 1984), total remaining oil reserves are estimated to be in the range of 1410–5280m tonnes, of which 480–3275m tonnes have yet to be discovered[39]. Figure 1.8 illustrates the extent to which Britain has become more than self-sufficient in oil over the past decade.

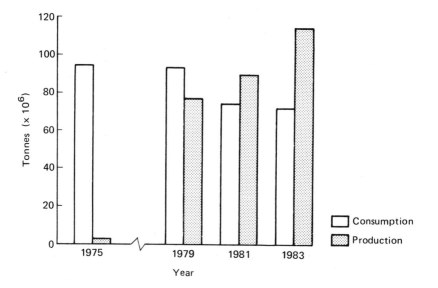

Fig. 1.8   Comparison of UK continental shelf and onshore oil production with oil consumption, 1975–83 (based on references 38 and 39)

The future

Vehicles

Any consideration of the future re the highway system in Britain must take into account the future of the motor vehicle, in the light of the many queries raised regarding the apparently limited future for oil. Currently, world crude oil reserves are estimated to have a life of anything from 50 to 150 years, depending upon the viewpoint of the prognosticator, the rate of discovery of new resources, and the projected rate of consumption (especially for non-transport purposes).

While, obviously, it is not possible for anybody to foretell the future exactly, there would appear to be a general consensus that the private car will continue very much as it is, well into the 21st century (see, for example, references 40-43). Changes there will be, however—of this there can be little doubt. In the short term, the emphasis in the land transport scene will be on developments which will make the use of the vehicle more efficient and promote energy conservation. In the long term, the private capsule will most likely continue, but with a different power-pack.

Energy conservation
Table 1.10 summarizes possible conservation measures that could be applied to land transport. What this table really states, of course, is that there are gross inefficiencies in the present system, and that there is ample scope for the application of conservation policies which will cause minimal detriment to the socio-economic fabric of the nation. It also shows that there is no single 'answer' to oil conservation, but rather that there is a combination of measures that can be implemented (see also Chapter 3).

While many of the measures recommended in Table 1.10 are quite obvious and logical—and indeed many have already been initiated to a greater or lesser degree—it is likely that their overall implementation will require some considerable time. For some to be accepted by the people, a major public education programme emphasizing the critical importance of fuel conservation will need to be carried out. However, the need for gradual implementation is not just associated with what people are willing to accept; it also relates to limitations on the speed with which the transport industry itself, especially the vehicle manufacturers, can respond to change.

Future numbers of vehicles
Any prediction of the future is fraught with problems at any time, and it is no less the case when forecasting total vehicle numbers at the national level. Basically, there are two possible approaches to forecasting, viz. using extrapolatory methods and using causal models. Historically, the practice in Britain has been based on the use of extrapolatory models[12] which fit past data and assume both a relationship between car ownership and gross domestic product and an eventual 'saturation level' for car ownership. The extrapolatory approach has now been seriously questioned and greater emphasis will be given in the future to the use of causal models for vehicle and travel prediction, i.e. models which use data produced by the Family Expenditure Survey (see, for example, reference 45), and which seek to identify the causes underlying decisions about car ownership and to quantify their effect.

One set of projections which uses the Transport and Road Research Laboratory's extrapolatory approach, and is based on 'middle' assumptions regarding economic growth and motoring costs, is given in

Table 1.10 Some conservation measures applicable to land-based transport[44]

Conservation measure	Remarks
1 Improve technical efficiency of car operation	
a Reduce average car size	a Petrol consumption is directly related to car mass, e.g. consumption, litres/100 km = 31 + 58 M, where M = car mass in 1000 kg units
b Optimize gearbox ratio	b Better matching of engine and road speeds (perhaps by programmable gearbox) can give large savings
c Substitute diesel for petrol engines	c Diesel engines are typically 20–25 per cent more efficient
d Utilize less luxury accessories, e.g. automatic transmission, power steering and generator, air conditioning	d Total avoidable energy losses are of the order of 30 per cent
e Use steel-belted radial tyres	e Can yield economies of 3–10 per cent
f Utilize fuel injection systems	f Permits leaner air/fuel ratio, and fuel shut-off during deceleration
g Reduce aerodynamic drag	g Better body design can give fuel savings
h Improve engine design	h Stratified charge engines have considerable potential
i Adjust the quality of petrol	i There is an optimum octane number that minimizes fuel requirements; varies with type of crude, whether leaded/unleaded petrol, type of emission control system
2 Improve efficiency of commercial vehicle operation	
a Adopt appropriate technical improvements from (1) above	a E.g. radial tyres, aerodynamic design
b Increase vehicle load factor	b Reduce number of trips to delivery point
3 Better driver education	
a Better car care	a More frequent maintenance, including engine tune-ups, reduces fuel usage
b Reduce maximum speeds	b Use monitoring/metering device in car to determine optimum speed to minimize fuel consumption on open road (varies from 65–90 km/h according to car)
c Promote improved driver training/habits	c E.g. minimize rapid acceleration/deceleration, poor usage of gears; employ short warm-up periods
4 Improve traffic movement	
a Provide coordinated traffic signals and clearways in urban areas	a Advantages (including fuel economy) must outweigh costs
b Use traffic signals to meter access to congested roads	b As for (a)
c Provide priority lanes for commercial vehicles at particular locations, design roads in urban areas with greater consideration given to the needs of commercial vehicles	c As for (a)

5 *Promote more efficient public transport operations*
a Improve bus service
b Place greater emphasis on energy-saving equipment

c Eliminate redundant services

d Utilize high-technology, low-energy guideway systems at appropriate locations

e Improve fare collection procedures on buses

6 *Promote more efficient car and public transport occupancies*
a Create car-free precincts in high-activity centres

b Reduce car parking provision, enforce parking regulations, introduce parking fees in high-activity centres
c Modify legal restrictions which prohibit private cars from carrying fare-paying passengers, taxis from multiple hiring, and other paratransit systems from operating
d Use differential fares to increase off-peak public transport usage

e Use pricing mechanisms to limit ownership and/or usage of the private car, and promote greater occupancies of public and private transport vehicles

f Ration petrol

7 *Improve low-cost communication technology*
a Reduce 'trips in person' with aid of e.g. group video tele-conference facilities, or databanks with computer links and visual display units

8 *Use land use powers to promote more efficient movement systems*
a Redevelop parts of existing cities to have high-density clusters capable of being serviced more efficiently by public transport

b Redevelop existing cities, and develop new cities, so as to minimize trip lengths

a Provide priority for buses on congested roads
b E.g. more efficient bus motors; improved traction motor, lighter cars, regenerative braking for rail systems; better bus and rail maintenance
c Parallel and competing bus lines and competitive bus–rail routes should be minimized
d People-mover systems carrying large numbers of passengers at moderate speeds over short distances in high-activity centres such as central areas of large cities and airports
e Idling time at bus stops is minimized

a Will promote car-pooling and bus-pooling as well as higher public transport utilization
b As for (a)

c Potential for car-pooling can be better realized when fee-charging is possible; institutionalized paratransit modes have considerable potential
d Off-peak (unlike peak) public transport is sensitive to lowering of fares in cities
e Various ways, e.g. higher car registration fees, higher sales taxes, increase taxes on petrol, higher parking charges

f More democratic than pricing, but very cumbersome and promotes black-market activities

a Group tele-conferences obviate the need for certain business trips; computer access to databanks could significantly reduce trips that are primarily information-based, e.g. educational lectures, library visits, comparison shopping

a E.g. redevelop so as to increase residential densities about major public transport stops
b E.g. 'non-heavy' industry and service trade concentrations could be interspersed with housing within the income range of workers in these concentrations

Fig. 1.9. The centre curve in this figure refers to a car-saturation level of about 0.5 car/person in Britain in the distant future, while the alternative levels of 0.4 and 0.6 represent a likely range of uncertainty. Application of the projections given in this figure to population forecasts enables estimates for the total number of cars for various years in the future to be determined.

In contrast, the official vehicle predictions made available by the Department of Transport do not attempt to predict precisely what will happen in the future. Rather, the Department recognizes the uncertain nature of forecasts by using a range of assumptions giving a range of traffic flow estimates. Its forecasts for car traffic are a 'best judgement' compromise between results obtained from two models, one extrapolatory and the other causal. (For comparison purposes, the Department of Transport's proposed car-ownership levels, as predicted in 1976, are also shown in Fig. 1.9.) Its forecasts for goods vehicle traffic allow for changes in the output moved by road, the average length of haul, and the changing composition of the lorry fleet.

The low governmental forecasts given in Fig. 1.9 assume a growth in the gross domestic product per capita of 66 per cent, and an increase in real petrol prices to the motorist of 200 per cent, over the period 1976–2000. The high forecasts assume a growth in GDP per head of 108 per cent, and a real petrol price increase of 54 per cent, over the same period.

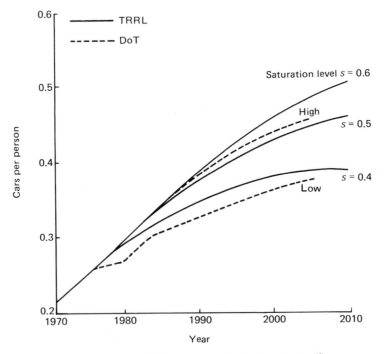

Fig. 1.9 National forecasts of future cars per head of population[12]

Table 1.11 Population forecasts for Britain ($\times 10^6$)

Year	Population
1975	54.38
1980	54.14
1985	54.47
1990	54.96
1995	55.37
2000	55.76
2005	55.91

Table 1.11 shows the population forecasts used in the development of the national data shown in Fig. 1.9. The inherent uncertainty in respect of projections into the future is reflected in the fact that these population forecasts are considerably down on those projected in previous years, e.g. in the late 1960s, Britain's forecasted population for the year 2010 was 68.21m.

The most recent national road traffic forecasts of growth in vehicle numbers[46] (long term, as of December 1984) are shown in Table 1.12. (See also Table 5.2.) Bus and coach numbers are not forecast; however, bus and coach traffic is forecast to maintain its 1982 level.

Table 1.12 Department of Transport long-term forecasts of growth in vehicle numbers in Britain (1982 = 100)

Year	Cars		Heavy goods vehicles		Light goods vehicles	
	High	Low	High	Low	High	Low
1982	100	100	100	100	100	100
1990	126	118	114	106	116	110
1995	138	126	118	105	132	117
2000	148	132	122	106	147	123
2005	157	138	126	105	162	129
2010	164	142	—	—	179	136
2015	170	146	—	—	198	143

Transport policy

In June 1977, after a lengthy consultative process, the Government published a major White Paper[47] which summarized its views in respect of transport developments for the future, with particular emphasis on public transport. The main conclusions of the fundamental review which preceded the policy statement were as follows.

(1) Expenditure in support of public transport and to moderate increases in fares should not be reduced (as had been planned) despite the declining share of public expenditure to be devoted to transport. This meant more support for buses, and less for road construction.

(2) More responsibility for planning transport to meet local needs

should be devolved to local government, since the most practical and democratic approach to coordination is local.

(3)   Greater emphasis should be placed on getting value for money, both in the financial regimes for the public sector industries and in the framework for decision-making on particular schemes and services.

(4)   There should be a more systematic and open involvement of people, Parliament, transport operators and unions in the continuing debate on transport and the formulation of policy, and this should be promoted by the regular provision of information by the Government and the transport industries.

(5)   Transport planning should be flexible so that it can face uncertainty and respond promptly to changing needs and developments in both the short and long term. Its assumptions and forecasts should be kept under review and revised as objectives and events require.

(6)   Decisions and planning as they affect people's needs for transport should recognize the changed realities in transport. For the future, the nation should aim to reduce its absolute dependence upon transport, see that transport makes its contribution to the conservation of energy resources, and key transport into the Government's wider policies for the environment and its views on how communities should develop.

The main policy decisions taken by the Government as a result of the above review are of considerable importance, and the reader is referred to the White Paper for details. In effect, they point the directions which transport developments will take into the 1990s. An interesting critique of the policy paper as a whole is available in the literature[48].

Highway policy

The highway policy decisions referred to in the statement of transport policy were elaborated upon in subsequent White Papers[49–51]. Basically, these policy documents indicate that the Government intended to stabilize spending on trunk roads in the 1980s at a level that it considered the nation could afford. Accordingly, after examining the road programme, it divided the proposals into the following three priority categories.

Roads which aid economic recovery and development   The first priority of the Government is national economic recovery, and it has judged the trunk road programme in that context. Thus it has given the highest priorities to highways which aid the economic effort. The prime example in this category is given as the M25 orbital motorway around London which will link up all major routes radiating from the capital, and help exports by improving transport links with Europe and with London's airports. In the same way, routes to major ports are of clear economic importance and are therefore noted as likely to receive a high priority for improvement. Another stated important aim is to build roads which remove restraints on industrial development, particularly where poor communications make it difficult to attract new firms to replace declining industries.

Roads which bring environmental benefit   There are still many towns and villages which are regularly choked by heavy traffic. Many of the worst problems are on major industrial routes and routes to ports, and these are given a high priority for both economic and environmental reasons. In addition, however, many towns on non-industrial routes suffer from congestion problems which could be very considerably alleviated by the provision of bypasses. Roads noted as falling into this category include bypasses of historic towns.

Preserving the investment already made Since the beginning of the 1960s, some 2000 km of motorway have been constructed in England at a cost (1980 prices) of about £4000m. There are a further 8000 km of trunk roads in England of which about 2250 km are dual carriageway. One of the Government's key aims is to ensure the proper protection of these national assets, a number of which (for example, sections of the M1, M5 and M6 motorways) are regarded as being in particular need of major renovation as they reach the end of their first phase of life.

The Government also stated that it aims to create a climate in which there will be fair competition between the various forms of transport. While it would continue to support public transport, it would not dictate to people how they should travel between cities or move freight. Public transport would continue to be important, particularly for long-distance travel, for journeys in the major urban areas, and for people who do not have the use of a car. However, for the majority of journeys outside the centres of conurbations, private cars were considered to provide the most convenient way of getting about.

Finally, the Government stated that with the completion of most of the remaining schemes in the inter-urban motorway and trunk road programme by the end of the 1980s, the emphasis would most likely shift: (a) from inter-urban trunk roads to principal local authority roads which complete the network, and (b) from inter-urban roads to those in towns and cities, especially London.

Selected bibliography

(1) Collins HJ and Hart CA, *Principles of Road Engineering*, London, Arnold, 1936.
(2) Pannell JPM, *An Illustrated History of Civil Engineering*, London, Thames and Hudson, 1964.
(3) Anderson RMC, *The Roads of England*, London, Benn, 1932.
(4) Belloc H, *The Road*, Manchester, Hobson, 1923.
(5) Hartman EH, *The Story of Roads*, London, Routledge, 1927.
(6) Jackman WT, *The Development of Transportation in M(.ern England*, London, Cass, 1962.
(7) Webb S and Webb B, *English Local Government: The Story of the King's Highway*, London, Longmans Green, 1913.
(8) Wilkin FT, The highland transport system—Genesis, growth and survival, *The Highway Engineer*, 1979, **26**, No. 2, pp. 9–19.
(9) Konski JL, The bicycle anarchy, Paper No. 10 120 in *Transportation Engineering Journal of ASCE*, 1973, **99**, No. TE4, pp. 757–766.
(10) Posthumus C, *The Story of Veteran and Vintage Cars*, London, Phoebus/ BPC, 1977.

(11) Prestwood-Smith P, The development of transport objectives, in Bonsall PW et al. (Ed.), *Urban Transportation Planning*, Tunbridge Wells, Abacus Press, 1977.

(12) Tanner JC, *Car Ownership Trends and Forecasts*, TRRL Report LR799, Crowthorne, Berks., The Transport and Road Research Laboratory, 1977.

(13) Department of Transport, *Roads for the Future: The New Inter-urban Plan for England*, Cmnd 4369, London, HMSO, May 1970.

(14) Scottish Development Department, *Scottish Roads in the 1970s*, Cmnd 3953, London, HMSO, March 1969.

(15) Welsh Office, *Wales: The Way Ahead*, Cmnd 3334, London, HMSO, July 1967.

(16) Department of Transport, *Transport Statistics, Great Britain, 1973-1983*, London, HMSO, 1984.

(17) Cundhill MA and Shane BA, *Trends in Road Goods Transport 1962-1977*, TRRL Report SR572, Crowthorne, Berks., The Transport and Road Research Laboratory, 1980.

(18) Smith WS, Urban transport co-ordination, *Traffic Engineering & Control*, 1966, **8**, No. 5, pp. 304-306.

(19) Watts PF, Stark DC and Hawthorne IH, *British Postbuses—A Review*, TRRL Report LR840, Crowthorne, Berks., The Transport and Road Research Laboratory, 1978.

(20) Martin PH, *The Harlow Dial-A-Bus Experiment: Comparison of Predicted and Observed Patronage*, TRRL Report SR256, Crowthorne, Berks., The Transport and Road Research Laboratory, 1977.

(21) Dinefwr Working Group, *Rural Transport Experiments: The Welsh Schemes*, TRRL Report SR507, Crowthorne, Berks., The Transport and Road Research Laboratory, 1979.

(22) Tebb RGP, *Differential Peak/Off-peak Bus Fares in Cumbria: Short Term Passenger Responses*, TRRL Report SR391, Crowthorne, Berks., The Transport and Road Research Laboratory, 1978.

(23) Buchanan CD et al., *Traffic in Towns*, London, HMSO, 1963.

(24) Sando FD and Batty V, Road traffic and the environment, *Social Trends*, 1974, **5**, pp. 64-69.

(25) Department of Science, *Final Report of the Committee on the Problem of Noise*, Cmnd 2056, London, HMSO, 1963.

(26) Millard RS et al., *A Review of Road Traffic Noise*, TRRL Report LR357, Crowthorne, Berks., The Transport and Road Research Laboratory, 1970.

(27) *Noise Insulation Regulations 1973*, Statutory Instrument 1973, No. 1363, London, HMSO, 1973.

(28) *Noise Insulation Regulations 1975*, Statutory Instrument 1975, No. 1763, London, HMSO, 1975.

(29) Sargent JW, Noise, Paper 1 in *Proceedings of the Colloquium on Highways and the Environment*, pp. 3-4, London, The Institution of Civil Engineers, 1978.

(30) Colwill DM, Pollution, Paper 3 in the colloquium cited in reference 29, pp. 7-8.

(31) *Lead and Health*, Report of a DHSS Working Party on lead in the environment, London, HMSO, 1980.

(32) Dawson RFF, Visual effects, Paper 2 in the colloquium cited in reference 29, pp. 5-6.

(33) Van Rest DJ and Hearne AS, Agricultural severance, Paper 5 in the colloquium cited in reference 29, pp. 11-12.

(34) Departments of Transport and of the Environment, *Report on the Review of Highway Inquiry Procedures*, Cmnd 7133, London, HMSO, April 1978.

(35) Report of the Ad Hoc Committee on Energy, *Journal of the Institution of Engineers, Australia*, 1975, **47,** Nos 7 and 8, pp. 25-30.

(36) Clegg MW, A brief survey of energy resources, in *Proceedings of the Conference on Transport for Society*, pp. 31-36, London, The Institution of Civil Engineers, 1976.

(37) Department of Energy, World energy resources: present position and prospects, in *Proceedings of the National Energy Conference held at London*, Energy Paper No. 13, London, HMSO, 1976.

(38) Department of Energy, *Energy Policy: A Consultative Document*, Cmnd 7101, London, HMSO, February 1978.

(39) Department of Energy, *Development of Oil and Gas Resources of the United Kingdom 1984*, London, HMSO, 1984.

(40) Hooper PO and Mullen P, The effect of increased fuel prices on car travel, *Traffic Engineering & Control*, 1974, **15,** No. 16/17, pp. 728-731.

(41) Mogridge MJH, The future of the car, *Traffic Engineering & Control*, 1976, **17,** No. 1, pp. 101-104 and 115.

(42) Ellis JR, The future of the vehicle, *The Highway Engineer*, 1976, **23,** No. 5, pp. 20-25.

(43) *Energy and Transport*, London, The British Road Federation, 1980.

(44) O'Flaherty CA, Energy, technology, and movement, in *Resolving Conflicts in Transport*, Volume 1, pp. 149-172, Melbourne, Victoria, The Chartered Institute of Transport, 1977.

(45) Tanner JC, *Saturation Levels in Car Ownership Models: Some Recent Data*, TRRL Report SR669, Crowthorne, Berks., The Transport and Road Research Laboratory, 1981.

(46) *National Road Traffic Forecasts 1984*, London, The Department of Transport, April 1985.

(47) Department of Transport, *Transport Policy*, Cmnd 6836, London, HMSO, June 1977.

(48) Beesley ME and Gwilliam KM, Transport policy in the United Kingdom; a critique of the 1977 White Paper, *Journal of Transport Economics and Policy*, 1977, **21,** No. 3, pp. 209-223.

(49) Department of Transport, *Policy for Roads: England 1978*, Cmnd 7132, London, HMSO, April 1978.

(50) Department of Transport, *Policy for Roads: England 1980*, Cmnd 7908, London, HMSO, June 1980.

(51) Department of Transport, *Policy for Roads: England 1983*, Cmnd 9059, London, HMSO, September 1983.

2
Highway finance and administration

The public roads of Britain are a government responsibility. They are supported by taxes raised from the people and maintained and operated by the central government and the local authorities through their civil servants. This responsibility is a necessary one, due not only to the magnitude of the highway network, but also because of the unique role that it plays in the national economic effort. Indeed, the influence of the road network is now such that major, and sometimes minor, improvements inevitably affect the lives, and even the fortunes, of the people living and working nearby.

This governmental relationship has caused a particular type of administration to develop in order to manage the highways and cope with the problems which they create. Since the highways necessitate the expenditure of vast sums of money, while at the same time they often impinge on the basic civil rights of the individual, much legislative effort has been exerted to protect the interests of the public. As a result, the young highway engineer entering upon a professional career and putting his or her special technical qualities to work is often confounded by the multiplicity of administrative and legislative factors which must be taken in before a position of executive responsibility can be attained.

The function of this introductory chapter therefore is to give a brief perspective of highway administration and legislation, and the manner in which they function to provide the highway transportation service. Intimately bound up with administration and legislation are the methods by which the highways are classified and financed, and so these are discussed also.

Legislation

The basis of all highway administration and finance lies in the legislative process which brought them into being. In formulating the laws which enable finance to be raised and allocated, and administration to function, Parliament indirectly as well as directly set out and defined the objectives of performance, thereby outlining the processes now in existence. The following is therefore a brief summary of the legislative history which underlies the operation of the present system. No attempt is made here to refer to all the laws that are of interest; these are literally too numerous and too detailed to cover within the limitations of this text. Instead, emphasis is laid on those laws which, it is felt, have had particular significance in setting in motion events or restrictions that are of particular importance to highways and their traffic today.

Finally, it should be mentioned that, although the Highways Act of 1959 and the Road Traffic Act of 1960 together consolidate most of the pre-1960 laws relating to the creation, management and use of highways and bridges, a significant volume of legislation remains scattered in a variety of statutes. For quick reference to this earlier legislation, the reader is referred to an excellent aide-mémoire that is available in the technical literature on highways[1].

Pre-World-War-I legislation

Prior to 1888, the roads of Britain were administered by highway boards, the area of their responsibilities being determined by the grouping of a series of parishes, each of which had the right of electing a member to such boards. The roads under the control of these bodies were all rural in character, while those passing through built-up areas were directly maintained by county borough councils, municipal corporations or boroughs, and local boards which, in addition to having the duties and powers of maintaining the highways, were the local authorities for such purposes as public health and public utility services. At that period, relatively negligible grants were made to the highway boards from the central government towards the cost of many local services, including the cost of through roads.

In 1888 and 1889, two most important Acts, the Local Government (England and Wales) Act of 1888 and the Local Government (Scotland) Act of 1889, brought into existence local authorities known as county councils which were given far-reaching duties dealing, amongst other matters, with public health, police, lunacy, small holdings, contagious diseases of animals, and 'main' roads. In addition to being the highway authorities for the erstwhile Turnpike Trusts, the county councils had the power to declare as main roads other roads which in their opinion should be so declared by virtue of their through traffic and the total volume of traffic. The obvious intention of these laws was to ensure that the maintenance of main roads should become a county, as against a district or local, charge. Under these Acts, the county boroughs were to maintain all roads within their districts.

The reform in highway administration initiated by the Local Government Acts of 1888 and 1889 was extended by the Local Government Act of 1894. This legislation abolished the highway boards and highway parishes and transferred the duty of maintaining roads other than main roads, in rural areas, to rural district councils.

What may well be described as the legislation which signalled the emancipation of the motor car was passed in 1896 in the form of the Locomotive on Highways Act. This Act abolished the requirement that a road locomotive had to be preceded by a man carrying a red flag, and gave to it the right to travel upon the highway at a speed not exceeding 22.5 km/h.

The year 1909 saw the beginning of official government recognition that special measures were needed on a national level to ensure adequate road facilities. This was signified by the passing of the Development and Road Improvement Funds Act of 1909 which constituted a Road Board with powers to use money raised from vehicle and petrol taxation to build new roads and to carry out schemes to widen and straighten existing roads, to bypass villages, and to allay the dust nuisance. This Act, for the first time, brought a central government department into being that really had power to define which roads should be given special assistance, and what form the improvements should take. Money for these purposes was raised by a petrol tax and a graduated scale of motor vehicle licence duties based on a horsepower rating defined by the Royal Automobile Club and paid into a fund called the Road Improvement Fund; in time this became known as the Road Fund.

The year 1909 also saw the passing of the Housing, Town Planning, Etc., Act, the purpose of which was to ensure that land in the vicinity of towns should be developed in such a way as to secure proper sanitary conditions, amenity and convenience. This legislation is of significance in that it disclosed for the first time the gulf which, unfortunately, exists in many instances to this day between the highway engineer and the town planner. The emphasis in the Act was placed upon securing proper public health considerations, within which the social implications raised by the construction of a new road or road system were not at all taken into account. As a result, and more by default than any other reason, the planning and design of roads were regarded as primarily the responsibility of the engineer and only secondarily as part of the town planning process.

Legislation between the world wars

In 1919, for the first time in its history, the Government created a Ministry of Transport to which the powers and duties of the Road Board, together with its officers and staff, were transferred. The Ministry of Transport Act empowered the Minister of Transport to devise and implement a road classification system after consultation with the Roads Advisory Committee and the local authorities.

Following this legislation, the country was divided into administrative divisions, each under the jurisdiction of a Divisional Road Engineer who was expected to study the road and bridge problems in an area, and to pronounce upon proposed future developments as they were submitted by the local highway authorities.

The year 1920 saw the passing of two important pieces of legislation, the Finance Act of 1920 and the Roads Act of 1920. The Finance Act created a simple system of taxing cars which, in essence, remained in operation until after World War II. For the first time also, goods vehicles were taxed more severely than passenger cars. The Roads Act established the form of registration and licensing which is in use to this day. The county and county borough councils were made responsible for registering the vehicles and collecting the duties imposed. An important byproduct of the Roads Act was that for the first time it was possible to obtain accurate data regarding the numbers of the different types of vehicles in use.

The Roads Act of 1920 also provided that contributions for the maintenance of existing roads could be granted from the newly constituted Road Fund, which took the place of the Road Improvement Fund.

The Road Improvements Act of 1925 gave to highway authorities the power to prescribe building lines along roads and, at bends and corners, to control the height and character of walls, fences, hedges, etc., and to restrict the erection of new buildings where adequate sight distances were required. The Act also authorized the Ministry to conduct research and experimental work; this led eventually to the establishment of what is now known as the Transport and Road Research Laboratory.

The Finance Act of 1927, for the first time since mechanically propelled vehicles were taxed, regarded a portion of the receipts from ordinary motor car taxation as a luxury tax, and used that portion as a contribution to the general exchequer expenses of the country.

The Local Government Act of 1929 recognized that the maintenance and repair of roads was in the hands of too many minor local authorities and set about reducing this number. Prior to this Act, by far the greatest amounts of roads, known as 'district roads', were under the control of some 600 rural district councils. With the passing of the Act, these bodies were eliminated as highway authorities and their roads, equipment, plant and officers were transferred to the county councils; the maintenance and improvement charges for these roads then became county-at-large charges. In like manner, municipal boroughs and urban districts with populations less than 20 000 also ceased to be highway authorities and the roads in their areas became county roads and the responsibility of the county councils. The larger urban districts in England and Wales were left in undisturbed possession of their roads with one major proviso: where a road not previously designated as a county main road was classified by the Ministry of Transport, then that road became a county main road for the expenditure on which the county council and the Ministry of Transport became responsible. The position of county boroughs remained unchanged. In Scotland, the small burghs were also dispossessed of their road maintenance powers, and the residents of these localities became for the first time county ratepayers for highway purposes.

In 1928, a Royal Commission on Transport, which was established to consider problems arising out of the growth of the motor vehicle population and usage, concluded that some regulation of public transport was necessary in the public interest. The result was the passing of the Road Traffic Act of 1930 under which control of public service vehicles was vested in Area Traffic Commissioners appointed by the Ministry of Transport. Their function was to fix and maintain fares, to sanction routes and timetables, and to eliminate any unnecessary services. Every operator of a public service vehicle was required to obtain a licence for the vehicle, and upon renewal had to justify the service provided against any rival provider of transport who might object. The net effect of these requirements was to protect established public transport operators; this was at the price of tying them down to fixed lines and thereby reducing their flexibility. This Road Traffic Act also revised the speed limits for certain types of commercial vehicles, while it swept away all limits for motor cars and motor cycles, thereby making it possible for the first time for highway engineers to design officially for high vehicle speeds. It made third party insurance obligatory and set standards and penalties for careless and dangerous driving.

The year 1933 saw the passing of the Road and Rail Traffic Act which attempted to regulate with respect to goods vehicle licensing and to ensure that commercial traffic paid a greater share of the cost of the road system. A second Road Traffic Act was passed in 1934 which imposed an overall speed limit of 48 km/h in all built-up areas, i.e. roads in areas where systems of street lighting were in existence. At the time, it was considered that this was the universal answer to the road accident problem; this, unfortunately, as is well known today, was not the case. The second major feature of this legislation was that it introduced a driving test for all drivers who did not hold a licence before 1 April 1934. A special driving licence was also required for drivers of heavy commercial vehicles.

The Road Traffic Act of 1933 also permitted advances from the Road Fund to local authorities for the purpose of erecting traffic signs. In addition, the Minister of Transport was empowered to make regulations regarding the use of pedestrian crossings, and the county councils were enabled, under certain conditions, to provide street lighting.

In 1935, the Government awoke to the dangers of ribbon development and its effects upon the free movement of vehicles on the road. As a result, it passed the Restriction of Ribbon Development Act which made it unlawful to construct, form or lay out any means of access to or from a classified road without the consent of the highway authority. This Act also incorporated a new system of defining the land needed for future roads and road improvements.

The following year another most important legislative Act was passed. This was the Trunk Roads Act of 1936 which for the first time made the Ministry of Transport a highway authority in its own right. This was done by giving the Ministry complete financial responsibility for 7249 km of major through routes outside the boundaries of London, the county boroughs in England and Wales, and the large burghs in Scotland. The Act also made provision for delegating these highway authority powers to the local authorities so that they could act as agent authorities for the Ministry for these roads.

The final important pre-war piece of legislation was the Finance Act of 1936. This Act required the Ministry of Transport to draw its revenue for highway purposes from sums voted by the House of Commons. Whereas, before, part of the receipts from vehicle taxation automatically went to the Road Fund where, *as a right*, they were devoted to road improvement, now the Ministry of Transport had to budget for its monies in a manner similar to, and in competition with, other ministries.

Post-World-War-II legislation

The first major road legislation after the war was the Trunk Roads Act of 1947. This Act transferred another 5929 km of Class I roads to the responsibility of the Ministry of Transport, together with all private bridges on the trunk roads. On this occasion, the Minister became responsible also for main roads designated as trunk roads which passed through county boroughs and through London. In addition, and for the first time, provision was made for the two carriageways of a dual carriageway road to diverge for reasons of engineering or property so that the land in between need not be considered part of the roadway. Furthermore, the Minister was required to keep the national system of through routes under constant review, and to reorganize it as changing circumstances demanded.

In 1947, the Town and Country Planning Act was passed and this had the effect of requiring planning authorities to prepare plans showing the intended use of all land within their areas. This Act resulted in the coordination of highway and land use planning since road proposals now had to be shown on the statutory development plans, and their relationship to land use and development was apparent to all. Furthermore, it meant that road proposals would be kept up-to-date since the development plans had to be revised every five years.

The year 1947 also saw the passing of the Transport Act which set up a British Transport Commission for the purpose of securing an efficient and properly integrated system of certain public transport and port facilities within Britain. This Act, which gave considerable powers to the Ministry of Transport in that it enabled the Minister to give directions to the Commission as to the exercise and performance of its functions in relation to matters which, in the Minister's view, were of national interest, marked the first major move towards having a coordinated public transport system in Britain.

The motorway era in Britain can be said to have begun with the passing of the Special Roads Act of 1949. This Act is of major importance in that it marks the departure from the previously held principle that everyone normally had the right of free access along the highway; now it became possible to construct roads reserved for the exclusive use of certain classes of traffic. By a government decision made in 1946, motorways were considered to be reserved for the purpose of facilitating the movement of long-distance motor traffic and of connecting some of the main centres of population.

The Transport Act of 1953, which provided for the winding-up of the Road Haulage Executive of the Transport Commission and the transfer of the Railway Executive's powers to area authorities, signified the enfranchisement of road goods transport.

The first of what were to be two major traffic management legislative measures was passed in 1956 in the form of the Road Traffic Act. Under this Act, compulsory annual tests were introduced for all vehicles aged ten years and older, as were stricter regulations on the sale of unroadworthy vehicles and the renewal of driving licences. In a further attempt to improve road safety, the penalties for dangerous driving were increased and extended to cover pedal cyclists, and pedestrians became legally bound to obey the directions of police on traffic duties. In addition, the Minister of Transport was empowered to approve local authority schemes fixing parking places and charges, and to sanction experimental traffic management schemes by the police. These latter provisions were brought about by the need to ease traffic congestion within the London metropolitan area.

The second bill of particular importance with respect to traffic management was the Road Traffic and Road Improvement Act of 1960. This Act provided the Minister of Transport with the power to take more effective action in dealing with the traffic, parking and roadworks problems in Greater London. Local authorities were also given new powers to make Orders for the purpose of a general scheme of traffic management in a specified area, as well as in connection with the provision of off-street parking facilities. In addition, provision was made for the police to appoint traffic wardens, while members of the public were given the option of paying fixed penalties instead of being prosecuted for certain traffic offences.

About this time also, a number of most important measures were taken in order to simplify and bring up-to-date the law with respect to roads and their traffic. These measures involved the passing of a number of consolidating Acts, namely, the Highways Act of 1959, the Road Traffic Act of 1960, and the Vehicle Excise Act of 1962.

In 1961, the Highways (Miscellaneous Provisions) Act was passed which made it possible for the first time to bring an action against a highway authority in respect of damage resulting from its failure to maintain a highway maintainable at the public expense. In 1962, another Road Traffic Act, which was primarily concerned with promoting road safety, granted more flexible powers to impose speed limits; a penalty system for traffic offences was also presented.

The Transport Act of 1962 provided for the reorganization of nationalized transport. The main change was the dissolution of the Transport Commission, and the division of its functions between four separate boards, appointed by the Minister of Transport, for British Railways, London Transport, British Transport Docks and British Waterways. A Transport Holding Company was appointed to take over all the commercial activities, i.e. bus companies, road haulage and travel agencies, run in company form.

The Road Safety Act of 1967 was of particular importance in that it clarified the situation regarding drinking and driving. This Act made it an offence to drive or attempt to drive if 100 ml of blood contained more than 80 mg of alcohol, and gave the police the power to impose selective roadside breath tests and to require blood or urine samples if the tests were positive. The existing offences of driving or being in charge while under the influence of drinks or drugs were retained. The Act also increased governmental powers with regard to roadside checks on commercial vehicles.

The Transport Act of 1968 marked a major breakthrough with regard to governmental support for public passenger transport. It gave the Minister of Transport and the Secretaries of State for Scotland and Wales the power to designate any area outside London as a Passenger Transport Area, with both an Authority and an Executive to integrate and develop public transport services. The governmental National Bus Company acquired the Transport Holding Company's bus operating and manufacturing companies, except for those vested in the Scottish Bus Group (responsible to the Secretary of State for Scotland). The Act also increased the fuel grant for stage-carriage services, provided 50:50 exchequer:local authority grants for uneconomic bus and ferry services and 25 per cent grants to buy new buses, as well as empowering the central government and local authorities to make grants towards approved capital expenditure on public transport facilities. The purposes for which local authorities could make Orders regulating traffic were widened to help buses, pedestrian movement and amenity; the authorities were also enabled to restrict parking meters to specific classes of users, and to use surplus meter revenues for purposes connected with public transport or road provision.

Whilst it had been realized for many years that roadway decisions could be the cause of major detriments to adjoining properties, it was not until the passing of the Land Compensation Act of 1973 that adequate machinery was available to compensate those whose properties were adversely affected by road and redevelopment schemes. Under this legislation, the local authority is required to consider the design of new road schemes in conjunction with the treatment of adjoining areas, i.e. as a single environmental unit. Thus, for example, the combined approach may require the construction of noise abatement barriers between the properties and the road or the installation of noise insulation in adjacent buildings; however, in the event that these measures are not sufficient to protect the properties from noise damages which result in lower property values or a reduced quality of life, compensation must be paid in lieu.

The Road Traffic Act of 1974 amended a number of previous Acts relating to the regulation of traffic on the public highway. Its provisions ranged from increases in fines for various traffic offences, to the elevation to a duty of the requirement that local authorities should promote road safety, to making it legal to install road humps on carriageways to control the speeds of vehicles, to changes in driver licensing which mean that licences are now generally valid to age 70 years without renewal.

The cost of land acquisition is often the decision-maker as to whether a highway scheme proceeds at all, is delayed or not, whether one route is preferred to another, or indeed whether some other solution is chosen for implementation. The Community Land Act, passed in 1975, is intended to ensure that the development value in land is transferred to the Government; it does this by requiring the ownership of the land to pass through the hands of the governmental authority prior to development. When this happens, the basis of compensation (eventually) will be the current use value of the land rather than its full market value. From the highway engineer's viewpoint, this Act should ensure that highway land, especially in urban areas, can be purchased at a lower price than would otherwise be the case.

Most of the provisions of the Transport Act of 1978 stem from decisions taken by the Government in 1977 in relation to transport policy. For example, the Act requires non-metropolitan county councils to publish and up-date annually a five-year public transport plan, makes legal the operation of regular car-pools for payment, and authorizes local authorities to license privately-operated public car parks.

It is widely accepted that two-wheeled motor vehicles are a comparatively dangerous form of transport. Riders of these vehicles are younger on average than drivers of other motor vehicles and tend to be more inexperienced. The Transport Act of 1981, and associated regulations, therefore sought to reduce two-wheeled vehicle accidents by encouraging proper rider-training, by the following means: (a) introducing a two-part test for learner motor cyclists, (b) limiting the length of time for which a provisional licence may be held to two years, and (c) reducing the maximum engine size of motor cycle which learners may ride from 250 cc to 125 cc.

Other regulations made under the Transport Act of 1981 made seat-belt wearing compulsory for front-seat occupants of cars and light vans from 31 January 1983. The objective in this instance is to reduce the severity of the injuries incurred in accidents.

The Transport Act of 1985 provided for the reorganization of road passenger transport in the public sector, and related passenger transport functions to the local authorities described in the Local Government Act of 1985. This Transport Act is of significance in that it suggests a loosening of governmental controls, and a search for greater accountability, in the public transport arena, e.g. it makes provision for the return of the National Bus Company to the private sector, and for the transfer of other bus undertakings to companies limited by shares registered under the Companies Act of 1985. These companies—which in metropolitan areas outside London are owned by Passenger Transport Authorities and operated by Passenger Transport Executives, whilst in non-metropolitan areas they are owned and operated by councils—are required under the Transport Act not to inhibit competition between persons seeking to provide public passenger transport services in their jurisdictional areas. The Act also provides for the establishment of a Disabled Persons Transport Advisory Committee to advise the Minister of Transport regarding the public passenger transport needs of disabled people.

Classification and nomenclature

Britain has one of the most comprehensive and penetrating roadway systems of any country in the world. The statistics show that there are about 345 000 km of surfaced roads at this time on the island. Table 2.1 gives the breakdown of the different types of road forming this total.

Trunk roads

Following the Trunk Roads Acts of 1936 and 1947, the central government assumed responsibility for 13 178 km of main road which were to form the nucleus of a national system of routes for through traffic. As is suggested by the statistics in Table 2.1, these early trunk roads still form the core of Britain's main road system.

Table 2.1   Composition of the highway system in Britain[2]

Road type	Kilometres of roadway
Trunk roads	
motorway	2606
non-motorway	12321
Principal roads	
motorway	103
non-motorway	34587
Other roads	295361

The term 'trunk road' does not guarantee that a particular level of service will be provided on any highway bearing the title. It is an all-embracing term which includes roadways ranging from very busy motorways to ones less than 5.5 m in width and carrying very low volumes of traffic. In general, however, the level of service provided by any given trunk road is a relative function of its national and regional importance.

By means of enabling legislation which became law in May 1949, special roads can be reserved exclusively for particular classes of traffic, e.g. motor vehicles, or pedestrians, or pedal cycles, or other prescribed traffic. The most important of these is the motorway, usage of which is restricted to motor vehicles and certain types of motor cycles. Apart from its high-quality geometric design standards, the most significant features of a motorway are that there is complete control of access and complete separation of all conflicting traffic movements. The fact that a roadway is constructed as a motorway does not, however, automatically mean that it is a trunk road.

It might also be noted here that the Department of Transport has designated a network of routes as 'primary' routes. These comprise the most important traffic routes, and include nearly all trunk roads and many principal (non-trunk) A-roads.

Principal and classified roads

Prior to 1967, Britain's classified non-trunk roads were divided according to whether they were Class 1, 2 or 3 roads. Class 1 roads were primarily of regional importance in that they connected large towns or were of special importance to through (non-town) traffic. Class 2 roads were not of sufficient importance to justify Class 1 status, yet formed important secondary connections with the smaller population centres. Class 3 roads, which were mostly local distributors, composed the remaining classified roads.

This governmental classification system was abolished in 1967, and a new system created which, in essence, made all Class 1 roads into principal roads. Table 2.1 shows that there are 34 587 km of principal roads in Britain, only about one-tenth of which are within urban areas.

'Other' roads

These are literally the remaining surfaced highways within the national (urban and rural) network. They are composed of 108 974 km of classified non-principal roads and 186 387 km of unclassified collector and access roads.

Highway nomenclature

As will have been gathered, the *classification* of a highway is related to its general importance within the national system; it also has implications with regard to the manner in which roads are funded. It is important not to confuse classification with *nomenclature* which simply refers to the numbering system by which given roads are known.

A glance at any road map will show that all roads other than motorways (M-roads) are first divided according to whether they are A-roads or B-roads. The A-roads—which tend to be the old trunk and classified roads—are then generally subdivided into three, indicated by the number of digits after the letter, whilst all the B-roads and some A-roads have four digits after the letter.

There are nine one-digit primary roads, six of which are centred on London, and the remaining three on Edinburgh. London-based roads are the A1 to Edinburgh and (proceeding in a clockwise direction) the A2 to Dover, A3 to Portsmouth, A4 to Bristol, A5 to Holyhead, and A6 to Carlisle. The Edinburgh-based roads are the A7 to Carlisle, A8 to Glasgow, and A9 to Wick.

Between each of the primary roads, there are other important roads which are given double numbers, starting with the initial primary number. Thus these roads between, say, the A3 and A4 start with the number 3, e.g. the A30 between London and Penzance.

Other lesser A-roads have three digits after the letter, and these start with the appropriate figure for the sector, e.g. the A483 between the A4 and A5, and joining the A48 with the A5. All B-roads and some A-roads of lesser importance have four figures, starting with the primary digit of the district in which they are located, e.g. the A3072 between Bude and Bickleigh, or the B2080 between Tenterden and New Romney.

Finance

Highway grants

In the financial year ending 1984, about £4.3 billion was spent by governmental authorities on transport in Britain[2], as follows: trunk roads, 17.0 per cent; local roads and car parks, 31.4 per cent; public transport (a) central government subsidies, 34.1 per cent, and (b) local investment, 5.9 per cent; ports, shipping and civil aviation, 2.7 per cent; local authority administration, 4.0 per cent; central government administration, 1.0 per cent; other expenditure, 3.9 per cent. Further examination of this national expenditure shows that more than half of the total figure is spent on roads. Also, about 47 per cent of the total is spent by the central government; 53 per cent is local authority expenditure.

The central government (via either the Secretary of State for Transport in England, the Secretary of State for Scotland, or the Secretary of State for Wales) provides all the funds for the planning, design, construction and maintenance of trunk roads. The decision as to whether or not a road is classified as a trunk road is made by the central government itself.

Prior to 1967, the central government paid for 75, 60 and 50 per cent of the capital costs of major, and certain minor, improvements on the (then) Class 1, 2 and 3 roads, respectively. In 1967, the decision was taken to change the system so that the central government would pay up to 75 per cent of the cost of specific improvements to (including new sections of) principal roads; all other expenditures, including maintenance costs of principal roads as well as maintenance and improvement costs of non-principal and non-trunk roads, had to be paid for out of local authority funds. To alleviate hardship which this change might cause to local authorities, the central government simultaneously introduced a non-specific or block rate support grant (RSG), i.e. a calculated sum given to the local authority from the national exchequer on the basis of approved expenditure estimates, but which the local authority could then divide between its local services as it saw fit.

In 1975, the local authority grant system underwent a major change, and the specific grant element of the old system was replaced by a transport supplementary grant (TSG), provided for by the Local Government Act of 1974. The specific grant system had been subjected to the criticism that it biased resource allocation towards projects eligible for grants (particularly capital-intensive projects which attracted the highest rates of grant), that only projects of one particular type (e.g. road improvement schemes) were evaluated against each other, and that local government did not have enough say in resource allocation. The objectives of the change therefore were to encourage the development and execution of comprehensive transport plans, to eliminate bias towards particular forms of expenditure, to ensure a better distribution of the central government's grant to reflect the needs of individual areas, and to reduce the amount of detailed central government supervision of individual schemes.

Under this system, support is still given through the rate support grant, but the transport supplementary grant is distributed to local authorities in England and Wales with accepted estimates of expenditure on capital works for highways and traffic regulation above a prescribed threshold level (50 per cent in 1985-86). In order to obtain a transport supplementary grant, the appropriate councils must submit annually to the Department of Transport a transport policies and programme document (TPP) which sets out the aims of their transport policies and provides costed five-year programmes to achieve them. In Scotland, all aid is still provided through the rate support grant.

A useful discussion of the development and efficacy of the TPP-TSG system of financing is readily available in the literature[3].

Sources of revenue

As is indicated in the previous discussion, the two main sources of highway revenue are central government grants and local government rating funds. The following discussion is therefore primarily concerned with the financial means by which these monies are raised.

Central government revenues

The road and transport grants provided by the central government are financed from the general revenue of the State and, following allocation by Parliament, they are distributed via the Ministerial system.

The general revenue of the State comes, of course, from many sources, not least of which is the taxation revenue from motor vehicles, i.e. the motor taxation contribution to the national exchequer is typically 11-12 per cent of the national tax revenue. Since the income from motor taxation by far exceeds the national expenditure on roads, it will be assumed for the purpose of this discussion that the money raised by this taxation is in fact the money used to finance the road system.

There are three main forms of motor vehicle taxation. These are vehicle excise duty, value added tax, and motor fuel duty.

Following the Development and Road Improvement Funds Act of 1909, the same graduated scale of *vehicle excise duty* was imposed on all motor vehicles. In 1921, the duties for cars and goods vehicles were separated and this practice has continued ever since. The duties imposed in 1985 are a flat rate of £100 per year for all cars, and a graduated rate for goods vehicles which varies according to unladen weight from £179 for a two-tonne vehicle to £1138 for an eleven-tonne vehicle. Taxis and buses pay a flat rate of £50, although in the case of buses £1 is added per seat over twenty. Motor cycles also have a tax scale which varies according to engine size, viz. £10 (< 150 cc), £20 (150-250 cc) and £40 (> 250 cc).

A *purchase tax* of $33\frac{1}{3}$ per cent was first imposed on cars in 1940, and since then the tax rate has varied at irregular intervals, depending upon national economic needs (e.g. $66\frac{2}{3}$ per cent in 1951, 25 per cent in 1972).

Other than during the period 1950–59, purchase tax has never been levied on goods vehicles. With the introduction of the *value added tax* (VAT) in Britain in 1973, purchase tax was abolished and replaced with a standard VAT plus a special car tax. At this time, the car tax amounts to 10 per cent of the wholesale price, while the VAT is 15 per cent of the retail price (including the car tax).

A *motor fuel duty* was first levied upon motor fuels in Britain in 1909, when a rate of tax of 3d per gallon was imposed on light oils used for motoring purposes, a 50 per cent rebate being given to commercial vehicles. In 1916, this tax was raised to 6d per gallon, and this was levied until 1921 when the motor fuel tax was repealed entirely. In 1928, however, a new petrol tax at the rate of 4d per gallon was again established; this was eventually raised (June 1985) to its current value of 81.56p per gallon (17.96p per litre). Value added tax accounts for an additional 26.46p per gallon (5.74p per litre), so that the total tax on petrol amounts to 108.02p per gallon (23.79p per litre), which is 53 per cent of its retail price. The equivalent total tax on diesel-engined road fuel, derv, is 94.83p per gallon (20.89p per litre), which is 48 per cent of its retail price.

Governmental arguments put forward for increasing the fuel tax have tended to fall broadly into two categories: firstly, the desire to encourage the use of public transport and, secondly, the need to conserve energy.

Local government revenues
Apart from the funds obtained from the central government, a local authority's main source of revenue for roads is the rates levied on non-agricultural property. Rates are actually a form of property taxation which attempts to spread over the property within a rating area the cost of the services expended within that area. In an attempt to spread these costs in a fair manner, the rates for a particular property are normally based on the 'annual value' of the property. This value reflects the ordinary letting value, the 'reasonably expected rent' or the rent that the property will command in the open market. This rent is estimated by a local valuation officer of the Board of Inland Revenue. When the local authority is informed of the annual value of the property, it then levies an assessment against the property and this is called the rate for the property. Responsibility for levying and collecting rates lies with the district councils and, in London, with the borough councils and the Corporation of the City of London.

'Key' loans, which must be approved by the central government, are other means by which local authorities can raise money for roads (usually principal roads). Another minor method of financing particular stretches of new roadway is to charge frontagers for the cost of constructing the new facilities. Thus, for instance, if a new housing estate is being constructed, the roads will be built by the estate contractors and the cost of so doing is then added onto the cost of the houses serviced by these roads.

Brief discussion relative to highway revenue

Roads cost a lot of money to construct and maintain; of this there is no doubt. What is very much in doubt, however, is the equitableness of the presently used methods of raising the required revenue. If road costs were to be charged in direct accordance with the benefits received, then the expenditure on each road or road system should ideally be assessed against the following three main beneficiaries: the general public, adjacent properties, and motorists. That this principle is accepted would appear to be generally reflected in the form in which highway finance is now carried out. Unfortunately, however, the manner in which the costs are practically allocated between the beneficiaries is not based on any scientific evaluation but rather has grown up as a result of practical expediency and need.

Payments by the general public
This concept requires payments to take place in the form of grants from the central government towards the construction and maintenance of highways that are in the national or general interest. Under this concept, the funds allocated for this purpose would come from finance raised by non-motor-vehicle taxation sources.

Roads falling into this national category are undoubtedly the non-urban trunk roads since their function is to serve the national economic needs for long-distance vehicle movement. Central government support for urban motorways should probably also be included since, in a densely populated island country the size of Britain, it can probably be said with justification that in cities where urban motorways are necessary their influence is indirectly, if not directly, national; motorways in such cities as London or Birmingham, to name only two, can certainly be included within this user classification.

It must be pointed out, however, that not since 1932 has the general public as a whole apparently contributed additionally towards the cost of the highway system in Britain. A simple comparison of recent motor vehicle taxation revenues with central government highway expenditure shows quite clearly that no non-motor-vehicle taxation monies have been used for road purposes, but that in fact it is the property owners/ occupiers and the motoring public who have contributed the entire financial support to the road system.

Payments by property owners/occupiers
The second group of people to benefit from the road system are the property owners/occupiers who use it to obtain access to their homes or places of business. Roads falling into the category for which the ratepayers should be primarily responsible can be divided into two types: roads which are used for the direct purpose of providing access to houses and land—the most obvious example is a cul-de-sac—and roads which provide internal communication within the local community, e.g. the streets of a town. The former type of road should ideally be financed by

means of direct assessments upon the owners/occupiers of the homes or land being serviced, while the funds to support the latter roads should be provided by the community's properties as a whole, i.e. from local government taxation.

The financing system in Britain generally attempts to apportion the road costs to the ratepayers according to the benefits derived. Whether the present apportionment to the ratepayers is the correct one, is, of course, not known. However, it is probable that the current apportionments reflect the general order of importance of these roads to the local communities who pay for their upkeep.

Payments by the motorists
It follows that since streets and highways exist for the primary purpose of allowing vehicles to travel upon them, then the road users should contribute a substantial amount towards their construction and upkeep. This principle is accepted not only by most highway administrators but also by the motorists themselves. It is at this stage that agreement ends, however, since it is not possible to say, firstly, just how much of the cost of the road system should be borne by the motorists and, secondly, how the portion to be borne by them should be allocated amongst the various users of the highways.

It is doubtful whether a formula based on scientific investigation will ever be derived that will equitably apportion the road costs between the motorists and the other beneficiaries of the road system; inevitably this apportionment will be—and perhaps should be—dictated by national policy needs. What is more likely, however, is that eventually sufficient evidence will become available to permit the allocation of costs in such a way that those vehicles which require highly designed facilities, or which cause greater wear and tear, have to pay a proportionately greater part of the costs than those smaller vehicles which cause less wear and tear and can be adequately served by more lightly constructed facilities.

Various theories have been suggested as to how highway costs should be allocated between different classes of road users. The following are the most widely known.

Differential-cost concept This concept, also known as the increment-cost concept, seeks to apportion the costs on an increment-mass basis. Thus the first costing is made on the basis of what the highways would cost if all the vehicles were passenger cars and light commercial vehicles. The additional costs needed to adapt the highways to meet the structural and geometrical requirements of larger commercial vehicles and buses, e.g. thicker pavements, wider carriageways, more gentle hill slopes, and additional maintenance, are next calculated and these extra costs are then allocated to the vehicles falling into this category. Finally, the extra costs involved in adapting the highways to meet the heaviest vehicles are calculated and allocated.

The total costs are then apportioned in the following manner: all vehicles pay the first or basic increment of cost; all vehicles except the passenger cars and light commercials also pay the second increment; while the final increment is paid only by the heaviest vehicles. In this way, the total costs allocated to the motoring public are distributed in an equitable manner.

Mass–distance concept   This theory assumes that the cost of using the highways should be directly related to the masses of the vehicles and the distances which they travel. The mass relationship is used to reflect the amount of wear and tear and the need for sturdier facilities, while usage is reflected in the distance tally. Thus, in concept, a passenger car of mass one tonne and travelling 5000 km per year might be expected to pay only one-tenth of the taxation paid by a five-tonne commercial vehicle which does 10 000 km in the year.

A sophisticated variation of this concept has been introduced into New Zealand[4]. Under the NZ Road Users Charge Act of 1977, a prepaid distance licence has to be obtained by the majority of heavy vehicles, with the distance-related fee taking into account the vehicle mass and axle spacing, as well as the type of vehicle. Rate-scales have been calculated for twenty-four different classes of vehicle with powered units being treated separately from trailers. For each class, a scale of charges relates the distance fee to the maximum gross vehicle mass. All vehicles in the scheme have to be fitted with distance recorders. Petrol-powered vehicles with a gross mass of 3.5 t or less are exempt because of petrol tax which, it is argued, produces a broadly equivalent charge. Off-road vehicles are also exempt; instead, they are required to buy 'time licences'.

Operating-cost concept   This taxation theory proposes that vehicles should be charged for using the highways on the basis of scientifically determined knowledge of their operating costs. Thus the taxes paid by commercial vehicles could be considerably greater than those paid by passenger cars since their operating costs are greater. Proponents of this method of charging argue that the operating cost of a vehicle is proportional to the value of the benefits derived from its use on the highway.

This method actually results in a lower assessment on commercial vehicles than is levied by the mass–distance method. While the cost of operation of a commercial vehicle increases as the mass of the vehicle increases, it actually decreases on a mass–distance basis as the load increases.

Differential-benefits concept   This theory proposes that motor vehicle taxation can be determined on the basis of estimates of the savings incurred by the different classes of vehicle due to the highway facilities being improved. Thus, prior to every taxation period, an economic analysis would be carried out on the effects of the current improvement programme on the operating costs of the different classes of vehicles, and taxes would then be levied in direct proportion to the savings incurred by each class.

With the present limited state of knowledge regarding road user costs, this method is not a practical proposition at this time.

Area-occupied concept   Another name for this might be the 'congestion tax theory' since it attempts to assign taxation on the basis of the carriageway space occupied by the different classes of vehicles. Thus commercial vehicles would again have to pay a higher rate of taxation than private cars due to their larger bulk.

This method of taxing road users has its most fruitful application in urban areas where the principal problem is one of motor vehicle congestion. Details of how a tax of this type might be used to reduce congestion in built-up areas are given in Chapter 7.

Present system
It is often suggested that motor vehicle taxation in Britain is unjust. Criticisms of the existing system can be divided into those which refer to the total share of the highway costs borne by the motorists, and those which refer to the manner in which the motorists' allocation is shared between the various classes of road users.

Total motor vehicle taxation At the present time, the motorist appears to be taxed very heavily in comparison to the highway benefits received in the form of direct road expenditure: the direct contributions to the national exchequer in the form of motor vehicle taxation are, and have been for some time, substantially in excess of not only the central government expenditure but also the total national expenditure on roads. In fact, the motorists' share is even greater than is illustrated by the central government taxation revenue since road users are, more often than not, property owners/occupiers as well and hence they have to contribute an additional amount in the form of rating towards the cost of supporting the non-trunk roads.

In discussing the contributions to the national exchequer, the major consideration is whether the motoring taxes are levied for the primary purpose of supporting the highways, or in order to help to finance general government expenditure. If this latter consideration is the one accepted— and in Britain it is so—then the Government is entitled to raise revenue by whatever legal means are available in order to provide the country with the administration, defence and social services deemed necessary. From this aspect, the motor vehicle is just another fruitful source of taxation, just as are cigarettes and drinks. Hence it can be argued that there is as little justification for the entire revenue derived from motor taxation to be used for the construction and maintenance of highways as there is for all the beer and spirit taxes to be used to build bigger and better public houses.

Even if this latter approach is accepted, it can still be argued that there is a strong case for having in existence some consistent relationship between the benefits received by the motorist in the form of improved road facilities and the finance provided in the form of motor vehicle taxation. From the highway administrator's viewpoint this is a desirable concept, since highway improvements are best carried out on the basis of a planned, orderly schedule coupled with the assurance of a steady income.

Differential apportionment It is indeed very difficult to say what should be an individual motorist's rightful contribution to the motor taxation revenue. For instance, there is no doubt that, as a means of charging for the use of the roads, purchase tax at least is indefensible[5]. It can really only be considered as a form of luxury tax which bears no relationship whatsoever to the use that a vehicle makes of the highways. This is most obviously demonstrated by the fact that at the time it was applied in Britain commercial vehicles, which cause the greatest amount of wear and tear to the highway system, paid no purchase tax at all. A further disadvantage of the purchase tax mechanism is that it encourages the road user to keep a vehicle on the road for a longer period of time in order to get better value for money. In fact, the only beneficial function of the purchase tax is the short-term negative one of restricting the entry of new vehicles onto the roads until better roads can be provided.

The motor fuel tax is a more equitable method of charging for highway usage since it at least partially reflects road maintenance costs due to wear and tear, as well as congestion costs. The more a vehicle uses the highway to move about, the greater is its fuel consumption and hence the greater is its contribution in the form of taxation. In addition, when a vehicle is travelling on a congested roadway it is forced to move at a low speed which, when coupled with numerous stopping delays, also increases the rate of fuel consumption. Nevertheless, the fuel tax in its present state can only be considered as a rather haphazard method of relating the charge to the usage. Neglecting for the moment any implications in terms of energy conservation, there are some very obvious inequalities, in that a car with a high rate of fuel consumption pays more than one of similar size with a lower consumption rate, or a petrol-driven goods vehicle pays more than a diesel-powered one. Also petrol consumption on poor roads is greater than on high-quality ones so that the anomaly occurs where the charge for using the low-quality roads is actually greater than that for the better ones.

Apart from its usage as a means of raising further revenue, the registration of vehicles is necessary for police and other social purposes. The administrative cost of providing for the registration of a vehicle is so small that it leaves a considerable margin for the Treasury when the existing rates are applied. What these rates should be is literally anybody's guess. They can only be regarded as service charges for using the highway facilities; hence they are not truly reflective of the wear and tear caused to the roads or the costs of congestion in particular locations. Thus, for instance, a household's second car doing perhaps 5000 km per year pays the same tax as the business supporting a commercial traveller's car on the road for, say, 30 000 km in the year. While the registration tax on goods vehicles at least attempts to differentiate between the various types, there are still many other anomalies within the system which make it unsatisfactory.

In summary, therefore, it can be said that the present methods of raising revenue by means of motor vehicle taxation are not at all reflective of the extent to which vehicles make use of the roadway. It is also true, however, that practice in Britain in this respect is not at variance with that in most other countries. Certainly an equitable method of motor taxation for highway purposes has yet to be devised and proved in practice.

Highway administration

As described previously, the system of highway administration in Britain was not specifically created to direct and manage the roads and their traffic. Rather, it is the outgrowth of the need to handle the problems created by the roads at various intervals in history. Some of these have already been chronicled in Chapter 1, as well as in the brief summary of highway legislation at the beginning of this chapter, and so will not be further described here. Instead, this brief discussion will be primarily concerned with highway administration as it exists today.

Highway authorities

In 1970, the total number of highway authorities in Britain was 1195: England—839, Scotland—233, and Wales—123. As might be expected, this multiplicity of authorities meant that there was an inefficiency built into the system of highway administration. Too much will not be said about this, however, as the system underwent a drastic change as a result of the local government reorganization in England and Wales in 1974 and in Scotland in 1975. Subsequently, for various reasons[6], the Government decided again to reorganize local government in London and the metropolitan counties; this reorganization was effected with the passing of the Local Government Act of 1985. Under this Act, effective from 1 April 1986, there is provision for the highway and traffic functions previously carried out by these councils to be transferred: (a) from the Greater London Council to the London borough councils and the Common Council of the City of London, and (b) from metropolitan county councils to metropolitan district councils.

Local authorities

In England, the six main conurbations outside Greater London, viz. Greater Manchester, Merseyside, South Yorkshire, Tyne and Wear, West Midlands and West Yorkshire, have populations ranging from 1.2m (Tyne and Wear) to 2.7m (West Midlands). Each conurbation is divided into compact districts (36 in all) with populations ranging from 0.164m to 1m.

Greater London has an administrative area of about 1580 km² and a resident population of about 7m. It comprises 32 boroughs (population 136 000–321 000) and the City of London. The 2.6 km² historic City of London only has a resident population of about 8000, although some 400 000 people travel to work in it every day.

The 39 non-metropolitan counties on the English mainland range in size from 0.289m (Northumberland) to over 1.450m (Hampshire). The Isle of Wight (population 0.114m) is also a county. These counties are subdivided into 296 districts with populations broadly in the range 75 000–150 000.

Wales and Scotland also have two-tier local government systems. Wales has eight counties, with populations ranging from 106 000 (Powys) to 538 000 (Mid-Glamorgan), and 37 districts (207 000–278 000). Mainland Scotland has nine regions, i.e. equivalent to counties in England, with populations varying from 0.1m (Borders) to 2.5m (Strathclyde), and 53 districts (9100–810 000); because of their isolation from the mainland, Orkney (18 000), Shetland (21 000) and the Western Isles (30 000) have a single all-purpose local authority.

Whilst the local planning authorities for the Greater London conurbation are the individual boroughs, the Local Government Act of 1985 provides for the establishment of a Joint Planning Committee to advise these authorities and the central government on matters of common interest relating to the planning and development of Greater London. Typically, it might be expected that this Committee will be concerned with strategic aspects of planning, e.g. structure plans, development plan schemes governing the preparation of local plans, and certain types of proposals for development. In the other conurbations, these functions will be carried out by the metropolitan district councils and, in non-metropolitan counties, by the county councils.

In non-metropolitan counties, traffic, transport and highway functions are the responsibility of county councils, except that the Minister of Transport deals with trunk roads. In the metropolitan conurbations and in Greater London, these functions are the responsibility of the district councils and of the borough councils, respectively; however, the Minister of Transport again deals with trunk roads, and has significant traffic regulation powers, in these metropolitan areas.

In each conurbation, a metropolitan county Passenger Transport Authority, comprising members from the constituent district councils, is responsible for providing the public transport service. In London, this responsibility is vested in a holding body, London Regional Transport, which is accountable to the Minister of Transport[7].

Departments of Transport and of the Environment
In 1970, a comprehensive central government department, the Department of the Environment, was brought into being; this was an amalgamation of the (then) Ministries of Transport and of Housing and Local Government. The intention of this amalgamation was to provide the opportunity for a combined approach to land use and transport planning at the national level. Government departments are constantly reorganizing in order to meet continually changing demands and, in 1976, this very large department was abolished, and two new ministries established instead, viz. a Department of Transport and a (new) Department of the Environment.

At the head of the Department of Transport is the Secretary of State for Transport, who is a politician appointed by the Crown on the advice of the Prime Minister. In England, the Minister (who is supported in his or her work by a staff of permanent civil servants at whose head is a Permanent Secretary) is responsible[8] for land, sea and air transport, including sponsorship of the nationalized airline, rail and bus industries; airports; domestic and international civil aviation; shipping and the ports industry; navigation lights, pilotage, HM Coastguard and marine pollution; motorways and trunk roads; oversight of road transport, including vehicle standards, registration and licensing, driver testing and licensing, bus and road freight licensing, and road safety; oversight of local authorities' transport planning, including payment of the Transport Supplementary Grant.

The Secretary of State for the Environment is responsible in England for a wide range of functions relating to the physical environment in which people live and work. As such he or she is responsible for planning, local government, new towns, housing, construction, inner city matters, environmental protection, water, countryside affairs, sport and recreation, conservation, and historic buildings and ancient monuments. The Minister is also advised on ways in which the quality of the built environment might be improved by an Environmental Board which comprises both outside experts and senior officials of the Departments of Transport and of the Environment.

Some very important services are shared between the Departments of Transport and of the Environment. These include the coordination of policy planning, transport and environmental protection statistics, transport economics and traffic forecasting, transport planning and research, and energy aspects of transport. The two departments also have eight joint regional offices located across England, each under the charge of a Director (a civil servant, normally of Under-Secretary rank) who deals with detailed local problems. Each Regional Director has at least two Regional Controllers, one of whom is always concerned with roads and transportation. It is through the office of the Regional Controller (Roads and Transportation) that most of the dealings with local authorities regarding highways take place. There is also very close liaison between the various Regional Controllers where it is apparent that developments in the transport field interact with planning and environmental aspects, e.g. when comprehensive transport demand studies are to be carried out.

Finally, it should be appreciated that, other than in respect of trunk roads for which the Department of Transport is entirely responsible, the central government does not control the detailed transport decisions of local authorities. Rather, it influences them through general statements on transport policy, provisions of financial support, and advice from the Department of Transport.

Planning and design of major trunk roads Although the Department of Transport is the highway authority for trunk roads, it does not, in practice, carry out any physical work on them but instead delegates the responsibility for improvement and maintenance. In the case of maintenance and the design of smaller improvement schemes, this responsibility is delegated to a local authority which acts as the Department's agent authority at an appropriate fee. When the improvement scheme—which can be a new motorway or simply a realignment of an old trunk road—is a large one, the responsibility for its design and supervision of construction is delegated to a county council or a consultant.

Prior to 1967, county councils or consulting engineers acted as agent authorities for all trunk road improvement programmes. This did not prove too satisfactory for the larger schemes, partly because some of the (then) local authorities lacked the expertise to plan, design and supervise the construction of motorways, and partly because the central government, although publicly accountable for road construction, then lacked sufficient personnel with experience of motorway construction who had the capability of carrying out thorough appraisals of the schemes put forward. To overcome this, the Government established, between April 1967 and March 1968, six Road Construction Units in England with the avowed intention of achieving three main objectives: to concentrate scarce design and engineering resources into larger units, to streamline administrative decision-making through improved delegation to regional offices, and to provide opportunities for central government personnel to gain first-hand experience of the design and supervision of road construction schemes. Within each road construction unit area, a number of sub-units (16 in all) were established to deal with the detailed design and supervision of major schemes; these were based on counties that were well staffed in terms of road design and construction expertise.

The six regional headquarter units of the RCU organization now deal mainly with the overall direction and control of trunk road schemes, advise Ministers, and handle the statutory procedures. In 1980, the Government decided that the sub-units should be phased out, with much of their work being transferred to consultants (with the aim of strengthening the consultants' home base for competing overseas[9]), while some schemes would be transferred to county councils who would act as agents for the Department of Transport.

Road research   Highway and traffic research has a very strong and active influence upon both governmental decision-making and road design, and so it is appropriate to comment here on the manner in which it is carried out in Britain. Without doubt, the greater part of this research is undertaken by the Transport and Road Research Laboratory, which services both the Departments of Transport and of the Environment. Other important research is carried out by the Greater London Council and at certain major universities.

In the 1920s, the Ministry of Transport decided to experiment with the construction of concrete roads, which were then being built in the USA at the rate of more than 3000 km per year. Most of these roads, which were built in the south of England where gravelly materials were readily available, contained experimental features which were aimed at the development of a general specification for concrete pavements[10]. In 1930, the Ministry decided to construct a small experimental station at the village of Harmondsworth, Middlesex, adjacent to one of the experimental roads (the Colnbrook Bypass), at which the testing of the concrete cubes could be carried out. In 1933, the title of this station was changed to that of the Road Research Laboratory, and its scope was widened to include research into highway engineering, soil mechanics, and bituminous and concrete technology. Over the years, this organization grew tremendously, and today the Laboratory has an international reputation for excellence in applied research into virtually all aspects of road planning, design, construction, maintenance, and traffic control (including road safety).

The results of the research carried out at the Transport and Road Research Laboratory (the name was changed from that of Road Research Laboratory in January 1972) are made known by the publication of technical papers and, particularly, a series of research reports coded as Laboratory Reports (LR series) and Supplementary Reports (SR series). In addition, papers are published by staff of the Laboratory in the journals of the learned societies or the technical press. The library contains most of the important literature dealing with highways; its resources are available to road engineers and other interested research workers.

Scottish Development Department/Welsh Office
Since 1956 the Secretary of State for Scotland (via the Scottish Development Department and the Scottish Economic Department) and since 1965 the Secretary of State for Wales (via the Welsh Office) have had powers and responsibilities in Scotland and Wales, respectively, which are comparable with those of the Ministers of Transport and of the Environment in England.

No Road Construction Units were created to deal with trunk roads in either Scotland or Wales. Instead, reliance was placed upon the local authority organizations or, where these may have inadequate resources, upon consulting engineers.

Programming trunk road schemes

In the mid-1960s, certain slippages occurred in the motorway programme that were attributed to inadequate forward planning procedures. Thus, in 1967, the system was changed to make the forward programme more flexible by providing a continuous reservoir of work, viz. by dividing the trunk road programme into two parts—a 'firm' programme and a 'preparation' pool. As is implied in the title, a scheme is included in the firm programme when a definite decision has been taken regarding its construction. Schemes considered by the Minister of Transport as likely to get into the firm programme are first added to the preparation pool where they are very carefully assessed re costs and likely benefits; they are then included in the firm programme when appropriate in the light of the available resources and priority needs.

Against the background of movements towards more openness in government, which is reflected in increasing demands for information about the Government's operations generally, the steps involved in the preparation of a typical trunk road scheme now take at least ten years, from the time the project is conceived until traffic actually runs on the road. If, as a result of considerations arising from the public consultation or public inquiries described below, the design of the scheme is radically altered during the preparation process, completion of the project can take fifteen or more years.

The following is an outline of the evolution of a typical trunk road scheme[11].

The problem is first identified; this may be by congestion, accidents or environmental problems suggesting that a road improvement is needed. If a strong case exists, an addition is made to the trunk roads preparation pool. Traffic and soil surveys, as well as economic and environmental assessments, then commence. Possible solutions are outlined for the preliminary work on the highway's design. A provisional estimate of cost is made. Local authorities (and affected public service bodies, e.g. electricity, gas and water undertakings) are consulted about their own plans.

Possible routes are identified to prepare for public consultation on the choice of route. This (typically about two years after the initiation of the project) is the first chance that the public has to become directly involved in the road planning process. It is a non-statutory process where the Department of Transport prepares an exhibition of possible solutions and seeks the views of individual members of the public, local authorities, and national and local interest groups. Comments and a questionnaire response are invited. The views expressed are analysed and included with the results of preliminary assessments of traffic, environmental and economic factors, and engineering design work in the Department's preliminary report. In the light of this report, the Minister of Transport reaches a decision and announces the preferred route (year 3). This decision enables design work to be concentrated on a particular route, and frees other routes from blight. Large-scale drawings of the design, and a detailed assessment of the scheme, are prepared prior to the publication of draft Orders.

Draft statutory Orders are usually prepared and published about $4\frac{1}{2}$ years after the initiation of the project; these define the preferred line of the road (the line Order) and the alterations that will need to be made to other roads (the side road Order). In a densely populated country such as Britain, the laying down of new roads inevitably affects, and necessitates alterations to, the existing pattern of public highways, footpaths and private accesses; statutory authority is needed for these alterations in the same way as for the line for the new road.

Concurrently, the Department prepares a firm programme report which examines the overall justification of the scheme in greater detail than does the preliminary report. Local authorities and interested groups also put forward their views on the proposals, as part of the formal statutory processes. Once the firm programme report is approved, then if there are objections to the proposals, there may need to be public local inquiries (year 5) into the line and side road Orders.

The purpose of the public inquiry is to inform the Ministers of Transport and of the Environment of the weight and nature of objections to the road scheme. An independent Inspector, nominated by the Lord Chancellor, is appointed to conduct the inquiry, to report on the objections, and to make recommendations to the Ministers on the proposals; however, it is the Ministers, not the Inspector, who jointly *decide* whether the Orders should finally be made, or made with modifications, or rejected.

Not only does the inquiry cover the preferred route, but it also deals with any alternative routes which might be considered as being of near-equal merit, as well as with variations of the preferred route[12]. Reports to the Departments from both the Landscape Advisory Committee and professional consultants are normally open for inspection and discussion at the inquiry. Pre-inquiry procedural meetings involving the objectors and representatives of the Departments, and chaired by the Inspector, may be held to streamline the operation of the inquiry, and make it as informal and non-legalistic as possible, as well as less costly to both the objectors and the State.

If the eventual decision of the Ministers is to go ahead with the road scheme, the next step is the acquisition of land. This cannot start until the design of the road is sufficiently far advanced to show the lateral limits of the land that will have to be obtained. When the design has reached this stage, plans can be prepared showing the land needed from each of the various properties affected by the road scheme; these plans are accompanied by reference schedules giving the descriptions, sizes and details of ownership and occupation. The District Valuer then enters into negotiations (year 6) with the owners and occupiers and seeks agreement to enter upon the land on the required date and the amount of compensation to be paid.

If, as usually happens on large schemes, some particular plot cannot be acquired by agreement, then compulsory powers of acquisition have to be sought; this is done via a compulsory purchase Order, which must first be published as a draft Order, describing in detail the land that is required; this can be published with the line and side road Orders or at a later stage. If, at the end of a given period, there are unresolved objections to the draft Order, on any grounds other than the amount of compensation, another public inquiry may need to be held. Unresolved disputes as to the amount of compensation are referred to the Land Tribunal, a procedure which does not delay the making of the Order or the use of the powers of entry given under it; often, indeed, the compensation cannot be finally assessed until the roadworks are completed and the actual loss or injury caused can be accurately described.

If, following the Inspector's report, the decision is taken to make the compulsory purchase Order, with or without modification, the road scheme then enters the final steps of preparation. When the Order is operative, entry to the land can be enforced within a minimum period by service of further notices, unless certain classes of land are involved, e.g.

common, open space or ecclesiastical land, when additional procedures must be followed.

The works commitment stage (typically about $7\frac{1}{2}$ years after the start of the project) sees the completion of the proposals and, after a final check that the scheme provides value for money, and that the necessary resources are available, contract documents are prepared and tenders are invited. The procedure for inviting tenders is mostly by what is known as 'selective tendering', whereby tenders are invited from selected contractors judged to have the requisite technical and financial resources, and to be able to submit a realistic tender for the particular contract.

After the contract is let, construction can begin. Typically, most trunk road contracts in Britain take about two years to complete before the road is opened to traffic.

Selected bibliography

(1) Hyde WS, Highway legislation, *Journal of the Institution of Municipal Engineers*, 1961, **88,** No. 5, pp. 169–174.
(2) *Basic Road Statistics 1983*, London, The British Road Federation, 1983.
(3) Naylor AE, TPPs through the years—a personal view, *Municipal Engineer*, 1984, **1,** No. 3, pp. 225–231.
(4) Starkie DNM, Charging freight vehicles for the use of roads, *Australian Road Research*, June 1979, **9,** No. 2, pp. 3–17.
(5) Foster CD, *The Transport Problem*, Glasgow and London, Blackie, 1963.
(6) Department of the Environment, *Streamlining the Cities*, Cmnd 9063, London, HMSO, October 1983.
(7) Department of Transport, *Public Transport in London*, Cmnd 9004, London, HMSO, July 1983.
(8) Civil Service Department, *Civil Service Yearbook*, London, HMSO, 1983.
(9) Department of Transport, *Policy for Roads: England 1980*, Cmnd 7908, London, HMSO, June 1980.
(10) Croney D, *The Design and Performance of Road Pavements*, London, HMSO, 1977.
(11) Department of Transport, *Policy for Roads: England 1978*, Cmnd 7132, London, HMSO, April 1978.
(12) Departments of Transport and of the Environment, *Report on the Review of Highway Inquiry Procedures*, Cmnd 7133, London, HMSO, April 1978.

3
Highway and traffic planning

Transport in general, and particularly highway transport, plays an essential role in the life of any community today. Good highway transport facilities are the result of sound planning—and more and more it is now being recognized that transport planning cannot be, and must not be, isolated from land use planning.

Any land use context, whether on the national, regional or local scene, may be broadly considered as composed of two types of housing concentration, viz. low-density and high-density housing. This pattern of land usage gives rise to the following three types of transport movement:

(1) connection of low-density to high-density areas, and vice versa, e.g. suburbia to central area travel,
(2) inter-communication between and within low-density areas, e.g. village to village travel, suburbia to suburbia travel,
(3) interconnection of high-density areas, e.g. intercity travel.

Over the course of the past thirty years, and most particularly in the last fifteen years, a study methodology and a variety of strategies have been developed with the ultimate objective of 'solving' the transport problems and tasks arising from these movements. The purpose of this chapter is to describe these approaches, and to explain the factors underlying their development, with emphasis on the problems of urban and intercity travel.

The presentation adopted in this chapter is to give some basic information regarding travellers and their usage of transport facilities, with a view to providing an introductory background to further discussion in relation to the development and application of transport demand studies. It is suggested that the latter part of Chapter 1, which deals with significant post-World-War-II transport developments, also be re-read in this introductory context.

What this chapter does not do, and cannot even attempt to do, is describe how to deal with the real problems of urbanization which are the root cause of the most difficult transport problems in Britain, viz. urban migration and population growth, changes in wealth and its distribution, industrial development and technological change, and their attendant land use pressures and developments. *Solving* these fundamental aspects of the task is beyond the capability of the highway and traffic engineer. While it is easy to point out their implications, the engineer must work closely with personnel from other professional disciplines, e.g. city administrators, economists, sociologists, transport operators and urban planners and geographers, if his or her proposals to alleviate these problems are to be accepted.

Transport and travel data

National expenditure

In the United Kingdom in 1983, it has been estimated that users spent a total of £50 475m on inland surface transport (see Table 3.1); 96 per cent of this total was spent on road transport. The £26 510m motoring expenditure—over one-third of which went on taxes—was accounted for as shown in Table 3.2.

Table 3.1　National transport expenditure by users in 1983[1]

Passenger	£m	Freight	£m
Buses/coaches	2100	Road	19200
Motoring (private and business)	26510	Rail	645
Taxis/hire cars	495	Inland waterways	80
Rail	1445		
Total	30550	Total	19925

Table 3.2　National motoring expenditure by users in 1983[1]*

Item	£m	%
Purchase of vehicle	10984	41.4
Petrol	8702	32.8
Oil	202	0.8
Repairs, servicing, etc.	3865	14.6
Vehicle and driving licences	1964	7.4
Insurance	516	1.9
Other costs (e.g. tests, driving lessons, garaging)	277	1.1
Total	26510	100

*21.3 per cent by business users, 78.7 per cent by private users

Public transport

As is illustrated in Table 1.4, and also in Fig. 3.1, public transport usage in Britain has been declining over the past few decades. This can be generally attributed to real increases in national wealth (and thus of car ownership), real increases in fares, and decreases in the level of public transport service provided.

At the same time as patronage was declining, the governmental subsidy for public transport was being increased. Table 3.3 summarizes the growth in subsidy and grant since 1969.

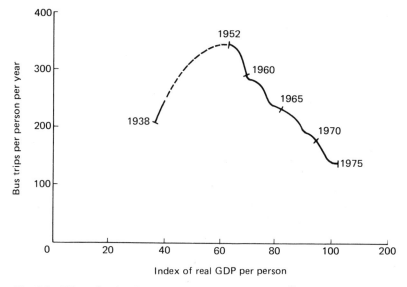

Fig. 3.1 Effect of national wealth upon trip-making by bus[2]

While the subsidies shown in Table 3.3 reflect deliberate government policy decisions in respect of public transport, they have nevertheless been criticized[3]. Specifically, the criticism has been twofold. Firstly, whereas transport users devote virtually all of their transport expenditure to roads, very significant governmental resources are nonetheless devoted to rail, where there is over-capacity and falling demand, instead of to roads, where there exists under-capacity (congestion) and growing demand. For example, in 1977, there were eleven bus journeys for every rail journey, yet British Railways received about twice the subsidy received by all bus operators. Secondly, in terms of grant per passenger journey, the bus passenger is subsidized to the extent of about one-quarter of the total cost of a journey, whilst the rail passenger is subsidized to the extent of one-half of a trip cost.

Table 3.3 Governmental subsidies and grants for public transport, 1969–83[1] (£m)

Transport mode	1969	1974	1983
Rail			
British Railways Board	76	391	1033
London Transport Executive	13	31	114
Road passenger transport	24	177	1003
National Freight Corporation	16	—	6
British Waterways Board	2	6	41
Total	131	605*	2197

*Equivalent to £1938m at 1983 values

Income and transport

Figure 3.2 shows in summary form the distribution of households with and without cars throughout Britain. In respect of this diagram, it might be noted that Britain does not at this time rank very high amongst Western European countries in terms of cars per head of population. For example, in the following countries in 1982, viz. Austria, Belgium, Britain, Denmark, the Federal Republic of Germany, France, Italy, Luxemburg, the Netherlands, Norway, Sweden and Switzerland, the numbers of motor vehicles per 1000 population ranged from 365 (Britain) to 540 (Switzerland), whilst the number of cars per 1000 population varied from 307 (Britain) to 391 (Germany)[1].

Many factors go into the decision as to whether a household should or should not have a car. Not least of these is job location and the availability of alternative forms of transport, e.g. one analysis[4] which

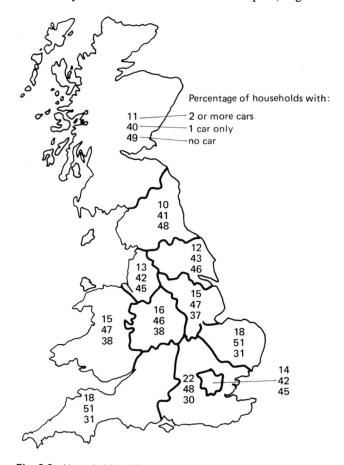

Percentage of households with:

11 ———— 2 or more cars
40 ———— 1 car only
49 ———— no car

Fig. 3.2 Households with and without regular usage of a car in 1981, divided according to economic planning regions (based on data in reference 1)

estimated that 71 per cent of Britain's working population lived in car-owning households, also found that the proportion of car-owning households varied according to location as follows: (a) city/town centre = 58 per cent, (b) suburban = 70 per cent, and (c) village/rural = 82 per cent. Another important factor influencing car ownership is household structure, e.g. a household with many male workers of driver-licence age will normally have more cars than will a single-income family with young children. An obvious overall factor affecting car ownership is household income. However, there is evidence[5] to suggest that, at high levels of household income, the relationship applying is that between the numbers of cars per adult and the income per adult, rather than that between the number of cars per household and the income per household.

A major practical concern of the transport planner is to make a reasonable mathematical estimate of how the likelihood of having a car in a particular locale may vary with household income. The general shapes of British car-ownership probability functions are shown in Fig. 3.3; these shapes have been confirmed by all transport studies which have investigated the relationship between car ownership and income. The term 'income' as used in these studies has been normally interpreted as meaning gross income (i.e. including tax and national insurance contributions, as well as unemployment benefits and other social security payments if appropriate). Whether the income variable thus obtained is truly 'gross income' is a moot point: some respondents (particularly those paid weekly) may only know what their take-home pay is, while other households may be extensively subsidized, e.g. one report[7] estimated that about 30 per cent of the total expenditure on car purchase is allowed as a

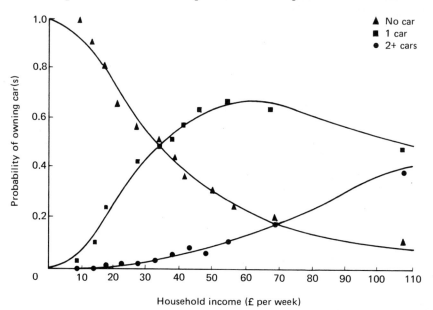

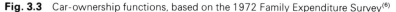

Fig. 3.3   Car-ownership functions, based on the 1972 Family Expenditure Survey[6]

business expense, while another reported[8] that of the business cars in use in 1973 (possibly 1m-1.25m), over 90 per cent were allocated to individual employees who were allowed to use them for private purposes.

Notwithstanding the above, it is generally accepted that any attempt to get more accurate income data would have a negative effect on the response rate to that question in transport studies. For this reason, the 'gross' definition has been accepted as standard, even though it most likely results in an underestimate of household income, particularly for the wealthier households.

Public transport

In 1983, the average household spent 14.9 per cent (£20.79) of its weekly expenditure on transport and travel, as follows: £7.33 on buying motor vehicles, spares, and accessories, £10.16 on running and maintaining motor vehicles, £0.95 on road public transport fares, £0.94 on rail fares, and £1.05 on other forms of transport, e.g. air, taxis, boats, ferries and bicycles.

The relatively low amount spent on public transport is as might be expected in the light of previously discussed data re growth in private motoring and decline in public transport patronage. This is also confirmed by the data in Fig. 3.4. The following brief discussion is based on the report[2] from which this figure is extracted, which contains a wealth of data regarding the factors influencing public transport usage.

Users of buses tend to be 'captive' passengers (i.e. people who do not have cars normally available for their use), and are spread fairly evenly throughout the income spectrum. In contrast, the users of urban rail fall mostly into the upper income group, and belong to car-owning

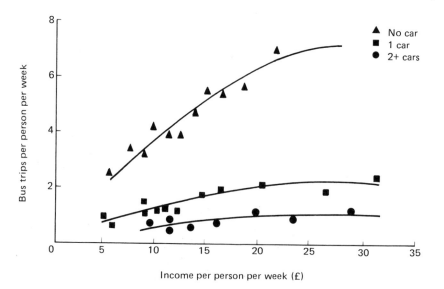

Fig. 3.4 Effect of income and car-availability upon trip-making by bus[2]

households. The predominant use of urban rail is for the journey to work for both men and women; for urban bus travel the main usages are work for men, work and shopping equally for women, and school for children. Women are the main users of buses, partly because fewer women have full driving licences; however, this latter situation is in the process of rapid change.

For buses, the elasticity with respect to change in fare levels is typically −0.3 (i.e. a 10 per cent increase in fares results in a 3 per cent fall in patronage) within a range of −0.1 to −0.6. These values seem to be stable with time, similar for different countries, and are largely insensitive to the size of the fare change, and whether it is an increase or decrease. Off-peak fare elasticities are likely to be twice those for the (relatively insensitive) peak period, while short-distance elasticities are also likely to be twice those for journeys of several kilometres. Bus patronage can be *apparently* increased by reducing fares. However, most of the new journeys are likely to be made by existing bus users, rather than by transfers from car travel.

For underground rail travel, the fare elasticity seems to be only about half of that for buses. However, the evidence suggests that for above-ground commuter rail, the elasticity is likely to be greater than that for underground rail.

The information available regarding the elasticity effect of level of service is generally inadequate and contradictory. Overall, however, it would appear that bus-kilometre elasticity will be greater for low-frequency services than for high-frequency services, while the elasticity with respect to train-kilometres is probably very low, especially if the rail services are reliable.

Journey time elasticity is estimated to be about −0.4 for buses and −0.6 for rail. Comfort is also important, e.g. the inability to get a seat may increase the 'cost' of the time spent travelling by the user by as much as two-thirds. Having to interchange public transport vehicles effectively adds the equivalent of 3 or 4 minutes waiting time to the actual waiting time caused by the interchange; this 'penalty' will be less where interchange conditions are relatively pleasant, and more where they are unpleasant, e.g. bus to bus without cover in inclement weather.

The journey-to-work trip

The journey to work is of critical interest to the highway planner and designer. The well-known tendency for work trips to happen at the same time results in the creation of peak period travel demands which are usually the focus of attention during the planning and design processes. It is essential therefore to have a clear understanding of the considerations surrounding this type of trip. The following brief summary is based on an excellent review of the subject available in the literature[9].

Modal choice

As discussed previously, the motor car has clearly become the principal mode of travel to work in Britain; its emergence in this role has also thrown into focus the attendant decline in public transport usage. However, the proportion of people using bicycles, motor cycles, or walking to work has remained reasonably constant over the past two decades. The change in modal split from public transport to the car may be attributable not only to the growth in car ownership, but also to an increase in the popularity of car-sharing, and to changes in peoples' attitudes towards the use of the different modes of transport.

Changes in the choice of travel mode are very often associated with relocation of residences and/or workplaces—hence the need for good land use planning data when carrying out transport studies. For example, one study of travel behaviour amongst employees of offices that were decentralized concluded that 80 per cent of the previous car users continued to drive to work, half of the employees who had previously travelled as car passengers transferred to other modes (28 per cent became car drivers), 50 per cent of previous bus users changed to other modes (mostly car passengers and trips on foot), and, amongst previous train users, 60 per cent changed to private modes.

Inconvenience factors which discourage the use of public transport include distance of residence from bus stops, long waits at bus stops, changing buses, and the longer time taken to travel by bus. These factors become, however, less important when the job location is in the city centre where congestion and difficulty of parking are important considerations against using a car.

Nationally, there appears to be a relationship between socio-economic group and modal split. Thus the use of a car for the journey to work is very high (about 75 per cent) amongst professional and managerial groups. Intermediate non-manual and skilled manual workers are also high car users; however, with these two groups, bus use (12 per cent) and walking (15 per cent) are also important. Work trips by car are much lower (35–40 per cent) amongst junior non-manual and semi-skilled manual workers; in these two groups, 20 per cent use a bus and 25–30 per cent walk to work. Amongst unskilled workers, 23 per cent use a car, 20 per cent travel by bus, and over 40 per cent walk to work.

Journey length and time

The historical development in respect of the length of the journey to work has been for it to become greater, e.g. between 1921 and 1966, the average length increased at the rate of about one per cent per annum. Unfortunately, no data are available regarding journey times over the same period. In 1973, the average journey-to-work length was 7.7 km; with the exception of social trips, this was the longest type of household trip.

Figure 3.5, which is from the 1975–76 National Travel Survey, shows the trip length distributions for the different modes. In respect of this diagram, it might be noted that over 50 per cent of train journeys were

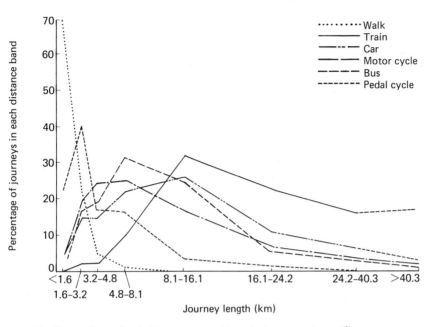

Fig. 3.5 National variation in journey-to-work length, by mode of travel[9]

longer than 16.1 km, whereas 80–90 per cent of car, motor cycle, and bus trips were less than this distance. Amongst cyclists, 62 per cent made trips that were less than 3.2 km, and more than 30 per cent made journeys of 3.2–8.1 km. Almost all journeys by foot were less than 3.2 km.

The trends for journey times in respect of train commuters, cyclists and walkers are generally as might be expected from their lengths, i.e. train commuters take the longest time (40 per cent make trips of over 60 minutes), while most cyclists and walkers spend less than 30 minutes travelling to work. However, although trip lengths are slightly shorter for bus users as compared with car travellers, the reverse is true with respect to journey times, e.g. 79.7 per cent of car commuters travel for less than 30 minutes, whereas only 47.7 per cent of bus users spend less than that time getting to work.

Spatial distribution of residences and workplaces
The evidence presented above suggests that people are now making longer trips to work, and possibly in shorter time. However, it should also be understood that the geographical distribution of journey lengths varies also with the part of the urban area being considered. In the period 1951–71, all metropolitan areas in Britain experienced a progressive outward shift of both residences and workplaces. The end result has been that the density of workplaces is still highest in the urban centre but it falls

sharply in the inner ring, and then declines gradually with increasing distance from the centre. In the case of residences, the density is now very low in the city centre, higher in the inner ring, and it then also declines gradually with increasing distance from the centre. As might be expected also, the residential locations of car users are now more widely distributed than those of public transport users.

The resultant effects of these movements have been threefold, viz.:

(1) residents in the inner areas still make shorter work trips than do residents in the outer areas,
(2) the increases in out-commuting by residents of metropolitan counties, and of their centres, are more than exceeded by the increased amounts of in-commuting,
(3) workplaces in inner urban areas attract longer work trips than do workplaces in peripheral areas, i.e. their catchments are greater.

While numbers of studies have been carried out which have examined the effects on the journey-to-work mode of a change in work location, no definite study appears to have been carried out yet on the reverse effects, i.e. on the role of accessibility on job choice. It is reported[10] that one study, which attempted to do so, determined that only 13 per cent of people who had changed jobs within the previous five years gave the work journey as a major reason for the change; factors such as pay, conditions of service, or redundancy were most commonly cited. It appears that when choosing a job, an individual generally takes mobility as given and looks for the best job available within the area thereby determined.

Non-journey-to-work trips

There are many other trips besides the journey-to-work trip that must also be taken into account in the highway planning process (see, for example, Table 3.4). There are a number of important points which can be made in relation to the data reflected in this table[10].

(1) Of the three main modes of travel (car, bus and walking) which together account for 92 per cent of all journeys, the car has the greatest potential for reaching a wide range of destinations. While the frequency of walking overall is nearly as great as for the car, it is only viable for very local journeys. The bus is chiefly used by people without the use of a car.
(2) The social groups with the lowest personal mobility (defined as the potential to travel freely) are young children and teenagers, non-working housewives, the elderly, and adults in low-income households. For each such social group, car use is relatively low, either because its members cannot drive (e.g. young people, some housewives, and many elderly people), or because a car is not available at certain times (e.g. most housewives during the working week), or because car ownership is low (e.g. the elderly and the poor).

Table 3.4 Some national statistics re trip purpose rates per day, distribution and modal split[10]: (a) trip purpose rate and distribution by social group and (b) percentage of trips by mode of transport for the whole population

Social group	Work		Shopping		Social		Recreation		Education		Personal business		Escort		Other		All purposes rate
	Rate	%	Rate	%	Rate	%	Rate	%	Rate	%	Rate	%	Rate	%	Rate	%	
All population	0.56	22	0.49	19	0.39	15	0.29	11	0.25	10	0.24	9	0.14	5	0.21	8	2.57
Working adults	1.31	41	0.38	12	0.41	13	0.26	8	<0.005	*	0.21	7	0.16	5	0.44	14	3.17
Young children	<0.005	*	0.29	13	0.30	14	0.32	15	1.03	47	0.17	8	*	1	0.05	2	2.19
Teenagers	0.07	3	0.35	13	0.38	14	0.45	17	1.07	40	0.27	10	*	*	0.07	3	2.67
Young adults	0.90	30	0.36	12	0.60	20	0.35	12	0.23	8	0.25	8	*	2	0.27	9	3.01
The elderly	0.09	6	0.57	36	0.30	19	0.23	15	<0.005	*	0.26	17	0.03	2	0.09	6	1.57
Non-working housewives	<0.005	*	0.87	38	0.46	20	0.26	11	*	*	0.28	12	0.34	15	0.07	3	2.29
Low-income adults	0.45	19	0.60	26	0.45	19	0.22	9	*	1	0.28	12	0.13	6	0.18	8	2.33

(a)

Mode of transport	Percentage by mode								
Car	51	35	58	40	13	45	67	63	45
Walking	19	46	26	46	60	38	29	26	35
Bus	15	16	11	8	13	11	2	5	12
Other	15	4	5	7	14	6	1	5	8

(b)

Note: Percentages are rounded and may not sum exactly to 100.
*Denotes small sample

(3)   Recognizing that bus patronage declines as car ownership increases, the less mobile groups may find over time that the locational opportunities accessible to them become fewer as bus services are contracted, and their relative disadvantage becomes worse, even though they themselves may make no changes in their travel patterns or requirements.

(4)   The highest journey rate for shopping is exhibited by housewives. Shopping journeys were on average about half the length of those made for most other purposes. People without cars made 5 per cent of their shopping trips by that mode, while those in two-car households made 68 per cent of their trips by car. Motorized travel becomes particularly important for trips to larger shopping centres. Where possible, the car is used in preference to public transport, even if this means deferring a trip until the weekend.

(5)   Seventy-six per cent of primary school children travelled to school by foot (average distance = 1.5 km), as did 46 per cent of secondary school children in the 11–16-year age range (average distance = 4.0 km), in 1974–75. Cycling was restricted less by the availability of a bicycle than by parental or school rules on usage. Bus travel tended to result in longer journeys and was less favoured for that reason; where contract coaches were used the timing of buses limited the flexibility open to a pupil to stay late or become involved in school activities. Although 69 per cent of children lived in car-owning households, the overall level of car use (13 per cent) was relatively low; however, it rose substantially in households with two or more cars. Car travel was considerably more common for primary, as compared with secondary, school trips and in the morning was about twice that in the afternoon. Only about one-fifth of the morning car journeys were made solely for the purpose of taking children to school.

(6)   Surprisingly, little definitive research has been carried out on trips for social purposes, but what has been done suggests strongly that when personal mobility becomes more limited, social travel is a journey purpose that is likely to be curtailed or rearranged. This applies both to families without cars and to communities which have had bus service withdrawals. Overall, the importance (and flexibility) of the car for social travel is reflected in the fact that for all of the less mobile social groups, journey lengths are significantly shorter than those for the population as a whole.

(7)   The use made of a particular type of recreation facility is very much influenced by its accessibility. Thus entertainment facilities such as cinemas and dance halls are usually located so as to be well served by buses as well as cars, if they are to remain commercially viable. Car ownership would appear to be the most significant factor influencing visits to the countryside, fell walking, etc.; the capability of people without cars to make such trips is severely curtailed. Children aged 5–14 who live within 0.5 km of a sports centre are ten times more likely to use it than those living over 1 km away; for the 15–19 age group the corresponding participation ratio is six, while for adults it is three.

(8)   A personal business trip refers to travel to facilities where an

individual receives some personal service, e.g. to banks, car repair garages, churches, cleaners and laundrettes, dentists, doctors, hospitals and solicitors. Frequently, these trips are combined with shopping or (professional) business trips, and hence they are sometimes underestimated in transport demand studies. Personal business journeys to any particular facility tend to be relatively infrequent, rather than regular, and this, combined with the fact that the various facilities differ considerably in nature, means that generalizations about this category of trip as a whole are difficult. One generalization that can perhaps be made is that when the per capita use of a facility is low, and the facility is large and remote, access for those without cars is difficult—for example, access by public transport can be time-consuming and costly.

Travel and the handicapped

A particular need not always taken into account by transport planners is that of the physically handicapped. Persons having musculoskeletal and central nervous system impairments, who together comprise some 56 per cent of the national handicapped population, experience the most difficulty in travel; by comparison, the least difficulty is experienced by those with respiratory impairments and, as would be expected, those without impairments of the legs.

Analysis of the urban deprivation factor with respect to the handicapped has shown the following[11].

(1) The most frequently used mode is foot travel, with a distance of about 0.4 km being the maximum that can be achieved by a large percentage of the handicapped.
(2) The second most widely used travel mode is the car and tricycle. Reliance upon this mode would be even higher if car transport were more available, i.e. nearly three-quarters of non-car users would appear not to have access to a car.
(3) Over two-thirds of the non-users of buses are prevented from doing so because of difficulties with the distance to the bus stop, the height of the entrance step, or the fear of falling while on them, e.g. buses moving off before invalid passengers are seated.

Transport and energy

Of all the energy used in the United Kingdom in 1981, 18 per cent was used on transport, 23 per cent by industry, and 20 per cent in households; the remainder was mainly used up via energy conversion losses, consumption by the energy sector, and distribution losses. Of the oil used for energy purposes, 43 per cent was used on transport—mostly (79 per cent) on road transport. A feature of road transport is the extent to which it is almost entirely dependent upon natural oil products. Furthermore, unlike other uses of oil, e.g. for heating, a substitute fuel will not be available to any significant extent for some time for transport vehicles.

Until relatively recently, energy considerations were of minor importance in the making of decisions about transport policy. The above data, and those summarized in Chapter 1, indicate why it is that they are clearly now given greater weight.

Comparative consumption efficiencies
Table 3.5 compares approximate energy utilization rates for alternative methods of vehicle propulsion. The energy consumptions are given in primary energy terms, i.e. the energy content of the unprocessed fuel input required to produce the final energy output; thus in the case of oil, this includes the refining processes and the transport of the refined products. Not taken into account in these calculations is the 'overhead' energy used by transport in the form of lighting and heating of related buildings, signalling, and on maintenance and repair work; not included either is the energy consumed indirectly in the construction of buildings,

Table 3.5   Primary energy consumption by passenger modes and types of traffic[8]

Transport mode	Megajoules per seat-km	Average load factor (%)	Average megajoules per passenger-km
Intercity			
Electric loco-hauled train	0.46–0.47	45	1.0
Diesel loco-hauled train	0.38–0.41	45	0.9
Express coach	0.20–0.26	65	0.4
Scheduled aircraft	2.20–3.00	65	3.9
Commuter			
Electric multiple unit train			
25 kV: 318 seats	0.30–0.45	25	1.6
750 V DC: 386 seats	0.24–0.32	25	1.1
Express bus (50 seats)	0.22–0.27	(25)	(1.0)
Urban			
Bus (70 seats)	0.15–0.23	25	0.8
Underground (LT)	0.20–0.24	14	1.6
Rural			
Diesel multiple unit train	0.29–0.36	(20)	(1.6)
Bus (45 seats)	0.19–0.23	15	1.4
		Average passenger load	
Car			
Motorway	0.75–0.80	2.0	1.6
Rural	0.80–0.83	(1.7)	(2.0)
Urban	1.00–1.20	1.5	3.1
Motor cycle			
Motor cycle (2 seats)	0.94 (average)	(1.1)	(1.7)
Moped (1 seat)	0.94 (average)	1.0	0.9

Notes:
(1)   Average load factors for commuter and urban services conceal large variations between peak and off-peak loadings.
(2)   Consumption figures in brackets are based on assumed load factors.
(3)   Energy consumption for London Transport underground is in terms of MJ/place-km.

vehicles, track, maintenance equipment and materials. Preliminary research has suggested that overhead and indirect uses of energy for car passenger transport could add about 30 per cent to that directly used as petrol, and perhaps about 50 per cent for railways as a whole.

Caution should be exercised in drawing strong conclusions from the relative energy efficiencies shown in Table 3.5, as the assumptions used in their preparation are approximate averages of variables that have wide ranges of actual values. Their major worth is in giving an indication of the average order of magnitude of difference between the various transport modes.

British road research

Research[12, 13] by the Transport and Road Research Laboratory in the energy area has indicated the following.

(1)   Obstructions to constant-speed driving, e.g. highway junctions, roundabouts, traffic lights, pedestrian crossings, and parked and moving vehicles, are the main reasons for energy consumption additional to that needed to propel vehicles at steady speed. Thus fuel consumption on urban roads is higher than that on all-purpose rural roads because these obstructions occur more frequently in built-up areas.

(2)   The effect of the increase in traffic congestion between off-peak and peak traffic periods in urban areas is to cause an average passenger car fuel consumption increase of 4.4 per cent in Greater London, and 3.4 per cent in the rest of the built-up areas of the country. These are due mainly to increases in fuel consumption in the non-central areas, i.e. the actual amount of travel in central areas amounts to only a small fraction of the total urban travel, thus making a relatively small contribution to overall fuel consumption. The estimated potential saving in road transport fuel through a reduction of traffic congestion from peak to off-peak conditions is about 1.5 per cent per vehicle per year.

(3)   People do have a good idea about what is meant by 'economical driving' so that if they were required to do so on a national scale, only about 80 per cent of the fuel now actually used would be required to drive the same distances. Most drivers interpret driving economically as driving more slowly (on average by about 15 per cent), which generally also results in an increase in journey time.

(4)   The effect of car shape (aerodynamic drag) on fuel economy in urban areas is negligible because of the low average speeds normally achieved. It is of much more significance at higher speeds, e.g. on rural all-purpose roads.

(5)   In typical mixed urban and rural driving, a small car with a 1.5-litre modern light-weight diesel engine can be expected to show something like a 50 per cent improvement in fuel consumption as compared with an equivalent petrol-engined version. This substantial energy benefit is currently obtained at the cost of a more expensive engine, and a slightly heavier car.

(6)   The optimal car journey speed for minimum fuel consumption is

about 50 km/h. Thus methods of traffic management which increase the average journey speed of urban traffic (up to about 50 km/h) tend also to reduce the amount of fuel used; this has given rise to a rule-of-thumb for urban traffic management that saving 10 per cent on journey time results in a saving of 6-7 per cent in fuel used.

(7)   The fine tuning of area traffic signal control strategies in urban areas can give small fuel savings without significantly increasing journey times.

(8)   The application of an 88.5 km/h speed limit on motorways would be unlikely to result in fuel savings greater than about 2.5 per cent, even with 100 per cent of driver compliance.

(9)   Driver technique (although not negligible) is less important with commercial vehicles than with private cars.

(10)   For both laden and unladen 16 t goods vehicles, the minimum fuel consumption occurs at an average running speed of about 55 km/h. On a motorway gradient of 3 per cent, fuel consumption is doubled as compared with level road conditions; at 4 per cent, fuel consumption is trebled.

(11)   The use of an air shield on a 16 t goods vehicle carrying a 2.44 m × 2.44 m container can reduce fuel consumption by about 10 per cent.

(12)   Running a lorry at 50 km/h on tyres that are under-inflated 20 per cent below the recommended pressure will increase the vehicle's fuel consumption by about 5 per cent.

Car-sharing

An often forgotten aspect of energy studies is the extent to which vehicle occupancies can affect energy utilization. Some data are available in the literature regarding occupancies and car-sharing as a whole[14].

(1)   The three National Travel Surveys for the years 1965-66, 1972-73 and 1975-76 show a decline in average car occupancies from 1.91 to 1.77 to 1.72. Assuming that the increased car ownership over the same period was the cause of reduced average occupancy, that the lost bus passengers transferred mainly to cars, and that both of these trends continue (as is likely), then car occupancies can be expected to decline further unless positive steps are taken to arrest the trend. In 1975-76, the average car occupancy for journeys during the morning peak period was 1.4.

(2)   In 1975-76 also, 10 per cent (all day) and 12 per cent (peak periods) of car passenger travel were made by people who obtained lifts in cars belonging to other households. These car passengers accounted for about the same amounts of passenger-kilometres as were travelled on local stage-bus services throughout the whole day and during peak periods, respectively.

(3)   If peak period car occupancies could be increased by 10 per cent (resulting from a corresponding reduction in car usage, with all occupants transferred to the remaining cars), there would be a saving of about 3.6×10^{14} litres of petrol per annum; this would amount to about 1.5 per cent of the petrol used for road transport.

(4)   Drivers who take regular turns to drive their own car to work and

give lifts to others in a group (car-poolers) account for only about 3 per cent of all car drivers travelling to work. Nevertheless they are estimated to save over 10^9 vehicle-kilometres per annum.

(5)   The families of car-poolers make limited use of the pool car on those days when it is left behind and is available for off-peak use, i.e. about 10 km per week on average. Off-peak bus trip rates are no less for the families of car-poolers than for families in which the car is driven to work every day.

Cycling as a transport mode

In recent years, there has been considerable renewed interest in cycling as a transport mode. It is difficult to be definite about the reasons for this upsurge of interest. Possibly it is related to environmental, ecological, health and recreational desires—and then again it may be the higher cost of petrol that is the basic cause. Whatever the reason, the highway and traffic engineer is now expected to take cycling into account in decisions relating to the planning, design and management of many transport facilities.

The change of interest in cycling is reflected in the historical travel statistics which show that between 1952 and 1974 the total distance cycled fell from 23×10^9 km to 3.2×10^9 km, after which it increased to 5.1×10^9 km in 1983. In percentage terms, bicycle-kilometres as a fraction of all vehicle-kilometres fell from about 25 per cent in 1952 to below 2 per cent in the early 1970s, and has remained at this level since then.

Research[15,16,17] by the Transport and Road Research Laboratory has brought together some very useful information about the users and usage of bicycles, as follows.

(1)   As might be expected, most cycle trips are relatively short, and carried out on urban roads. Figure 3.6 shows the distribution of journey lengths for various trip purposes.

(2)   Cycling amounts to about 5 per cent of the vehicular trips in urban areas of less than 0.25m population, and about 8 per cent of the people in such cities use a bicycle at least once in any one week. Cycling is less common in large cities.

(3)   Boys aged 11–15 years from higher social groups and older adults from lower social groups are most likely to be bicycle users.

(4)   About two-fifths of all cycle trips are for the journey to/from work, and some two-thirds of bicycle trips by adult men are for this purpose.

(5)   The seasonal variation in cycle use is similar to that for motor vehicles in urban areas, and less than that for motor vehicles in rural areas.

(6)   It is not possible at this time to estimate the amount by which cycling is likely to increase if special facilities for cyclists are provided, e.g. cycle paths. The limited evidence available to date would suggest, however, that few cyclists were influenced by the existence of cycle routes to take up cycling. What is known is that existing cyclists tend to do more cycling when special cycle routes are provided.

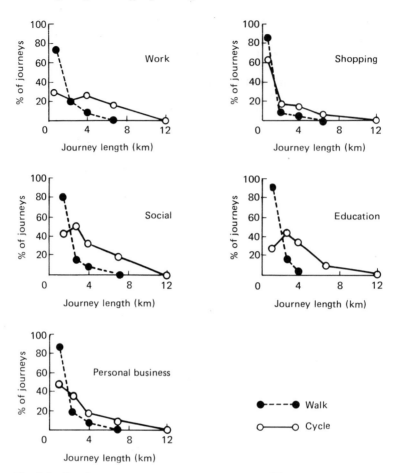

Fig. 3.6   The distribution of cycle and walk trip lengths[15]

Characteristics of the national highway network

Non-urban road system

From what has been said in Chapter 1, it will be clear that, other than during Roman times, there has never been a serious attempt, until recently, to create what might be termed a national road system. The road system which eventually evolved was composed of direct links from town centre to town centre (see Fig. 1.1). It is this inherited tortuous road system, with its concentration of traffic through urban centres, that has been the major cause of traffic congestion and associated environmental problems in Britain.

The creation, first, of the intercity trunk road system and then of the national motorway system, with its lateral spurs to various communities, marked determined attempts by government to create a national skeletal network on which long-distance traffic (particularly commercial vehicles) could be concentrated. This strategic highway network (see Fig. 1.5) has undoubtedly brought great benefits to the populace in general, in terms of shorter journey times, less driving stress, and significant release from traffic congestion in towns.

When it is considered that the lateral influence of a rural motorway-type road is spread through a width that is roughly equal to half its length, it can be seen that the strategic road system affects traffic on nearly every highway throughout the length and breadth of the land. Not only does it enable unnecessary travel through towns to be avoided, but it also has the very important 'byproduct' that many overloaded rural and urban roads are so relieved that they can again fulfil their more local linkage functions without any extra reconstruction being required.

Urban road systems

Essentially, all urban areas are communication centres for the exchange of goods, services and ideas, and the higher the level of these activities, the greater the city or town. Historically, urban development has been significantly affected by two transport-related factors, viz.: (a) the number of inhabitants of a city was limited by its connections to the hinterland, and (b) the city's population was strongly influenced by the capabilities of its own internal transport network.

Before the introduction of modern modes of transport, town dwellers generally had to live within walking distance of the town centre; this limited the urban radius to about 2 km, with a walking speed of about 4 km/h. The introduction of the railways substantially negated the importance of the first factor noted above because rail connections provided an urban centre with an increased economic (as well as population) capacity upon which to draw for its well-being. Consequently, the second of the above two factors grew in relative importance.

In the early rail age, with travel speeds via suburban railways rising to 20 km/h, the populations of many built-up areas increased by as much as tenfold as people moved outwards to new developments built near the suburban railway stations. The introduction of tramways and bus networks (see Chapter 1) resulted in a more dense urban development; whereas the railways encouraged urban growth along main radial routes, road-based public transport permitted the development of urban land away from the radial routes.

The introduction of the motor car enabled travel speeds in built-up areas to increase to about 30 km/h, and generally stimulated low-density residential infill—and the transformation of towns into polycentric agglomerations. As the populations of urban areas continued to grow, the lack of additional locational opportunities within city centres, and increasing traffic congestion, further stimulated the decentralization of both housing and employment. As these occurred, particular forms of internal urban road pattern were developed.

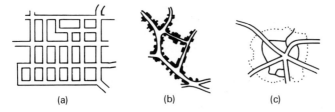

Fig. 3.7 Basic major road patterns in urban areas: (a) gridiron, (b) linear and (c) radial

Broadly speaking, three principal types of major road pattern were developed in urban areas. As shown in Fig. 3.7, they are the radial, the gridiron and the linear patterns. Of these, the radial pattern is the predominant type in Britain and so it will be discussed in greater detail than the other two.

Gridiron pattern
Originally favoured by the Romans, this major road pattern is adopted extensively throughout the USA. In one sense it is by chance that this pattern came into favour throughout American urban areas. The basic reason for its acceptance is that, when the counties and townships were being laid out in the early/mid-19th century, it was easier for the surveyors to set them out using straight lines and rectangular coordinates. In time, the pattern became the accepted one so that all newer towns and town expansions tended to be set out in a similar manner. This is particularly noticeable in the mid-western and western towns in the USA, whereas parts of the older eastern cities tend to be more uncoordinated and 'European'.

Although the gridiron pattern can produce monotonously long streets flanked by dull blocks of buildings, it has some considerable traffic-moving advantages. It encourages an even spread of traffic over the grid, and as a consequence the impact at a particular location is reduced. It facilitates the imposition of extensive one-way street systems since alternate streets of the grid can be made one-way in opposite directions. If there is a definite central business area in the middle of the grid, it is relatively easy for through traffic to bypass it since there are usually alternative bypass routes available in all four directions.

One objection to the gridiron pattern is that extra distances must be travelled when going in a diagonal direction. This has been remedied in many American cities by superimposing major diagonal routes upon the grid. While these diagonals have aided traffic movement, they also have an unfortunate aesthetic effect on the architectural development of adjoining frontages, producing in many instances acute-angled plots and awkward street intersections.

There are very few examples of towns in Britain being laid out on the gridiron pattern. Individual housing estates laid out in this fashion are, however, often to be seen.

Linear pattern

Historically, this type of urban road pattern developed as a result of local topographic difficulties. The most obvious example is that of a settlement which originally became established alongside a well-used trackway through a long valley. As time progressed and the settlement grew, the length of the town became too great for the (then) available modes of internal travel, and so the town began to grow laterally; the homes and other buildings were then located alongside feeder trackways connected to the major route. Today, many examples of towns which developed in this way are found in valleys such as those of the West Riding of Yorkshire and in South Wales.

From the point of view of traffic movement, the'linear pattern appears at first to have particular advantages. The most important of these is that the main traffic flow is canalized into one major roadway. In principle, this concept of canalization is a good one, as evidenced by the modern usage of the motorway. In practice, however, the channelling of all traffic into the main street of a town can create more problems than it solves.

Nowadays, the motor vehicle is used both for trips within the town as well as for those with destinations outside the town. As a result, the main street serves as a route both for the completely internal traffic and for the internal–external traffic, due to the lack of other suitable longitudinal routes. This traffic is in turn increased by the addition of external–internal traffic and through traffic.

The net result is that many of these linear towns are now literally bisected by the heavy flows of traffic concentrated on the central major street. In addition, since the central street was never designed as a thoroughfare, and in fact usually just 'grew', it is normally easily overloaded.

Radial pattern

As discussed earlier, the highway system in Britain developed in the form of a network of roads connecting town centre to town centre. Thus any given town had several roads radiating from its centre to the other towns and villages about it. Eventually, as the towns grew in size, they tended first to develop along the radials and then to fill the spaces in between. Thus the towns of today have a strongly marked radial system with the business area stabilized at the centre of a more or less symmetrical cartwheel plan.

The location of the main traffic generators within the central area, the system of radial roads converging on the main source of attraction and, usually, the lack of suitable bypass routes for through traffic eventually produced the belief that the cause of central area congestion was traffic on the radials and that the solution was to build ring roads to divert the traffic around the town centre or the town itself, as the case may be.

Usage of ring roads By definition a ring road is a highway that is roughly circumferential about the centre of an urban area, and which permits traffic to avoid the centre of this area. In practice, three forms of ring road have come into being: an inner ring road, an outer ring road, and intermediate ring roads. A town with a population of about 20 000 will tend to have a single inner–outer ring road, whereas a city of 500 000 might have both an inner and an outer ring road. Urban areas with larger population groupings tend to have one or more intermediate ring roads in addition to the inner and outer roads.

Perhaps the easiest way to visualize the basic urban road pattern in Britain is to liken it to a cartwheel. The spokes of the wheel are the radial routes. The hub of the wheel is the inner ring road, its function being to serve the needs of local and local through traffic. Its purpose is to promote the convenient use as well as the amenity of the central area by deflecting all vehicles which have no need to traverse it, while affording convenient means of entry for those to whom access is essential. Thus the location and design of the inner ring road is very closely bound up with the size, layout and usage of the central area. In practice, therefore, the inner ring road may be round, square or elongated, and may be incomplete on one or more sides, depending upon the conditions prevailing.

The outer ring road can be considered as the rim of the wheel. Although it is often used by through traffic in order to bypass a town, its original purpose was to serve the traffic of the town itself by linking up the outer communities and acting as a distributor between the radials. To serve this purpose, the outer ring road was generally located within the outer fringe of present and future urban development and not outside the limits of the town. The outer ring road has tended to be more formally circumferential than the inner ring road because of the greater availability of space at the edges of towns. Again, however, its completeness depends upon the needs at specific locations.

In the larger population centres, other reinforcing ring roads may be located with the outer and inner ring roads. These intermediate roads serve the needs of traffic, whether from a distance or of local origin, which desires to reach points between the outer and inner roads. They are normally located in areas already partially developed, and thus, to a larger extent than outer ring roads, they incorporate many existing local roads with all their diversity of use.

Advantages of ring roads   What was originally the principal advantage of the radial system, its provision of direct access to the town centre, is now often a liability. During peak periods especially, as the radial road is followed inward, the traffic volume builds up progressively until the central area is reached and dispersal to parking places is attempted. In the older towns, this has meant that the greatest accumulation of vehicles occurs where road conditions are most critical and the cost of additional road space is at a premium. In addition, it has been found that through trips via the radial routes not only often take longer because of central area congestion, but also they can mean greater travelling distances.

The principal advantage of a ring road lies in its ability to service a central area while circumventing it. This is suggested by the diagrams in

Route		Average distance
Direct		1.13r
Radial arc		−1.38r
Rectangular		1.44r
Radial		1.67r
Ring		1.90r (0.33r within the central area)

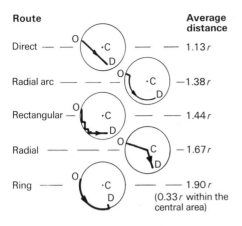

Fig. 3.8 Some possible routes for journey-to-work travel in the central area[18]

Fig. 3.8 which show some possible routes of travel from a point O outside a central area enclosed by a completely circular ring road of radius r, to destinations inside the ring road. If it is assumed that the destinations are scattered at random over the central area, then the *theoretical* average distances travelled from point O to a destination D within are as shown in the figure.

Although the ring route is appreciably longer than the other routes, it has the particular advantage that only 0.33r is travelled within the central area. The remainder of the journey, 1.57r is on the edge of the area, and thus, in theory, should be free of the central area congestion problems.

In practice, the advantage of ring routing is not as great as this, because towns are not circular, because it is not possible to join a ring route at any point, and because, for journeys wholly within the central area, the relative advantages of the routes are different. Nevertheless, as Table 3.6 shows, the ring road concept can be very saving of road space where it is most required.

Table 3.6 Effect of the number of commuters upon the road space required within a central area[18]

	Area of road per person (m²)		
		Ring route	
Number of commuters	Direct, radial arc or rectangular route	Total	Within central area
10 000	0.74	1.02	0.28
100 000	2.60	3.44	1.02
1 000 000	9.10	12.45	3.25

Note: peak period = 2 hours; 1.5 person per car

Highway traffic characteristics

It is necessary for the highway planner to be able to predict the highway volumes that can be expected on the roadway or network of roads being evaluated. Traffic volumes, however, are much heavier during certain times of the year and during certain periods of the day. Hence the engineer-planner must be familiar with the manner in which traffic behaves as a whole before attempting to conduct basic traffic prediction surveys.

Traffic volumes

There are three cyclical variations that are of particular interest to the highway planner: the manner in which traffic flow varies throughout the day and night, the day-to-day variation throughout the week, and the season-to-season variation throughout the year. In addition, the planner must also be aware of the directional distribution of the traffic and the manner in which its composition varies. A thorough understanding of the manner in which all of these behave is a basic requirement of any highway planning programme.

Hourly patterns

Typical hourly patterns of traffic flow, based on data collected at urban and rural census points throughout Britain, are shown in Fig. 3.9. There are some characteristics of particular interest illustrated in this figure.

First and perhaps most important is that the weekday pattern is basically unchanged from Summer to Winter. In both examples shown in Fig. 3.9, there is a peak about 8.30 a.m. followed by a drop until the middle of the afternoon, after which there is a further rise until an evening peak is reached about 5.30 p.m., after which the traffic flow falls again to its lowest point at about 4 a.m. The most marked difference between the two patterns is that in July a higher proportion of the day's traffic is carried in the brighter evening hours after 6 p.m.

The data in Fig. 3.9 also illustrate the fact that, although the Monday to Friday pattern is relatively consistent and is essentially as is to be expected, the patterns during the weekend can vary considerably. Not only do the Saturday and Sunday patterns differ from each other, but they also differ between themselves at various times of the year. The most marked variations are: (a) the lack of a morning peak, as such, during the weekends, (b) a more pronounced peak on Winter Sundays, and (c) the significant shift and spreading out of the weekend evening peaks from Winter to Summer.

A point which should be emphasized is that the patterns in Fig. 3.9 represent average data obtained at 49 and 39 national survey points in July and January, respectively. Thus the actual pattern on any individual road may vary significantly from these data. For instance, not only are traffic volumes heavier on urban roads than on rural roads, but they are also more concentrated during the peak periods of the day. Peak hour

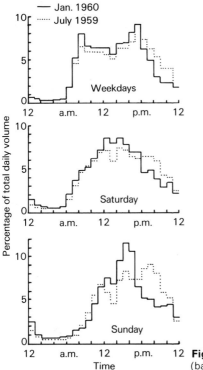

Fig. 3.9 Hourly patterns of traffic flow in Britain (based on data from reference 19)

volumes are more pronounced and directional on urban radial streets. Urban ring roads, however, do not have such sharp peaks. The durations of peak flows also vary considerably; some roads have high sharp peaks, whereas others have peaks that may not be so high but are more critical because of their length.

A most important point to note here is that data gathered at the same sites on later occasions reflect the same general pattern illustrated in Fig. 3.9. In other words, although traffic volumes may grow, the relative *percentages* of traffic at the different hours of the day in, say, July are quite consistent year after year.

Daily patterns

Figure 3.10 gives estimates of the daily travels starting at 8 a.m. for the year 1960. The diagram classically illustrates that distances travelled vary considerably with the time of the year. In general, however, the weekday flows behave in a surprisingly consistent manner as do, generally, the Saturday flows, whereas Sunday traffic varies considerably from week to week. During the Winter, the weekend traffic in general and Sunday traffic in particular is less than during weekdays; however, as the weather improves, travel increases on the weekends, especially on Sundays. Again, it is of interest to point out that, although the measurements shown in

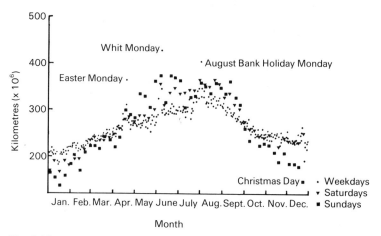

Fig. 3.10   Daily vehicle-kilometres for Britain in 1960[19]

Fig. 3.10 were gathered in 1960, the same general trends are reflected today.

Note also that the traffic on the holiday Mondays is in each case higher than on any other day within a week in either direction. Furthermore, the Saturdays and Sundays of the holiday weekends have more traffic than do the preceding and following Saturdays and Sundays. In addition, the weekdays preceding and following the holiday weekends also have more traffic than do the normal weekdays at those times of the year.

It must be emphasized that these figures represent average country-wide rural and urban values. The pattern at any particular highway may vary considerably depending upon its location and purpose (see, for example, reference 1). For instance, a seaside resort highway carrying heavy holiday traffic will have a pronounced difference between Summer and Winter values. Furthermore, a rural road which primarily acts as a service road for a farming area may show less variation between traffic on the different days of the week and months of the year than, say, a rural highway which is used regularly by long-distance traffic. Table 3.7 shows some daily and seasonal traffic flow patterns, as determined for various types of road, from a more recent national 50-point traffic census.

Monthly and yearly patterns
Figure 3.11 illustrates the monthly and yearly variation in car and light goods vehicle travel between January 1972 and December 1978. For comparison purposes, the monthly index for October 1972 is taken as 100. In general, this figure shows the very consistent seasonal trend in respect of cars, whereas the light goods (and also heavy goods) vehicle trends are much less marked.

These data are quite interesting from two viewpoints. Firstly, they show that similar monthly and yearly patterns can be relied upon for traffic prediction purposes. Secondly, they show that the monthly

Table 3.7 Average daily flows for each month expressed as a ratio of the average annual flow for that day of the week: (a) Monday to Friday, (b) Saturday and (c) Sunday[20]

Group*	January	February	March	April	May	June	July	August	September	October	November	December
1	0.91(4.3)	0.94(3.4)	0.98(3.0)	1.01(2.0)	1.03(2.4)	1.03(3.0)	1.02(5.3)	1.02(5.5)	1.04(2.1)	1.03(1.2)	1.01(1.4)	0.98(2.6)
2	0.88(3.7)	0.88(4.3)	0.96(3.1)	1.01(3.0)	1.04(2.7)	1.07(3.4)	1.10(3.6)	1.07(6.5)	1.06(4.1)	1.02(3.0)	0.98(3.3)	0.94(3.2)
3	0.82(5.6)	0.86(5.3)	0.93(4.0)	1.02(3.5)	1.03(4.5)	1.07(2.5)	1.16(6.1)	1.20(8.0)	1.10(4.3)	1.01(2.9)	0.92(5.4)	0.88(6.2)
4	0.63(18.9)	0.66(19.1)	0.73(15.9)	0.86(10.0)	0.97(4.6)	1.30(10.2)	1.57(15.7)	1.79(18.3)	1.25(9.2)	0.84(10.3)	0.71(17.9)	0.68(18.8)

(a)

Group*	January	February	March	April	May	June	July	August	September	October	November	December
1	0.88(6.6)	0.92(4.2)	0.97(4.7)	1.00(2.6)	1.02(2.4)	1.06(3.2)	1.05(8.1)	1.04(7.3)	1.05(2.6)	1.05(2.0)	1.00(2.0)	0.96(4.6)
2	0.81(9.3)	0.84(9.7)	0.91(7.7)	0.98(7.8)	1.06(3.9)	1.13(6.7)	1.18(14.1)	1.21(18.3)	1.08(4.7)	1.02(5.9)	0.92(8.5)	0.86(7.5)
3	0.69(13.9)	0.75(13.5)	0.85(11.3)	0.91(6.1)	1.03(4.7)	1.21(7.0)	1.36(15.5)	1.41(15.9)	1.20(7.6)	0.99(7.8)	0.82(13.2)	0.78(11.9)
4	0.58(20.7)	0.63(19.2)	0.72(13.5)	0.87(8.0)	1.07(7.2)	1.37(10.3)	1.61(15.4)	1.75(14.2)	1.25(8.5)	0.86(10.3)	0.67(19.4)	0.62(23.1)

(b)

Group*	January	February	March	April	May	June	July	August	September	October	November	December
1	0.83(8.0)	0.90(5.1)	0.99(3.7)	1.03(3.1)	1.08(3.6)	1.09(3.9)	1.08(7.8)	1.07(6.8)	1.07(3.4)	1.04(1.6)	0.94(5.2)	0.88(5.7)
2	0.77(7.4)	0.83(7.6)	0.97(2.9)	1.11(8.7)	1.13(5.7)	1.14(4.8)	1.16(8.2)	1.15(5.5)	1.08(4.4)	1.02(3.9)	0.86(6.8)	0.81(11.1)
3	0.67(10.1)	0.77(8.8)	0.91(6.6)	1.01(4.6)	1.11(7.0)	1.23(3.3)	1.31(7.7)	1.31(8.6)	1.18(6.4)	1.03(7.2)	0.78(8.6)	0.69(10.6)
4	0.50(23.0)	0.57(19.2)	0.72(12.6)	0.88(8.3)	1.19(11.2)	1.47(10.1)	1.72(10.9)	1.82(13.8)	1.26(8.6)	0.82(8.9)	0.57(20.7)	0.48(26.1)

(c)

Note: The coefficients of variation are given in brackets.

*Group 1 = urban/commuter roads; Group 2 = low-flow (<1000 veh/day) non-recreational rural roads; Group 3 = rural long-distance roads; Group 4 = recreational roads

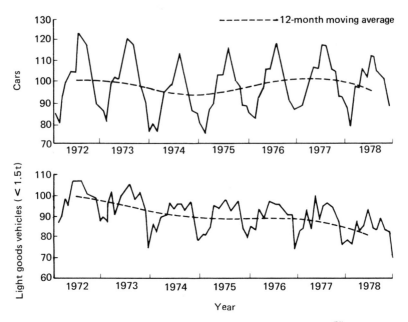

Fig. 3.11   Monthly indices of vehicle-kilometres per licensed vehicle[21]

distances travelled by cars have remained relatively static throughout the 1970s, whereas they reflected a gentle but continuous increase (about one per cent per year) in the late 1950s and the 1960s[22]. (In 1983, the average annual distances travelled by cars and taxis, motor cycles, buses and coaches, and goods vehicles were 13 400, 4400, 47 700, and 24 500 km per year, respectively.)

Effect of fuel prices
As noted in Chapter 1, it can be expected that the real price of oil (net of tax) will rise significantly by the turn of the century, so it is useful to examine the effect that the price of fuel to the vehicle user has upon travel. With the large fluctuations in the price of fuel that have taken place since 1973, it is now possible to obtain reliable estimates of the elasticity of demand for travel with respect to petrol prices after they have been corrected for inflation.

The corrected-for-inflation monthly indices for fuel prices at the service station from 1972 to 1978 (with the October 1972 index taken as 100) are shown in Fig. 3.12. It can be seen that, whereas the post-World-War-II trend had been a falling of real petrol prices, there have been significant fluctuations (increases and decreases) in the service station prices of both petrol and diesel fuel (derv) since the beginning of 1974. After taking into account such factors as changes in real public transport fares, the industrial production index, the real earnings index, gross domestic product per head, adult unemployment, and weather variables, the analysis on which Figs 3.11 and 3.12 are based concluded that when

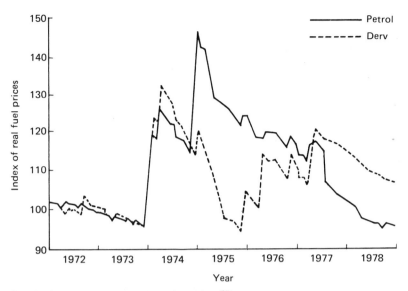

Fig. 3.12 Monthly indices of real fuel prices[21]

real petrol prices at the service station were increased by one per cent, there was a decrease in vehicle-kilometres per licensed petrol-driven vehicle of between 0.10 and 0.17 per cent for all vehicles. Light goods vehicles ≤ 1.52 t unladen mass formed an exception to this, i.e. one type of analysis for light goods vehicles produced similar and significant petrol elasticities to those estimated for cars and all vehicles, whereas a different method of analysis failed to find significantly non-zero elasticities. No significantly non-zero elasticities with respect to derv prices were obtained for heavy goods vehicles by any type of analysis, i.e. the elasticity for these vehicles appears to be close to zero, at about $+0.05 \pm 0.10$.

Effect of weather
The data already shown point up partially the effect that the state of the weather has upon the flow of traffic. During the Summer months the distances travelled are high, while they are down during the Winter months. Just as important, however, is the fact that, at any time of the year, months that are wetter than normal tend to have less traffic than normal; the difference is statistically significant but not very large.

Directional distribution
A most important consideration in the planning of any highway is the directional distribution of traffic. The importance of this factor can be gauged by considering the case of a dual carriageway highway designed to carry 4000 vehicles per hour. If the directional distribution of the traffic results in a 50:50 split, then two traffic lanes in each direction may well be sufficient to handle the traffic. If, however, some 70 per cent of the traffic flow is in a given direction at a given time, then three lanes in each

direction may be needed to supply the equivalent level of service to the traffic on that highway.

While this example of a 70:30 distribution may seem to be extreme, it is not so, in any sense. For instance, a major radial road in Leeds has a consistent traffic distribution of 70:30 and 40:60 during the Monday to Friday morning and evening peak hours, respectively.

The actual distribution to be used for design purposes can only be determined by field measurements. If an existing road is to be reconstructed, then the field studies can be carried out on it beforehand. If the highway is to be an entirely new facility, then the measurements should be made on adjacent roads from which it is expected that traffic will be diverted.

A most important feature of the directional distribution is that it is also relatively stable and does not change materially from year to year. Hence relationships established on the basis of current traffic movements are normally also applicable to future movements.

Traffic composition

Passenger cars, buses, lorries, motor cycles and bicycles all have different operating characteristics and hence, for comparison purposes, may be given different weightings to allow for their varied effects within the traffic stream upon traffic movement. In estimating the design volume of a road, it is therefore necessary that the percentages of the different classes of vehicle present during the design period should be determined.

In 1983, there were 20.216m motor vehicles *registered* in Britain, of which private cars, light vans and taxis accounted for 84.9 per cent, 0.6 per cent were buses and coaches, 2.8 per cent were goods vehicles, 6.4 per cent were motor cycles, and the remaining 5.3 per cent included tractors, crown and exempt vehicles[23]. While national figures such as these are generally assumed to represent the number of vehicles *in use*, this is not necessarily the case as they do not include (illegal) vehicles which are not licensed.

The above national proportional distribution of vehicles should also not be assumed to reflect the traffic composition at any particular location or time of day. As a general rule, the composition of the traffic stream varies according to the locality being served. For instance, where the locality is highly industrialized, the numbers of commercial vehicles will be relatively great, whereas they will be low in non-industrialized areas. In either case, the percentage of commercial vehicles present in the traffic stream will be less during the peak periods than during other hours of the day, since the number of passenger cars on the road is greatest at these times.

Table 3.8 summarizes peak and off-peak traffic flow data gathered on main roads in London, the principal cities of five other conurbations (population 0.3m to 1.1m), and eight towns (population 0.08m to 0.56m) which illustrate the manner in which traffic composition can vary in built-up areas. Actual values for planning and design usage are best determined from field studies. Generally, the traffic composition determined from

Table 3.8 Weekday traffic speeds, flows and compositions in built-up areas in 1976: (a) for peak periods and (b) for off-peak periods[13, 24]

Location	Average speed (km/h)	Average flow (veh/h)	Composition (%)			
			Cars and taxis	Light commercial vehicles	Heavy commercial vehicles	Buses and motor cycles
London						
Inner	n.a.	1900 ⎫	79.0	7.0	9.5	4.5
Outer	n.a.	2010 ⎭				
Conurbations						
central area	20.4	1485	77.5	8.8	5.0	8.7
non-central area	35.5	1640	79.3	8.8	6.8	5.1
Towns						
central area	20.8	1570	78.0	9.3	4.7	8.0
non-central area	31.4	1490	78.9	9.5	5.7	5.9

(a)

Location	Average speed (km/h)	Average flow (veh/h)	Cars and taxis	Light commercial vehicles	Heavy commercial vehicles	Buses and motor cycles
London						
Inner	20.7[25]	1360 ⎫	67.3	13.0	15.3	4.4
Outer	n.a.	1220 ⎭				
Conurbations						
central area	21.4	1150	67.3	14.9	10.9	6.9
non-central area	41.0	1120	66.0	14.3	15.9	3.8
Towns						
central area	25.0	1190	69.7	14.4	9.9	6.0
non-central area	36.1	980	68.3	14.8	12.6	4.3

(b)

current traffic figures is assumed stable when used for future design purposes.

Highway traffic measurements

When planning a new or improved road or road system, it is necessary to know the distribution and performance of the traffic on existing roads. Not only is this of use in predicting future traffic behaviour, but it is also of value in assessing whether alterations are justified, and in deciding priorities for road improvement.

In almost all highway and traffic planning studies, measurements of vehicle flows and speeds are needed. Often measurements are also needed of stopped times and the frequency with which these stops occur. As detailed information on these is readily available in the literature[26, 27], the following discussion is only summarily concerned with the manner in which the data are obtained and analysed.

Speed studies

The term 'traffic speed' is often used very loosely when describing the rate of movement of traffic. To the highway engineer, there are many different types of speed, each of which describes the rate of traffic movement under

specific conditions and for a specific purpose. The vehicle speeds of most
interest are spot speeds, running speeds, and journey speeds.

Spot speed
This term is used to describe the instantaneous speed of a vehicle at a
specified location. Spot speeds have a variety of uses. They can be used as
evidence regarding the effect of particular traffic flow constrictions such as
intersections or bridges. Since spot speeds at ideal sections of highway are
indicative of the speeds desired by motorists they can be used for
geometric design purposes on improved or new facilities. In addition, as
will be discussed in Chapter 7, spot speeds can be used to determine
enforceable speed limits.

The location at which the spot speed measurements are taken depends
upon the purpose for which they are to be used. Whatever the purpose of
the study, it should be conducted so as to reduce to a minimum the
influence of the observer and measuring equipment upon the values
obtained. Hence the observer and equipment should be located as
inconspicuously as possible. If it is not possible to measure all vehicle
speeds, vehicles should be selected at random from the traffic stream in
order to avoid bias in the results. Thus, for instance, every tenth vehicle
or those whose registration numbers end in a preselected digit should be
measured, rather than leaving it to the observer's discretion to select the
test vehicles.

Methods of measurement   One simple method of manually collecting spot
speed data is to observe the time required by a vehicle to cover a short
distance of roadway. With the *direct-timing* procedure, two reference
points are located on the roadway at a fixed distance apart and the
observer starts and stops a stopwatch as a vehicle enters and leaves the
test section. While this is a most uncomplicated way of collecting spot
speed data, it has the obvious disadvantage of being subject to error
because of the parallax effect. This error is of particular importance in
before-and-after studies, where the observer may have to change
observation position.

With the *pressure-contact-strip method*, two contact strips, usually two
pneumatic tubes (each 12.7 mm ID/25.4 mm OD), are laid across the
carriageway at a fixed distance apart. When a vehicle passes over the first
tube, an air impulse is sent instantly along the tube which activates a
time-measuring instrument in the hands of the observer. When the second
tube is depressed by the same wheels of the vehicle, the timer is
automatically stopped, and the reading noted either visually by the
observer or by automatic data recorders.

Provision can usually be made with these devices for switching the
direction of the start and stop detectors so that speeds in either direction
of travel can be measured in the course of a particular study. To avoid
invalid measurements due to overtaking manoeuvres occurring at the
contact strips on a busy road, an observer may have a superimposing
ready-switch which, until activated, does not allow any readings to be

taken. Advantages of this system are that it is relatively inexpensive, portable, requires no specialized personnel for operation and maintenance, gives very accurate speed results, and can have a relatively long life on low-volume roads. Disadvantages include its vulnerability to wear and tear by trailing vehicle parts, street cleaners and snow ploughs, as well as to vandalism; also, its installation can disrupt traffic for excessive periods of time because of the safety precautions necessary at the time.

In a manner similar to pressure-contact strips, *inductive loop detectors* can also be used to measure spot speeds. In this instance, two wire loops are inserted in the carriageway pavement at a known distance apart, and radio-frequency (RF) energy at 85–115 kHz is fed to circuits tuned to avoid local electrical interference, whose inductive elements are the loops. When a vehicle passes over a loop, a change is caused in its inductance which in turn causes either an amplitude impedance or a frequency or phase shift, thus recording the vehicle's presence. The vehicle's speed is then determined by measuring the time spacing between successive output pulses from the two detectors.

Advantages of the inductive loop system are that it is fairly easy to install, is not subject to damage because of its in-pavement location, and is capable of detecting small vehicles. A further advantage is that it is relatively inexpensive to abandon use of the loops for a given period of time and reuse the amplifier at a new location. Disadvantages include the relatively high cost of loop installation, and the safety need to close traffic lanes at the time of installation; also, the loops can be difficult to tune to detect both small and large vehicles.

A very common way of measuring spot speeds is by means of a *radar speedmeter*. When this is in operation, a beam of very high frequency is directed from the radar speedmeter to the moving vehicle. The waves are bounced back from the vehicle but, because of the Doppler effect, these reflected waves have a slightly different frequency from that of the transmitted waves. This difference, which is directly measurable, is proportional to the speed at which the vehicle is moving. Provided that the beam of the transmitter is contained within a cone of about 20 degrees of the line of movement of the vehicle, the spot speed can usually be measured to an accuracy of about 2 per cent with the normal type of radar speedmeter.

The radar speedmeter may be permanently located, e.g. fastened to a pole (above the carriageway), or portable and operated from near the edge of the carriageway at a height of about 1 m above ground level. Its operating zone extends a distance of about 45 m and so it can measure all speeds within this zone, whether they be approaching or receding. A drawback associated with using this speedmeter is that individual vehicle speeds can be difficult to obtain in heavy traffic flows where vehicles are likely to overtake/mask each other. No very definite information is available on this, but it appears that it is very difficult to operate the speedmeter efficiently on two-lane roads when the traffic flow is heavier than about 500 vehicles per hour. American experience suggests that positive identification of all individual speeds becomes impossible at traffic volumes greater than about 1000 vehicles per hour on multilane highways.

The major advantages of a radar speedmeter are its ease of use, mobility (the portable type), freedom from vehicular damage, and immunity from electromagnetic interference. Its disadvantages are its relatively high capital cost, and the need for experienced personnel to install (permanent type) and maintain the equipment.

Sonic detectors are sometimes also used to measure spot speeds. In a manner analogous to that for radar speedmeters, they operate on the principle of transmitting a beam of ultrasonic energy at 18–20 kHz towards the roadway and sensing the shift in the frequency of the ultrasonic tone when it is reflected from a moving vehicle. Their output is a direct current voltage that is proportional to vehicle speed. The advantages and disadvantages of this type of detector are as those for the radar speedmeter.

On very crowded highways, i.e. motorways or busy city streets, spot speeds can be very accurately obtained by means of photographs obtained with a *time-lapse camera*. This camera takes photographs at fixed intervals of time, thus obtaining a permanent record of all vehicle movements within the camera vision. After the film has been developed, the speed of any type of vehicle present in the traffic stream can be obtained by comparing its carriageway position on successive exposure frames[28].

The main disadvantages associated with using the camera method are the relatively considerable time required and the expense involved in processing the film and analysing the data (unless sophisticated analysis equipment is available). The major advantage of this system is, of course, that a permanent visual record is obtained of vehicular behaviour at the site under examination.

Analysis   Normally, the spot speeds measured at a particular location will vary considerably, the degree of variation depending upon the number and type of vehicles being measured and the condition of the roadway. In analysing the data, therefore, it is necessary to consider carefully beforehand what information is required. Some of the desired values can most readily be obtained by graphical interpretation, whereas others can be easily calculated directly from the field data.

Graphical analysis   Two methods of graphically interpreting the data shown in Table 3.9 are illustrated in Fig. 3.13. Figure 3.13(a) contains examples of both a histogram and a frequency curve, each illustrating the number of occasions upon which the different speeds occurred. Figure 3.13(b) is an example of a cumulative spot speed distribution curve for the same data.

The frequency curve is a very useful preliminary guide to the statistical normality of the data. If the measurements are normally distributed, then the frequency curve will be bell-shaped. The *modal* speed, which is the spot speed which occurs most often, can be detected easily at the peak of the frequency curve.

The cumulative curve shown in Fig. 3.13(b) was obtained by plotting

Table 3.9 Illustrative grouped spot speed data

Speed class (km/h)	Average speed (km/h)	Number of vehicles	Frequency (%)	Cumulative frequency (%)
10–14.9	12.5	3	1.5	1.5
15–19.9	17.5	10	5.0	6.5
20–24.9	22.5	21	10.5	17.0
25–29.9	27.5	31	15.5	32.5
30–34.9	32.5	54	27.0	59.5
35–39.9	37.5	43	21.5	81.0
40–44.9	42.5	21	10.5	91.5
45–49.9	47.5	10	5.0	96.5
50–54.9	52.5	5	2.5	99.0
55–59.9	57.5	2	1.0	100.0

the cumulative percentage versus the upper limit of each speed group shown in Table 3.9, and then drawing a smooth S-shaped curve through the points obtained. This curve is most useful in determining the speed above or below which certain percentages of vehicles are travelling.

The *median* speed is the middle or 50th percentile speed. It is the speed at which there are as many vehicles going faster as there are going slower.

The *85th percentile* speed is the speed below which 85 per cent of the vehicles are being driven. This speed is often used as the criterion in establishing an upper speed limit for traffic management purposes. Some

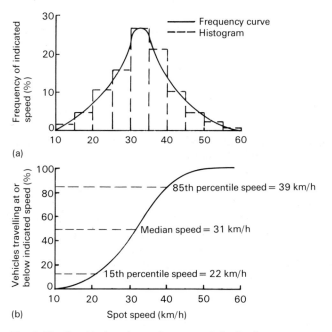

Fig. 3.13 Graphical analyses of spot speed distribution

authorities urge the use of the 85th percentile speed of a highway as a measure of the design speed to be selected on similar new highways.

The *15th percentile* speed is generally considered to be the speed value which should be utilized as a minimum speed limit on major highways such as motorways. Vehicles travelling below this value on high-speed roads are potential accident hazards because of their obstructive influence upon traffic flow.

Mathematical analysis   In a spot speed study, probably the most often obtained statistic is the *arithmetic mean* or *average* spot speed. This is the sum of all the variable speed values divided by the number of observations. Mathematically this is expressed as follows:

$$\bar{x} = \frac{\Sigma X_j}{n}$$

where $\bar{x}$ = average spot speed, X_j = jth spot speed, and n = number of observations.

More usually, to avoid excessive computations at a later stage, the entire data are grouped into speed-class intervals in the manner indicated in Table 3.9. A speed-class interval of convenient size is easily obtained by using the following equation:

$$CI = \frac{R}{1 + 3.322 \log n}$$

where CI = class interval, R = range between largest and smallest speed values, and n = number of observations.

For the data in Table 3.9, the fastest speed was 59 km/h, while the slowest was 12 km/h. Thus

$$CI = \frac{59 - 12}{1 + 3.322 \log 200} = 5.5$$

It is more convenient to use integers, so a class interval of 5 was used to group the data in the table.

When the data are grouped into speed-class intervals, the average value is obtained from the following equation:

$$\bar{x} = \frac{\Sigma f_i x_i}{n}$$

where $\bar{x}$ = average spot speed, x_i = average speed of the ith speed interval group, f_i = frequency of the ith group, and n = number of observations.

A statistical measure of the dispersion of the spot speeds is given by calculating the *standard deviation* of the set of observations. This is estimated by first obtaining the variance of the sample, and then taking the positive square root of the variance. Thus

$$s^2 = \frac{\Sigma(X_j - \bar{x})^2}{n} \quad \text{and} \quad s = +\sqrt{s^2}$$

where $s=$ estimate of the standard deviation of the distribution, $s^2=$ variance of the sample, $X_j = j$th spot speed, $\bar{x}=$ average spot speed and $n=$ number of observations.

For grouped data, the standard deviation is estimated as follows:

$$s = \left(\frac{\Sigma f_i(x_i)^2 - (\Sigma f_i x_i)^2/n}{n} \right)^{1/2}$$

Thus, for the data in Table 3.9, $s=8.62$ km/h.

The usefulness of the standard deviation lies in the fact that the arithmetic mean plus and minus one standard deviation contains 68.27 per cent of the data, plus and minus two standard deviations contains 95.45 per cent, and plus and minus three standard deviations contains 99.73 per cent.

A statistic which indicates the confidence with which the arithmetic mean of the sample can be considered as the actual mean of the entire traffic speed population on that section of the highway is the *standard error of the mean*. This is determined by first calculating the variance of the mean and then taking its square root. Thus

$$s_{\bar{x}}^2 = \frac{s^2}{n}$$

and $\quad s_{\bar{x}} = \dfrac{s}{\sqrt{n}}$

where $s_{\bar{x}}=$ standard error of the mean, $s_{\bar{x}}^2=$ variance of the mean, $s=$ standard deviation of the sample, and $n=$ number of observations.

Therefore, for the data in Table 3.9,

$$s_{\bar{x}} = 8.62/\sqrt{200} = 0.61 \text{ km/h}$$

This statistic enables it to be said with 95.45 per cent confidence that the actual mean of all the spot speeds at this particular location lies between $33.25 \pm 2(0.61)$ km/h, i.e. between 32.03 and 34.47 km/h.

The standard error of the mean is also a most useful statistic in determining whether the speed differences at different locations or in before-and-after studies are significant. This significance can be tested by calculating the *standard deviation of the difference in means*. Thus

$$\hat{s} = \sqrt{(s_{\bar{x}B}^2 + s_{\bar{x}A}^2)}$$

where $\hat{s}=$ standard deviation of the difference in means, $s_{\bar{x}B}^2 =$ variance of the mean of the before study, and $s_{\bar{x}A}^2 =$ variance of the mean of the after study.

If the difference in mean speeds is greater than twice the standard deviation of the difference in means, that is,

$$\bar{x}_A - \bar{x}_B > 2\hat{s}$$

where $\bar{x}_A$ and $\bar{x}_B$ are the mean speeds of the after and before studies, respectively, then it can be said with 95.45 per cent confidence that the observed difference in mean spot speeds is statistically significant. In other words, the difference in speeds is a true change reflecting the changing road conditions, and is not due to chance.

The standard error of the mean is also very useful in estimating the number of vehicles that must be sampled in order to ensure a certain degree of accuracy. For example, let it be assumed that it is desired to know how many vehicles must be sampled at the site at which the data in Table 3.9 were collected in order to ensure an error of less than 0.5 km/h in the average speed obtained. Thus

$$\frac{2\sigma}{\sqrt{n}} < 0.5 \, \text{km/h}$$

where n=desired number of observations and σ=standard deviation of the entire population of spot speeds. In this case it is assumed that $\sigma=s$, the previously determined estimate of the population standard deviation. Therefore

$$n > 4\sigma^2/(0.5)^2$$
$$> 4(8.62)^2/0.25$$

i.e. $n > 1188.9$

Since n is an integer, this means that the number of vehicles sampled must be greater than 1189 to ensure the required accuracy.

In conducting before-and-after studies, care must always be taken that similar measurement procedures are used in both stages of the study; otherwise the speed differences may not be true reflections of the change in road conditions. For instance, if the spot speeds are measured instantly by means of a radar speedmeter or pressure-contact strips placed closely together, the average speed is given by

$$\bar{x}_t = \frac{\Sigma X_j}{n}$$

where $\bar{x}_t$=time-mean speed, and X_j and n are as defined before. Note that the average spot speed is here called the *time-mean* speed.

However, if a time-lapse camera is used to obtain the data, the average mean speed is obtained by

$$\bar{x}_s = \frac{\bar{l}}{t}$$

where $\bar{x}_s$=space-mean speed, $\bar{l}$=average distance travelled by each vehicle, and t=time interval between two successive exposures (a constant). Note that the mean spot speed calculated in this way is called the *space-mean* speed.

The basic difference between the two mean speeds is that the spot speed for each individual vehicle is calculated prior to the determination of the time-mean speed, whereas the average time (in the above illustration this is the fixed time interval t) is first determined when calculating the space-mean speed. Both methods have been examined and it has been shown that

$$\bar{x}_t = \bar{x}_s + \frac{\sigma_s^2}{\bar{x}_s}$$

where $\bar{x}_t$ = time-mean speed, $\bar{x}_s$ = space-mean speed, and σ_s^2 = variance of the space distribution of speeds.

Thus the time-mean speed is always greater than the space-mean speed unless the very unlikely happening occurs that there is no variation at all amongst the speeds. The difference is usually of the order of 12 per cent.

Running and journey speeds

Although spot speeds are most useful in measuring fluctuations in speed at particular locations, they give no information regarding fluctuations in speed throughout a route as a whole. Statistics that are of value in this situation are the running and journey speeds.

Running speed is defined as the average speed maintained over a given route while the vehicle is in motion. Thus, in determining the running speed, the times along the route when the vehicle is at rest are not taken into account in the calculations. Normally, only the average running speed and the standard deviations are the variables determined.

Knowledge of the running speed can be very useful to the highway planner. For instance, the running speed can be used as a measure of the level of service offered by a highway section over a long period of time.

The motorist about to set out on a journey is primarily concerned with the time required to complete the trip. Hence the speed in which there is most interest is the average journey speed, i.e. the distance travelled divided by the total time taken to complete the distance. This total time includes both the running time, when the vehicle is actually in motion, and the time when the vehicle is at rest, i.e. at traffic signals, in traffic jams, etc.

Knowledge of the journey speed and, in particular, the journey time is most useful in highway planning. For instance, it is an excellent direct measure of traffic congestion and the general adequacy or inadequacy of a road or road system. Highway economic studies normally utilize journey times and speeds in their analyses. In addition, journey times are used as criteria on which to base decisions regarding the diversion and assignment of traffic to new and improved highway facilities.

Methods of measurement   The following three methods are those most used in obtaining the running and/or journey speeds of traffic streams.

Registration number method   With this method, observers with synchronized watches are stationed at both ends of the highway test section. As each vehicle passes an observer, its registration number and time of passage are noted and recorded. At a later time, the registration numbers are matched and the individual vehicle times determined. Knowing the distance over which the vehicles have travelled, the mean running speed of traffic can then be determined.

Unless traffic is very light, it is usually necessary to employ two observers for each direction of travel; the first observer notes the registration numbers and the times, while the second observer records them. Two observers can accurately record at the rate of about 300 vehicles per hour in this manner. If the traffic flow is heavier, an unbiased sample of the vehicles can be obtained by noting only the licence numbers ending in certain digits.

Besides being laborious and time-consuming, the registration number method has the very obvious disadvantage that it can only be used on highway sections having minor or no intersections, as otherwise vehicles may leave the test section or stop along the route unknown to the observers. It is therefore most suitable for highway sections in rural locations.

Elevated observer method   With this method, two or more observers are stationed on top of a high vantage point located adjacent to the road section being evaluated. These observers then select vehicles at random from the traffic stream and observe and time their progress over the test section.

While this method has the advantage that any vehicle delays can be noted, it has the disadvantage that it is very often difficult to secure suitable vantage points. Of necessity, also, the road test section will usually have to be relatively short, so that the selected vehicles remain within easy sight of the observers. As a result, this method is most useful on relatively short sections of city streets.

Moving observer method   This method[29] of determining the journey speed can best be illustrated by considering a stream of vehicles moving along a section of road of length l in such a way that the average number of vehicles q passing through the test section per unit time is constant. The stream can be regarded as consisting of flows q_1 moving with speed v_1, q_2 with speed v_2, etc.

Suppose that an observer travels with the stream at a speed v_w and against the stream at a speed v_a. When travelling against the flow q_1, which is moving at a speed v_1, the rate at which vehicles pass is greater than if the observer were stationary, i.e. they pass at the rate of $(v_a + v_1)/v_1$, and the number of vehicles that pass per unit of time is equal to $q_1(v_1 + v_a)/v_1$. Similarly, when travelling with the traffic stream, the number of vehicles going in the same direction, as counted by the observer, i.e. the number of overtaking vehicles minus the number of overtaken vehicles, is given by $q_1(v_1 - v_w)/v_1$.

If t_w denotes the journey time l/v_w which corresponds to speed v_w, t_a denotes the journey time l/v_a which corresponds to speed v_a, x_1 denotes the number of vehicles with speed v_1 met by the observer when travelling against the stream, i.e. in time t_a, and y_1 denotes the number of vehicles which overtake the observer when travelling with the stream minus the number the observer overtakes, i.e. in time t_w, then

$$x_1 = q_1(t_a + t_1)$$

and $$y_1 = q_1(t_w - t_1)$$

Similar results hold for all the other speeds, so that if q denotes the flow of all vehicles in a given stream, i.e. $q_1+q_2+q_3+\ldots$, $\bar{x}$ denotes the total number of vehicles met in the section when travelling against the stream, y denotes the number of vehicles overtaking the observer minus the number the observer overtakes when travelling with the stream, and $\bar{t}$ denotes the mean journey time of all vehicles in the stream, i.e. $(q_1t_1+q_2t_2+\ldots)/q$, then summing overall gives

$$x=q(t_a+\bar{t})$$

and $y=q(t_w-\bar{t})$

Therefore

$$q=(x+y)/(t_a+t_w)$$

and $\bar{t}=t_w-y/q$

The space-mean speed is then given by $l/\bar{t}$. It must be remembered that the times used in the equations may (journey times) or may not (running times) include the periods during which the test vehicles are actually stopped.

In practice, when using the moving observer method to study traffic conditions on a highway section, a car (or preferably a pair of cars, one starting from each end and each carrying three observers) is started along the test section. One observer starts a stopwatch which is left continuously running and records the time history of the vehicle along the highway. A second observer counts the vehicles in the opposing traffic stream that are met on the journey, totals being recorded by the first observer en route at points which tend to divide the length of highway being studied into sections of reasonably homogeneous traffic conditions. Meanwhile the third observer counts the number of overtaking and overtaken vehicles, the totals being recorded at the same points as before. The values of x, y, t_a and t_w then used in the above equations are the averages of all the individual values obtained. If it is desired to know the average speed of certain types of vehicles only, the procedure is the same, except that the only vehicles taken into account in the calculations are those of the type being considered.

This procedure has the particular advantage that it is most economical in personnel—a small team of observers with one or two cars can collect reliable data over considerable roadway distances in a few weeks. More important is the fact that it allows much information to be collected at the same time, i.e. journey times and speeds, delays due to intersections or parked cars, and other relevant information.

It is generally held that 12–16 runs in each direction along a test section will be sufficient to give reasonably consistent speed estimates, using the moving car observer method. In general, the heavier the traffic volume on a given type of road, the less the number of test runs required. However, the method is quite sensitive to variations in the traffic stream, so that at low traffic volumes (e.g. <250 veh/h in a given direction on a two-lane road) the number of test runs required to achieve a given degree of accuracy on short sections of road can be so great as to render the normal use of the method unpractical[30].

In the case of a one-way street, it is not possible to utilize the moving observer method in the normal way since no vehicle runs against the traffic stream are possible. In this case the practice is to vary the solution method slightly, and carry out two sets of runs with the stream at test vehicle speeds that are considerably above and below the average space-mean speed. The best results are obtained by making as large a number of runs as possible and choosing the two speeds so that they differ by a factor of at least two[31]. Alternatively, it has also been suggested that accurate results can be obtained by substituting a pedestrian walking (on the footpath) for the vehicle which has to travel against the traffic stream[32].

Delay studies

Delay studies can be carried out separately at particular locations or in conjunction with studies determining running and journey speeds. They are of considerable value to the highway planner since they enable locations where conditions are unsatisfactory to be pinpointed, as well as determining the reasons for and extent of the delays. This information can be used to indicate the urgency of need for improvement and the extent to which the improvement should be carried out. Highway economic studies normally utilize delay times in their calculations.

Studies into the cause and extent of delays on highways must take into account the fact that there are two forms of delay. The first of these, termed *fixed delay*, occurs mostly at roadway intersections. It is the result of some fixed roadway condition, and hence it occurs irrespective of whether or not the highway is crowded. Typical highway fixtures causing this type of delay are traffic signals, railway crossings, roundabouts and stop signs. The second type of delay is called the *operational delay* and this is primarily a reflection of the interacting effects of traffic on the highway or street. Operational delays can be caused by parking and unparking vehicles, by pedestrians, by crossing and turning vehicles at uncontrolled intersections, as well as by vehicles stalling in the middle of the traffic stream. Internal frictions within traffic streams can be another cause of operational delays. For instance, vehicle volumes in excess of capacity will cause traffic congestion and result in considerable delay to traffic. Intersections adjacent to each other and carrying heavy turning movements can be the cause of a considerable amount of weaving within the traffic streams as vehicles attempt to enter and leave the main roadway; these manoeuvres inevitably reduce the mean speed of traffic and lower the general efficiency of movement.

Methods of measurement   The moving observer method is most useful in determining the cause and extent of delay encountered along a route. Before the experiments are started a survey of the route is made and a journey log prepared. Then as the test vehicle moves along the route all locations and times of stopping and starting are noted, as well as ancillary data. Analysis of the journey logs easily points up the locations and extent of any delays to traffic.

While the moving observer method will determine the locations and extent of any delays, it will not always provide sufficient information by itself on which remedial action can be based. For instance, at intersections it will be necessary to obtain additional detailed information such as the traffic volumes on the different approach roads, the traffic management methods in use at or adjacent to the intersection, the accident statistics, and the physical dimensions and geometric layout of the intersection, before decisive action can be taken.

Volume studies

The terms traffic flow and traffic volume are used interchangeably to define the number of vehicles that pass a given point on the highway in a given period of time. As such it is probably the statistic that is of most value to the highway planner; to attempt to plan a modern high-speed highway without knowing the traffic volumes that can be expected upon it is like trying to design a structure without knowledge of the loads that it will have to carry.

The type of traffic volume data collected at any given time and location depends upon the use to which the data will be put. Similarly, the method of collecting the data is dependent upon its usage. In either case it is most important to remember that, notwithstanding the timing, location and method of data collection, traffic flow studies provide only very limited information regarding existing traffic on the roads. For instance, the 24-hour data in Fig. 3.14—the widths of the traffic flow bands in this diagram are proportional to the measured traffic on each roadway—clearly show that the traffic in Greater London is diffused over many roads. This can be attributed to the fact that at the time this survey was carried out, most main roads in London provided similar moderate traffic capacities, so that motorists tended to seek indirect alternative routes, thereby distributing the demand more or less evenly on all main roads. However, other than pointing up which roads are more used than others, this figure by itself tells little; no information at all is given about the origins and destinations of vehicles and whether the volumes of traffic on each road accurately reflect its importance, i.e. whether motorists are using a particular route because it is less crowded or whether it is actually the most direct and desirable route between the various origins and destinations.

Nevertheless, despite the obvious limitations associated with the data obtained from traffic flow studies on their own, they have many and varied uses. In rural areas well served with highway facilities and in smaller urban areas, volumetric measurements can very often be used to interpret the relative importance of individual roadways and thus justify particular highway improvements or traffic management measures. When

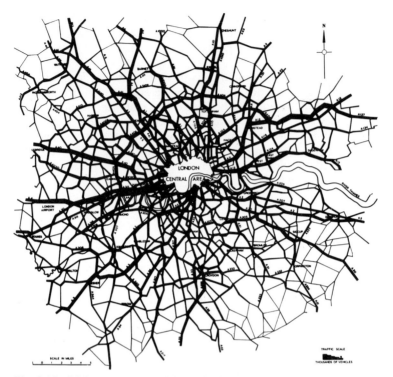

Fig. 3.14   24-hour average weekday traffic flow pattern within the London traffic survey area[33]. Courtesy of the Greater London Council (Crown copyright in map reserved)

carried out in conjunction with vehicle classification counts, they have a variety of uses, e.g. in capacity determinations and highway economic studies, in establishing correction factors to be applied to automatic traffic counts, and in establishing geometric design and traffic control criteria, particularly at intersections.

Figure 3.15 illustrates one example of the type of information that can be obtained regarding traffic flow through an intersection. This method of presenting data on traffic volumes and turning movements is very informative; the relative importance of each turning movement is discernible at a glance. If a vehicle classification count is carried out at the same time, then 'pie' diagrams can be drawn to show the composition of each traffic stream.

Volumetric counts conducted at cordon locations can provide valuable information regarding the accumulation of vehicles within an enclosed area. If vehicle occupancy counts and pedestrian counts are conducted as well, information is obtained regarding the accumulation of people within the cordon area, and the number of people using different transport modes.

As was discussed earlier, the manner in which traffic volumes and traffic patterns behave throughout the year and from year to year is

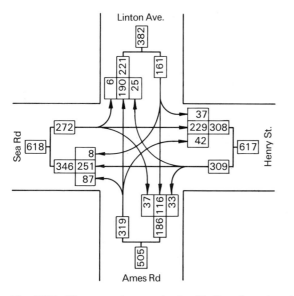

Fig. 3.15 Diagrammatic example of traffic flow through an intersection

predictable on similar highway types. Thus volumetric data collected over a relatively short period of time on individual highways can often be scaled to allow for daily, weekly, monthly, seasonal or yearly variations, as required[34].

Methods of measurement There are many different methods of obtaining traffic volume data. These include the use of permanent devices which can automatically count and record the data; manual counts where traffic observers do the counting and recording; a combination of mechanical and manual methods, e.g. multiple pen recorders where observers actuate pens which mechanically record the data; moving observer methods; methods involving the use of photography. Of these, the methods most widely used are the automatic methods, the manual methods, and the moving observer technique. Whichever is used at any particular location depends upon the facilities available and the type of information required.

The greater part of traffic counting, especially when data are required over long periods of time, is carried out using automatic equipment. Typically, this equipment consists of a device which detects the passage of a vehicle and a counting mechanism which records the detection pulses.

All of the detectors described previously in this chapter in respect of spot speed determinations can also be used to carry out vehicle counts—except that, of course, in the case of contact strip and inductive loop detectors, only one device needs to be laid on/in the carriageway. They also have the same advantages and disadvantages.

Contact strip devices can be of the pneumatic tube or electrical variety. In the case of the *pneumatic strip*—used mostly for temporary vehicle counts—a single length of rubber tubing is stretched across the roadway and fastened to the carriageway by clips or saddles of canvas and brass at about 1 m spacings. One end of the tube is sealed (except for a tiny air-hole whose purpose is to minimize the reflection of air waves in the tube) and the other end is attached to a pressure-actuated diaphragm switch. The passage of a wheel causes a pulse of air to travel along the tube so that it moves the diaphragm outwards against a contact point, thereby completing an electrical circuit which activates a recorder housed at the roadside.

The permanent *electrical contact strip* is a device consisting of a steel baseplate beneath a vulcanized rubber pad holding a strip of suspended spring steel, which is inserted in the carriageway pavement. The gap between the two metal contacts is filled with a dry inert gas during the assembly and sealed as a unit during the vulcanizing process. As each vehicle axle crosses the treadle pad there is a positive electrical contact closure which actuates a recorder. This contact strip detector has the advantage of being an entity which is inserted in each traffic lane, and thus has the ability to count vehicles by lane.

Inductive loop detectors are very widely used in Britain, especially at permanent counting stations on heavily-travelled roads. They are designed to count vehicle units not axles, and thus, unlike data obtained by the previous detectors, the data obtained with this system need not be adjusted for this factor. They have a very long life, e.g. the replacement of on-carriageway pneumatic tube counters by inductive loop installations has been shown[35] to produce a tenfold extension of detector life before any sort of failure is experienced. Typically, inductive loop detectors have a life of about 5.8×10^6 vehicle countings before component or tuning failure occurs. The problem of the overtaking vehicle is also obviated with the use of this detector, i.e. when two or more loops are installed in different lanes of the same carriageway, the loss of count due to the coincident passage of two or more vehicles is prevented by the incorporation of a 'logic/serializer module' between the detector module (to which the loop is connected) and the roadside counter.

Another form of in-carriageway vehicle counter is the *magnetic detector*. This consists of a coil in the form of a large number of wire windings on a ferromagnetic core (encased in a metal housing) which detects the residual magnetic field carried by a moving vehicle as a result of its travelling in the earth's magnetic field. When a vehicle travels over the coil, the earth's lines of flux passing through the coil are disturbed and a voltage is induced which, after amplification, trips a relay to register a count. This detector is generally easy to install or relocate and has low maintenance costs. A major disadvantage of this system is that it is subject to false calls when located near large direct current lines. Furthermore, its usage is not recommended on multilane highways.

A fixed *radar detector* which is suspended over a roadway and continually transmits/receives a radio signal of known frequency is sometimes used to count vehicles, e.g. on heavily-travelled roads in urban areas with many overbridges. Whenever the frequency of the reflected wave differs from the transmitted one, a vehicle is detected in the lane covered by the detector. Although of high initial cost, this type of detector is relatively simple, very reliable, and very accurate.

Vehicles can also be detected by *ultrasonic detectors* of either the pulsed or resonant type. The pulsed type detects vehicles by measuring the time required for the echos from bursts of ultrasonic energy transmitted towards the road surface to bounce back to a detector suspended above the carriageway. When a vehicle passes beneath the unit, the sound waves are reflected from its roof and so the received echo occurs earlier because of the shortened path length. In contrast, the resonant type of ultrasonic detector requires the receiver and transmitter units to be mounted opposite each other so that vehicles can pass between them and break the beam of ultrasonic energy.

There are various forms of automatic counting units utilized in conjunction with the detecting devices just discussed. The types most useful are as follows.

(1) A simple accumulating counter—this is read directly and as necessary by an observer. No printed record, other than the accumulated total, is available at any specified interval of time.

(2) A more sophisticated accumulating counter—upon actuation by a timing mechanism, this prints the results out on a paper tape at required intervals, say every 15 minutes or every hour.

(3) A counter using a circular chart recorder—this can record traffic volumes of up to 1000 vehicles, utilizing counting intervals of 5, 10, 15, 20, 30 and 60 minutes, for periods of 24 hours or 7 days. The principle of this counter is a very simple one: the circular graph paper rotates at a uniform speed and, as vehicles move across the detector, a recording pen moves out on the graph and records the traffic volumes.

(4) A punched-tape recorder—this enables tape punched automatically at the roadside to be processed through a translator which, when connected to a key-punch machine, produces tape which can be input to a computer for analysis. The operation of this counter consists of an input-impulse memory, an interval timer period, and the punching and reset mechanism.

The normally used automatic counter can provide only very limited information on its own regarding the numbers of vehicles that pass by the recorders. It may be that more detailed information is required, such as the exact number of vehicles instead of the number of axles or the turning movements at an intersection, or simply that no automatic devices are available or data are required which cannot be measured by a physical device. In such cases recourse is made to *manual counting procedures* utilizing field personnel who record the required data on previously prepared tally sheets or on manually-operated counters. Manual observers are also needed to determine the different types of vehicle present in the traffic stream.

The advantages of manual counting are that the counts are more accurate, very specific information is obtained and, in general, office work is simplified. It is usually more expensive to obtain data in this way, however, and so the use of manual methods is normally limited to short periods of time; typically, field observers can count up to about 1200 veh/h over a 15-minute interval with less than a 2 per cent counting error.

The *moving observer method* has already been described with reference to obtaining running and journey times and speeds. The basic method requires that the test cars pass up and down the highway test section for a prescribed number of times while travelling at approximately the estimated average speed of traffic and noting the travel times and the number of vehicles overtaking and overtaken.

If it is desired to express the flow in vehicles per hour, this is given by the following formula:

$$q = \frac{60(x+y)}{t_a + t_w}$$

where q = traffic flow in one direction (vehicles per hour), x = number of vehicles met while driving against the flow to be estimated, y = number of vehicles overtaking minus the number of vehicles overtaken while driving with the flow to be estimated, t_a = travel time while driving against the flow to be estimated (minutes), and t_w = travel time while driving with the flow to be estimated (minutes).

Department of Transport data system

The Department is currently responsible for carrying out a variety of traffic censuses and surveys, most of which have been established for many years. The main 'continuous' programme of work is most easily described in the first instance in terms of its principal underlying aims[36], as follows.

(1)   The statistical aim of providing a numerical description of the traffic characteristics of the national road network (e.g. vehicle-kilometres by type of vehicle)—this objective, which is most important for general policy and research purposes, is undertaken by the use of large-scale but infrequent 'benchmark' censuses at intervals of several years. These are supplemented by smaller surveys on a core of fixed sites which are undertaken several times a year in order to monitor trends and the seasonal flow variation of different classes of vehicle.

(2)   The technical aim of providing traffic flow data related to specific road links, which can be used as part of the input to the design, assessment and management of highway schemes—this is partly provided by a programme of more intensive traffic flow measurements that are undertaken on a very large sample of road links covering all the main types of road, and tackled consecutively over a period of years.

(3)   The functional aim of obtaining some continuous year-round traffic flow measurement on various types of highway in order to establish patterns of hourly flow and seasonal variation—this is necessary in order to calculate factors which enable the transformation of short-period,

manually-recorded counts gathered in the main censuses into such quantities as peak flow, annual average daily flow, etc., that are required, for example, for the design and assessment of highway schemes.

The prime framework for the above traffic data collection is formed from the Department's General Traffic Census, 200-point census, 50-point automatic census, and 1300-point (benchmark) manual classified census[37].

The *General Traffic Census* is a system whereby vehicle data are recorded manually and classified into various vehicle classes, viz. motor cycles, cars, light vans, buses and coaches, several sub-classes of heavier goods vehicles (2-axle, 3-axle rigid, 4-axle rigid, 3-axle articulated, 4- or more axle articulated), and (optionally) pedal cycles or cars with trailers. This is an 'annual' census which covers motorways, trunk roads and principal roads, the actual points sampled being selected so as to be representative of road links in the national system. Some 6250 points are currently sampled with the measurement points being rotated so that each is counted once every four to five years. The counts are always carried out on both Sunday and Monday for 16 hours, most usually in August; however, 25 per cent of the points are recounted as a seasonal check in the following April or May.

The *200-point census* is similar to the General Traffic Census, except that pedal cycles are always counted, and mopeds and scooters are counted separately from motor cycles. The essential feature of this manual census is that it is carried out monthly on a randomly selected sample of sites. Its prime use is in the estimation of national vehicle distance-travelled trends; however, the data are also used to estimate seasonal variations of flow for the various vehicle classes.

The *50-point automatic census* is the only truly continuous national census. As the title suggests, this census currently utilizes magnetic loops in the carriageway, or pneumatic tubes fixed to the highway surface, for continuous automatic traffic counting. The census points are fixed, albeit they were originally randomly selected. It is this census that is the standard source of data on traffic flows that are disaggregated in terms of hours, days, weeks and months throughout the year; it is also used to provide an independent check on, and updating of, overall vehicle-kilometre levels and traffic trends as determined from the 1300-point census.

The vehicle classes separated out in the *1300-point (benchmark) census* are the same as for the 200-point census. In fact, the 200-point census sites (which are only on classified roads) form a subset of the 1300-point census, which is carried out about once every six years at a random sample of sites representative of the whole network. Its primary purpose is to provide 'benchmark' estimates of vehicle-kilometres by vehicle class, region and highway type (including unclassified roads).

In addition to the above major census requirements, the Department of Transport and local authorities also gather data via various ad hoc studies, e.g. surveying vehicle speeds for traffic planning and management purposes, monitoring before-and-after vehicle separation distances to determine the effects of highway safety advertising campaigns, and measuring commercial vehicle axle loadings to provide research information for highway pavement design purposes. Other data regularly collected in the course of traffic and transport surveys include: journey data (e.g. origins and destinations, journey speeds and times, trip purposes, commercial vehicle loads, and private vehicle occupancies); vehicle axle spacings; junction flows; delay and queue length measurements; road lane vehicle concentrations; pedestrian flows.

As can be gathered from the above, many if not most of the national traffic data are obtained by manual means, and this can be very expensive and tedious in respect of both the gathering of data and their analysis. Looking to the future, it can be expected that there will be an increasing shift towards the gathering and analysis of data on an automatic basis[38]. While, obviously, certain data can never be collected non-manually, there would appear to be no technical reason why much of the information now gathered manually cannot be collected automatically. Ultimately, what can be measured will depend upon the sensors utilized, but by combining the output from several it should be possible to obtain quite complex and detailed information about both the total traffic and individual vehicles. The key to the practical realization of such installations will be the inclusion of microprocessors as integral parts of the on-site instrumentation[39].

A microprocessor-controlled system has the inherent advantage of being extremely flexible, e.g. it can accept simultaneous inputs from a number of sensors, carry out appropriate computations, and produce a data output which can be in a computer-compatible form for further analysis if required. A single piece of equipment can be used for a variety of dissimilar tasks merely by changing the control program. The form, content and timing of the output can be directly under the control of the equipment user, so that only data actually required by the study are collected. Data storage can be on magnetic tape, memory, or on a terminal printer. The printer is especially useful for on-site calibration; it may also be conveniently employed for short-duration collection of data, and the validation of magnetic tape recording. The processor can also output data to a BT line for transmission to a central computer, which should enable substantial cost and time savings to be made, if only by reducing errors produced during the data punching and recording processes.

Environmental studies

Nowadays, transport proposals need to be evaluated on environmental as well as economic and operational grounds. A number of environmental factors are difficult to evaluate in other than qualitative terms, e.g. visual effects and severance; others such as noise, air pollution and vibration can be examined quantitatively.

Noise measurements

Sound can be defined as the sensation produced in the ear as a result of fluctuations in air pressure. This definition can be applied equally to noise, which may be described as unwanted sound.

The basic unit of noise and of sound pressure is the decibel (dB). The reference pressure (zero dB) is the level of the weakest sound at 1000 Hz which can be heard by a person with good hearing in an extremely quiet location. By itself, however, the decibel is inadequate as a measure of noise as it takes no account of the ear's decreasing response at low and high frequencies. To overcome this, sound meters are usually fitted with three internationally-defined frequency-weighting filters, the 'A', 'B' and 'C' filters. The A-weighted filter (and the A-weighted decibel, dB(A)), which discriminates against low-frequency sounds, is that which is most frequently used in noise studies.

Figure 3.16 indicates the levels of some common sounds on the dB(A) scale.

The basic noise level meter is composed of: (a) a microphone to transform sound-pressure changes into corresponding electrical signals, (b) amplifiers to raise the microphone output to a useful level, (c) calibrated attenuators to adjust the amplification to an appropriate level for readout, and (d) a detector and meter to provide a visual indication of the detected sound levels. To obtain a permanent record of the data gathered, an output connector is used to supply the noise data to other equipment such as graphic level recorders or magnetic instrumentation tape recorders. Graphic level recorders produce hardcopy time histories of the noise level fluctuations, while the tape recorders retain the signal information on magnetic tape for further analysis.

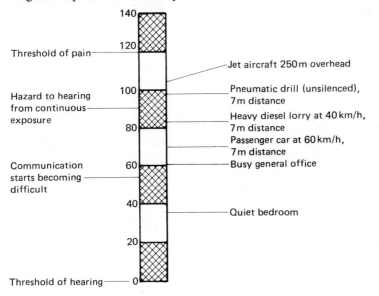

Fig. 3.16 Levels of common sounds on the dB(A) scale

Most environmental study measurements consider the entire traffic stream rather than individual vehicles as the noise source[40]. The variability of traffic noise over time is such that statistical analyses are often used to obtain indices which reflect the noise levels exceeded during certain percentages of the measured time, e.g. social surveys have shown that annoyance correlates well with L_{10} which is the noise level that is exceeded 10 per cent of the time. Another widely used index is L_{eq} which relates to the average equivalent sound energy over the period measured.

Normal practice when measuring L_{10} is to tape record the noise level on a microphone every hour from 6 a.m. to midnight. On high-volume highways with free-flowing traffic, each recording should last for 3 minutes or for the time it takes 100 vehicles to pass, whichever is the longer; for normal urban roads and lower traffic volumes, a recording period of at least 15 minutes is required[26]. The less continuous and orderly the flow, the longer will be the sampling period required for accurate measurements.

Experience has shown that errors in noise measurement are more dependent upon the operator than upon the instrument; the latter normally accounts for no more than ± 1 dB(A), whereas the placement of the microphone could contribute ± 5 dB(A). Good practice for general outdoor measurements is to place the microphone at a height of 1.2 m; measurements at building facades are usually taken 1 m from the facade. Additional data usually gathered at the time of noise measurement include traffic flow, highway gradients, conditions of the highway surface, information re adjacent areas (e.g. between the microphone and the highway), wind direction and velocity, air temperature and humidity, place of measurement, and conditions of measurement (e.g. whether reflected or background noise is present).

Air pollution

The major components of the exhaust emissions from motor vehicles are nitrogen, oxygen, carbon dioxide and water. About 3 per cent of a typical unfiltered emission contains unburnt petrol (hydrocarbons), organic compounds (mostly hydrocarbons), carbon monoxide, nitrogen oxides and lead compounds. At this time, it can be said that only the carbon monoxide emission has been definitely shown to be a health hazard. Under conditions of very bright sunlight and still air—not normally experienced in Britain—the hydrocarbons and nitrogen oxides can combine to form secondary products (peroxyacetylnitrates and ozone) which are often referred to as photochemical smog. This smog can cause eye irritation, as well as being injurious to plants.

As the main air pollutants are dissimilar in their chemical properties, each must be measured in a different way. The physical and chemical methods most often used to measure the most common pollutants are summarized in Table 3.10. The physical methods noted in this table have the advantage that they are easily automated and continuous records can be therefore obtained. A computer-based method of predicting air pollution is described in the literature[41]; this procedure, which uses data

Table 3.10 Methods currently used to determine the concentration of pollutants in the air[26]

Pollutant	Method	Type
Carbon monoxide	Infrared gas analysis	Physical
Hydrocarbons	Flame ionization detection	Physical
Nitrogen oxides	Chemical absorption	Chemical
	Chemiluminescence	Physical
Lead	Chemical or physical absorption	Chemical

inputs of traffic flows, vehicle speeds, highway layout and meteorological conditions, assumes a Gaussian-type dispersion of the pollutants.

Pollutant levels are normally measured at or within 100 m of the edge of the carriageway. Note is usually taken at the time of measurement of the following weather conditions, as they can affect atmospheric quality: wind direction and velocity, temperature, and amount of sunlight. When monitoring the effects of various transport schemes on air pollution, it is usually necessary to take before-and-after measurements at a number of reference points, to determine not only where improvements result but also where conditions worsen.

Vibration

A moving vehicle generates vibrations in the ground as a result of the fluctuations of wheel contact loads that are caused when the vehicle travels over irregularities in the road surface, each wheel in turn producing a short train of vibrations. When a road surface is in poor condition, vibrations may be generated by heavy commercial vehicles, which are noticeable by people, and there may also be a risk of structural damage to adjacent buildings. Tests have shown[42] that the levels of vibration generated at a given location are dependent upon the size of the carriageway surface irregularity, the type of vehicle suspension system, and the speed of the vehicle and its axle loading. Soil transmission properties and the dynamic characteristics of a building are also factors which influence the dynamic response of the building.

Various techniques have been used to measure vibration levels; these have ranged from measurements of displacement through particle velocity to acceleration. The accurate recording and interpretation of vibrations requires the use of high-quality equipment and a high degree of professional experience.

Basic origin and destination surveys

When new or improved traffic routes and facilities are being planned, it is necessary to estimate where they should be located so as to attract or relieve most traffic, and how much traffic they will actually carry when constructed. To do this properly, it is necessary to determine the pattern of the journeys that people make. A basic origin–destination (O–D) survey shows what amounts of travel there are between various locations.

It does not normally reveal much about the actual routes travelled between particular origins and destinations; instead it emphasizes the travel desires rather than the actual routes.

The scope of an origin and destination survey may be limited to one particular route in an urban or rural locality, or may be extended to include any part or all of an urban, conurbation or rural area. The larger comprehensive surveys can cover a city as large as London[43] or indeed a country as a whole[44].

While O–D surveys can be carried out on existing facilities and within existing towns, they are obviously not practical where new towns are being planned. Instead, some method of predicting the traffic-generating potential of houses, factories, commercial establishments and the multitude of other forms of land use is required. Moreover, if this information is available it will also have application to existing urban and rural areas. Another major purpose of comprehensive O–D surveys, therefore, is to determine what relationships do exist between the amount and type of travel and the traffic-generating factors. Of particular interest in this respect is the databank established as a result of the *Regional Highway Traffic Model Project*[44]. Before discussing the main forms of O–D survey and when they are used, it is useful to describe the RHTM study because of the impact it is likely to have on future data collection.

The RHTM project is the largest O–D study carried out in Britain, both in scale and in expenditure. Between 1975 and 1979 some £7.4m (November 1978 prices) were spent on this project; there were in addition the Department of Transport's internal costs, and those incurred by local authorities. About two-thirds of the cash cost was incurred in collecting and analysing survey and planning data, and one-third was spent on developing models.

The specific objective behind the RHTM study was to develop a trip assignment model for England as a whole for the year 1976, which could be used to forecast zone-to-zone private passenger and road freight traffic flows, averaged for all weekdays (24 h periods) for the years 1981, 1986, 1991 and 2001. In the event it was not possible to achieve the objective, and efforts to calibrate the model were stopped in 1979. However, the data obtained have been banked in a computer and are now available for general use in transport planning studies[45].

As part of the study, Britain was divided into 3613 'regional zones', the boundaries of which corresponded with local authority and ward boundaries. These zones were not required to be uniform in population or area, but were chosen to be reasonably homogeneous in character, and to provide an adequate representation of traffic patterns.

Home-interview and roadside-interview techniques were used to collect data about the pattern of trips taking place on an average weekday. As the information obtained had to be representative of the whole country, 50 869 home interviews were carried out in 21 areas that were judged typical of six types of settlement ranging from conurbations, through large and small towns, to rural areas; these represented about 0.77 per cent of the households in these areas.

Household information gathered included the number of people in each home, and their ages and sex, the number with driving licences, the number of employed people, how many vehicles were available, and the household income. The people interviewed were also asked about all the journeys they had made during the previous 24 h period, including information about the origin, destination and purpose of each trip.

The roadside interviews also provided information regarding the origins and destinations of drivers, and on the number making journeys on a typical day between each pair of the 3613 zones. 1021 survey sites were used to establish screenlines and cordons which intercepted 70 per cent of the 13m possible zone-to-zone trips. At each site that was surveyed, the number of vehicles of various types passing every 0.5 h was counted manually, and about one in ten was stopped at random and the driver interviewed regarding his/her origin, destination and trip purpose. In all, about one-million drivers were interviewed. The national postcode system was used to allocate each origin and destination to its proper postal zone, and then (by computer) to its corresponding RHTM zone, i.e. the Post Office has divided the country into 1.5m postal zones and has published directories giving, for each address in Britain, the corresponding postcode zone.

It might be noted that the standard definitions of vehicle type and trip purpose used in the RHTM study, as well as the system used to distribute addresses to zones using postcodes, provided for the first time a nationwide basis for collecting and storing origin and destination data. To ensure compatibility between this and other survey results, it is desirable therefore that these definitions be normally used in future transport studies in Britain, so as to ensure a continuation of the national approach.

Also included in the unique RHTM computer bank are basic data which describe the national highway network. The total RHTM network for England comprises 12 266 separate stretches of highway totalling 42 705 km of trunk (all) and principal (95 per cent of all) road; 6034 km of highway in Scotland and Wales are also included to allow for journeys entering England's network from beyond its national boundaries. The computer's data files contain, for each of the stretches of road, its identity number and national classification, length, number of lanes and carriageways, and estimated average traffic speed. In addition, each highway section is classified according to its character (i.e. rural, suburban or urban), 'hilliness', and 'bendiness'.

Survey area zoning and networks

All origin and destination surveys begin with the definition of the survey area. Whilst it would be desirable to analyse every journey individually, it is impracticable to attempt this, and hence origins and destinations are grouped by zones. Thus all trips produced in a particular zone are assumed to have their origins at the *centroid* of the zone, whilst all trips attracted to the same zone are assumed to have their destinations at the centroid also.

Internal O–D zones can be either large or small; what is more important is that each should include land uses that are as homogeneous as possible. Thus, to maximize homogeneity, zones in urban areas will generally be smaller the closer they are to the central area, whereas zones in residential suburban districts can be considerably larger. Furthermore, zones outside the urban area, i.e. external zones, can be quite large and increase in size as the distance from the survey boundary area increases. In practice, zone boundaries are often dictated by geographical or physical limitations such as rivers, hilly topography, embankments or other such features that inhibit traffic movement and help to create 'traffic sheds'. Preferably, a major highway should be towards the centre of each zone.

National Grid references have also been used to delineate the boundaries of individual zones; typically, a 1 km reference is adequate for inter-urban or regional studies, whilst a 100 m grid reference may be required for detailed studies of urban or country areas[46]. As noted previously, in reference to the RHTM project, postal zones are also used in transport and traffic studies. The main advantage of such zones is that they allow maximum flexibility in the subsequent reuse of the data gathered—albeit at some additional cost incurred in the initial setting up of the zone files.

Data gathered in the Registrar General's population census are invaluable in transport studies. The National Census is taken at 10-year intervals, during the first year of each decade, i.e. 1971, 1981, etc., and is updated on the sixth year of the decade by means of a 10 per cent sample. Data collected[47] include household structure by sex, age and marital status, as well as data re occupation and workplace; also gathered are travel information regarding the journey to work and car availability. The obvious value of these census data to transport study preparation, and the checking of survey information, has caused the trend in recent years to be the selection of zones whose boundaries are compatible with those of the enumeration districts used in the census.

However the zones are defined, trips into and out of them are made via the highway and public transport *networks*. The networks are normally composed of the major traffic routes; these are divided into *links*, which in turn are limited by *nodes*. All links are identified by the coded nodes at either end. A node is normally a point on the traffic route at which there is a major change in the character of the route. For example, in the case of a highway network, a node will normally be located at a major junction, or where a single carriageway becomes a dual carriageway; in the case of a public transport network (which is different from the road network, even though the same routes may be utilized), the nodes are usually located at major access points for passengers, e.g. route interchanges or major bus stops. Access to and from a zone centroid is via a *centroid connector* that is linked with a node or series of nodes in the transport network. The travel characteristics of the centroid connector are normally reflective of the average for the trips starting/terminating within the zone.

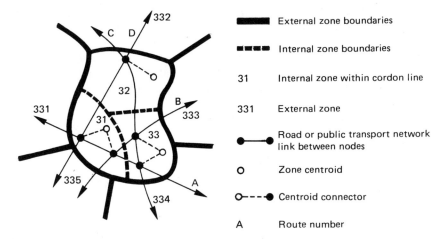

Fig. 3.17   A simplified survey area and transport network

Figure 3.17 is a diagrammatic illustration of a basic survey area and transport network.

O–D methods

There are many and varied procedures for carrying out origin and destination surveys. The personnel involved can vary from just a few observers to many hundreds of interviewers and analysers. The analysing equipment required may vary from pencils, paper and a pocket calculator to the most modern electronic computing equipment capable of processing extensive and complicated data accumulations.

Whatever the type of survey, it is essential that its aims and objectives are clearly enunciated in advance of the design of the survey. Many of the problems associated with, in particular, large-scale travel surveys can be related to multiple objectives that were not clearly described, so that consequent conflicts could not be satisfactorily resolved. Thus, for example, some objectives[48] which could be the cause of differing home-interview survey designs are as follows: (a) to establish or up-date general travel parameters, (b) to develop and test models of travel behaviour, (c) to estimate patronage levels on particular modes, (d) to reveal any particular transport-disadvantaged groups, and (e) to determine attitudes to transport.

Roadside-interview method

This is a direct-interview O–D method whereby motorists are stopped and questioned regarding origins and destinations and other journey data as required. Normally, this interview takes place on the highway, the motorist being stopped en route to his/her destination. A variation of this method, however, requires the drivers to be interviewed just as they have parked their cars; this usually occurs in conjunction with town centre parking studies. In either case, interviews are confined to the drivers of the motor vehicles and normally do not provide detailed information regarding passengers.

This method is capable of supplying detailed and accurate information, since all the information is gathered directly from the motorist and there is no need to infer anything from his or her behaviour.

Interview sites   The location of interview sites depends primarily upon the particular problem being studied. Thus if the survey is concerned only with trip data on a single isolated minor highway, a single midpoint location may be adequate. With town studies, however, sites may be located on all main radial roads so that perhaps 95 per cent of the traffic entering or leaving the survey area can be intercepted. In the case of a survey for an isolated major highway, interview stations may also be sited on all other highways expected to be affected by the construction or reconstruction of the new facility.

In general, interview sites should be located as near as possible to the boundaries of zones. If they are not, then apparently absurd answers can be obtained when motorists moving through survey points have both origins and destinations within the same zone.

A primary consideration in locating interview sites is the safety of both the interviewer and the motorist. Thus all sites should have a minimum of about 230 m of unrestricted sight distance available on both sides, and should be well clear of any intersections. Signs explaining the purpose and extent of the stoppages should be located well in advance of each site in order to give motorists time to adjust speed accordingly.

Interview team   Each interview team should consist of at least two members: one to make a classified count of all vehicles passing in the direction being studied, and the other to conduct the actual interviews. More usually, however, an interview team surveying traffic on a fairly busy two-lane road and interviewing in both directions will consist of a party chief, two recorders, six interviewers and two police officers. The police officers are necessary as it is only they who have the power in law to stop and question a person on the public highway.

Sampling   It is not usually possible to stop and question all motorists, and so sampling procedures are normally used. This is necessary not only to avoid traffic delays and maintain good public relations, but also because congestion may cause local motorists to detour around the interview site, thus distorting the traffic pattern. To avoid this, one of three sampling methods is commonly used.

The first of these requires a fixed number of vehicles to be stopped. Thus, for example, three vehicles may be stopped out of every twelve and the next nine allowed to proceed, in order to give a 25 per cent sample.

The second method requires all vehicles to be stopped that arrive at the site during a predetermined period of time. Thus all vehicles would be stopped every alternate half-hour in order to obtain a 50-per-cent-by-time sample.

The third procedure is similar to the second in that no attempt is made to maintain a fixed relationship between the number of drivers interviewed and the total number of vehicles on the road. Instead, a variable sampling fraction is obtained by the interviewer, after completing an interview, signalling to the police officer to stop the very next but one vehicle irrespective of what it is. This has the practical advantages that the interviewer is able to work at a constant rate, and is never too busy or too slack, traffic congestion and delays are kept to a minimum, and bias is not introduced by always selecting a slow-moving vehicle leading a platoon. It has the statistical advantages that it is never left to the discretion of either the police officer or the interviewer as to which vehicles should be stopped or whether interviewing should cease because of congestion. In addition, because of the variable sampling procedure, the fraction of total traffic interviewed is greater in light traffic and lower in heavy traffic; this has advantages over the other two methods from a statistical viewpoint.

Just as it is not possible to stop all vehicles on the road at a given time, so also it is not normally possible to carry out a direct-interview survey throughout the entire year. Instead, the survey can be carried out during a period when the traffic pattern is considered to be representative of the whole year. Alternatively, the survey can be conducted during the period when the problem is most acute. In either case, the data are adjusted later on the basis of knowledge obtained from continuous automatically-obtained volume data. It is the second method that is most commonly used in Britain, the surveys on rural roads being often conducted during the normally busy weeks of August.

Normally, O–D surveys are carried out only during part of the 24-hour day, usually from 6 a.m. to 10 p.m. Furthermore, it is not usually necessary to carry out the survey at all stations on the same day. Instead, two or three stations can be covered in one day, and another two or three on other similar days, and so on. In fact, if the number of interviewers is limited, a team can work the early shift on one day and the late shift on another similar day at the same site without causing any bias to the data.

The direct interview   The exact questions asked at the interview vary with the needs and objectives of the survey. Typical information that might be recorded for each interview is as follows: survey-station number, date, hour, vehicle type, direction of travel, number of passengers, origin and destination and purpose of trip. With respect to the O–D questions themselves, a trip is considered to commence at the last place of call and end at the next place of call.

When collecting O–D data by roadside interview, it is usual to exclude scheduled buses from any interview. Instead, they are recorded in the classification count, and information about frequency of service and routes is obtained at a later time.

Considerable attention should be paid beforehand to the framing of brief unambiguous questions. If proper care is taken in this respect, and the interview itself is carried out quickly, courteously and efficiently, then most forms of bias can be eliminated from the questioning.

Analysis   The first stage in the analysis of most O–D survey data is the classification of the data by means of code numbers. In this way vehicles are grouped into different classes, and origins and destinations are grouped by zones. This coding is normally carried out in the office with the interviewers coding their original field sheets. At this stage, it is usual to find some journeys that do not apparently make sense, and so their real meanings and the rectification of mistakes are best noted by the interviewers themselves.

It is usually impracticable to attempt to analyse the data by manual means, and so the coded information is normally transferred onto a computer for later sorting.

The next step in the processing of the data is the production of expansion factors, one for each period of the survey. An expansion factor is the number of trips of each vehicle category represented by each interview during that particular period. For instance, if during a one-hour sample period 13 passenger cars are interviewed out of a total of 72 cars, then the expansion factor for these journeys is 72/13. Similar expansion factors are determined for each of the categories used.

The next step is to apply the appropriate expansion factors to the data on file, so that these can then be regarded as representing an actual number of trips. At this stage also, an additional adjusting factor may be taken into account to allow for, say, the time of year when the surveys are carried out.

The final step in the basic analysis of O–D survey data is the sorting of the data in order to determine the total number of trips for all classes of vehicle and between any pair of zones.

The basic results of an O–D survey are usually expressed in the form of *desire-line graphs*. One simple form of desire-line graph, obtained from a very limited survey involving only a vehicle count at an intersection, is that shown in Fig. 3.15. For large-scale surveys, however, the form of desire-line graph used is shown in Fig. 3.18. Desire-line charts are usually drawn to represent numbers of trips between zones, the greater the band thickness the greater being the number of trips. They take no account of the routes taken by motorists but instead reflect their desires for ideal routes. Thus, for instance, the desire-line graph in Fig. 3.18(b) illustrates the general pattern of sector-to-sector movements across the River Thames within the internal survey area of the London traffic survey.

Desire-line charts can be presented in many forms. Separate graphs may be drawn to show desire-lines for through trips, internal trips, and trips between internal and external survey zones. This last type is illustrated in Fig. 3.18(a). Often separate charts are drawn to show different types of trips between zones. Important zones such as central areas, or industrial districts, may require special diagrams to determine their exact influence. Again, depending upon the survey, major desire-lines

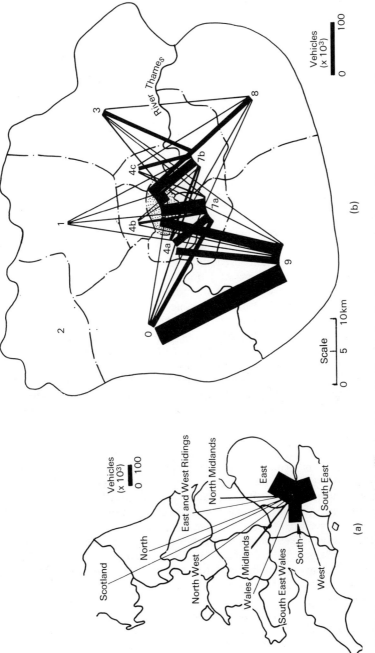

Fig. 3.18 Examples of desire-line graphs[33]: (a) origins of traffic entering the London survey area and (b) sector-to-sector Thames-crossing traffic desires

may be drawn to reflect the traffic requirements of particular survey sectors, i.e. groups of zones with common characteristics.

Mail-return surveys

There are two main methods of carrying out origin and destination surveys using mail-return questionnaires. The two methods may be carried out separately or they may be used to complement each other.

The first method involves posting return-stamped and addressed questionnaires to all registered motor vehicle owners within the survey area. Each recipient is normally asked to record all trips of his or her motor vehicle made on the day after the questionnaire is received; the day chosen is usually a weekday.

The second method involves handing out the questionnaires to motorists as they pass slowly by selected sites. If carried out in conjunction with the first type of postal survey, the sites are normally located on a cordon-line about the survey area, so that information can be obtained about both external and internal trips. If carried out in isolation, the sites are normally where traffic is so heavy that vehicles cannot be stopped to enable direct interviews to take place.

In either case, the success of a questionnaire survey depends to a very large extent upon the willingness of motorists to complete and return the questionnaires. A good return is best achieved by mounting an extensive publicity exercise prior to and during the survey so that motorists will become conscious of the importance of returning the questionnaires. Even so, the returns may not exceed 50 per cent, with a typical response being of the order of 33 per cent.

The reliability of the results obtained from questionnaire surveys has been criticized on the grounds that they may be heavily weighted in favour of facilities of particular interest to the motorists interviewed, that the results may be distorted because of better cooperation by motorists who use writing as a regular part of their daily lives as compared with manual workers who may not bother to complete the questionnaires, and that the returns from through trippers will be especially low and unrepresentative. However, other research[49] has concluded that provided the mail-return questionnaire is mainly factual, and has a short and easy-to-fill format, the respondents will form an unbiased sample of the population surveyed.

Major advantages of mail-return questionnaire surveys are that all data are obtained at the one time for the one period of travel, and untrained personnel can be used to hand out or send the questionnaires by post. In addition, there is little interference to traffic at congested road locations.

Registration number method

Again there are two variations on this method, each of which has its particular usage. The first and more often used procedure requires observers with synchronized watches to be stationed about and/or within the survey area[50]; then as a vehicle *passes* an observer its passage is timed and its registration number is noted. (This is similar to the method described for speed and journey time studies.)

At the end of the survey, the records at all the observation sites are compared, and each vehicle's trip through the survey area is traced. For the purpose of the study, the vehicle's origin is assumed to be where it is first observed, while its destination is assumed to be its last observation point. By noting the entry and exit times, the journey time of each vehicle can be estimated; when this is compared with known journey times for the same trip, the vehicle can be classified as stopping or non-stopping within the survey area.

The principal advantage of this method is that it can be used where traffic is very heavy and it is not desired to stop vehicles for questioning. In addition, since the motorist is unaware of being scrutinized, the results obtained will not be biased because of poor motorist cooperation. It has the disadvantage that, because of large personnel requirements, its use is limited to single highway facilities or to small survey areas with few exits and entrances; furthermore, all observation sites have to be staffed for the same one day, and observations must be continuously carried out at every entrance and exit point.

The second method of conducting a registration number O–D survey is to record for a given day the registration numbers of all vehicles *parked* within the survey area. These are then compared at a later time with the motor vehicle registration lists, and the origin of each vehicle is assumed to be where it is registered, while its destination is taken to be where it was parked.

The principal advantages of this method are its simplicity, low cost and freedom from bias. Its main disadvantage is that the results obtained are of limited value in that little or no information is obtained about public transport and commercial vehicle travel, all vehicles are assumed to come directly from the registered address to the parking location, and no data are obtained regarding through traffic trip purpose or time of journey. Furthermore, its use is again confined to small survey areas because of the numbers of observers required.

Tag or sticker method
This is also a moving vehicle method that is not dependent upon the complete cooperation of the motorist. The usual procedure is for an observer to stop each vehicle briefly at the entrance to the survey area and to place a precoded tag or sticker, bearing the time and place of entry, under the windscreen wiper. At the same time, the driver may be handed a card requesting his or her cooperation and explaining the nature of the survey. As the vehicle leaves the survey area, the tag is reclaimed and the time, station, direction of travel, and any other possible information are recorded on it.

The main advantage of this method is that the path of a vehicle can be traced through the survey area by having intermediate observers note the colour and/or shape of the tag on each vehicle. Thus the need for a bypass can be determined by comparing the vehicle journey times with subsequently determined trip times over the same routes. A commonly used criterion assumes that a vehicle which takes less than 1.5 times as long as the average minimum time normally required to make the trip is a potential bypass user.

Home-interview method

Large urban areas attract traffic from great distances outside their boundaries as well as generating intensive traffic in their own right. Within cities, narrow streets, well-developed and highly-valued property, constrictive topography, and concentrated traffic volumes create obstacles which have a considerable influence on the manner in which people travel and the routes they take to their destinations. Many variables are involved and so it is not normally possible in the larger urban areas to decide by merely observing the existing traffic flows where highway improvements should be located. Even if it were possible, it is generally not practical to obtain data within heavily-trafficked areas by stopping vehicles in the streets. Not only does the possibility of costly congestion make this unrealistic, but the immense road lengths within urban areas also make it impractical to attempt to interview a representative sample of vehicles. Thus a more sophisticated method of data collection must be used to obtain the comprehensive but detailed information required by highway planners in order to be sure of the need for new or improved facilities, as well as their correct location and adequacy for the future.

The US Bureau of Public Roads was the first to develop a detailed procedure for making extensive urban traffic studies that are based on a home-interview sampling technique[51]. This procedure is still widely used as a model for the home-interview phase of the comprehensive transport demand studies carried out in urban areas today.

The home-interview study utilizes a sampling process similar to that used in public-opinion polls. A representative sample of homes within the survey area is selected and the inhabitants questioned regarding their travel during a particular weekday, usually the day before the interview. If the data obtained in this study are combined with those from a taxi and commercial vehicle study and from a roadside-interview study on all roads that cross the survey boundary, an almost complete record can be obtained of a typical weekday's travel within the survey area.

Sampling   As has been discussed previously, travel is an expression of individual behaviour and as such it has the fascinating characteristic of being habitual. Since it is a habit it tends to be repetitive and the repetition occurs in a definite pattern. Thus it is not necessary to obtain travel information from all residents of an urban area nor is it necessary to interview individuals over long periods of time in order to ascertain their travel habits or need for travel facilities.

The existence of this habitual pattern therefore enables statistical sampling methods to be used with confidence in determining travel patterns within urban areas. A sample will be representative if the persons included in it are distributed geographically throughout the survey area in the same proportion as the whole population is distributed. Consequently, sampling based on the place of residence is the most practical means of assuring that the proper geographical distribution is obtained. In Britain, the residences are normally selected either from the *Register of Electors*, which contains a list of the names and addresses of all qualified electors, or from the *Valuation List*, which is a register of all rated units as held by each local authority.

The size of the sample selected is variable and depends upon the population of the survey area, the purpose of the survey and the relationships which it is desired to establish, the accuracy required and, to a certain extent, the population density within the survey zones. Table 3.11 provides a simple guide to the sizes of samples required by new comprehensive surveys in particular urban areas. Note that this table shows that the larger the urban area, the smaller the sample that may be utilized. More detailed guides regarding the sampling approaches to adopt in comprehensive transport studies are available in the literature[52, 53].

One point that should be emphasized is that proper population sampling is fundamental to the success of any home-interview survey. Thus it is desirable to obtain the advice of a competent statistician when designing the survey.

Table 3.11 Typical sample sizes for conducting home-interview O–D studies[51]

Urban population	Sample size
<50 000	1 in 5 dwelling units
50 000–150 000	1 in 8
150 000–300 000	1 in 10
300 000–500 000	1 in 15
500 000–1 000 000	1 in 20
>1 000 000	1 in 25

The home interview   The actual interview itself is the most important part of the home-interview study. The interviewer provides the only first-hand link between the survey and the general public, thereby determining whether or not the public response is to be cooperative. The completeness of the data, and thus the validity of the decisions based on them, depends upon the willingness of the members of each household to provide travel data.

The interviews are conducted within the selected domiciles and questions normally refer to travel by each member of the household above the age of five on the day prior to the interview. Since the accuracy of the survey depends to a large extent upon obtaining the desired information from preselected dwellings, most interviews are conducted in the evening because of the greater chance of finding people at home. In order to maintain the unbias of the sample, 'call-backs' must be made rather than the alternative of interviewing the next-door neighbour when there is nobody at home at a preselected residence. Each interview usually requires about one hour; this time includes an allowance for travel time and call-backs.

Two types of information are normally obtained during a home interview: household structure data and household travel data. Typical household structure data include the size of the household and its gross income, the age, sex and occupation/employment status of each resident, and the number of cars available to members of the household, together with any other socio-economic data that may be deemed appropriate to the survey area. Travel data collected for each person normally include the origins and destinations of all trips, start and finish times of journeys, journey purpose, mode of travel (car driver or passenger, by bus and/or rail), number of occupants in car (if car trip), public transport route and cost (if public transport trip), and details of car parking utilized, if appropriate.

Commercial, taxi and public transport vehicles   Normally, surveys of commercial vehicle travel are carried out in conjunction with the home-interview study. If the complete pattern of travel within the survey area is desired, this phase of the survey will need to be carried out with the same care as, and in a manner similar to, the home-interview study. The samples of vehicles are usually selected at random from numerical or alphabetical registration lists; customarily, the sample rate for lorries and similar heavy vehicles is at least twice that in the home-interview survey. The normal approach adopted is to seek the cooperation of goods vehicle fleet operators in encouraging vehicle drivers to keep logs of all trips (i.e. a trip being defined as a journey from one essential stop to another essential stop) made on the survey day.

Desirably, the taxi sample should amount to at least 20 per cent of the total number of taxis registered within the survey area; at the very least the percentage of taxis should be as great as that used for the goods vehicle study. In either case, the desired information is gathered by visiting the company offices where the vehicles are registered and obtaining the vehicle records for the previous day.

Information needed concerning public transport vehicles can be obtained from the records of the companies involved. The locations of the existing routes, schedules of operation, and the numbers of passengers carried on an average weekday in the course of the survey are examples of typical public transport data collected. At screen- and cordon-line station points, a manual volumetric count should normally be made of all public transport passengers crossing the lines, as well as counts of the vehicles.

External-cordon study   A roadside-interview study is usually carried out in conjunction with the home-interview study in order to obtain information about those travellers who pass through the internal survey area, but live outside it, and who have not therefore been sampled in the home-interview survey. In carrying out this phase of the study, a cordon-line is drawn about the survey area and motorists are stopped and interviewed as they cross the cordon-line.

In order to determine the travel patterns of the non-residents of the survey area, and to obtain corroboratory information regarding the travel patterns of the survey area residents, it is usual practice to sample the traffic streams in both directions of travel at each interview station on the cordon. It is evident that some duplication of information will occur at this stage. This duplication will need to be eliminated, by removing the appropriate trip information from the internal survey data, before the analysis is undertaken.

Evaluation of survey accuracy   It is most important that the home-interview data should be checked for completeness and accuracy. Three methods are most often used to check the reliability of the data; they involve comparisons with data obtained at screen-lines, at control-points, and/or at the external cordon. If these comparisons reveal considerable differences, then the survey data are usually adjusted accordingly.

Screen-line comparisons provide the best means of checking the reliability of the trip data collected in both the internal and external phases of the study. A screen-line commonly follows natural or artificial barriers such as rivers, canals, railway tracks, ridges or parks. The purpose of a screen-line is to divide the internal area so that the number of vehicles moving from one part to another can be determined by direct count at all screen-line crossings. In order to function properly, the screen-line should extend across the survey area from cordon-line to cordon-line and coincide with zone-lines. The value obtained at the screen-line count is then compared with the number indicated by the home-interview study.

Ideally, every screen-line crossing volume should be obtained using permanent counters. Rarely, however, are these data available and consequently most screen-line counts are limited to one or two days at each location, and sampling techniques must be used[54] to ensure that the data gathered meet predetermined accuracy criteria. It is normally considered that the main internal- and external-interview studies are sufficiently complete and reliable if the expanded interview results account for 85 per cent or more of the total volume as determined by the direct counts at the screen-line.

Control-point comparisons are used only when the routes of trips are obtained in the course of the home interviews. Two or three points, well-known to motorists and preferably in different quadrants of the internal survey area, are selected as control-points before the survey is begun. Viaducts, large bridges, underpasses and other points through which large volumes of traffic are funnelled are excellent for this purpose. As with the screen-line count, a comparison of the traffic volumes obtained from the internal and external surveys and the actual traffic volumes at these control sites gives a measure of the accuracy and completeness of the data.

Control-points should never be adjacent to the external cordon-line. Although both external and internal travel data would be collected, the primary purpose of these checks is to measure internal travel reliability, and it is likely that the amount of traffic near the outskirts of the survey area would be too small to provide a satisfactory check. The control-points should not be so close to the central area, however, that the reliability of the data may be in question because of the passing and repassing of vehicles looking for places to park.

If it is not possible to use a screen-line, and if a few good control-points are available, a control-point comparison can be a most useful method of checking home-interview data reliability. The main drawback associated with this method is that it entails extra questioning during the internal and external studies.

Cordon-line comparisons can be the easiest type of comparison since they may not require any extra data. In this context, the cordon counts represent screen-lines drawn around specific areas such as the survey area as a whole (external cordon). In order to obtain data for comparison purposes, it is necessary to derive from the internal survey the car, public transport and commercial vehicle trips that cross the external cordon-line drawn about the survey area. These volumes are then compared to similar trips across the cordon-line made by internal survey residents, as determined from the external-cordon survey.

The principal advantage of the cordon-line method is that the comparison is made from data gathered as an integral part of the regular survey. However, the external-cordon check obviously cannot verify the reliability of all data, and hence it is normally considered as complementary to the screen-line or control-point comparisons and not as a substitute for them, i.e. the trips recorded in the internal survey which pass through the cordon-line are not normally representative of internal travel within the survey area as a whole.

Selection of O–D survey method

The choice of which origin and destination method to use in a given situation is primarily a function of what information is required. Unfortunately also, however, the type of survey utilized is often controlled by the size of the available funds. This is to be deplored since, as has been determined in so many instances, data obtained on the basis of inadequate surveys can often be more misleading than if the surveys had never been carried out at all.

The registration number and tag-on-car methods are normally easy to organize and carry out. However, if they are not to become unwieldy in analysis they are restricted to localized situations such as determining the manner in which vehicles move through a complicated intersection or in determining the need for a bypass about a small town.

If more detailed information is required, such as for the location of a new river crossing, then a basic direct-interview survey on existing bridges, tunnels and ferries will adequately decide the most suitable location. Similarly, the location of a new bus terminal can be determined by direct interviews at the other existing terminals or on the buses en route.

In urban areas and on major rural roads, more complete evaluations of drivers' needs are required. In particular, detailed origin and destination surveys are required in order to estimate highway needs in urban areas. The following five systems of O–D surveys may be considered appropriate for collecting travel data in urban areas.

External-cordon survey An external-cordon O–D survey on its own is usually adequate in built-up areas with populations of less than about 5000. In free-standing urban areas of this size, it is the traffic from outside which usually exerts the major influence on the traffic patterns. Furthermore, any traffic movements not recorded during the course of this survey are normally not sufficiently stable to have any significant impact on long-range highway plans.

This type of survey can also be used in urban areas with populations between 5000 and 75000 when the predominant flow of traffic is on through routes and where problems associated with public transport or the movement of traffic on the overall street system are not important considerations.

The external cordon itself is an imaginary line set up about the survey area at which motorists are stopped and asked questions about their trips. The cordon-line should be situated sufficiently far beyond the built-up areas so as to intersect a minimum number of roads, yet not so far out that it takes in too much rural area. Ideally, it should encompass the general area of daily commutings and at the same time allow for future urban growth.

External–internal-cordon survey This type of survey is most useful in urban areas containing populations of between 5000 and 75000 where most of the traffic is oriented towards the central area and there are no major deficiencies in the street system outside the central area.

In this type of survey, roadside interviews are conducted at two cordon-lines. The external cordon is located outside the edge of the built-up area, while an internal cordon is placed about the central area. The set-up of the interview stations on both cordon-lines is essentially the same as for the external cordon alone and the interviews are geared to obtain the same basic type of information.

Information given by this type of survey can be used to establish the pattern of passenger car and commercial vehicle movement fairly comprehensively in compact motorized communities. However, urban areas which rely heavily upon public transport, or have public transport routing problems, may require special additional studies. For instance, direct interviews with bus passengers or mail-return questionnaire studies may very usefully be employed to obtain any additional information required.

External-cordon-parking survey This type of O–D survey has been shown to be most useful in free-standing urban areas with populations between 5000 and 75000 where the principal traffic destinations are within the central area, and where a parking problem also exists within the central area. Thus the survey is composed of an external-cordon study, as described before, and a comprehensive parking study within the central area of the city.

In essence, the information obtained is similar to that obtained with the external–internal–cordon survey, with the additional bonus that data are obtained about the parking needs of the central area itself. Its principal disadvantage is that no information is obtained about trips that are completely internal and which do not terminate within the town centre. Thus if it is expected that a substantial amount of non-stopping traffic may pass through the central area, or if one of the major purposes of the survey is to determine the need for an inner ring road, this method of origin and destination survey should not be utilized.

External-cordon–controlled-questionnaire survey   A survey of this type can be utilized in urban areas of up to about 200 000 population. The mail-return questionnaire method of obtaining information is used for the internal survey phase of the study, while the external-cordon study takes care of the trips originating externally.

The results obtained should be checked for accuracy by means of a screen-line comparison.

External-cordon–home-interview survey   This type of survey can be used successfully in cities of any size. However, because of the very large cost of conducting the survey and analysing the data, it is primarily applicable to larger urban areas and is rarely justified in cities with less than 75 000 population.

As described before, the home-interview data are supplemented with data on trips originating outside the survey area, obtained from roadside interviews conducted at the external cordon-line. Internal commercial vehicle movements are determined by sample interviews of vehicle operators and/or drivers. The combined data, when expanded, should give a composite of practically all trips made within the survey area on a typical weekday during the period of the survey.

Constituents of the highway design volume

The final step in the traffic analysis leading to the design of a highway or highway system is the determination of the volumes of traffic which will have to be handled. This can be a most complicated process, particularly in urban areas. Why this should be is most easily illustrated by considering the basic ingredients which make up the volume finally selected for highway design purposes.

Horizon year

One of the first problems facing the highway planner is the decision as to what year in the future should be used for planning purposes. At the present time, it is Department of Transport policy that a trunk road should have sufficient capacity to cater for traffic volumes 15 years after its opening. Added to this is the fact that typically 10–15 years may elapse between a highway proposal being conceived and the highway actually opening for traffic. In practice, therefore, the 'design' period may vary from 15 to 30 years, depending upon when an analysis is carried out prior to the opening of the highway. In carrying out the prediction calculations, however, it should be recognized that their credibility is apt to decline with increasing time into the future, e.g. due to possible changes in land use and population, or regional economy, or even in the methods of transport.

In urban areas, the problem of choosing a horizon year is much more complex. Once a major highway is constructed within a city, extensive building development can be expected alongside, and so there is much less scope for changing the geometric design in the future. The evidence for this statement can be seen in any crowded city or suburban centre today. Since modern buildings are normally considered to have an economic life of at least 30–50 years, it is not uncommon to think *strategically* in terms of this length of time when planning a new major highway system, particularly in the larger urban areas. Practically, however, it is now becoming more common to design-plan in terms of a more immediate future, say 15 years also, while ensuring that the design-plan does not contradict long-term strategic developments. In other words, rather than planning for the long-term future on the basis of a 'one-off' transport study (with all of its consequent risks), current thought is turning towards transport planning in a 'rolling programme' context, i.e. where one designs for a relatively short (and predictable) period ahead, and then constantly revises the transport and land use plan as time progresses, further data become available, and policies become clearer.

Traffic prediction components

Design traffic volumes for some future date are derived from knowledge of current traffic and estimates of future traffic. The basic constituents of the design volume for an individual highway are illustrated graphically in Fig. 3.19.

Current traffic
By current traffic is meant the number of vehicles that would use the new roadway if it were open to traffic at the time the current measurements were taken. Current traffic is composed of reassigned traffic and redistributed traffic.

Reassigned traffic is the amount of existing traffic that will immediately transfer from the existing roadways which the new highway is designed to relieve. *Redistributed traffic* is that traffic which already exists on other roadways in the regional network, but which will transfer to the new highway because of changes in trip destination that are brought about by its attractiveness.

Depending upon the type and location of the new highway, current traffic can be estimated from traffic counts on existing roadways that are

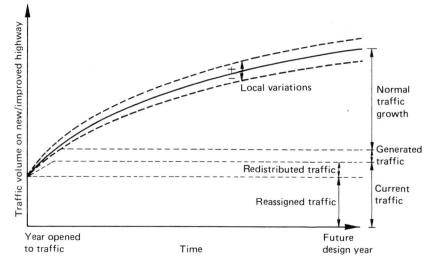

Fig. 3.19   Constituents of a highway's design traffic volume

likely to come under its travel influence and/or from roadside-interview origin and destination studies. In large urban areas, more comprehensive transport demand studies will normally be required.

On low-volume highways in rural areas, use of classified traffic count data alone is often sufficient to evaluate the current traffic volumes. In this case, the amount of attracted traffic can often be estimated adequately by a highway engineer having a thorough knowledge of local conditions. The importance of local knowledge must not be underestimated since the amounts of reassigned and redistributed traffic are dependent upon the attractiveness of the new highway as compared with the existing adjacent roadways. For instance, when the existing roads are of low quality, but are (relatively) heavily travelled, the percentage of attracted traffic will obviously be greater than when there is little congestion on the roads. The extent of this changeover can best be estimated by a local engineer familiar with the area rather than by a highway engineer from 'outside'. It must be emphasized here that this method of estimating the attracted traffic volumes is of course very crude and should be limited to usage on low-cost, low-volume roadways and where the cost of any underestimation or overestimation is minimal.

On high-volume rural roads or roads through the smaller urban areas, a combination of classified traffic counts and roadside interviews is at least required in order to obtain the data on which to base estimates of the current traffic volume. Practically, however, additional information regarding journey times is normally also required to estimate the amount of traffic that might be attracted to a major new or improved facility. Intuitive practical recommendations based on local knowledge are certainly inadequate in this instance, and more sophisticated methods

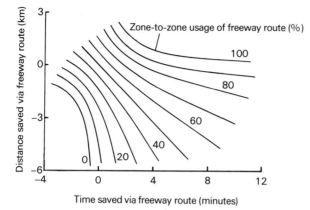

Fig. 3.20 Percentage of traffic attracted to Californian freeways, based on distance–time criteria[55]

must be employed if the large investments in the new highway are to be justified. This is most simply illustrated by the pioneering data in Fig. 3.20, which were developed to suit Californian traffic conditions. This figure suggests that: (a) when the travel time and distance remain unchanged, about 35–40 per cent of the traffic will be voluntarily attracted to a new motorway, primarily because of its higher quality of design, (b) when the distance is shortened significantly, but journey time remains the same as a result of traffic congestion on the new facility, a greater proportion of traffic will still be diverted, (c) if the distance is unchanged, and the journey time is shortened slightly, a much greater amount of traffic will be attracted to a new motorway, and (d) the greatest percentage of vehicles will be diverted when either the travel time or the distance savings, or both, are significant.

The pioneering study[56] in Britain in relation to the factors which influence motorists to divert to a new high-quality highway is that which preceded the opening to traffic of the London–Birmingham motorway in November 1959. In this study, roadside origin and destination surveys (at 23 stations) and journey time measurements were carried out on the 2900 km highway network which was deemed to fall within the zone of influence of the 110 km main section of the new motorway and its 10 km of spur roadways. The 16-hour traffic volumes estimated in this study for a typical weekday in June/July 1960 on the M1 are shown in Fig. 3.21, together with the average weekday flows that were actually achieved on the motorway during the same period. (It should be noted that whilst the estimated traffic volumes consisted entirely of current traffic, the measured flows probably also included some generated traffic.)

Simple diversion curves such as are shown in Fig. 3.20 are inadequate for studies of highway networks in urban areas. In such instances, recourse will need to be made to comprehensive transport demand studies such as are described later in this chapter.

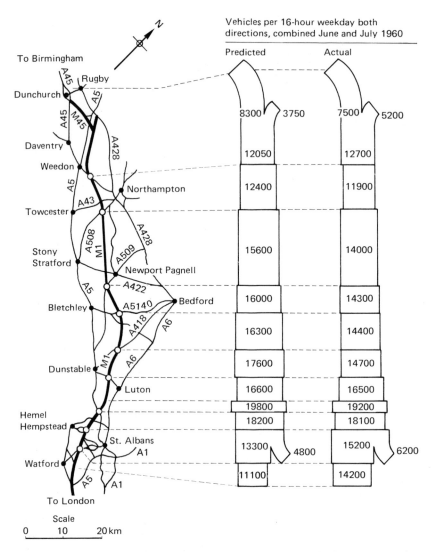

Fig. 3.21   Actual and predicted traffic flows on the London–Birmingham motorway in 1960[57]

Normal traffic growth

This is the increase in traffic volume due to the general normal increase in the numbers and usage of motor vehicles. The very evident desire of people for the mobility and flexibility offered by the motor vehicle, coupled with the fact that the motor vehicle industry has assumed such economic importance in the country as a whole, makes it inevitable that further substantial increases in motor vehicle ownership can be expected in Britain (see Chapter 1). Care should be taken, however, when utilizing national projection figures for particular design purposes, as they may not reflect the growth rate in the area under consideration. For example, one study[58] concluded that for a national saturation level of 0.45 car/person, the saturation levels given in Table 3.12 would apply to different environmental locations. It should be noted, firstly, that the larger the urban area, the lower is the predicted car-ownership value, and, secondly, that rural areas exhibit the highest number of vehicles per head of population.

A further important factor which should be taken into account is the possible growth in motor vehicle usage. Between 1974 and 1984, for example, the average distance travelled by cars and taxis in Britain increased from 12 000 km to 14 000 km per annum, whilst that for heavy goods vehicles changed from 2800 km to 3600 km per annum. With a continually rising standard of living, coupled with the present trend towards a shorter working week and longer leisure periods, there will be continued pressures to increase car usage; working against this, however, probably will be a long-term real increase in the cost of motoring.

Table 3.12 Suggested car-saturation levels in different environmental areas, for a national saturation level of 0.45 car/person

Location	Saturation level
Central cities of conurbations ⎱	0.25–0.30
Inner London ⎰	
Other large cities (formerly county boroughs)	0.30–0.45
Other areas	0.45–0.60

Generated traffic

As the term is used here, it refers to future motor vehicle trips that are generated anew as a direct result of the provision of a new highway. Generated traffic has itself got three constituent components, each of which should theoretically be estimated separately.

The first type of generated traffic can be entitled *induced traffic*. This can be traffic which did not exist before in any form and which results entirely from the introduction of the new highway. Another type of induced traffic is that composed of extra journeys by the same vehicles, as a result of the increased convenience and reduced travelling time via the new roadway.

A measure of the amount of induced traffic that will occur can be obtained by examining the estimates for redistributed traffic. It can be postulated that where the maximum attraction occurs the maximum amount of induced traffic will also occur. However, where adequate road capacity is already available or where the travel time ratio (i.e. the time via the new highway divided by the time via the quickest alternative route) is high or where trip-end parking facilities are poor, then the amount of induced traffic is likely to be low.

The second category of generated traffic can be termed *converted traffic*. This traffic is created as a result of changes in the usual method of travel. Thus the building of a motorway may make a route so attractive that traffic which formerly made the same trip by bus or railway, or even airway, may now do so by passenger car or lorry.

The amount of converted traffic is dependent upon relative journey times, and comparisons of convenience and economy. These in turn depend upon the dynamics of the various transport technologies, their management, the extent of governmental regulation and investment, and operating and maintenance costs. In practice, preliminary indicators of the likely amounts of converted traffic can be obtained from examination of the records of the public transport authorities who have experienced similar changes elsewhere.

The final category of generated traffic may be described as *development traffic*. This is the portion of the future traffic volume, due to improvements on land adjacent to the highway, that is over and above that which would have taken place had the new highway not been constructed. Increased traffic due to normal development of adjacent land is a part of normal traffic growth and so is not considered a part of this development traffic. Experience with highly improved roadways indicates that lands adjacent to them tend to be developed at a more rapid rate than normal, and it is the extra traffic generated that is therefore considered to be development traffic. The exact amounts generated will, of course, depend upon the extent to which regional and local authorities allow this development to take place.

For this reason, estimates of development traffic require consulting the planning authorities involved and determining the proposed land usage alongside the new roadway. The numbers of zone-to-zone trips expected to be produced by each type of development can then be estimated on the basis of previously obtained travel habit data.

Comprehensive transport demand studies

Development of the process

Planning for highways in Britain generally presents a difficult proposition because of the highly urbanized nature of the country. The process is at its most complex within urban areas, where the influences of, and upon, a new major roadway system can be immense.

As well as being complex, planning for highways is also relatively new and still undergoing changes in methodology and analysis. For example, prior to World War II, highway planning did not exist in the context of the sophisticated methodologies used today. Road decisions were simply taken on the basis of engineering experience, political support, and whatever funds were available. With the onset of the prosperous era following World War II, the mass advent of the motor car—with the consequent freedom which it brought to the ordinary person with respect to home and work locations, and play capabilities—and the realization that there was a direct interaction between land use development and trip generation, a much more sophisticated approach to highway planning, particularly in urban areas, became essential; with the mass development of the electronic computer, with its capability for handling masses of travel data, it became possible.

The seductive quality of highway planning as practised worldwide in the 1950s and early 1960s was its emphasis on the projection of 'numbers', i.e. its implicit promise to remove uncertainty with regard to decision-making in the whole field of transport, and most particularly roads. The (then) acceptance of this approach, however, resulted in plans for urban motorways in practically every important city in Britain. Presentation of these proposals was usually accompanied by dire prophecies to the effect that cities would grind to a standstill if the plans were not implemented. These plans were usually developed in the context of an economically *optimum* transport (meaning road) master plan which both theoretically and practically was intended to serve the 'common interest' of society. They assumed that a commonly held hierarchy of goals and objectives for the future could be agreed upon by the technical planners, political leaders, and the public in general, that a unique plan could be drawn up to achieve these goals and objectives, and that a staged programme of implementation would receive the consistent and continuous political and financial support necessary to ensure the achievement of the resultant master plan. However, as the highway proposals usually involved major changes to the urban fabric, as well as large-scale disruptions to the homes and social environments of many inhabitants, they were faced with strong opposition from those affected by and/or opposed to change—and the highway planners' concept of the common interest faded, and many such plans were relegated to the filing cabinet.

The lack of confidence in motorway plans as the answer to, in particular, urban transport problems coincided with the rise of the 'public transport solution'. At its simplest, the hypothesis was that people needed to travel, and, if urban motorway solutions were not acceptable, then obviously ones incorporating major public transport improvement schemes would need to be implemented instead, if the predicted travel demands were to be met. The result was that, in the mid-late 1960s, urban transport studies which resulted in policies and plans which were heavily oriented towards public transport were better able to win acceptance and substantial government funding.

Complementary to the growing awareness of the interactive effects of decisions taken in respect of land use, road transport, and public (bus and rail) transport was the realization in the late 1960s and early 1970s that greater attention needed to be given to the formulation of transport policies and objectives, and that traffic management and control was another key factor in the transport planning process. Thus, in the 1970s and early 1980s, urban transport planning proposals have given greater emphasis to the policy objectives underlying the proposals, and integrated road, public transport, and traffic control strategies became recognized as a means of implementing them.

Coincident with these changes in the philosophical approach to transport planning, and hence to the carrying out of transport studies in urban areas, a number of other significant developments also took place. Firstly, the technical aspects of the planning process were considerably improved as a greater understanding was obtained of the factors which influence movement. Secondly, transport modelling methodologies were developed to a very high degree of sophistication. Thirdly, the transport planning process as a whole began to become more flexible, pragmatic and responsive to current community needs, and less concerned with the determination of *unique* optimum long-term answers, i.e. it began to be recognized that what was required was a continuously-changing evolutionary or iterative process whereby the emphasis could be placed upon the preparation of credible, achievable shorter-term proposals[59, 60], instead of upon 'ideal' long-term master plans that might never be implemented.

Overall it can be said that one major message has emerged from the transport studies carried out over the past twenty-five years, viz. that complete reliance upon past and present trends to predict long-term decision-making re highways and their traffic can be very hazardous. Such elusive considerations as governmental policy decisions regarding, for example, available finance for transport facilities, industrial relations, urban development, land use controls, energy, or perhaps protection of the environment have very significant impacts on practical decision-making relative to transport study proposals. Add to these the general uncertainties regarding changing life-styles, especially changes in the organization of work and leisure, new developments in communication technology, suspicions regarding the long-term supply and cost of fuel oils, not to mention the ever-varying effects of natural economic forces, and it can be understood why absolute reliance upon unique proposals based on long-term travel predictions can be hazardous—even though the predictions are based on the most sophisticated transport planning methodologies and processes. Engineering judgement and commonsense are basic requirements in the final analysis and preparation of transport proposals.

Whatever the philosophical approach adopted at a given time, the general transport planning process has nonetheless retained its basic identity since its development in the 1950s. In essence, this can be divided into five main steps:

(1)   the carrying out of inventories of local and national goals, objectives and standards, present travel data, present traffic facilities, public transport services, and transport (including parking) policies, present and future land uses and populations, and appropriate present and future economic, environmental and employment data,

(2)   the determination of the existing interzonal travel patterns, and the derivation and calibration of mathematical models to represent them,

(3)   use of the travel models to predict future trips, and the development and evaluation of realistic transport options to meet future needs,

(4)   selection of the optimum acceptable proposal, and the detailed development of this option,

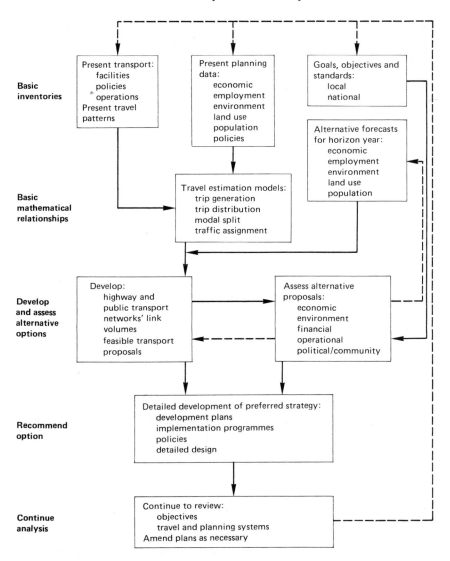

Fig. 3.22 A simplified representation of the transport planning process

(5)   continued analysis and replanning of the transport system.

The relationship between these various steps is reflected in the simplified flow diagram in Fig. 3.22.

Basic inventories

The first step in any comprehensive transport study, and possibly the most difficult to describe in real terms, is the definition of community goals and objectives. The pervasive influence that transport systems have in shaping, for example, urban areas means that it is most important for the goals and objectives to be set and (difficult though it may be) agreed as early as possible, if the study's final recommendations are to have reasonable hopes of acceptance. Unfortunately, the terms 'goal' and 'objective' are often used synonymously so that attempts to develop them are often frustrated by people having different interpretations of what is being attempted.

A *goal* may be defined as a broad comprehensive statement of an ideal result that may never be achieved, of an end state which is ever-sought. Its purpose is to provide the transport planner with overall guidance regarding the directions in which to proceed in the course of study analysis and proposal development. Goals are usually found in the statements of, and following consultations with, political leaders and advisory civic groups, in previously accepted comprehensive plans, and in the reports of public and judicial inquiries.

An *objective* may be defined as a sub-goal that is capable of being measured. As well as being more tangible, an objective should be comprehensive to the public and command its respect, and be capable of being achieved within a specified period of time. Objectives are normally prepared in support of goals, with one objective possibly relating to several goals, and several objectives relating to one goal.

A *standard* may be defined as a lower order objective which can be quantified and its achievement measured.

At the same time as transport goals, objectives and standards are being defined and clarified, work is also initiated on the collection of information regarding the present physical and operating characteristics of all transport systems within the study area. *Transport system information* normally gathered in respect of highways, on a link-by-link basis (usually from major junction to major junction), includes road-link lengths and classification, effective widths and capacities, journey times during off-peak and peak traffic periods, accident rates, and traffic classifications and volumes.

On- and off-street car parking data are also sought in central areas, as well as at other locations where parking problems have already been identified; information usually gathered includes location, number and usage of spaces, parking durations, whether the facilities are privately or publicly operated, types of control and the regulations by which they are enforced, and the charges levied by the operator. Additional parking information gathered relates to the availability and usage of on- and off-street loading facilities for goods vehicles.

Public transport information collected typically includes the definition and description of interchange points with other transport services, and (on a link-by-link basis also) vehicle speeds, headways and travel times between nodes, capacities and passenger volumes carried, fares charged and total revenues collected, and the condition of rolling equipment, track (if rail), terminals and maintenance facilities.

The main sources of data on present-day *travel patterns* are home-interview, roadside-interview, and taxi and commercial vehicle origin and destination surveys. As these types of survey have already been discussed earlier in this chapter, they will not be further discussed here.

A major component of the basic inventory phase of a transport study is the detailed determination of the *planning factors* known to influence trip generation and attraction, e.g. education, employment, entertainment, shopping and wholesaling. First and foremost, therefore, there is a need for the identification and measurement of all land uses, on a zone-by-zone basis, within the survey area. Most of these data can be obtained through the planning department of the local authority concerned, but they may need to be supplemented by field surveys, i.e. changes in land use are continuous and up-to-date reliable data may not be available in all instances. Appropriate information regarding the population, employment, retailing, wholesaling and educational activities in each zone are also gathered at the same time.

While current reliable planning information regarding the survey area is fairly easy to obtain, it is much more difficult to get forecast information, particularly if the horizon year is well into the future. The most difficult to predict are the economic trends, which in turn influence all future land use activities, as well as employment, car ownership and income levels. Expert economic help may need to be sought at this stage to ensure that the assumptions made at this phase of the study are compatible with the quality of the output desired from the study as a whole.

The prediction of future populations can be made on the basis of established demographic criteria relating to the size of family and the age distribution of its members, birth and death rates, the number of women within the child-bearing age range, and migration data on movements into and out of the survey area.

Travel estimation

The fundamental assumption underlying all transport planning studies is that there is some order in human travel behaviour that can be measured and described, and that present-day inventory data can be manipulated to provide the basis for formulating these measurements and descriptions. In a typical study, the most likely pattern of land use is predicted for the horizon year, and travel estimation models are derived on the basis of current land use; trip-making relationships obtained from the basic inventory data are then used to predict the transport demands generated by this future land use. Classically, therefore, the second phase of the transport planning study involves the derivation of acceptable trip generation, trip distribution, modal split and traffic assignment models.

Figure 3.23 shows a simplified example of the four classical steps in the travel estimation process, utilizing the basic zone system identified in Fig. 3.17. In practice, of course, the numbers of possible zones (or households) and trips are considerably greater and a computer is needed to carry out the very many calculations required to develop the prediction formulae.

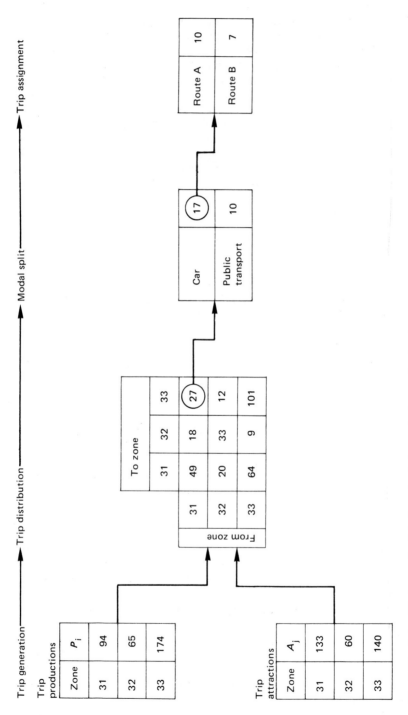

Fig. 3.23 A simplified example of the four classical steps in the travel estimation process

Trip generation

A trip is a one-way movement made between two places. Since by definition a trip has two ends, viz. its origin (where the trip is produced, e.g. home) and its destination (to which the trip is attracted, e.g. work), it is normally necessary to develop separate sets of generation equations to predict the trips starting from, and attracted to, each zone within the survey area (i.e. internal trips). In addition, equations must also be developed to predict the trips with one end within the survey area and the other outside (i.e. external trips), and those with both ends outside the survey area (i.e. through trips).

The two methods of analysis most widely used to develop trip-end generation models are zonal multiple regression analysis and cross-classification category analysis.

With the classical *multiple regression analysis* method, each zone of the survey area is treated as one unit of analysis, and the average number of trip productions/attractions which leave/enter the zone for a specified purpose is taken as one value of the dependent variable. Trip purposes for which such zonal trip-end models are derived typically include: (a) home-based work trips, i.e. between the home and the workplace, (b) other home-based trips (usually grouped for analysis purposes), e.g. from the home to/from major generators such as airports, educational institutions, hospitals or shopping centres, between the home and personal business destinations such as banks or doctors, and between the home and recreational destinations, and (c) non-home-based trips, e.g. work to business destinations.

Independent variables in the home-based work and non-work trip-end models might include some or all of the following: zonal population, average household size, average household income, average rateable value of house, residential density, average car availability, distance from the central business/shopping area (or an equivalent locational factor). The trip purpose equations so derived might be further subdivided according to whether they are related to car-owning or non-car-owning (i.e. captive to public transport) households, and whether they were determined for the peak travel periods, selected off-peak periods, and/or an average 24-hour day.

For the non-home-based trips, or for the derivation of trip attraction equations for non-residential zones, appropriate independent variables might include the number of employees in each standard industry category, volumes of sales, and/or commercial and industrial floor areas. These are often presented on a 24-hour basis, the results being factored back at a later time, if desired, to give the appropriate data for the peak or non-peak period.

The following example from Cardiff illustrates the applicability of the method:

$$Y_w = 0.097X_1 - 351X_2 + 0.773X_3 + 0.504X_4 - 43.6$$

where Y_w = zonal work trips by all modes, X_1 = population of zone, X_2 = number of households in zone, X_3 = number of employed residents in zone, and X_4 = number of cars owned in zone.

Similar equations for other types of trips were also developed so that, when the various *Y*-values were combined, ΣY gave the total number of trips associated with each zone for the year of the study. On the assumption that the relationships expressed by the various *Y*-equations remain stable, the future numbers of zonal trips were estimated by substituting appropriate future estimates of the independent variables.

Prior to the 1970s, zonal regression models were the most widely used of the trip generation models. With time and experience, however, they were rigorously criticized[61] on the grounds that equations produced for one study area were not consistent with those obtained for another, and thus were also unlikely to be stable over time into the future. Furthermore, it was shown that the greatest proportion of variation in trip-making behaviour can occur within zones, but the zonal models make no attempt to explain this. Lastly, while a significant reduction in the 'within zone' variance can result from reducing the zone size, particularly when the smaller zones are homogeneous in respect of socio-economic characteristics, small zones present problems of data accuracy and large survey samples may be required to estimate the zonal averages.

The above arguments led to a lessening of interest in the zonal multiple regression analysis approach, and to a concentration on household-based models where observed data on individual households could be used directly in the model-building procedure. Eventually, a cross-classification method known as *category analysis*[62] became very popular and is now widely used in urban transport demand studies in Britain. In concept, this procedure considers the household as the fundamental analysis unit, and assumes that the number of trips from a household is a stable function of three main parameters: household income, car ownership or availability, and family structure. In all, 108 different categories of household were initially defined as follows, and a trip rate established for each.

Income classes (£/annum) (as reported in 1967)	Ownership (cars/household)	Family structure (persons/household)
$\leqslant 500$	0	0 workers + 1 non-employed adult
501–1000	1	0 workers + $\geqslant 2$ non-employed adults
1001–1500	2+	1 worker + $\leqslant 1$ non-employed adult
1501–2000		1 worker + $\geqslant 2$ non-employed adults
2001–2500		$\geqslant 2$ workers + $\leqslant 1$ non-employed adult
>2500		$\geqslant 2$ workers + $\geqslant 2$ non-employed adults

Category analysis has major appeal as a method of predicting home-based trips. One obvious reason for this is that it is intuitively appealing to categorize households in a given zone, both now and for the future, and associate each with an expected trip-making behaviour. More important is the fact that its usage should eventually reduce considerably the amount of current home-interview data which need be gathered and analysed, since much of the household information required for any given urban area can be obtained directly from the National Census, while trip rates are available from previous studies (see, for example, Table 1.5), and only require sample checks for model-calibration purposes in the area being surveyed. A disadvantage of the category analysis method is the need to ensure that enough data are collected in the initial demand survey to permit the determination of meaningful average trip rates for each of the selected categories of household.

One final point that might be noted relates to corrections that need to be carried out before any trip generation models are calibrated. Whatever the type of model actually used to simulate trip-making, it is logical that the total number of trip productions (ΣP_i) internal to a traffic zone should approximate the total number of attracted trips (ΣA_j) over a 24-hour period. In practice, of course, this is rarely the case, so it is usual to adjust one of the sets of data. As the determination of P_i is usually more accurate than that of A_j, it is normally the attracted data that get adjusted on a uniform zone-by-zone basis.

Trip distribution
This is the term used to describe the procedure whereby the trips produced by/attracted to each zone are distributed in relation to every other zone. The output from the trip distribution analysis is a series of trip tables, each for a particular trip purpose, in which are tabulated the number of trips between all zone pairs. Three main types of distribution models tend to be used to develop these trips, viz. growth factor methods, gravity models, and opportunity models.

Growth factor methods Four different types of growth factor model have been developed: these are the uniform factor, average factor, Fratar, and Detroit methods.

The *uniform factor* method involves the determination of a single growth factor for the entire survey area; all existing interzonal trips are then multiplied by this factor in order to get the future interzonal flows. The expansion factor used is the ratio of the total number of future trip-ends to the existing total.

The *average factor* method is an iterative procedure which attempts to take into account in a simplified way the fact that rates of development in zones are normally different from the rate for the survey area as a whole, and this can be expected to be reflected in different interzonal trip growth rates. Thus it utilizes a different growth factor for each interzonal movement: this is composed of the average of the growths expected at each pair of origin and destination zones.

The *Fratar* method of successive approximations[63] is perhaps the most widely known iterative process. More complicated than the average factor method, it gives more rational answers, albeit at the expense of a considerable amount of computer time. A variation of the Fratar method which requires significantly less computer time is known as the *Detroit* method[64].

Although relatively easy to understand, the growth factor methods are now rarely used to predict the distribution of internal survey area trips over long time-horizons. The basic philosophy underlying these models, which were developed at a time when comprehensive transport study data were not available (only total trip-end and trip interchange data), is that existing interzonal trips can be expanded into the future on the basis of anticipated rates of growth in the zones' total productions and attractions. Major weaknesses of this approach are, for example, that it completely underestimates growths in existing trips which are at or near zero (even though the land use may change significantly in a growing town), while it assumes that the resistance to travel between zones remains a constant into the future (even though a new motorway might be opened).

In particular instances such as when there is a lack of detailed O–D data, the Fratar and Detroit methods are still used to predict future trips (e.g. internal-external or external-external trips), or to extrapolate origin and destination matrices over short time periods (<5 years), particularly in rural area studies. Examples of the use of these and other models are readily available in the literature[65].

Gravity model   As the name implies, the gravity model formulation used in transport studies is borrowed from Isaac Newton's 1686 law of molecular gravitation which attempted to explain the force between, and consequent motion of, the stars and planets in the Universe. After a long history of application in other fields, notably sociology, the gravity model approach was adapted to trip distribution in 1964[66]. Its utilization for this purpose is based on the premise that all trips starting from a zone are pulled by various traffic generators or land uses, and that the degree of pull of each 'magnet' varies directly with the form of the generator and inversely with the distance (or travel time) over which the attraction is generated.

The general form of the gravity distribution model as used today is

$$T_{ij} = \frac{P_i A_j F_{ij} K_{ij}}{\sum\limits_1^n A_j F_{ij} K_{ij}}$$

where T_{ij} = number of trips from zone (i) to zone (j) for a particular purpose, P_i = total number of trips produced in zone (i) for that purpose, A_j = total number of trips attracted to zone (j) for that purpose, F_{ij} = empirically derived friction factor which is an inverse function of the separation (i.e. travel time, distance or generalized cost) between zones (i) and (j), K_{ij} = empirically derived interzonal adjustment factor which takes account of socio-economic influences not otherwise included in the model, and n = total number of zones in the survey area.

What this model says is that the proportion of trips, P_i, produced at zone (i) which is distributed to destination zone (j) is a function of both the attractiveness of zone (j), A_j, and the deterrent separation, F_{ij}, between zone (i) and zone (j) *relative* to the same features for all other attracting zones. Thus a zone to which access is considerably improved as a result of, say, a new motorway (i.e. decreasing friction) would subsequently improve its relative 'pull' on the P_i trip productions.

Currently, the most widely used measure of the separation between zones is a trip decay function (usually a power or exponential function) based on the *generalized cost* of the journey from zone (i) to zone (j). By generalized cost is meant the cost to the traveller as he/she is considered to perceive it. Thus the perceived cost of a trip between any two zones by a particular mode can be expressed in a general way as

$$C_{ij} = a_1 t_{ij} + a_2 e_{ij} + a_3 d_{ij} + p_j$$

where C_{ij} = generalized cost, t_{ij} = in-vehicle travel time, e_{ij} = excess (i.e. outside-vehicle) travel time, d_{ij} = interzonal distance (for cars only), p_j = terminal cost at destination end of trip (for cars only), and a_1, a_2, a_3 = average perceived cost values of the trip components.

The gravity model is by far the most widely used trip distribution procedure. As well as recognizing the gravitational pull of land uses of differing attractiveness, and the effects of changes in, say, travel time, it also allows for the influences of differing trip purposes. What used to be its main disadvantage, the relative complexity of the iterative calibration process carried out on the basic inventory data, is now overcome through the development of well-established standard computer routines.

Opportunity model A major criticism that can be levelled at the gravity model is that it does not take direct account of individual behaviour patterns. An opportunity model, which is derived on the basis of probability theory, assumes that as trips which are generated in zone (i) move away from that zone, they incur increasing opportunities for their purposes to be satisfied at any zone (j), and therefore there is an increasing probability that they will not proceed beyond zone (j). Two forms of opportunity model have been developed, the *intervening opportunity* model and the *competing opportunity* model. The difference between these relates to the manner in which the probability function utilized in the distribution equation is calculated, viz. the intervening model requires trips to remain as short as possible and they only become longer if acceptable destinations cannot be found closer to zone (i), whereas the competing model assumes that the probability of a trip from zone (i) stopping in zone (j) is dependent upon the ratio of the trip destination opportunities within a given time boundary.

The basic intervening model is expressed as

$$T_{ij} = P_i [e^{-LD} - e^{-L(D + D_j)}]$$

where T_{ij} = number of trips from zone (i) to zone (j), P_i = total number of trips produced in zone (i), D = total number of destination opportunities closer in time to zone (i) than are those in zone (j), D_j = number of destination opportunities in zone (j), and L = probability that any given destination opportunity will be selected.

The opportunity model approach has been rigorously tested in a number of studies in the USA, and it has been shown that in general the intervening opportunity model can give results of a reliability equal to that of the gravity model. Unlike the gravity model, however, there is no possibility of including an 'adjustment' factor in the intervening opportunity model's formulation to ensure that the trip interchanges from the model balance with those actually measured at the inventory stage— so calibration becomes more difficult. In addition, the opportunity model tends to overestimate journeys made in the face of natural hindrances such as major rivers or hill ranges, i.e. the model is not sensitive to the actual amount of travel time or distance but responds only to the ordering of zones by time or distance.

As of this time, the opportunity model does not appear to have been reported in the literature as having been used in practice in Britain.

Modal split
The purpose of the modal split analysis is generally considered as being to determine the proportion of trips that can be expected to be made in the private car vis-à-vis those by public transport, i.e. walking and cycle journeys are usually omitted. In all such analyses, the assumption is made that travellers make rational choices between the modes available, and that these are based partly on the characteristics of the modes and partly on their own personal characteristics.

Classically, two types of modal split models have been developed, viz. trip-end models and trip-interchange models. A major difference between the two types of model is that trip-end models are developed *before* the trip distribution phase of the study (so that trip distribution is concerned with the allocation of vehicle trips), whereas trip-interchange models allocate person-trips to different transport modes *after* the trip distribution process.

The early trip-end models assumed that the main factors affecting modal split were factors such as car availability, residential density, and zone location relative to the central area of a town. Later the term 'accessibility' was added to describe the ease with which people with origins in a given zone could access relevant destinations in any other zone, and regression analyses were then carried out to predict the number of trips by the different modes, in a manner similar to that described for trip generation. With the development of category analysis, zone accessibility was added as a further household cross-classification factor.

There is ample evidence now to show that accessibility (of some form) is a factor which affects trip rates and activity participation rates, trip lengths, car ownership, the choice of (and satisfaction with) home location, and the rate of development of land for residential purposes. Nevertheless, only some of the accessibility indices which have been developed to date can be considered satisfactory. Reference 67 provides an excellent summary of the most common accessibility measures, and the applications to which they are put.

Trip-interchange models have tended to assume that the levels of service provided by the competing transport systems are the main factors affecting modal split. Proxies for level of service which have been used in various studies include relative travel times, excess travel times, travel costs, and (more recently) generalized travel costs. A typical travel time ratio is

$$R_{TT} = \frac{t_1 + t_2 + t_3 + t_4 + t_5}{t_6 + t_7 + t_8}$$

where t_1 = time spent walking to, say, the bus stop at the trip origin, t_2 = waiting time at bus stop, t_3 = time on bus, t_4 = time spent changing from one bus to another, t_5 = time spent walking from bus stop to destination, t_6 = time spent driving car, t_7 = time spent parking car at destination, and t_8 = time spent walking from parked car to destination. A typical excess travel time ratio is

$$R_{ETT} = \frac{t_1 + t_2 + t_4 + t_5}{t_7 + t_8}$$

where t_1, t_2, t_4, t_5, t_7 and t_8 are as defined before. A representative travel cost ratio is given by

$$R_{TC} = \frac{nf}{c_1 + c_2}$$

where n = average car occupancy, f = fare by public transport, c_1 = cost of fuel (petrol and oil) for car trip, and c_2 = parking cost. (Note that road tax, insurance or maintenance costs are not included in the R_{TC} formulation as they are not perceived by the motorist as being important when deciding about the mode for a particular trip.)

Whatever the type of proxy used, the usual practice is to develop diversion curves on the basis of the inventory data to express the percentage of person-trips likely to take place via the public transport system.

Unlike the other phases of the modelling stage of a transport study, there are no 'accepted' modal split models. Rather the general tendency in Britain has been to develop models which are deemed appropriate for the study in question and most of these predict the number of trips by, say, public transport by the extrapolation of trends.

It is recognized that there are two major travel populations that have to be considered in modal split analyses, viz. public transport *captives* who do not have a car available for their use and for whom therefore there is normally no question of mode choice, and the *choice* travellers who do have cars and therefore have a real choice between public and private transport. It is common practice to assign all or the great majority of future captive-passenger trips to the public transport system; these are usually separated out from other trips at the trip generation stage, and then distributed only to destinations served by public transport at the trip distribution stage. In the case of choice travellers, either trip-end or trip-interchange modelling procedures are carried out in order to estimate the number of choice public transport trips that must be added to the captive trips, to get the final modal splits for the various categories of travellers. Figure 3.24 shows some basic diversion curves derived as part of the London studies.

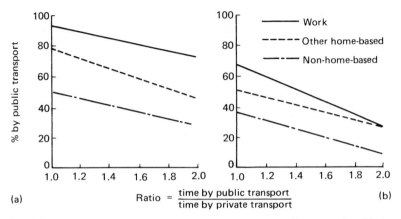

Fig. 3.24   Modal split diversion curves derived from London Transportation Study data[68]: (a) to central area and (b) to non-central area

A major criticism of present-day modal split models is that they are based on current levels of service, and these generally have a built-in bias in favour of the private car. Thus future estimates of public transport usage could err significantly on the low side should, for example, the bus service be significantly improved by the provision of reserved trackways in combination with a stringent parking policy. The reverse would obviously be true of a survey area in which the travel choice was biased in favour of public transport, whereas the decision might be taken to make significant improvements to the highway system.

Traffic assignment

The final step in the travel estimation process is the assignment of the distributed volumes of trip, i.e. public transport (captive plus choice), private car, taxi and commercial vehicles, to the different highway routes within the survey area. The output from the analysis therefore is expressed as the numbers of vehicles assigned to each link of both the public and private transport networks.

In real life, a motorist selects a journey route for a variety of reasons, e.g. time, distance, monetary cost, convenience, comfort and safety. In theory, therefore, there exists in each situation an optimum route which offers the driver the best combination of all of these. In practice, most assignment procedures assume that the optimum route is either the minimum travel time or the minimum travel cost route.

In the transport studies carried out in the 1950s and early 1960s, an approach was developed whereby all vehicular trips between the centroids of any two zones were loaded onto the single network path or 'tree' with the minimum travel resistance, usually specified in terms of travel time. The philosophy underlying this *all-or-nothing* assignment was that this volume represented the maximum traffic demand under free-flow traffic conditions, and facilities should be provided for it.

Another well-known procedure uses empirically derived *diversion curves* to assign traffic. This is similar to the all-or-nothing assignment process except that traffic is assigned to the two or more routes between the two zones with the lowest travel resistance.

With time, however, it began to be recognized in transport studies that it might not be realistic to assume that all network links could be upgraded to the standard necessary to maintain free-flow conditions, and that if too much traffic were loaded onto an inadequate link, then the speed on that link must inevitably be reduced, and traffic would probably be diverted onto other links. The value of speed–travel-time work carried out in 1952[69] then began to be better appreciated; in essence, this seminal research postulated that traffic on a network will distribute itself in such a way that the travel costs on all routes used from any origin to any destination are equal, while all unused routes have equal or greater travel costs. Or, putting it another way, traffic continues to distribute itself until no driver can reduce his or her travel cost by switching to another route.

Questions then began to be asked regarding the validity of the least-path trees used in the all-or-nothing and basic diversion curve assignments. Thus, instead of relying upon only one simplified exercise, route assignment procedures began to use a number of iterative *capacity restraint* techniques[70,71] to redistribute traffic volumes until an equilibrium situation is reached where the travel speed used in the calculation for each link of the network is relatively unaffected by the assignment process.

Examples of factors commonly included in current capacity restraint models are travel times, costs, comfort, and volume/capacity level of service, i.e. travel data already available from the inventory stage of the transport study. Standard computer programs have been written which enable these factors to be taken into account in the analysis in a fairly routine way. Advice regarding British practices in respect of trip assignment is readily available in the literature[72,73].

One final point may be emphasized about the practical use of assignment models. If the objective is to produce a strategic road network to accommodate all future road traffic demand, then an 'all-or-nothing' type of model is best used. However, if the ideal uncongested highway network is obviously neither practicable nor desirable, so that the objective becomes one of identifying overloaded links and critical junction bottlenecks for upgrading in the light of certain policy decisions, then capacity restraint assignment becomes a more useful technique.

Mathematical modelling: some comments
Realism in travel estimation has been an ideal long sought after by highway transport and traffic engineers. As with other aspects of engineering, this ideal became more practical with the advent of digital computers of sufficient size to manipulate the large arrays of data developed through the transport study process.

However, at this time, it cannot be said that this ideal is even near achievement. Some of the major criticisms of the current travel estimation process are as follows.

(1)   The process is inherently expensive, with its heavy emphasis on data collection.
(2)   It relies upon relationships determined at a particular point of time, and assumes that these relationships will still hold true for the future.
(3)   There are many arbitrary simplifying assumptions and 'fudge factors' included in the analyses which give an illusion of accuracy and objectivity which the process simply does not possess. Furthermore, analyses carried out by different agencies can produce significantly different travel estimates.
(4)   The planning models are unresponsive and insensitive to the policy questions which need to be asked nowadays.
(5)   The forecasting process is incomprehensible to the general public, who now seek to be involved in the decision-making which affects their daily lives.
(6)   The validity of the forecasting process has not been historically substantiated by before-and-after study results.

Notwithstanding the general validity of the above criticisms, it is essential to appreciate

'that, in spite of all their shortcomings, these four basic models represent the best available travel-forecasting process, although they have many serious problems and many inherent errors ... and that the planning they permit shows a much greater awareness of interacting systems, allocation of scarce resources, and the long-term impacts of present decisions, than anything before in the transportation field'

Whilst these words were published[65] in 1975, they are valid today and will remain so for many years to come.

Research developments in respect of travel estimation appear to be centred on three main approaches. One of these concentrates on *refining the existing modelling processes*, so as to get immediate improvements; differing examples of this approach are available in the literature[74,75]. Another is centred on the development of *econometric models* (see, for example, reference 76) in which travel choices are modelled simultaneously, although they still use data derived from geographical aggregates; these models would appear to be particularly useful in the analysis of short-term policy options, e.g. the effects of public transport fare increases. Third, and probably most important, is the emphasis currently being placed on the development of *behavioural models*[77,78]. By 'behavioural' models is meant the following three groups of methods, viz. disaggregate models, attitudinal models, and a recent group of studies (too formative to be termed models) which deal with household-level travel patterns.

Disaggregate models evolved from the literature on consumer choice theory in the late 1960s. The thrust of these models is: (a) the use of individual-level trip data, rather than aggregates of observations, as the basic building blocks of model construction, (b) the use of specialized calibration methods that are unfamiliar to most planners, and (c) the use of specific functional forms, e.g. logit. Whilst most research at this time appears to be concentrating on the problem of model choice, the mathematical evolution of the research methodology has rapidly outstripped its applications to date.

Attitudinal models have expanded rapidly since the early 1970s, building on the following concepts: (a) the use of individual-level data (as above), (b) perceptions of alternatives as reported by individuals and scaled by a variety of procedures taken from the literature on attitude measurement and psychometrics, (c) treatment of a wide range of qualitative attributes and, more recently, (d) detailed attitudinal models for separate individuals using conjoint and functional measurement models.

The newest trend, which focuses on household-level travel behaviour, is primarily concerned with: (a) attempts to integrate certain concepts of household structure from sociology and psychology, particularly family decision-making, role allocation, life-style and life-cycle, (b) development and applications of small group (e.g. family) games to investigate family decision-making and responses to changes in activity patterns and/or transportation systems, (c) recent work on household travel time and cost budgets and their use as constraints on travel, and (d) several analyses of learning and adaptation processes and their relationship to transport.

Urban highway and traffic options and strategies

Alternative proposals

Once the travel estimation models have been developed, the next stage in the transport planning process is their application to the preparation of a range of physical proposals, and the testing of the effects of various traffic management and public transport policies and strategies in relation to these proposals. As might be expected, this is a complicated process which requires not only considerable engineering planning experience to ensure the development of professionally acceptable proposals but also significant acumen in terms of what is likely to be acceptable to the public and the Government. This process is at its most complex in urban areas.

At its simplest, the development of the physical options can be viewed as comprising four main approaches, viz.: (a) do nothing, (b) develop a transport network that is heavily private-car-oriented, (c) develop a network that is heavily public-transport-oriented, and (d) develop a series of networks that are intermediate between (b) and (c).

The *do-nothing approach* is by definition a misnomer in that nothing, particularly urban development, ever stands still into the future. In reality, this approach is better described as the 'do-minimum with respect to highway or public transport capital works' option. On the whole, the do-nothing approach assumes that the present highway network will remain essentially unchanged, except for local roads that have already been committed, and adjacent national highways (trunk roads/motorways) that can be reasonably expected to be constructed within the planning period. Similarly, the numbers of, and charges for, public and private parking spaces will be neither increased nor reduced, unless policy commitments have already been entered into, e.g. the decision may already have been taken to raise parking charges to try to arrest the decline in public transport usage. In the case of public transport, the situation is more complex, i.e. the continued decline in public transport necessitates the making of some assumptions regarding the level of service to be maintained within the planning period and the scale of the operating subsidy to be provided.

In the case of a *private-car-oriented approach*, it is normally necessary to postulate significant improvements to the highway network, particularly those links leading to the town centre, in order to accommodate the inevitable increases in motor vehicle trips which will result during the planning period. Associated with this will be the need to provide extra car parking spaces in and about the town centre, at parking charges that will not discourage potential users. Correspondingly, the public transport system is allowed to decline to some base level of service, and an operating subsidy compatible with that base level is assumed.

With a *public-transport-oriented approach*, the emphasis is on capacity restraint in respect of both the highway network and town centre parking provision, and the development of proposals which result in significant improvements to the quality and quantity of public transport services. Typically, it results in a do-nothing approach being adopted with respect to the provision of highways which primarily serve the journey-to-work trip. Existing roadworks might normally be utilized to provide priorities for buses, and busways on separate rights-of-way could be proposed.

In practice, most transport plans nowadays are developed using an *intermediate approach*, with some new road construction and a significant emphasis on public transport. Associated with these is a series of integrated traffic management strategies that aim to influence the nature and scale of the transport gap created when the highway infrastructure provided is unable to meet the untrammelled vehicular demands of the urban area. The strategies commonly adopted in this type of situation are most easily described by dividing them according to whether their objectives are primarily:

(1)   to influence the time and place of trip-making,
(2)   to influence the choice of mode,
(3)   to improve traffic flow,
(4)   to reduce energy consumption,
(5)   to alleviate environmental problems,
(6)   to improve road safety.

Practically, of course, these strategies are all interactive and all affect, to a greater or lesser degree, mobility and accessibility as well as improving the environment and safety, and reducing energy consumption[26, 79].

Influencing trip-making

The two measures most widely used to influence trip-making are land use controls, which limit where and to what extent building (and therefore traffic generation) can occur, and travel time controls.

Land use control

Why land use decisions are in effect transport decisions, and vice versa, is illustrated in Fig. 3.25. This figure shows that for whatever purpose the land is used, its activities generate trips—and some activities obviously generate more trips than do others (1–2). These trips, in turn, point up the need for particular types of transport facilities (2–3–4). The extent to which the transport facilities are able to cope with the trip demands determines the quality or degree of accessibility associated with the land in question (4–5). This accessibility influences the value of the land since, logically, the land has no commercial value if people cannot get to it (5–6). Finally, it is the land value which helps to determine the use to which the land is put (6–1). Thus it can be seen that the control of land use is to a large extent the key to the control of movement.

It is within the metropolitan counties that the main land use (and transport and environmental) policies and strategies will be applied in the future. These conurbations have been described[80] as consisting 'of groups of urban settlements more or less in a state of coalescence which have developed over a period of time, usually but not always around a central dominant city'. London, in particular, has a structure which is strongly dominated by its central city, and this is emphasized by the form of its communication system which tends to be strongly radial and focuses on the centre. The West Yorkshire conurbation consists essentially of a conglomeration of free-standing towns and cities, whereas in the other conurbations, the central city theme is dominant, but less so than in London. In all of the conurbations, the trend over the past thirty years

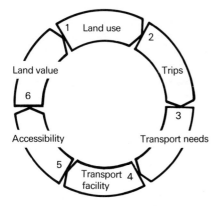

Fig. 3.25   The land use transport cycle

has been towards the central cities losing their residential populations while the traffic regions containing them have gained population.

Initiating area-wide changes in land use is not something that is easily done under the democratic governmental system in a heavily built-up country, and hence the full impact of major land use proposals may not be felt for many years to come. This is particularly so with respect to the planning of the large urban counties where the pressures are so great that the land use strategies which are particularly aimed at reducing traffic congestion tend to get overwhelmed by the sheer needs of more pressing current problems. In London, for example, businesses were encouraged for some years to leave the city centre in order to disperse jobs and ease the transport problem; then, however, concern for the decline in business activity led to a reconsideration of this strategy. Current government policy in relation to inner city areas is to strengthen their economies, and secure a new balance between the inner areas and the rest of the city region in terms of populations and jobs[81].

The manner in which strategic land use and transport planning in the conurbations should ideally develop has been outlined in concept form by Buchanan et al. in their study[82] of the region linking Southampton and Portsmouth. In this study, a rational hierarchy of travelways (in the form of a 'directional grid') was proposed which would accommodate different modes of transport having scales of operation which fall naturally into a graded order. Thus Buchanan postulated six categories of routes from '1' to '6', each successive route having a larger scale of operation in relation to distance (see Fig. 3.26). The categories proposed were such that the '1' and '2' routes correspond to paths and roads in housing areas, while '6' routes correspond to regional or national communication lines. '1' routes generally only connect with '2' routes, '2' with '3', and so on; thus at intersections, where routes cross, are concentrated the facilities which require accessibility from two consecutive categories of urban sub-systems served by these routes, e.g. shops which need to be accessible to a residential area at one level and to wholesale distribution at another, or industry which needs to be accessible not only to its employees but also to a regional freight route. In Fig. 3.26, the main urban facilities are shown grouped on alternate 'red' and 'green' routes which Buchanan called 'spines of activity'. Thus the 'red' routes accommodate public as well as private transport (the two systems running parallel); the 'green' routes are through routes (possibly also used by express public transport systems) through landscaped areas—they also serve for the longer-distance random movements more likely to be carried out in private transport vehicles.

In the smaller- and medium-sized towns, the land use strategy will probably continue along the lines by which these towns are now generally being guided, i.e. towards the development of the centripetal town of the type idealized in Fig. 1.4. This type of town has very many advantages from a movement aspect. It is only when they get too big that these towns become inadequate; in practice, it would be both undesirable and uneconomic for changes to be attempted at this stage.

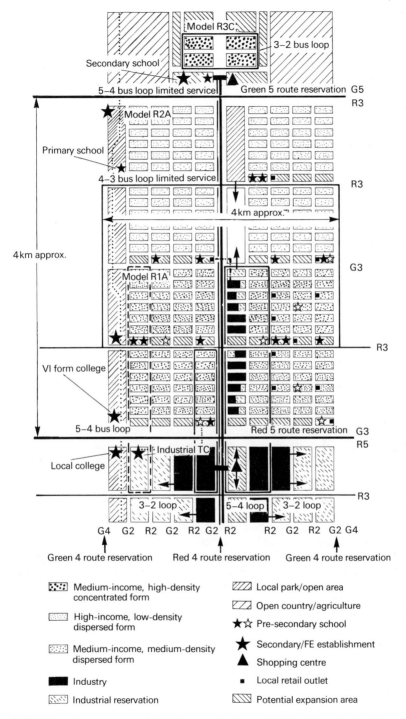

Fig. 3.26 Six routeways: a composite diagram showing an idealized directional grid-route structure in a conurbation-type area

Changing the time of travel

As noted previously in this chapter, the use of the urban transport system is very unevenly distributed in time. If the morning and evening traffic peaks could be 'flattened' and their volumes distributed over long periods of time, then much more efficient utilization of the public and private transport networks would be achieved. Strategies suggested to change the time of travel include: (a) rearranging the temporal pattern of society's activities by introducing flexible/staggered working hours, and (b) using peak period public transport fares to make travelling in peak periods relatively more expensive, and therefore less attractive.

Varying working hours   Both the staggering of the start/finish of work and the use of flexible working hours are now quite common industrial practices. In the case of staggered work periods, a number of large firms or governmental offices in a given area agree to stagger their start/finish times by, say, 15–30 minutes with no change in the total amount of hours worked per day or week. With flexible working hours, the employees (typically) start work at any time between 8.00 a.m. and 10.00 a.m. and finish at any time between 4.00 p.m. and 5.30 p.m., so long as no more than ten hours are gained (which may be taken later as 'flex leave') or five hours lost (to be made up in the following flex period) in any four-week period.

The overall effect of their introduction[83] is generally to reduce the number of private vehicle movements and the demand for public transport during the (normally) most heavily loaded 15–30 minutes in the morning and evening peak periods. Another effect can also be to increase slightly the proportion of car users vis-à-vis public transport users; however, this may well be offset by the public transport operator being able to reduce the number of peak period duplicate buses (and costs) without affecting the quality or frequency of service.

Varying fares   It is well recognized that the level of fares charged is one of the main determinants of travel by public transport[84]. It is often therefore suggested that by increasing the peak period fare vis-à-vis the off-peak fare, some of the peak passenger load would be encouraged to travel in off-peak time. In practice, however, this is not necessarily the case. While a differential fares scheme can be effective in reducing the imbalance between peak and off-peak passenger levels—and thus ultimately reducing public transport vehicle requirements—it will not influence public transport passengers to re-time their journeys to avoid the peaks[85,86]. This is particularly so if they are tied to normal (non-flexible) working hours, when in fact an increase in peak period fare is more likely to encourage the use of a household's private car. A beneficial effect of lowering the off-peak fare is to encourage greater usage of public transport by existing users; in addition, it may help to curb the increase in car ownership, especially of second cars.

Influencing choice of mode

In recent years, it has been government policy to favour public transport in preference to the private car mode in urban areas. Strategies used to favour public transport include: (a) establishing new public transport services, (b) improving existing services, and (c) making private car travel more difficult.

New public transport services

In most instances, providing new public transport services simply means operating buses along new routes. The very high capital cost of new underground or surface rail systems generally makes it impractical to consider rail developments in urban areas unless fairly exceptional circumstances exist. The only two new urban rail projects initiated in recent years are London Transport's rail extensions and Tyne and Wear's light rapid transport system[87].

Improving public transport services

Level of service factors which can affect decisions about whether to travel by public transport instead of walking or going by car, and whether to go to the central area instead of shopping locally, include the following: service frequency, route density and distance between stops, journey speed, timetable convenience, comfort on vehicle, and comfort at stops/stations. Much has been reported[2, 84, 88] about the relative importance of individual factors, sometimes with contradictory results. Overall, however, the general consensus appears to be that the scope for improving public transport patronage, simply by providing a better service on its own, is limited. On present evidence, the main beneficiaries of such improvements are existing public transport users; few new passengers are attracted from their private cars simply because the level of service is improved. What can happen, however, is that the decline of patronage can be arrested or slowed down as a result of level of service improvements; in addition, when combined with measures which discriminate against the private car, public transport patronage may be increased.

In recent years, particular attention has been paid in urban transport strategies towards improving bus regularity and reliability, and reducing overall journey time and cost. Measures used to implement these include limited-stop services, the use of bus control systems, and schemes which give priority to buses over private vehicles.

Limited-stop services These typically pick up passengers at a limited number of stops in a residential area and then travel non-stop to the city centre or the main industrial area. A flat fare is usually charged for the trip. The services are most heavily used when applied to peak period travel.

Bus control   Several procedures are used to maintain regularity of service in the light of traffic congestion, by providing immediate information on traffic conditions which enables the bunching of buses to be reduced or gaps in the services to be filled. The methods used range from the traditional manual methods involving route inspectors, mobile supervisor-dispatchers and other personnel, to fitting buses with radio telephones to allow direct communication between a control centre and drivers. Also developed are procedures which use radios in conjunction with sophisticated computer systems[89] to indicate on visual display units at the control centre the positions of buses on all major routes.

Bus priority   Whilst this concept has been in use for many decades, it was not until the mid-1960s that priority schemes began to appear in appreciable numbers, and since then their usage has increased enormously[90, 91]. The nature of bus priority has also changed over the years from the implementation of simple schemes (e.g. a short length of bus lane or an exemption from a turning prohibition) to the use of comprehensive schemes involving large-scale changes to whole areas (e.g. pedestrianization, road closures, highway construction and traffic management measures) in which improvements to public transport which involve bus priority are an integral part of the overall strategy.

Some bus priority measures in widespread usage at this time (see also Chapter 7) are as follows.

(1)   *With-flow bus lanes* are lanes reserved for buses travelling in the same direction as the general traffic.
(2)   *Contraflow bus lanes* are lanes reserved in a one-way street for buses travelling the 'wrong way', to avoid a detour or to serve the street.
(3)   *Reserved bus lanes on, and priority access to, urban motorways*—these are sometimes reversible for tidal-flow operation.
(4)   *Bus-only streets* are existing streets turned into bus roads.
(5)   *Busways* are segregated roads that are designed and constructed for buses only.
(6)   *Bus access to pedestrian precincts*—this takes passengers right into the central area, possibly avoiding congestion on an inner ring road.
(7)   *Selective bus detection at traffic signals* may be used when bus flows are low or road widths cannot accommodate a bus lane. A transponder on the bus activates an inductive loop in the roadway to adjust automatically the start/finish of the green period at a congested junction; also, the signal programmes in area-wide traffic control schemes may be redesigned to give favourable attention to bus flows.
(8)   *Other basic traffic management measures* may include traffic regulations giving priority to buses when leaving bus stops, parking restrictions near bus stops and on bus routes generally, and exemptions from prohibitions affecting turning movements at intersections.
(9)   *Comprehensive schemes* may be used (as mentioned above). For example, a scheme may divide the city centre into non-interconnecting segments to prevent the through movement of private traffic, restrict the

number of vehicles entering the central area by the use of traffic management techniques, pedestrianize large areas of the centre, use bus lanes and bus-only streets to provide direct routing between the central zones, and provide park-and-ride facilities to encourage motorists to transfer to public transport.

The most common method of giving buses priority over other traffic is the with-flow bus lane; this allows buses to bypass queues of vehicles, particularly on the approaches to signal-controlled intersections. The majority of bus lanes are thus less than 250 m long. Many such lanes operate only in the morning or evening peak periods when traffic congestion is at its height. In very many cases in Britain, taxis, works and contract coach services, emergency services, and cyclists are also allowed to use the reserved bus lanes.

A few older British towns, e.g. Reading and Oxford, have some streets in their central areas that are devoted to buses only, while a number of new towns, e.g. Irvine, Peterborough, Redditch, Runcorn and Washington, have bus-only roads. Washington has twelve sections of bus-only roads cutting through housing estates; all but one are only 20 m long. In contrast, the Runcorn busway is an 18 km segregated road system reserved for the exclusive usage of buses; a further 8 km of roadway is shared between buses and other mixed traffic. Runcorn new town is designed so that nearly all homes, factories and shops are within 5 minutes walk of a bus stop. The figure-of-eight bus route across Runcorn connects the new residential areas to the industrial estates, shopping centre, railway station and secondary schools more directly than would be possible if the buses were to use a traditional road pattern. The average speed on the Runcorn busway is about 31 km/h, which compares with about 19 km/h for conventional town bus services.

Restraint on car usage
Various strategies are used/proposed to discourage people from taking their cars into town centres. Most utilize a 'carrot-and-stick' approach whereby it is made economically difficult for the typical journey-to-work motorist to bring a car into the town centre except under certain conditions (e.g. car-pools), when improved public transport can provide a reasonable alternative. Restriction methods proposed for this purpose include parking control, controlled physical restraint, and road pricing.

Parking control The general traffic strategy adopted in most medium-sized and large British towns is to restrict highway traffic movements to artificially low levels by the deliberate imposition of parking restrictions. Control measures utilized are: (a) a limiting of the total number of central area parking spaces, with the objective of ensuring that the access road system is kept within its capacity, (b) a limiting of the length of time during which a vehicle may stay in a controlled parking space, so that the journey-to-work vehicle is kept out of the choice parking spaces, and (c) the imposition of a relatively severe parking charge, so that only those who wish to park in the central area for short periods of time will (economically) do so.

What is possibly the prototype parking control strategy for medium-sized cities is that implemented in Leeds throughout the 1970s[92]. In Leeds, the decision was taken in the late 1960s to limit the total quantity of long-stay parking about the highly pedestrianized central area, so that the number of commuter cars would be limited to providing for approximately 20 per cent of the 163 000 daily person-journeys anticipated in 1981. The decision to settle on a modal split of 20 per cent private car/ 80 per cent other mode was taken so that: (a) the peak hour demand would be within the highway capacity which could be provided in a realistic plan for the design year, and (b) the remaining demand—which was primarily (66 per cent) for bus transport—would be sufficient to support an efficient bus system at satisfactory frequencies.

In addition to the above long-stay parking spaces, approximately 8000 on- and off-street parking spaces were planned for short-stay shopper and business parking needs within the central area of the city, close to offices, shops and warehouses, and linked with the large pedestrian precinct within the city centre.

As a complement to the above intentions, the Leeds plan originally proposed (in addition to the normal inter- and intra-district bus services) express bus services which would use the primary road network and provide non-stop services between suburban community centres and the city centre, minibus services which would provide for short (shopping and business) movements within the central area, and park-and-ride services which would link outer suburbia interchange points with the central area.

The general outline of much of the basic Leeds plan is reflected in Fig. 3.27. Note that the highway system—part of which is an urban motorway—is designed to enable traffic not wishing to enter the central area to bypass it. The long-stay multistorey parking facilities are located at the fringe of the central area so as to intercept the commuter traffic, the intention being that the (20 per cent) commuters would leave their cars at these facilities and then travel either on foot or by bus to their work destinations. In contrast, the short-stay parking facilities and the bus terminals are generally located well within the central area, thereby ensuring that the users of these facilities are close to their destinations.

In 1979, the Leeds plan underwent a major revision in the context of the development of a county-wide plan for West Yorkshire as a whole[93]. The recommended strategy that emerged from this re-evaluation is a complex balance between the following interrelated elements:

(1) a continued restraint through central area parking controls to moderate the demands for travel by private cars in the peak periods to match the available highway capacity,

(2) a limited number of physical improvements to the highway network, cfl (3) an increasing emphasis on bus priority measures, low-cost environmental and traffic management measures, and urban traffic control,

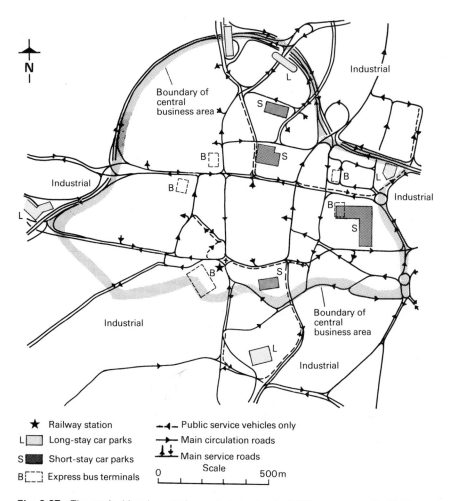

Fig. 3.27 The seminal Leeds central area strategy plan for 1981, as proposed in 1969

(4) further improvements to the bus services, particularly their routing,
(5) limited investment in new infrastructure on the existing rail system,
(6) greater coordination between the bus and rail systems,
(7) bus and rail system fares set at significantly higher levels,
(8) promotion of car-pooling programmes,
(9) proposals for the staggering of the hours of schools and workplaces,
(10) for the longer-term, the safeguarding of routes for a possible light rail transport system.

Limiting the use of the private vehicle in the peak periods through control of the supply and price of public long-stay parking spaces is still a key element of the new Leeds strategy proposed for 1995, in order to effect acceptable operating conditions for general traffic, acceptable operating conditions for buses, and some environmental protection.

Physical restraint   By physical restraint is meant the limiting of traffic volumes by the deliberate reduction of highway capacity on the approaches to the restraint area.

In 1972, the City of Nottingham (population = 0.3m plus 0.25m in the immediate surrounding urban area) abandoned its proposed £150m highway construction programme in favour of an alternative strategy which placed the emphasis on the improvement of public transport, combined with traffic management and control measures, to solve its movement problems. As part of the change, a number of streets in the central core of the city were made pedestrian-only, with the remaining streets forming a labyrinth suitable for maintaining access but offering no through routes except for buses; two free bus services were introduced in the central area; bus frequencies between suburbs and the city centre were increased by approximately 30 per cent; also, the number of long-stay commuter parking spaces on-street and in public car parks was reduced in the central area. Combined with these developments, a major transport and traffic planning experiment[94, 95] involving physical restraint using traffic signals was initiated in a sector on the western side of the city in August 1975.

Within this western sector, there is an outer ring road to deal with non-city-bound traffic, and three main radial roads which border two large residential zones that provide much of the peak period commuter traffic (see Fig. 3.28). The three radial roads converge at a common congested intersection on the fringe of the city centre. In the experiment, traffic wishing to leave the residential zones during the morning peak was restricted to certain exit roads, most of which were controlled by traffic lights; the traffic controls at these intersections were designed to limit the rate at which vehicles could join the main roads, so that buses and other main road traffic could run freely. Traffic lights on each of the six main radial roads formed a partial 'collar' of control signals about the city; these signals were timed to limit the amount of traffic entering and, particularly, passing through the inner city area.

To prevent bus passengers from being delayed in the 'control queues' expected at the junctions, priority bus lanes were provided at the approaches to all the collar signals and some of the zone exits. Buses were also able to bypass delays at other zone exits by making turns that were banned to private vehicles, or by using short sections of bus-only street. Travellers for whom the conventional bus services were inconvenient were encouraged to travel by car to specially provided peripheral park-and-ride sites and from there to continue their journey by buses operating 7.5-minute frequency, limited-stop services into the central area.

A year after Nottingham's zones-and-collar scheme was initiated, it was abandoned as no significant changes in the mode of travel by commuters were observed, and thus it was deemed that the experiment had not achieved its basic objectives of encouraging bus travel and

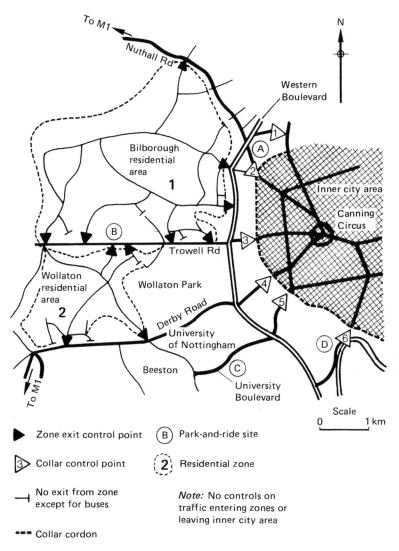

Fig. 3.28   Nottingham's experimental zones-and-collar scheme

discouraging private car travel. Some valuable conclusions drawn from the experiment were, however, as follows.

(1)   If modal change in favour of public transport is to be achieved using traffic signal restraint, substantial delays must be imposed on car drivers. (In the Nottingham study, bus journey times into the city centre were reduced by less than one minute and other traffic journey times were only increased by two or three minutes.) This requires the provision of very extensive queueing accommodation on the controlled roads (not available

at Nottingham), arranged so that queueing vehicles do not cause severe disruption to essential traffic on other parts of the road network.

(2)   A major drawback with this type of restraint is the large number of infringements that can be expected at the control/collar points and along bus lanes. (In Nottingham, about 2000 infringements per 2 h morning peak period were experienced, of which 25 per cent were vehicles passing traffic signals on red; further restraint could well have created unacceptable enforcement problems.)

(3)   Area-wide physical restraint using traffic signals to discriminate against the private car is not likely to be successful in causing people to transfer from their cars to buses in most British towns. Intensive control of parking is likely to be less objectionable, and thus more effective, in urban areas where traffic restraint systems are deemed necessary.

(4)   Park-and-ride schemes for commuters are unlikely to be effective without some form of subsidy and complementary parking control schemes within the central area. (In Nottingham, over 90 per cent of car commuters had free or subsidized parking at their destinations; a further 25 per cent of the commuter-drivers received financial assistance from their employers for the running of their cars.)

Road pricing   While parking controls are the most widely used form of traffic restraint in the central areas of cities, they do not affect through traffic movements at all and are only fully effective on traffic terminating at local-authority-controlled car parks. For example, whilst parking controls in Central London reduced peak period traffic to on- and off-street public parking spaces by 30 per cent between 1962 and 1974, traffic to private parking spaces and through traffic were both doubled (see Fig. 7.30). To overcome this latter aspect of the problem, another method of restraint often advocated is that of road pricing (see Chapter 7 for an overall description), whereby motorists would be charged an 'economic' price for the use of town centre roadways during certain times of the day.

A road pricing scheme was developed in considerable detail for London[96], but was abandoned without ever being implemented, on the grounds that it would be discriminatory against lower-income social groups. This study examined the effects of supplementary licensing charges for Central London ranging from (1973 prices) 20p to £1.20 per day, with charges at two and three times this level for large commercial vehicles, and lower rates for residents; charges at one-quarter and one-half of these levels were tested for Inner London.

An area licensing scheme, which was supplemented by a park-and-ride scheme and increased parking charges in the central area, was introduced into Singapore in 1975[97]. This scheme, which had the effect of increasing the monthly cost of commuting by 50 per cent, has been operating successfully since then[98]. In the Singapore context, a special supplementary area licence must be obtained and displayed if a motorist wishes to enter a designated restricted area (containing the central area) within which the objective is to reduce congestion between 7.30 a.m. and 10.30 a.m. The restricted zone, which covers about 500 ha and has 22 entry points, is flanked by the sea on one side. An area licence is not required by buses or two-axle commercial vehicles, in order to favour public transport and maintain the commercial activity of the restricted zone; commercial vehicles with more than two axles are banned during the restricted time period. Taxis are not exempt from the licence scheme. Motor cycles and car-pools (i.e. cars carrying four or more persons) do not require licences; these particular exemptions also help counter objections that driving into the central area is a luxury that only the rich can afford.

The basic objective of the Singapore scheme, which was to reduce the peak hour traffic, was quickly achieved; between March and October 1975, the total traffic into the restricted zone during the 7.30-10.30 a.m. restricted period fell by 44 per cent. The absolute number of car-pools increased by 60 per cent, while the number of private cars accessing the zone during the controlled period declined by 73 per cent. Travel times for people who did not change mode during the restricted period were relatively unchanged (a one-minute saving for both car and bus users); car drivers who changed to the bus took an average of nine minutes longer, however. About the same numbers of people from vehicle-owning households chose the options of changing to the bus, joining/forming car-pools, or making the trip outside the restricted time; however, people from non-vehicle-owning households did not change their travel behaviour. The park-and-ride scheme attracted very few patrons.

Overall, pedestrian movement in the central area of Singapore has now become easier, and air pollution has been lessened. Business leaders generally agree that the area licence scheme has not had an adverse impact on business, although they believe that the restrictions on car travel will accelerate the existing trend towards decentralization. The revenues from licence sales per month exceed operating costs (including special police for enforcement, and the printing and distribution of licences), so that the Singapore Government earns an annual cash return of more than 90 per cent of the total capital cost.

Special considerations that are thought to contribute to the success of the Singapore strategy include:

(1) competent transport management, with an organizational structure that fosters comprehensive policy-making and planning for all aspects of transport in the metropolitan area, including traffic management, traffic policy, bus services and motor vehicle registration,
(2) good design and implementation of the scheme,
(3) advanced education of the public as to the reasons for the scheme, its expected benefits (both short- and long-term), and the choices open to people,
(4) the fact that cars are used by a minority of central area commuters (not only were there therefore fewer opponents to the scheme, but the public transport system had a greater capacity to absorb those who changed travel mode),
(5) Singaporeans are culturally disposed to believe that the Government

acts in the general social interest and to accept rules and costs imposed on them,
(6)   laws that make enforcement easy.

Improving traffic flow

Streets and highways serve many functions, most particularly to provide for travel to local or long-distance destinations, and/or to provide access to homes, shops, places of work, etc. In many congested urban areas, strategies may be aimed at improving the hierarchical structure of the highway and street network with a view to regulating usage of the roads according to their main functions. By so doing, not only can vehicles be made to flow more easily as a result of different types of traffic being separated, but improved environmental districts can also be developed.

Road hierarchy
A well-planned and well-designed hierarchical urban road system should be able to accommodate traffic wishing to enter or leave a town rapidly and safely, or to circulate freely within it; it should also be able to provide for any through traffic not diverted to outer bypasses. This is achieved by the development of a network of high-capacity primary distributor highways which are linked to business, industrial and residential districts by separate district and local distributor roads. These link roads, in turn, are serviced by access street systems which enable vehicles to reach houses, factories, shops, car parks, etc.[99].

As their title implies, *primary distributors* form the primary network for the urban area as a whole, and all longer-distance traffic movements are canalized onto these highways. Ideally, primary distributors are designed with full control of frontage access; however, the extent to which this is applicable will depend upon the traffic volumes to be handled. In the large urban areas where traffic flows are heavy, they usually have grade-separated junctions, and motorway status may be necessary; in smaller urban areas, much lower design standards can be used. When existing routes are to be utilized (e.g. if funds are not available to construct the necessary grade-separated junctions), then all-purpose roads may be designated as primary distributors, and steps taken to restrict frontage access and street parking, and to minimize the number of turning or crossing movements; the loading and unloading of commercial vehicles is also normally banned during peak traffic periods.

The shape of any urban area's primary highway system will usually depend upon the needs and pattern of development of the town in question. Generally, however, the radial and ring roads in the larger British towns tend to evolve into primary distributor routes.

Outer primary distributors also serve as town bypasses. The data in Fig. 3.29 illustrate quite classically that the percentage of bypassable traffic can be related directly to the population of an urban area. A more recent examination[101] of the bypass needs of commercial vehicles gave

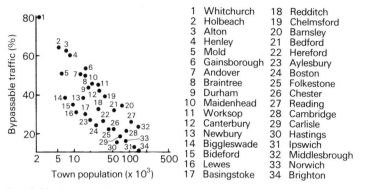

1 Whitchurch	18 Redditch
2 Holbeach	19 Chelmsford
3 Alton	20 Barnsley
4 Henley	21 Bedford
5 Mold	22 Hereford
6 Gainsborough	23 Aylesbury
7 Andover	24 Boston
8 Braintree	25 Folkestone
9 Durham	26 Chester
10 Maidenhead	27 Reading
11 Worksop	28 Cambridge
12 Canterbury	29 Carlisle
13 Newbury	30 Hastings
14 Biggleswade	31 Ipswich
15 Bideford	32 Middlesbrough
16 Lewes	33 Norwich
17 Basingstoke	34 Brighton

Fig. 3.29   Relationship between town size and percentage of bypassable traffic[100]

the following relationships with respect to towns of up to 0.13m population:

(1)   for heavy goods vehicles (>1.52 t unladen mass),

$$X_1 = 79.403 - 0.000\,4307 X_2$$

(2)   for light goods vehicles (<1.52 t unladen mass),

$$X_1 = 63.34 - 0.000\,4348 X_2$$

where $X_1 =$ percentage of bypassable heavy or light goods vehicles and $X_2 =$ population of the town. With smaller urban areas, especially those situated on main routes between large towns, the need for bypass facilities tends to be relatively great, even though the actual volume of bypassable traffic may be relatively small; in larger towns, the volume of through traffic may be sufficiently large to warrant a bypass route for its own sake, even though the percentage of bypassable traffic may be small.

In the same way as primary distributors serve the town as a whole, *district distributors* serve large environmental areas within the urban area, such as the town centre or large residential districts or industrial estates. Their function is to feed traffic from the primary road network to these localities. District distributors (and primary distributors) may border the environmental areas but not traverse them. In a typical medium-sized town, radial routes may be designated as distributor routes; in a smaller town, they may be omitted entirely, e.g. when the primary network surrounds a single environmental area such as the centre of the town.

The function of a district distributor is normally preserved by appropriate restrictions on frontage access and street parking, and the imposition of clearway operation during peak periods.

Local distributors are intended to penetrate into the environmental districts. As such they provide the link between district distributors and access roads. *Access streets*, in turn, give direct access to buildings and land. Depending upon the size and function of the environmental area being served, local distributors may or may not have restrictions on frontage access and parking. Public car parks are normally located so as to have easy access to both local and district distributor roads. A whole range of traffic management measures can be used to strengthen the hierarchical structure of the network (see Chapter 7 for details). Basically, these divert unwanted traffic away from roadways not intended to serve them. Thus management measures used on the lower hierarchy of roads tend to encourage low traffic speeds, whilst those applied to the upper end encourage higher speeds and seek to facilitate the most efficient movement of traffic that is possible.

Traffic management
In recent years, particular attention has been paid to the use of traffic signals to favour particular traffic flows, and thus influence the routing of vehicles in the urban road network[102]. There is a well-proven axiom of traffic engineering to the effect that a motorist will choose the route that minimizes his or her perceived journey time when making a regular trip such as the journey to work. With this objective in mind, several procedures have been developed which calculate traffic signal timings at the junctions of a road network controlled by coordinated fixed-time signals; as these signal timings determine the total delays experienced on particular junction approaches, they can be varied to influence the traveller's choice of route since the junction delays contribute significantly to the total journey time.

One-way streets, turning restrictions at junctions, tidal-flow operation, and closure of side streets are other well-tried and proven means by which the throughput of traffic can be increased in urban road networks. Basic to the success of these schemes is the implementation of proper road-signing systems which ensure that motorists, particularly in through-going vehicles, have no doubts as to which routes to follow.

Traffic management measures used to improve traffic flow—and to encourage the development of a hierarchical street system—are described in Chapter 7.

Energy conservation

There are a number of strategies that can be implemented which directly reduce motor vehicle fuel consumption. Table 1.10, for example, summarizes possible conservation measures; note that very many of these are included also in strategies previously discussed in this chapter in relation to the achievement of other urban transport objectives. Possible components of a strategy related to energy conservation that have not been discussed specifically so far, however, are paratransit activities such as car-pooling/sharing, demand-responsive buses, and park-and-ride schemes.

Car-pooling/sharing
The basic objective of any energy conservation programme of this nature is to increase the number of passengers per vehicle, thereby reducing the

number of vehicle-kilometres travelled and uneconomic operation in traffic congestion. Important secondary benefits include[103]: (a) a reduction of the peak of the peak period public transport demand, so that uneconomic extra buses can be withdrawn, (b) an acceptable cushion against the withdrawal of public transport services, (c) a reduction in the number of parking spaces required at journey termination, and (d) accessibility to a labour-pool without which a given employer (e.g. at an isolated site) might not be able to operate.

The current interest in car-pooling and car-sharing effectively dates from the 1973-74 'energy crisis' and the resultant concern to reduce energy consumption. Organized schemes have been in operation in the USA for many years[104] and a number of terms have originated there which have now come into accepted usage in the British technical literature[105, 106], as follows.

(1) *Car-pooling*—where a number of car owners agree to run each of their cars in turn, from adjacent origins to adjacent destinations, with the other owners as passengers. Usually no money changes hands, and any imbalance from disturbance of the routes is corrected in the rota.
(2) *Car-sharing*—a regularized form of lift-giving which usually involves a financial contribution by the passenger towards the running expenses of the vehicle.
(3) *Lift-giving*—any form of trip offered to a passenger in a private vehicle. It may or may not elicit a monetary contribution from the passenger.
(4) *Van-pooling*—refers to the use of a small personnel carrier or minibus that is privately owned or hired, often by a group of employees who use it in place of their cars. (In many cases in the USA, the vehicle is owned by a firm and leased to an employee on favourable terms, provided that it is used to carry a minimum number of other employees to work.)
(5) *Subscription bus*—a coach, usually hired by a group of employees, with or without employer help, for their journeys to and from work.

Up to now, most pooling/sharing arrangements in Britain would appear to have been the result of personal or business initiatives, and so no systematic investigative work has been reported in the literature regarding any of the above types of schemes. Information available from abroad, however, suggests the following as being factors which encourage successful pooling and sharing arrangements: (a) the provision of priority for cars of greater than a given occupancy on (long) urban bus lanes and busways, (b) preferential charges at tolls for entry to bridges and tunnels, and in public car parks, and (c) preferential spaces in car parks within town centres.

Most of the successful car-pool schemes appear to be those formed at the place of employment. The exact reasons for this are not known, but it may be that people feel more at ease in contacting workmates, or are more willing to travel with them than with unknown people from their home neighbourhood; generally, workplaces are also more compact than residential areas. Most of the successful schemes for matching potential car-poolers in the USA are operated manually by small- or medium-sized firms. Computer matching is utilized, but experience shows that the successful schemes have each involved somebody who was willing to carry out the coordination work required to get the pools into operation. US experience also suggests that van-pools and subscription buses are used mostly for long-distance commuting, e.g. >48 km; if this trend is transferable, it would suggest that their potential for major development in Britain is limited.

It is reported[105] that a sophisticated microsimulation model, calibrated on a 10 000-household survey in Yorkshire, produced forecasts which suggested that in a typical scheme only a marginal proportion (1.5 per cent) of the target population could be expected to become car-poolers/sharers, i.e. the model predicted that some 8 per cent of the target population would apply to be included in a matching system, that about 90 per cent of these could be provided with lists of potential partners, but that only 20 per cent of these would find compatible partners on their list. The predicted 1.5 per cent increase was associated with a 0.3 per cent reduction in work-trip vehicle-kilometres and a 2.0 per cent reduction in peak period work-purpose public transport patronage.

Demand-responsive services
Whereas car-pooling/sharing strategies are generally considered to be most effectively implemented during peak period traffic conditions, demand-responsive 'public transport' operations which use shared taxis, demand-responsive buses, jitneys and daily and short-term shared car rentals have been suggested[107] as being appropriate for use in lieu of conventional public transport when the travel demand is low, e.g. in small towns or during off-peak hours in larger urban areas. Of these, the demand-responsive bus systems have received the most consideration in industrialized countries where, according to their proponents, they are supposed to provide a quality of service somewhere between that of the private car and a stage-bus, at a lower cost. In addition, these bus systems may also provide for the needs of the elderly and the handicapped in a way that cannot be matched by conventional public transport.

Demand-responsive buses   The demand-responsive bus systems may be divided[108] into three main groups, graduated according to their flexibility and responsiveness, viz: corridor services, zonal many-to-one or many-to-few services, and area-wide many-to-many services, without transfers.

By a *corridor service* is meant a bus service which is given space and/ or time flexibility in its mode of operation. Space flexibility is provided by allowing the bus driver to undertake minor pick-up/drop-off deviations from an otherwise fixed route. Time flexibility is provided by the use of special communication devices located along the fixed route which allow potential bus users to call the next bus and thus reduce their waiting time. Corridor services are less expensive than zonal or area-wide systems, and do not depend upon private telephone ownership for their successful operation.

Zonal dial-a-bus services, which require telephones at home for their successful operation, have been given trials in a number of locations in Britain[109, 110]. With this type of operation, the area to be served is divided into sections, each of which is serviced by one or more small buses which collect passengers at their homes following phone requests, and deliver them to a single destination (many-to-one operation) or to two or three possible destinations (many-to-few operation). The buses may operate for all or only certain times of the day, depending upon need; they may have fixed or semi-fixed schedules, and/or operate on the basis of advance bookings only.

In general, 'successful' dial-a-bus services are those that have been kept in operation because they are considered less expensive to operate than alternative fixed-route buses, and/or the quality of service offered is higher than other kinds of public transport operation, and/or the service is heavily subsidized and thus can be offered at low fares, for political or marketing reasons. In relation to these factors, it should be kept in mind that labour costs, which are already high in conventional systems, may be further increased by 10–15 per cent with dial-a-bus systems as they can be expected to need at least two dispatchers/telephonists; furthermore, dial-a-ride buses tend to have low vehicle utilization rates, and the fares charged are generally quite low in comparison with the quality of service offered.

Area-wide demand-responsive systems, which are the most sophisticated type of dial-a-bus service, operate on the basis of picking up passengers on request at many origins and delivering them to many destinations (many-to-many operation). In essence, the service provided may be considered as close to that of a taxi, but at a much lower fare level. As might be expected, extensive area-wide systems are likely to be very costly and require computer control for their successful operation.

Park-and-ride schemes

Energy-saving schemes of this nature are commonly proposed as components of urban transport strategies which have other objectives. However, their implementation has not always been judged successful.

In practice, their usage is most applicable to large cities which are capable of supporting high-speed public transport systems on their own rights-of-way. Essential to the operation of a successful park-and-ride scheme is the provision of ample parking spaces at public transport terminals located in the suburbs at nodes of major feeder roads and inbound bus/rail routes. If the scheme is to be effective, the advantage to the motorist of using the combination of car and public transport modes must significantly outweigh that of using the car alone, as otherwise the motorist will stay in the car for the entire trip.

Further details of the factors which influence the successful operation of a park-and-ride scheme are given in Chapter 4.

Protecting the environment

Environmental programmes are mostly designed to alleviate detrimental effects of noise and air pollution resulting from traffic. Strategies which help to reduce these aspects of environmental pollution include noise alleviation through statutory vehicle regulations, good highway design, traffic management, noise barriers and the insulation of dwellings, and the reduction of air pollution by statutory vehicle regulations and traffic management. Also of importance are strategies involving heavy goods vehicle movement.

Noise alleviation

The noise level at any one instant in time on a roadway depends upon the following main parameters: individual vehicle noise, traffic volume, composition and speed, and highway gradient and surface. The level at a given location away from the highway is then dependent upon the distance to the reception point and upon the nature and scale of any intervening barriers.

Statutory regulations   Individual vehicle noise sources are the engine, gearbox and transmission, exhaust, bodywork, and the road–tyre interface; the effect of each is then a function of the vehicle's operating characteristics and age. The Motor Vehicles (Construction and Use) Regulations of 1973 state that the maximum noise level limits permitted for various classes of vehicle when in use are as follows: cars, 88 dB(A); goods vehicles, 92 dB(A); coaches and buses, 92 dB(A); motor cycles (>125 cc), 89 dB(A).

A strategy of further quietening vehicles by regulation is obviously an attractive concept. However, there would appear to be a practical limit to its likely effectiveness for the foreseeable future. For example, in the case of a typical highway traffic flow of 2000 vehicles per hour containing, say, 20 per cent heavy commercial vehicles, over half of the commercials would need to be quietened by about 10 dB(A) in order to achieve a 2 dB(A) reduction in the traffic stream's L_{10} noise level. Practically, it is likely to be many years before vehicle manufacturers are able to change their engines to such an extent, and even then the noise reduction might not be significant until much of the existing fleet is replaced.

Improving traffic flow   That traffic volume and noise are interrelated is well established. Typical L_{10} measurements taken 4 m from the edge of a carriageway carrying a traffic stream with 20 per cent heavy commercials, and flowing at 80 km/h, show that the noise level rises sharply with increasing volume up to about 1200 veh/h; beyond 1200 veh/h, the noise level increases again, but at a much less rapid rate.

From a noise aspect, heavy commercial vehicles can be defined as large vans, buses, coaches, and lorries of unladen mass >1.52 t. As these vehicles are normally noisier than light vehicles, the extent to which they are present in a traffic stream is of obvious importance. Typically, doubling the heavy vehicle content of a traffic stream that is freely flowing at 50 km/h will result in an L_{10} noise level increase of between 1.2 and 2.2 dB(A); the corresponding change at 20 km/h is 1.9 to 2.5 dB(A), with the greater noise level reflecting the increased importance of the traffic stream's higher heavy vehicle content under low-speed, congested-flow conditions.

When the traffic stream is moving at speeds greater than about 50 km/h, free-flow conditions generally prevail and the noise level increases more or less steadily as the speed is increased, e.g. doubling the stream speed from 50 to 100 km/h typically increases the L_{10} level by 3.0 to 4.5 dB(A). In contrast, when the traffic flow is slower and interrupted (as in most urban streets), the noise–speed relationship is more complicated. At low speed (e.g. in congested conditions), the major noise sources are vehicles' engines; thus, since movement in low gear is noisier than in high gear, the L_{10} level tends to decline slightly as traffic speed increases from congested conditions up to about 35 km/h, after which it starts to increase again.

Overall, therefore, it can be seen that any traffic strategy in urban areas which results in a reduction of the heavy commercial vehicle content of the traffic stream, and ensures more free-flowing traffic conditions, will help to alleviate the noise pollution at the location in question.

Highway design The effect of highway gradient on traffic noise is important, albeit complex. Typically, the noise level associated with traffic on an ascending gradient will increase as vehicles climb the hill; however, the descending traffic stream on the same slope will tend to emit reduced noise. As vehicles climb the hill, they also tend to lose speed and consequently the increase in noise level due to the ascending gradient is lessened by this reduction in speed (and vice versa for descending traffic). Overall, however, the tendency is for a highway gradient to cause an increase in the L_{10} noise level.

All other factors being equal, the difference in noise level associated with a change in road surfacing is relatively small. Of much more practical importance is whether the carriageway is dry or wet, e.g. the noise generated by a wet carriageway is typically 3–4 dB(A) greater than when it is dry, and it can be as much as 10 dB(A) higher. The exception in respect of a dry carriageway is a grooved concrete pavement; in this instance, the tyre–road interaction can be the cause of a significant increase in noise level, the extent of which depends upon the percentage of heavy vehicles present in the traffic stream.

Whether a highway is at-grade, elevated or depressed is also of considerable significance in relation to noise propagation to adjacent properties. Thus buildings which front onto an at-grade roadway will bear the full brunt of the traffic noise; they largely act as noise barriers which shield the second and further rows of buildings behind them, even though there may be street gaps or alleyways in the front row. If the roadway is elevated, the front row of houses—if they are really close-in—will normally experience a reduction in noise level as a result of being in the noise shadow of the highway structure; in contrast, the buildings in the second and third rows may experience noise increases, as the front row is not effective as a screen in this case. If the highway is depressed, the noise level at buildings adjacent to the cutting will be decreased, with the greatest reduction being experienced at ground level. However, the magnitude of the reductions measured adjacent to a vertical-sided cutting can be lessened if there is noise reflection from the farside retaining wall; this may be avoided by increasing the angle of batter of the retaining walls, and/or lining them with a sound-absorbent surfacing material.

Noise barriers   Roadside screens such as purpose-built fences or earth mounds (on open ground locations) can be used to reduce noise levels by up to 20 dB(A) just immediately behind them. In the case of screens, the maximum acoustic effectiveness is achieved by a structure that is substantially free of gaps and of density at least 10 kg/m². Whilst trees and shrubs may be useful for the visual masking of traffic, they are relatively ineffective for noise screening purposes.

The extent of the protection provided by any given barrier constructed as part of an environmental protection strategy depends upon the geometry of the situation. The relevant parameters in this instance are: (a) the length difference between the direct path from the noise source to the reception point and the path from one to the other via the top edge of the barrier ($a+b-c$ in Fig. 3.30(a)), and (b) the sizes of the angles of view subtended by the screened (θ_1) and unscreened (θ_2, θ_3) sections of the roadway (see Fig. 3.30(b)). The noise level at the reception point R in Fig. 3.30 is therefore the sum of the noise levels from the screened source S and from the separate unscreened sources subtended by the angles θ_2 and θ_3.

Procedures used in the design of noise barriers are described in Chapter 6.

Noise insulation   Another form of screening is that provided to the inside of a building by its own facade. For most buildings, however, it is the noise insulation provided by the windows which determines the internal noise level. When window insulation has been improved, the other noise

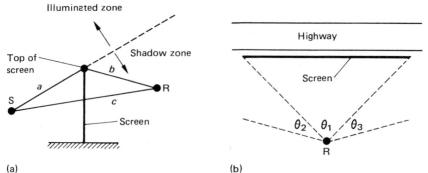

(a)   (b)

Fig. 3.30   The screening effects of a noise barrier: (a) path difference, shown in elevation, and (b) angles of vision, shown in plan

paths then start to become more important, e.g. chimneys, ventilators, and gaps about doors.

The difference in noise levels associated with the attenuation properties of various window types is given in Table 3.13.

Table 3.13   Effect of window type on noise attenuation within a building

Type of window	Difference in noise level (dB(A))
Single window	
open	5–15
fully closed	15–20
sealed (mechanical ventilation)	25–30
Double window	
staggered openings	20–25
closed (mechanical ventilation)	30–40

Air pollution

Air pollution from motor vehicles is another major environmental impact of road traffic. Whilst it is now generally accepted that the level of toxic materials (lead and carbon monoxide) found near busy roads is not sufficiently high to constitute a health hazard in Britain, it has also been shown that the exhaust components that can be detected by people (e.g. hydrocarbons by their odour and smoke by sight) cause considerable annoyance and concern.

A good indicator of the level of air pollution at any given highway site is the concentration of carbon monoxide. Studies[111] show that the concentration of this gas at any given location is heavily influenced by the traffic flow and windspeed.

Strategies used to reduce air pollution are generally of two types, viz. legislation which limits the exhaust gas emissions from motor vehicles, and traffic management measures which result in reductions in traffic flow. The legislative approach is most used in Japan and the USA, e.g. legislation was passed in the USA in 1973 which imposed severe limits on vehicle manufacturers in respect of exhaust emissions. In Britain, legislation has been used to curb the amount of lead emissions by placing limits on the lead content of petrol. All of the traffic management strategies discussed previously which limit traffic flow also reduce air pollution. Alternatively, if it is necessary for the highway system to carry heavy volumes of traffic, then the smoother the traffic flows, the less the pollution emitted.

Heavy commercial vehicles   As noted previously, heavy goods vehicles cause a number of problems in urban areas, mainly because of the size of the vehicles, their operational characteristics, their high noise level, and their effects upon other vehicles in the traffic stream. The environmental damage caused by these vehicles generated intense public debate during the late 1960s and early 1970s, and led often to demonstrations and to the blocking of highways[112]. This concern was eventually given national focus with the passing of the Heavy Commercial Vehicles (Controls and Regulations) Act of 1973—it came into operation in July 1977—which placed a statutory responsibility on local authorities to prepare plans to regulate the use of roads in their areas by HCVs, so as to preserve or improve amenity. Since then the establishment of commercial vehicle control measures has become an integral part of the planning procedures followed by most highway authorities, albeit the methods adopted by different authorities may vary somewhat. The most comprehensive strategy is probably that in London which is based on overnight on-street parking restrictions, local street controls on moving traffic, and lorry-routing combined with area-wide restrictions on moving commercial vehicle traffic[113].

Heavy goods vehicles parked at night in residential streets are physically dominating (and also pose a security problem). Such parking is due partly to the lack of provision at operators' premises, and partly to drivers using their vehicles for personal transport. The banning of night-time parking on streets by vehicles $> 2.54\,t$ unladen mass was first introduced into London in 1971, and has been extensively extended since then. In controlled areas, a high level of observance has been obtained with 85 per cent of the on-street lorry parking being displaced; half of those displaced use strategically located public car parks, whilst the others use their own yard space or make arrangements to park at the premises of another operator.

Local street controls have also been developed in London to discourage heavy commercial vehicles from using minor residential roads as alternative routes to the main highway network during, particularly, the peak traffic periods. The most successful type of control has been that of width restriction; this involves using barriers to prevent the passage of vehicles over $2.15\,m$ wide on residential streets. This method, which allows cars and small vans to pass, but effectively bars heavy commercial vehicles, has the advantage of being self-enforcing. Where necessary, special provision can be made for the passage of emergency vehicles and buses.

Where many different alternative routes exist to the same standard as the controlled route, a width restriction may be of little value if traffic is simply transferred to other unsuitable roads. This can be overcome by the use of well-signed heavy commercial vehicle routes, combined with area-wide restrictions of traffic movement—provided that an alternative main road network exists for the use of the displaced vehicles, which is adequate in both environmental and geometric design terms. An example of the successful operation of this strategy is the $15.5\,km^2$ area of Central London, which has very little industrial activity and is a focus for tourists. In 1973, large commercial vehicles $> 12.19\,m$ long were made subject to control, and required to carry LONG VEHICLE plates. A through-movement ban was then imposed using traffic signs on the approaches to the area and, despite a lack of police presence for enforcement purposes, an 85 per cent level of compliance was obtained. Flows of 'long vehicles' increased by about one-third on the perimeter routes, whilst smaller increases were recorded on other routes up to 1.5 km or more from the control area; these routes are generally in commercial areas, however, and so very few complaints of additional traffic nuisance were received.

Improving road safety

It is an aim of practically every transport and traffic plan that it should have as one of its objectives the reduction of road accidents. Strategies employed for this purpose normally include:

(1) localizing activities, so that traffic volumes and traffic conflicts are reduced,
(2) separating motor vehicle traffic from pedestrian and bicycle traffic in time and space,
(3) differentiating traffic of different character, so that each stream is as homogeneous as possible,
(4) providing clarity, simplicity and uniformity in highway design and signing, so that road user decision-making is facilitated.

Detailed factors influencing the implementation of these strategies are discussed in Chapters 6 and 7.

Public participation in highway planning

It is difficult nowadays to visualize any highway or transport proposal involving significant physical change which would not attract some public concern. In the 1950s and for much of the 1960s, the highway planner could generally assume that his or her professional expertise and unbiased presentation of the 'facts' would find reasonable acceptance at the final decision-making stage in the development of a transport/highway plan; the late 1960s and the 1970s, however, saw this comfortable feeling challenged by members of the general public, often in very vociferous and determined ways. The professional's positive response to this product of rising levels of affluence and education and people's desire to have more direct influence on decisions that affect their lives has been to develop processes whereby the public could become more directly involved in the plan development prior to its formal presentation for decision-making. This is known as public participation.

Reasons why

The overall objective of public participation programmes has been described as being[114] 'to facilitate responsive and timely decision-making by those in authority through maximum intercommunication with the public'. In concept, it can be considered as a democratic ideal, i.e. with its accent on such democratic essentials as open discussion and debate, face-to-face personal contact, and respect for people's diversity and the protection of their rights, it can be argued that public participation is part of a process of assuring the continuance of democracy in an increasingly complex society.

There are many practical and political benefits to be derived from the participation process, as follows.

(1) It provides a mechanism whereby planners can get a direct and timely understanding of community problems and values, and perceptions of alternative operations. Being outside the formal planning process, members of the public are able to provide this information from a position that is less constrained by political limitations and precedence.

(2) Potential opponents of the plan are involved in the decision-making process, thereby strengthening the planner's ability to implement the answer eventually decided upon, and to avoid the delays and costs (both economic and political) which would otherwise occur at a later stage.

(3) Public participation is educational for both the public and the highway and transport planner. The public learns more about the planning process, and thus is better able to cope in the long term with other problems of this nature. Planners benefit both professionally and personally by gaining skills and perspectives that are not traditionally part of their training, e.g. improvement of skills in working with people, increased understanding of the other disciplines involved in the transport planning process, and better job satisfaction as a result of getting answers that are capable of implementation.

(4) The participation process supplements the inability of normal parliamentary representation to cater adequately for people's preferences on single issues. People get a 'good feeling' from being able to have a direct influence on decisions which affect their lives and, in many instances, dignity.

Some programme requirements

A successful public participation programme is the result of a series of complex and interrelated deliberations and decisions[115]. Most of the successful programmes are carried out with a good measure of careful planning, commonsense, a supportive governmental climate, simple techniques, and the basic skills possessed by public participation practitioners who are sensitive to people. There are no great secrets or formulae guaranteed to work in every instance; however, openness, flexibility and a willingness to deal with people in a constructive, non-defensive way are important attributes. Successful developers of programmes take advantage of lessons learned in past exercises, are open to useful new ideas, and use well-thought-out mixes of participation techniques that are tailored to the constraints and opportunities at hand.

An explicit statement of objectives—and these will usually vary from project to project—is the framework on which any effective public participation programme is based. This is an essential requirement if the participation activities and techniques are to be linked with the transport planning process. These objectives should express the motives, assumptions and desires underlying the programme, and guide its development throughout the planning study. Furthermore, the objectives should be achievable; while glowing promises of unrealistic objectives may pacify people for a time, or even cause a temporary surge of enthusiasm, they usually lead to eventual disenchantment with the programme by both the planners and the public.

There are three essential components of any successful public participation programme, as follows.

(1) The members of the public who should be involved in the planning process must be identified. A variety of techniques may be used for this purpose, e.g. the 'snowball' technique whereby initial discussions of project issues and community concerns are begun with people known to be interested, and these are then asked to contribute the names of other people who should be involved.

(2) A two-way communication flow must be established whereby the planner gives information on the planning process, requirements that have to be met, public participation events, the public's role in working towards decisions, and alternative plans and their impacts, whilst the public participants give information on community conditions, contribute ideas and opinions on transport problems and potential solutions, and express their concept of values pertinent to the planning issue.

(3) Positive interaction must be encouraged using techniques which get people to work together in a positive way on a shared concern. Small meetings such as workshops and briefings, which are informal, personal and intensive, create a natural setting for building positive interaction between the planner and the public. The provision of information via large group meetings (one-way communication) is of limited value as a means of having the public truly influence the content of a transport plan.

Table 3.14 summarizes some techniques used to identify participants and to promote two-way communication flow and positive interaction in public participation programmes.

Timing

However a public participation programme is devised, the schedule set for its implementation is crucial to its success. The most sophisticated technique will fail if it is used too early or too late in the planning process, e.g. before the local community is ready for it, or after important review points are past.

Participation programmes should be initiated at the very start of the planning process, and be continued through all decision points. Furthermore, the professionals organizing the participation activities should be actively involved in the overall design of the transport/highway planning study, so that public involvement can be keyed into such normal planning milestones as setting objectives, producing alternative proposals, and making important technical and political decisions.

The implementation schedule for the public participation phase of the overall programme will vary according to the needs of the overall

Table 3.14 Participation techniques and the needs satisfied by them[(116)]

Techniques	Attributes	Needs satisfied
Group		
1 Small groups	6–10 people; homogeneous group already existing, locally organized or identified by planner; not highly structured discussion, but directed to particular issue or problem	Need of those not normally attracted to participation to express a view on an issue of special concern; planners' need for detailed and pertinent information, e.g. on how particular groups are affected; need of potentially affected people for individual consideration and counselling; political need to identify points of conflict before they become polarized and particular solutions become advocated
2 Public meetings	Usually more than 20 people, although depends upon level of interest in subject being investigated; self-selected by open advertised invitation; formalized proceedings aimed at presenting information to large audience or demonstrating support for community cause	Need of vocal sections of the community for forum to express their views; can satisfy planners' need for evidence of community interest, but not necessarily their need for a rewarding dialogue
3 Search conference	Usually 20–30 people; participants selected to be heterogeneous in important respects but sharing identifiable interest, often in a similar general environment; staged discussion aimed at identifying broad cross-section of views on variety of issues	Planners need to identify community attributes to assist remainder of study and participation programme design; planners need to gain understanding of all relevant issues at outset of planning; need of those with general concern about the community's future to participate in planning and be educated about other people's perspectives on the problems; planners and participants need to develop and refine ideas on planning issues that have not already been discussed
4 Workshops	Sub-groups of 8–15 people; selected on the basis of skills or specialized interest; structured sessions aimed at producing plan or programme of recommendations	Need of local experts or lay specialists to contribute in actual processes of planning; planners or institutions need to expand resources of study team
5 Committees	Approximately 15 members; members elected or appointed by planners or authorities; set up to provide advice on community views or specialist advice	Planners' and institutions' need for advice, for comment on developing programmes, and for a barometer of community feeling; lay specialists need to contribute to and monitor specific planning advice
6 Forum	Representatives nominated by existing groups and associations; set up to facilitate exchange of views amongst these groups and relevant authorities	Provides existing groups with more informed and united base from which to lobby authorities and decision-makers; institutional need for rationalized system of interacting with community groups

Individual			
7	Individual discussions	Selected by planners by random, snowball or other sampling techniques; loosely structured but aimed at gaining information about relevant issues and participants' views on them	Snowball sample discussions can satisfy planners' need for quick and efficient means of identifying the range of issues; random sample discussions can provide planners with information or views that represent the broader community; provide interested participants with an undemanding opportunity to express personal views directly to study team representatives
8	Submissions	Oral or written but often do not demand any dialogue between submitting groups and planners; openly invited, but generally attract organized groups or individuals with a well-defined position	Political and institutional need to demonstrate commitment to open planning; provide a focus for groups to organize a basis from which to lobby; provide planners with some information on positions of key authorities and groups
9	Surveys	Means of gathering information about objective characteristics or attitudes in a community; usually involve administering formal questionnaire to selected sample with varying levels of interest; minimal discussion	Study team's and institutions' need for hard data to document probable effects of proposal; political need to gauge likely public reaction to proposal
10	Participant observation	Means of gathering information and establishing contacts in a community through planner residing in area	Planners' need for thorough understanding of a community as preparation for further contact and participation; study team's need for credibility; need amongst certain community groups to feel confident of planners' understanding of area and comfortable interaction
Publicity			
11	Displays	Means of disseminating information to the community at critical stages of study and can be designed to elicit feedback; mobile or continually changing permanent exhibition	Planners need to ensure that all those who are interested have the opportunity to be informed; study team needs to get some direct feedback and discussion on issues; opportunity for some study team members to have direct contact with members of community; need of some community groups to keep abreast of developments; need of some individuals to speak directly to members of the study team with special expertise; undemanding opportunity for some individuals to register a view; institutional and political need to demonstrate commitment to participation
12	Site offices	Provide temporary accommodation for members of the study team in an approachable location in the study area; source of information and counselling advice for members of the community	Provide study team members with a convenient base from which to work and establish contact in study area; satisfy the need of some members of the community for individual attention to their views or problems
13	Media releases	Information dissemination through printed and electronic media; can be aimed at information generating interest and feedback	Satisfy political and institutional need to ensure that basic information is provided; satisfy the need of some community groups to be kept informed; provide an opportunity for some groups to contribute who might otherwise not be contacted

programme, although commonsense and experience suggestions are as follows.

(1)   It is easier to keep to a schedule that uses a master calendar to key the public participation elements into benchmarks in the overall planning process.

(2)   Frequent team meetings help to assure a smooth interface between the participation programme schedule and that of the technical planning elements.

(3)   The participation programme should be flexible so as to cater for unforeseen issues that develop quickly, or for conditions within the community that change dramatically for reasons unrelated to the study.

(4)   Consultation with a few key members of the public before any major activity is firmly scheduled will help to ensure that it is carried out at an appropriate time for the community.

British practice

In Britain, public participation is required by statute in the preparation of development plans, by government direction in the development of trunk road schemes, and by general acceptance in county highway design and transport study work. An indication of the stages in which the public is typically involved in these various procedures is given in Fig. 3.31.

Development plans

Public involvement in development plan work is governed by the Town and Country Planning Acts of 1968 and 1971 which established the two-part framework of current development plans, as follows.

(1)   *Structure plans*—these deal with the planning authorities' intentions in respect of all land uses. As such they cover overall policies and strategies for transport, diagrammatic representations of the primary and secondary route network, and broad programmes for its enhancement.

(2)   *Local plans*—these deal with more specific issues and, in essence, fill in the details of the structure plan framework for a particular area.

With both of these types of plan, the public is guaranteed that adequate publicity will be given to reports of survey and to proposals contained in the draft plan, that opportunities will be provided to make representations on these proposals, and that these representations will be considered by the governing authority.

Structure plans, which contain broad policies and strategic proposals, generally do not encourage participation on specific highway proposals. Local plans by contrast contain proposals in sufficient detail to identify individual properties likely to be affected by highway schemes, and can be expected to become an important arena for public debate—and to justify significant public involvement programmes in their preparation.

The primary reason for seeking public participation in local plan work is to obtain a broad spectrum of view and information on all aspects of

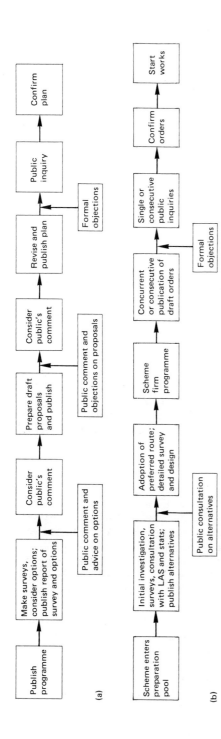

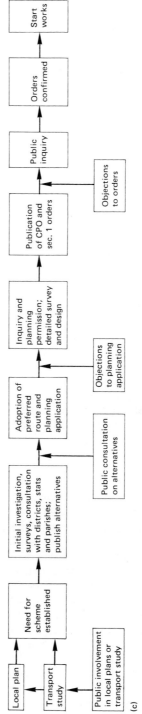

Fig. 3.31 Current British practice in relation to public participation[114]. (a) local plans (structure plans are similar). (b) trunk roads and (c) county roads

an area's future development—housing, industry, conservation, transport and leisure—from the local population. Whilst this level of participation is quite adequate for the establishment of highway corridors and strategies, it is inadequate for their detailed location. Thus, if the adopted local plan is regarded as only conferring 'outline planning approval' on the highway schemes that it contains, a subsequent stage of participation is required during the detailed design of the highway schemes to ensure that further inputs and information are obtained which enable decisions on exact alignments to be made and necessary ameliorating work to be properly carried out.

Trunk roads
Prior to 1973, the public's only opportunity to make representations on trunk road proposals was at or immediately preceding the public inquiry; the purpose of the inquiry was to examine the justification for the various Orders which, if confirmed, resulted in the establishment of the line of the highway and gave authority to alter side roads and acquire land. In the early 1970s, however, considerable dissatisfaction was expressed by some public groups at the limited opportunity for change provided by this approach, and so a new procedure was devised in July 1973 which has the objective of 'informing people about practicable alternative routes for trunk road projects, and for obtaining their views about them'.

The new procedure begins with the publication of a timetable for the consultation, and this is followed by the distribution of a comprehensive consultative document. This publication contains a statement of the needs that the scheme is expected to satisfy, a description of alternative alignments and an assessment of their relative efficiency, together with details of cost, layout, side road diversions and earthworks. The document also contains a rigorous assessment of the environmental, economic and physical effects of the proposed highway upon all nearby properties.

The consultative document—which is published as an attractive brochure with maps of alternative routes, and a prepaid reply slip for an expression of public views—is circulated to all whose personal interests are likely to be in any way affected by the alternative routes, and to local authorities and amenity and conservation groups. A comprehensive exhibition of the proposals is also mounted locally, and this is staffed by representatives of the Department of Transport or the agent authority. The presentation at these exhibitions is normally of a high standard, and considerable effort is made to convey as much information as possible on the proposals to the public. The need for a scheme or alternative means of meeting that need without extensive highway construction are, however, specifically excluded from the debate, since these are taken by the Department to be matters for parliamentary decision.

The public inquiry is the ultimate stage of the public's participation in the preparation of trunk road proposals. The basis for this is the Highways Act of 1959 which states that any member of the public may register an objection to the Department's proposals and appear at a public inquiry to press the objection.

Transport studies
Public participation in transport studies tends to follow a pattern similar to that for local plans. Inevitably, a general consultative approach must be employed, and thus the same pros and cons relate to this process as to participation in local plans. Indeed, it can also be argued that transport options are so dependent upon development proposals that public participation in comprehensive transport studies, as currently carried out, is little more than an information exercise unless it is carried out in the local plan context.

County highways
The pattern of participation carried out during the preparation of county highway proposals is similar to that for trunk roads, except that there is a greater involvement of locally elected members. This in turn has resulted in a greater emphasis on public meetings as the channel for two-way communication.

Selected bibliography

(1) Department of Transport, *Transport Statistics Great Britain 1973-1983*, London, HMSO, 1984.
(2) Webster F V, *Urban Passenger Transport: Some Trends and Prospects*, TRRL Report LR771, Crowthorne, Berks., The Transport and Road Research Laboratory, 1977.
(3) *No Time to Stop: The Case for the Road Programme*, London, The British Road Federation, 1979.
(4) Mouncey P and Harrison P, Mobility and the role of the car, *Traffic Engineering & Control*, 1977, **18,** No. 10, pp. 460-463.
(5) Tanner JC, *Saturation Levels in Car Ownership Models: Some Recent Data*, TRRL Report SR669, Crowthorne, Berks., The Transport and Road Research Laboratory, 1981.
(6) Bates JH, Gunn H and Roberts M, *A Disaggregate Model of Household Car Ownership*, London, The Department of Transport, 1978.
(7) Mogridge MJH, The future of the car, *Traffic Engineering & Control*, 1976, **17,** No. 1, pp. 101-104 and 115.
(8) Advisory Council on Energy Conservation, *Passenger Transport: Short- and Medium-term Considerations*, Energy Paper No. 10, London, HMSO, 1976.
(9) Das M, *Travel to Work in Britain: A Selective Review*, TRRL Report LR849, Crowthorne, Berks., The Transport and Road Research Laboratory, 1978.
(10) Town SW, *The Social Distribution of Mobility and Travel Patterns*, TRRL Report LR948, Crowthorne, Berks., The Transport and Road Research Laboratory, 1980.
(11) Feeney RJ, Ashford NJ, Morris A and Gazely D, *Travel and the Handicapped: A Project Summary*, TRRL Report SR480, Crowthorne, Berks., The Transport and Road Research Laboratory, 1979.
(12) Waters MHL, *Research on Energy Conservation for Cars and Goods*

Vehicles, TRRL Report SR591, Crowthorne, Berks., The Transport and Road Research Laboratory, 1980.

(13) Gyenes L, *Assessing the Effect of Traffic Congestion on Motor Vehicle Fuel Consumption*, TRRL Report SR613, Crowthorne, Berks., The Transport and Road Research Laboratory, 1980.

(14) Vincent R A and Wood K, *Car-sharing and Car-pooling in Great Britain: The Recent Situation and Potential*, TRRL Report LR893, Crowthorne, Berks., The Transport and Road Research Laboratory, 1979.

(15) Stores A, *Cycle Ownership and Use in Great Britain*, TRRL Report LR843, Crowthorne, Berks., The Transport and Road Research Laboratory, 1978.

(16) Mitchell CGB, Cycle use in Britain, in *Cycling as a Mode of Transport*, TRRL Report SR540, Crowthorne, Berks., The Transport and Road Research Laboratory, 1980.

(17) Quenault SW, *Cycle Routes in Portsmouth, III—Attitude Surveys*, TRRL Report LR875, Crowthorne, Berks., The Transport and Road Research Laboratory, 1979.

(18) Smeed RJ, The traffic problem in towns, *Town Planning Review*, 1959, **35**, No. 2, pp. 133–158.

(19) Tanner JC and Scott JR, *50-point Traffic Census—The First Five Years*, Road Research Technical Paper No. 63, London, HMSO, 1962.

(20) Bellamy PH, *Seasonal Patterns in Traffic Flows*, TRRL Report SR437, Crowthorne, Berks., The Transport and Road Research Laboratory, 1978.

(21) Oldfield RH, *Effect of Fuel Prices on Traffic*, TRRL Report SR593, Crowthorne, Berks., The Transport and Road Research Laboratory, 1980.

(22) O'Flaherty CA, *Highways*, Volume 1: *Highways and Traffic* (2nd ed.), London, Arnold, 1974, p. 77.

(23) *Basic Road Statistics 1983*, London, The British Road Federation, 1983.

(24) Marlow M and Evans R, *Urban Congestion Survey 1976: Traffic Flows and Speeds in Eight Towns and Five Conurbations*, TRRL Report SR438, Crowthorne, Berks., The Transport and Road Research Laboratory, 1978.

(25) Nicholl JP, Journey times by car and bus in central London, *Traffic Engineering & Control*, 1977, **18**, No. 12, pp. 581–583.

(26) OECD Research Group TSI, *Traffic Measurement Methods for Urban and Suburban Areas*, Working Paper RR/TSI/78.3, Paris, OECD, November 1978.

(27) Gerlough DL and Huber MJ, *Statistics with Applications to Highway Traffic Analyses* (2nd ed.), Westport, Conn., The Eno Foundation for Transportation, 1978.

(28) Sumner R, Measurement of vehicle following times and speeds, *Traffic Engineering & Control*, 1975, **16**, No. 7, p. 327.

(29) Wardrop JG and Charlesworth G, A method of estimating speed and flow of traffic from a moving vehicle, *Proceedings of the Institution of Civil Engineers*, Part 2, 1954, **3**, pp. 158–171.

(30) O'Flaherty CA and Simons F, An evaluation of the moving observer method of measuring traffic speeds and flows, *Proceedings of the Australian Road Research Board*, 1971, **5**, Part 3, pp. 40–54.

(31) Bennett TH, Use of the moving car observer method on one-way roads, *Traffic Engineering & Control*, 1975, **16**, No. 10, pp. 432–435.

(32) Hewitt RH, Sandilands NM and Seed IWM, Moving observer surveys in one-way streets, *The Highway Engineer*, 1976, **23**, No. 1, pp. 22–27.

(33) Freeman, Fox and Partners et al., *London Traffic Survey: Existing Traffic and Travel Characteristics in Greater London*, Volume 1, London County Council, 1964.

(34) Phillips G, When to mount a traffic count, *Traffic Engineering & Control*, 1980, **21**, No. 1, pp. 4–6.

(35) Kember Smith J, *Life of Automatic Punched Tape Traffic Counter and Detector Installations*, TRRL Report SR385, Crowthorne, Berks., The Transport and Road Research Laboratory, 1978.

(36) Fry R E, Setting the scene—data uses present and future, Paper 1 in *Traffic Data Collection*, London, The Institution of Civil Engineers, 1978.

(37) Searle G A C, Setting the scene—present methods of collection and analysis—an international comparison of data collection means, Paper 2 in *Traffic Data Collection*, as reference 36.

(38) Ham R, Future proposals for the automatic collection and retrieval of traffic data—its potential and problems, Paper 10 in *Traffic Data Collection*, as reference 36.

(39) Hillier J A and Moore R C, Trends in traffic data collection and analysis, Paper 9 in *Traffic Data Collection*, as reference 36.

(40) Lassiere A, *The Environmental Evaluation of Transport Plans*, London, The Department of the Environment, 1976.

(41) Hickman A J, Colwill D M and Hughes M R, *Predicting Air Pollution Levels from Traffic Near Roads*, TRRL Report SR501, Crowthorne, Berks., The Transport and Road Research Laboratory, 1979.

(42) Leonard D R, Grainger J W and Eyre R, *Loads and Vibrations Caused by Eight Commercial Vehicles with Gross Weights Exceeding 32 Tons (32.5 Mg)*, TRRL Report LR582, Crowthorne, Berks., The Transport and Road Research Laboratory, 1974.

(43) Greater London transportation survey, *Traffic Engineering & Control*, 1975, **16**, No. 5, pp. 212–214 and 219, No. 6, pp. 270–272, and No. 7, pp. 332–334.

(44) Standing Advisory Committee on Trunk Road Assessment, *Forecasting Traffic on Trunk Roads: A Report on the Regional Highway Traffic Model Project*, London, HMSO, December 1979.

(45) The RHTM data bank, *Traffic Engineering & Control*, 1979, **20**, No. 5, p. 274.

(46) Douglas A A, Grid reference coding; application to transportation studies, *Traffic Engineering & Control*, 1976, **17**, No. 11, pp. 456–459.

(47) Department of Social Services, *1981 Census of Population*, Cmnd 7146, London, HMSO, July 1978.

(48) Dumble P L, Improving home interview travel survey techniques, *Proceedings of the Australian Road Research Board*, 1980, **10**, Part 5, pp. 252–266.

(49) Falin D, How reliable are mail return questionnaires for transportation engineering surveys?, *Traffic Engineering & Control*, 1980, **21**, No. 7, pp. 370–372.

(50) Clarke R H and Davies D H, Origin and destination survey by registration number method, *Traffic Engineering & Control*, 1970, **11**, No. 12, pp. 600–603.

(51) Bureau of Public Roads, *Conducting a Home-interview Origin and Destination Survey*, Procedure Manual 2B, Washington DC, US Public Administration Service, 1954.

(52) *Consequences of Small Sample O–D Data Collection in the Transportation Planning Process*, Report FHWA-RD-78, Washington DC, The Federal Highway Administration, 1976.

(53) Di Renzo J F, Ferlis R A and Hazen P I, Sampling procedures for designing household travel surveys for statewide transportation planning, *Transportation Research Record*, 1977, No. 639, pp. 37–43.

(54) Roark A L, Levinson H and Guhin J S, A methodology for measuring

screen-line traffic volumes, *Traffic Engineering & Control*, 1975, **16**, No. 10, pp. 423–425.

(55) *A Policy on Arterial Roads in Urban Areas*, Washington DC, The American Association of State Highway Officials, 1957.

(56) Coburn TM, Beesley ME and Reynolds DJ, *The London–Birmingham Motorway*, Road Research Technical Paper No. 46, London, HMSO, 1960.

(57) Leitch Sir G et al., *Report of the Advisory Committee on Trunk Road Assessment*, London, HMSO, October 1977.

(58) Tanner JC, *Car Ownership Trends and Forecasts*, TRRL Report LR799, Crowthorne, Berks., The Transport and Road Research Laboratory, 1977.

(59) O'Flaherty CA et al., *Intertown Public Transport: Alternatives for Canberra*, Canberra, The National Capital Development Commission, 1976.

(60) Pak-Poy PG and Associates, *Canberra's Short-term Transport Plan Study*, Canberra, The National Capital Development Commission, 1977.

(61) Douglas AA, Home-based trip end models—a comparison between category analysis and regression analysis procedures, *Transportation*, 1973, **2**, pp. 53–70.

(62) Wootton HJ and Pick GW, Trips generated by households, *Journal of Transport Economics and Policy*, 1967, **1**, No. 2, pp. 137–153.

(63) Fratar TJ, Vehicular trip distribution by successive approximations, *Traffic Quarterly*, January 1954.

(64) Hill DH and Von Cube HG, Development of a model for forecasting travel mode choice in urban areas, *Highway Research Record*, 1963, No. 38, pp. 78–96.

(65) Stopher PR and Meyburg AH, *Urban Transportation Modelling and Planning*, London, Lexington Books, 1975.

(66) Philbrick AT, A short history of the development of the gravity model, *Australian Road Research*, 1973, **5**, No. 4, pp. 40–54.

(67) Jones SR, *Accessibility Measures: A Literature Review*, TRRL Report LR967, Crowthorne, Berks., The Transport and Road Research Laboratory, 1981.

(68) Tressider JO, Meyers DE, Burrell JE and Powell TJ, The London Transportation Study; methods and techniques, *Proceedings of the Institution of Civil Engineers*, 1968, Paper 7092, pp. 433–464.

(69) Wardrop JG, Some theoretical aspects of road traffic research, *Proceedings of the Institution of Civil Engineers*, 1952, Part 2, **1**, pp. 325–378.

(70) Chu C, A review of the development and theoretical concepts of traffic assignment techniques and their practical applications to an urban street network, *Traffic Engineering & Control*, 1971, **4**, No. 13, pp. 136–141.

(71) Van Vliet D, Road assignment, *Transportation Research*, 1976, **10**, pp. 137–143, 145–149 and 151–157.

(72) Black JA and Salter RJ, A review of the modelling achievements of British urban land-use transportation studies outside the conurbations, *Chartered Municipal Engineer*, 1975, **102**, No. 5, pp. 100–105.

(73) Lai FK and Van Vliet D, Trip assignment techniques current in the UK, *Traffic Engineering & Control*, 1979, **20**, No. 7, pp. 348–351.

(74) OECD Road Research Group, *Urban Traffic Models: Possibilities for Simplification*, Paris, OECD, August 1974.

(75) Wilson AG, The use of entropy maximising models in the theory of trip distribution, mode split and route split, *Journal of Transport Economics and Policy*, 1969, **3**, No. 1.

(76) Shepherd LE, An econometric approach to the demand for urban passenger travel, *Proceedings of the Australian Road Research Board*, 1972, **6**, No. 2.

(77) *Behavioral Demand Modelling and Valuation of Travel Time*, Special Report 149, Washington DC, The Transportation Research Board, 1973.

(78) Hartgen DT, *Behavioral Models in Transportation: Perspectives, Problems and Prospects*, Preliminary Research Report 152, Albany, NY, The New York State Department of Transportation, 1979.

(79) OECD Road Research Group T13, *Integrated Traffic Management*, Working Paper RR/T13/77.2, Paris, OECD, September 1977.

(80) Buchanan CD and Partners, *The Conurbations*, London, The British Road Federation, 1969.

(81) Department of the Environment, *Policy for the Inner Cities*, Cmnd 6845, London, HMSO, June 1977.

(82) Buchanan CD and Partners, *South Hampshire Study: Report on the Flexibility of Major Urban Growth*, London, HMSO, 1967.

(83) Department of Transport Traffic Advisory Unit, Some effects of flexible working hours, *Traffic Engineering & Control*, 1978, **19,** No. 1, p. 39.

(84) Stirling WN, McTavish AD, White PR, Kilner AA and Stark DC, *Factors Affecting Public Transport Patronage*, TRRL Report SR413, Crowthorne, Berks., The Transport and Road Research Laboratory, 1978.

(85) Tebb RGP, *Differential Peak/Off-peak Bus Fares in Cumbria: Short-term Effects*, TRRL Report SR368, Crowthorne, Berks., The Transport and Road Research Laboratory, 1978.

(86) Tebb RGP, *Differential Peak/Off-peak Bus Fares in Cumbria: Short-term Passenger Responses*, TRRL Report SR391, Crowthorne, Berks., The Transport and Road Research Laboratory, 1978.

(87) *Tyne and Wear Public Transport Impact Study: Study Definition Report*, TRRL Report SR478, Crowthorne, Berks., The Transport and Road Research Laboratory, 1979.

(88) Chapman RA, Gault HE and Jenkins IA, The operation of urban bus routes, *Traffic Engineering & Control*, 1977, **18,** No. 9, pp. 416-419.

(89) Wheat MH and Cohen NV, Bus monitoring and control experiment and the assessment of its effects on bus operation, in *Traffic Control and Transportation Systems*, London, North-Holland, 1974.

(90) *Bus Priority Systems*, NATO CCMS Report No. 45, Crowthorne, Berks., The Transport and Road Research Laboratory, 1976.

(91) Ball RR and Brooks RJ, Planning and road design for bus services, *Chartered Municipal Engineer*, 1976, **103,** No. 7, pp. 117-124, No. 8, pp. 129-137, and No. 9, pp. 155-160.

(92) Leeds City Council et al., *Planning and Transport—The Leeds Approach*, London, HMSO, 1969. (This policy was reappraised in a leaflet entitled *Leeds, Public Transport and the Environment*, published in 1973.)

(93) West Yorkshire transportation studies, *Traffic Engineering & Control*, 1978, **19,** No. 12, pp. 536-540, and 1979, **20,** No. 1, pp. 27-31, No. 2, pp. 82-85 and 87, No. 3, pp. 111-116, and No. 5, pp. 269-273.

(94) Vincent RA and Layfield RE, *Nottingham Zones and Collar Study—Overall Assessment*, TRRL Report LR805, Crowthorne, Berks., The Transport and Road Research Laboratory, 1977.

(95) Department of Transport Traffic Advisory Unit, The Nottingham zones-and-collar experiment, *Traffic Engineering & Control*, 1978, **19,** No. 5, pp. 240-241.

(96) May AD, Supplementary licensing: an evaluation, *Traffic Engineering & Control*, 1975, **16,** No. 4, pp. 162-167.

(97) The impact of traffic policies in Singapore, *Traffic Engineering & Control*, 1977, **18,** No. 4, pp. 152-157, No. 6, pp. 299-302, and No. 7/8, pp. 357-361.

(98) Holland EP and Watson PL, Traffic restraint in Singapore, *Traffic Engineering & Control*, 1978, **19**, No. 1, pp. 14–22.

(99) Ministry of Transport, *Roads in Urban Areas*, London, HMSO, 1966.

(100) Reynolds DJ, *The Assessment of Priority for Road Improvements*, Road Research Technical Paper No. 48, London, HMSO, 1960.

(101) Mackie PJ and Urquart GB, *Through and Access Commercial Vehicle Traffic in Towns*, TRRL Report SR117C, Crowthorne, Berks., The Transport and Road Research Laboratory, 1974.

(102) Charlesworth JA, Control and routing of traffic in a road network, *Traffic Engineering & Control*, 1979, **20**, No. 10, pp. 460–466.

(103) Bonsall PW, Car-sharing—where are we now?, *Traffic Engineering & Control*, 1980, **21**, No. 1, pp. 25–27.

(104) Plum R and Edwards J, *Car-pooling: An Overview with Annotated Bibliography*, Minneapolis, Minn., University of Minnesota Center for Urban and Regional Affairs, 1977.

(105) Green GR, *Car-sharing and Car-pooling—A Review*, TRRL Report SR358, Crowthorne, Berks., The Transport and Road Research Laboratory, 1978.

(106) Bonsall PW, *Car-pooling in the USA: A British Perspective*, TRRL Report SR516, Crowthorne, Berks., The Transport and Road Research Laboratory, 1979.

(107) Kirby RF, *Paratransit—A State-of-the-art Overview*, Transportation Research Board Special Report 164, 1976, pp. 37–44.

(108) OECD Road Research Group UT2, *Transport Services in Low Density Areas*, Working Paper RR/UT2/79.1, Paris, OECD, 1979.

(109) *Symposium on Unconventional Bus Services: Summaries of Papers and Discussions*, TRRL Report SR336, Crowthorne, Berks., The Transport and Road Research Laboratory, 1977.

(110) Tunbridge RJ, *A Comparison of Optimal Minibus, Dial-a-bus, and Conventional Bus Services*, TRRL Report SR409, Crowthorne, Berks., The Transport and Road Research Laboratory, 1980.

(111) Hickman AJ, Bevan MG and Colwill DM, *Atmospheric Pollution from Vehicle Emissions: Measurements at Four Sites in Coventry, 1973*, TRRL Report LR695, Crowthorne, Berks., The Transport and Road Research Laboratory, 1976.

(112) Ratcliffe BG, The problem of the heavy goods vehicle and the highway system, *Chartered Municipal Engineer*, 1976, **103**, No. 11, pp. 201–208.

(113) Hansell BB, Foulkes M and Robertson JS, Freight planning in London—(3) reducing the environmental impact, *Traffic Engineering & Control*, 1978, **19**, No. 4, pp. 182–185.

(114) Bergg JA, Public participation in highway proposals—the pros and cons, *Chartered Municipal Engineer*, 1980, **107**, No. 12, pp. 345–354.

(115) *Guidelines on Citizen Participation in Transportation Planning*, Washington DC, The American Association of State Highway and Transportation Officials, 1978.

(116) Sinclair A, *Public Participation in Transport Planning (in Australia)*, Bureau of Transport Economics Occasional Paper No. 20, Canberra, Australian Government Publishing Service, 1978.

4
Parking

There are few towns in today's highly developed countries—and in many newly developing ones as well—which do not experience traffic congestion and car parking problems. Parking, in particular, is such a sensitive issue that parking policies in many communities often appear to be both conflicting and indecisive. Nevertheless, there is no doubt but that the public has now become reconciled—albeit reluctantly—to the need to exercise firm controls on parking in town centres, whilst civic leaders, in turn, have come to recognize the need to develop strong parking policies and strategies for the urban area as a whole.

The importance of this latter point cannot be overemphasized: *parking control is now, in many towns, the key to proper traffic control and transport policy implementation.*

Parking is a commodity that is subject to the basic laws of economics. Thus, for example, if a parking policy is developed which results in a reduction in the number of long-stay parking spaces in a town centre, then a new equilibrium point will be established which will result in a higher price to the consumer for parking; this higher price, in turn, will make it less attractive to make the journey-to-work trip to the town centre by car. Thus parking policy affects how people will travel—and good parking management and control can lead to some or all of the following[1]: higher car occupancies, decreasing person-trips, faster travel times and less travel delays, greater public transport usage, decreasing congestion, and reduced air and noise pollution.

Until the late 1960s, relatively little regard was paid by local authorities to the strategic implications of providing parking in town centres. Even in the 1970s, whilst greater attention was given to reducing the amounts of on-street parking, and to placing the balance of the publicly-operated parking (on- and off-street) under various forms of control, relatively little was done to regulate off-street commercial parking facilities—the operators of which, not unnaturally, felt free to attract parkers on the basis of the best economical return, and without much regard for the strategic parking needs of the central area. This major loophole in parking control was only closed nationally in 1979 when local authorities were given the power[2] to require privately-operated public off-street car parks in central areas to be licensed, and to impose conditions in the licence regarding: (a) the maximum number of spaces, (b) the proportion of spaces to be available for different categories of parking, (c) the scale of charges for the various types of parking, and (d) the times of opening and closing.

The role of parking policy and strategy within transport planning has already been discussed in Chapter 3, and so will not be repeated here. Rather the approach adopted in this chapter is to discuss the detailed development of the town centre parking plan, including the technical factors underlying the location and operation of on- and off-street parking facilities. Firstly, however, some of the detrimental effects associated with the parking problem are discussed.

Traffic and parking: some associated effects

Aesthetics

An environmental 'disamenity' which receives little attention is the aesthetic deterioration often associated with the parking of vehicles. The simplest way of describing this detriment is to quote from what is possibly the most significant document published so far this century regarding the role of the motor vehicle in the urban area, i.e. the Traffic in Towns Report[3] has this to say regarding the visual consequences of the motor car:

'... the crowding out of every available square yard of space with vehicles, either moving or stationary, so that buildings seem to rise from a plinth of cars; ... the dreary formless car parks, often absorbing large areas of towns, whose construction has involved the sacrifice of the closely-knit development which has contributed so much to the character of the inner areas of our towns. ... visual intrusion is a serious matter to which society, perhaps after some false starts and bitter experiences, will be bound to pay serious heed. ...we are concerned with the future when, unless positive policies are adopted and implemented, the number of vehicles will be so great as to dominate the visual scene entirely.'

To overcome these and other problems associated with the motor vehicle, the Traffic in Towns Report advocated the development of parking policies based on the premise that 'it is the liability of the owner or driver of a stationary vehicle to dispose of it off the highway.... From this it follows that parking on the highway, or any form of publicly subsidized parking, are in the nature of concessions which should be zealously safeguarded by the public authority.'.

Business, accessibility and parking

Commercial interests consider that they are directly affected by the parking situation and, even in this day, many merchants and business people in city centres tend to regard packed kerbs as necessary visual evidence of trade prosperity. Hence the reason why regulations to control parking in towns are viewed with suspicion by Chambers of Commerce, and usually accompanied by new demands for off-street multistorey parking facilities.

Table 4.1 Factors influencing the decision whether to shop in the central area or in a suburban shopping centre

Location	Factors	
	For	Against
Central area	Greater choice of competitive goods	Parking is difficult
		Can be very crowded
	Opportunity to do many types of errands on one trip	Traffic congestion
	Ease of accessibility by public transport	
	Often lower prices	
Suburban centre	Closer to home with consequent shorter travel time	Fewer types of businesses, with consequent more limited selection of competitive goods
	Easier parking	
	More convenient shopping hours	Poor accessibility by public transport
		Prices can be higher

Attempts have been made to relate the prosperity of a town centre to the availability of parking spaces (see, for example, reference 4), but no conclusive results have been derived to this effect. Although inadequate parking is widely believed to be one of the main drawbacks of the central area of a large town, analysis suggests[5] that, when other factors are taken into account, it may not greatly affect the shopping orientation of persons (see, for example, Table 4.1). At the same time, however, it must be noted that another survey[6] reported that between one-tenth and one-fifth of the respondents in the households queried (belonging to Automobile Association members in Britain) stated that they avoided going to town centres wherever possible because of parking difficulties.

Anything that can be done to reduce congestion, and allow people to travel to the town centre in a shorter time, will make the central area more accessible—and thus will help people to decide to shop there as against in the suburbs or out of town. It is government policy[7] that inner city areas should be revitalized, and local authorities are now urged to concentrate revenue support on public transport for actual and potential journeys to work in these locations, to give weight to the implications for local firms when designing traffic management schemes which improve access for central area traffic, to ensure efficient loading arrangements for goods vehicles, and to provide adequate and convenient parking for shoppers.

One simple way of reducing traffic congestion is to eliminate or control parking at the sides of roads approaching the town centre, e.g. parked vehicles reduce speeds simply by occupying road space that could be used by moving vehicles, thus reducing capacity. (This is especially noticeable if the parked vehicles are at, or close to, an intersection.) Table 4.2 shows that a small number of parked cars can have a quite significant effect upon effective roadway width, although the relative effect diminishes as the parking intensity increases. If the volume of traffic is fixed, then parked

Table 4.2   Effect of parked vehicles upon useful roadway width (based on data from reference 8—indicative only)

Number of parked vehicles per km (both sides added together)	Effective loss of carriageway width (metres)
3	0.9
6	1.2
30	2.1
60	2.6
125	3.0
310	3.7

vehicles will also reduce traffic speed and, hence, increase journey time which, in turn, reduces town centre accessibility.

Accidents

Although vehicles parked at, or manoeuvring into or out of, kerb parking spaces can be an important cause of accidents, the number of studies which have been carried out in order to produce an understandable and accurate description of the role of parking in accident occurrence is relatively few. What is probably the classic study of this nature—it was carried out in the USA in 1946[9]—found that 17 per cent of city accidents and 10 per cent of rural road accidents involved vehicles which were parked, manoeuvring into or out of a parking position, or stopped in traffic. In Britain in 1983, 20.7 per cent of the pedestrians involved in road accidents were deemed to have been masked by stationary vehicles; two-thirds of these were children or people of retirement age[10].

Apart from other general statistics of the type given above (e.g. 17 per cent of the urban area accidents in the American State of Connecticut were estimated[11] to involve vehicles which were parked, manoeuvring into or out of a parked position, or stopped in traffic), there is only one major study reported in the literature which has attempted to look deeper into the relationship between kerb parking and traffic accidents. This study[12], which was carried out in 1965–66 in the USA, examined in detail (including statistical analysis) some 11 620 road accidents on 152.36 km of arterial and collector streets in 32 cities, representing 17 States and the District of Columbia. The main results of this most comprehensive study—which while confirming some previous concepts of traffic engineers re this problem also raised some questions regarding others—are as follows.

(1) An average of 18.3 per cent of all accidents studied involved parking, either directly or indirectly.
(2) Almost 90 per cent of the accidents involving parking were a direct result of parking activity—either a parked vehicle or a vehicle entering or leaving a parked position—while in slightly more than 10 per cent, parking was a factor only.

(3) There was no significant difference in parking accident experience between street segments on which parking was prohibited and those on which parking was restricted to less than two hours. A significant increase was found in parking accidents where the parking operation was unrestricted.

(4) Parking accidents have a higher rate in residential areas than in either commercial or industrial areas.

(5) Accidents involving parking have a tendency to decrease slightly as the roadway width increases.

(6) Parking accidents in town centres have a somewhat higher rate than in intermediate or outlying areas, the latter two showing rates which were practically identical.

(7) In 92 per cent of all parking accidents, cars only were participants. This compared with about 86 per cent of the total vehicle-kilometres operated by such passenger vehicles.

(8) Almost 40 per cent of the total number of vehicles involved in accidents during the parking operation were attempting to drive forward into the kerb. In slightly over 20 per cent of the cases, vehicles were backing into the kerb whereas, in about 24 per cent, they were leaving the kerb in a forward direction.

(9) Forty-six per cent of the vehicles in motion were attempting to drive straight ahead on the street in question and collided with a vehicle which was parked or was involved in a parking or unparking operation. Vehicles attempting to park were involved in a little over one-third of the total parking accidents.

(10) In almost 94 per cent of the accidents involving kerb parking, the vehicles were reported as legally parked.

(11) In slightly over 77 per cent of the parking accidents, male drivers were the participants. This is slightly higher than the percentage of drivers who were men.

(12) Almost 13 per cent of drivers involved in parking accidents were under 20 years of age, as compared with approximately 8.5 per cent of the total national driving done by persons in the same age group.

(13) The month of December was significantly higher than any other month in both the number and the percentage of parking accidents experienced.

(14) Almost as many parking accidents occurred during hours when the road surface was either wet or covered with ice or snow as compared with the periods when the surface was dry—even though it may be reasonable to assume that the carriageway surface was normal during the greater percentage of the total time.

(15) Night-time was not a major factor in accidents involving parking operations.

(16) Of the almost 2180 parking accidents examined in these urban areas, only one (this involved a pedestrian stepping from behind a parked vehicle) resulted in a person being killed. Very few injuries were recorded and, in most cases, these also involved pedestrians who stepped from behind parked vehicles.

Surveys

Parking surveys are carried out in order to obtain the information necessary to provide an assessment of the parking problem in the area(s) being studied. The objective of any such study is to determine facts which will provide the logical point of departure in relation to indicating parking needs. Common to all types of parking studies, whatever their scale, are parking supply and parking usage surveys.

Parking supply survey

Parking supply surveys are concerned with obtaining detailed information regarding those on- and off-street features which influence the provision of parking space, the existing situation with regard to parking space, and how it is controlled. A typical survey would require an inventory of the on-street accommodation, and of all off-street car parks and parking garages serving the traffic area being studied. Whether the survey area involves the central area of a town (which may be broadly defined as the area in which all streets have business frontage), or a suburban shopping area, or a hospital or university, it should also include the surrounding fringe area where vehicles are parked by persons with destinations within the survey area.

A parking supply survey can be considered as being composed of three main parts, viz. an on-street space inventory, a street regulation inventory, and an off-street space inventory. Data usually collected in each of these survey phases are listed below; they are obtained by simple inspection of the survey area.

On-street (including alley) space items:

(1)   footpath crossings and accesses to premises,
(2)   loading bays,
(3)   bus stops,
(4)   taxi stands,
(5)   pedestrian crossings,
(6)   visibility splays at junctions,
(7)   one-way streets,
(8)   private streets,
(9)   service and rear access alleys,
(10)   vacant or unused land suitable for temporary or permanent parking space,
(11)   carriageway widths,
(12)   other local factors (e.g. areas of special amenity).

Street regulation inventory items:

(1)   controlled parking: (a) by regulation (including discs, with each separately classified) or (b) by meters (classified by type of control),
(2)   parking prohibited: (a) always or (b) during peak hours only,
(3)   controlled loading and unloading,
(4)   uncontrolled parking (the remainder).

Off-street inventory items:

(1) type: (a) surface only, (b) multistorey or upper level only, or (c) underground only,

(2) ownership and use: (a) publicly owned, for public use, (b) privately owned, for public use, or (c) private use only,

(3) commercial vehicles only,

(4) payment: (a) fee-charging (subdivided by rates of charge) or (b) free,

(5) time limit, for example: (a) up to 2 h only, (b) up to 4 h only, or (c) over 4 h or no limit,

(6) number of spaces provided,

(7) size of parking spaces,

(8) number and location (on map) of entrances and exits.

For summary purposes, these data may be marked on a map with a scale of, say, no more than 1:2500 (more usually 1:1250), using figures to indicate the number of spaces available in various sub-area locations, and letters, symbols and colours to differentiate between parking classifications. Tabulations are usually also prepared showing, for each sub-zone within the survey area, the total number of on- and off-street parking spaces available by classification, together with the space available for the loading of lorries and vans, or reserved for buses, taxis, etc.

The information obtained from the parking supply survey is an invaluable reference in connection with many of the routine decisions which have to be made in connection with the development and implementation of a parking plan. When questions arise as to what are the kerb or off-street conditions at any particular location, the answers are immediately available. The supply study also provides the basis for evaluating the available space-hours of parking as well as the parking turnover at any particular location.

Parking usage survey

The parking phenomenon is very much based on the law of supply and demand, whereby supply is the total number of spaces available within a designated area, and demand is the desire to park based solely on the location of the trip destination. However, unlike the true supply and demand situation, there is a third variable—*usage*—associated with parking which reflects the desire to park close to the destination, *but within the limitations imposed by the available supply*, as well as the desire to park at a reasonable cost. In other words, demand is a constant, reflecting the desire to park at the trip destination, whereas usage is a variable that depends upon the conditions at the terminal area and upon the characteristics of the trip as well as of the trip-maker.

If the parking supply is in excess of the parking demand, then the *true* demand can be determined by means of a parking concentration survey— usually carried out in conjunction with a parking duration survey. If the parking supply is less than the demand, then an *indication* of the demand may be obtained by means of a direct-interview parking survey; however, the *true* demand, which might include 'suppressed' vehicles not able or willing to find parking within the survey area, can only be properly estimated from comprehensive land use/transport surveys of the nature noted previously in this text (see Chapter 3).

Concentration survey

The purpose of a concentration survey is to determine not only where vehicles do park, but also the actual number parked at any given instant at all locations (on- and off-street) within the survey area. The street phase of the survey is usually carried out by dividing the survey area into 'beats', each short enough to be toured on foot or by car within a predetermined time interval. Information regarding the numbers of vehicles parked (legally and illegally) at each street and alleyway location is then noted on prepared forms or spoken into a tape recorder for transcribing at the end of the day. Each beat normally forms a closed circuit so that no time is lost by the observer(s) in returning to the starting point for successive inspections.

The selection of a suitable length of beat—or, more properly, the 'trip interval'—depends upon the accuracy required, and upon the amount of time, money and personnel available for the survey. A long trip interval, whilst cheaper in total survey time, cost and analysis, can overlook a significant proportion of short-term parkers, thus completely invalidating the survey. Whilst a method to compensate for the error experienced is described in the literature[13], it is usually considered preferable to carry out continuous observation on a sample of streets on each beat prior to the survey in order to determine an acceptable trip interval, as well as to estimate the correction factors which will need to be applied to the data subsequently obtained. Whilst experience suggests that trip intervals of 30 minutes and one hour may be satisfactory for many on- and off-street parking surveys, respectively, this may not always be so in the parts of town centres where parking durations are short and where sharp fluctuations in demand are likely to exist.

Where the survey personnel available is limited, it is generally better to carry out the on-street survey on a sample basis using short trip intervals, rather than to attempt to cover the whole area using longer intervals. If sampling is used, care must be taken, however, to ensure that it is representative, covers all types of parking, i.e. commuters, visitors, shoppers and residents, and that the results can be scaled up to give a picture of the whole area.

The off-street phase of the concentration survey can be carried out in the same manner as the on-street survey, i.e. by counting the number of vehicles parked at regular intervals in each facility. However, if traffic counters are available, it is simpler to use one of these at each entrance and exit in order to monitor the traffic—in essence, to carry out a cordon count of the vehicles entering and leaving the car park, from which the accumulation of vehicles can be calculated.

The selection of time and season for a concentration survey (as for a duration survey) is primarily dependent upon the characteristics of the town, and upon the extent of the variation in parking usage likely to be experienced throughout the week. In most towns in Britain, seasonal demand is at an 'average' level in September and October, and a normal weekday then (from, say, 7 a.m. to 7 p.m.) may well be convenient for the survey. Days immediately preceding or following holidays, special shopping days, or days when shopping hours differ from the usual should not be generally considered as suitable survey days. However, there may well be a case in a particular town for carrying out a special study on, say, the market day or, in the case of holiday resorts, at special times of the year.

Where the survey area includes fringe residential areas, there should be a special survey so as to isolate the residents' parking usage.

Duration survey
As the term implies, the primary purpose of a duration survey is to determine the lengths of time that vehicles are stored within the survey area. In so doing, data normally collected during a concentration survey can also be collected (in essentially the same way, in the case of the on-street survey) during the duration survey.

As with the concentration survey, the most accurate way of carrying out a duration survey is to make continuous observations on parkers at all possible locations. For practical reasons, this is not usually possible; instead, the duration information is obtained in a more economical fashion by an observer noting the first three numbers or four letters/ numbers of each vehicle's registration number while patrolling the beat (or possibly from parking tickets in the case of appropriate off-street facilities). As with the concentration survey, the trip interval utilized during the duration survey is again critical, e.g. if a trip interval of 30 minutes is utilized, a car which is observed but once during the survey may possibly have been parked for as little as one minute or for as long as 59 minutes, while a significant number of cars may have been parked for periods up to 29 minutes without being observed at all.

Since they also obviously measure the degree to which the existing parking regulations are observed, duration surveys should always be carried out as inconspicuously as possible. The accuracy of any survey's results could be seriously impaired if parkers thought that the check being made was one which might be used for law enforcement purposes, e.g. a study[14] of parking meter spaces limited to 2 h found that as a result of 'meter feeding' associated with the presence of observers conducting a licence plate survey, only about half of the cars recorded by a hidden time-lapse camera as having exceeded 2 h were actually identified in the survey as having parked for longer than that duration.

Parker interview survey
This, the most expensive—and comprehensive—of the parking surveys, normally involves interviewing motorists at their places of parking and questioning them regarding the origins of the trips just completed, the

primary destinations whilst parked, and the purposes of the trip. Particulars regarding parking duration and concentration (of the type already noted) may also be gathered during this survey and, if required, information may be obtained regarding the number, sex and age of each vehicle's occupants.

With this type of survey, the survey area is usually divided so that each sub-area can be covered for the desired period(s) of the day by the interviewing team. Each individual interviewer is then given a specific section and is responsible for interviewing each driver and recording each parking incident in that area; while each section will obviously vary in size according to the parking concentration and durations, typically it is unlikely to exceed 75–100 m of kerbside space in central areas. At off-street facilities, the duration phase of the survey is simply carried out by interviewers stationed at entrances or exits.

When the survey personnel available is limited, and particularly in

Table 4.3   Categories of parking: based on parking durations[15]

Category of parking	Approximate duration	Examples
Stop-and-go	0–5 minutes	Drop-off of shoppers and others; pick-up of persons and packages; parcel drop-off and pick-up; taxi loading and unloading; bus loading. (*Note:* Most of this category of short-term parking is at the kerb; the driver remains in the vehicle, but a parking space is occupied.)
Errand	0–15 minutes	Brief shopping and business errands or pick-up; bank deposits and withdrawals; drop-off of dry-cleaning; buying cigarettes or newspapers; paying bills; taxis awaiting fares; dashing in for a cup of coffee. (*Note:* These are mostly of short duration but averaging longer than the above; the driver typically leaves the vehicle; mostly but not entirely kerb parking.)
Convenience	0–30 minutes	Purchasing of convenience goods other than large orders; purchasing or ordering equipment for business or consumer use; deliveries. (*Note:* These are split between kerb and off-street parking.)
Services	0–1 hour	Trips to doctor, dentist, lawyer, travel agents, etc.
Basic	0–4 hours	'True shopping' for major purchases or group of purchases; multiple-purpose trips; entertainment (movies, etc.); dining-out; cultural purposes; salesman parking; loading and unloading of large trucks; parking by repair and service vehicles; tourists
Employee	0–8 hours	Parking by bosses, employees, some business visitors, hotel guests, tourists, conventions
Night-time	0–15 hours	Resident parking; hotel guests; tourists; entertainment parkers; work vehicles; out-of-service vehicles; buses and taxis. (*Note:* These are long-term, off-hour parkers.)

areas subject to long parking durations, information normally obtained by a direct-interview survey may instead be obtained by means of a mail-return questionnaire survey. In this case, questionnaires are either handed to the motorists as they park or else placed under the windscreen wipers of all vehicles parked at the kerb or off-street within the survey area, with the request that each driver answer the questions posed and then return the completed questionnaire by post. The problem with this type of survey is that expansion of the data obtained is statistically difficult; whilst a good survey may result in, say, one-third of the questionnaires being returned, little is known about the extent to which these returns are representative of the population being sampled, e.g. 'satisfied' drivers are less likely to return the questionnaires than those having difficulty in obtaining parking places, and parking habits such as duration differ according to trip purpose and type of driver (see Table 4.3). The result is that the data obtained from the mail-return questionnaire survey, unless treated with considerable care, may be such as to result in entirely incorrect conclusions being deduced.

Information provided
Parking usage surveys can provide very useful data regarding existing parking characteristics within the survey area. In the first place, the variation in the concentration of parked vehicles with time of day may be obtained. Figure 4.1(a) shows, for example, the variation in the number of vehicles parked on the streets in the central area of Edinburgh[16]. Note that the parking peak in this city occurs at about 11.30 a.m. and that a second major peak occurs at 3.30 p.m.

If p is the number of vehicles parked in a street, or group of streets, and P is the number of kerb spaces available (obtained from the parking supply survey), then a parking index defined by

$$PI = \frac{100p}{P}\%$$

may be used to assess the relative amounts of parking at particular locations. It is usual to show these *PI* values as numbers on a map of the survey area. Alternatively, the supply may be compared to the concentration in order to determine to what extent the occupancy values exceed the supply values; in addition to indicating where extra capacity is required, this also shows the areas where illegal parking, e.g. double parking, takes place, thereby implying the need for enforcement procedures at these locations.

The duration survey provides very useful information regarding the lengths of time that people park at particular locations. Figure 4.1(b) shows, for example, that of the 77 970 vehicles parked on the streets of the central area of Edinburgh at the time of the survey, approximately 71 per cent parked for 2 h or less—but they only occupied 28 per cent of the 21 856 parking space usages that day. Figure 4.1(a) also shows the percentage of short-term parkers in the central area of Edinburgh at any particular period of the day.

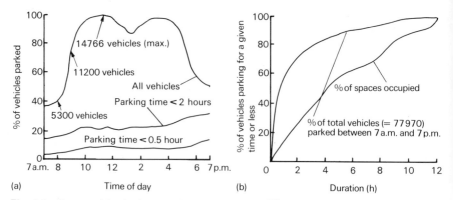

(a) Time of day
(b) Duration (h)

Fig. 4.1   Street parking in the central area of Edinburgh[16]: (a) time-of-day characteristics and (b) duration characteristics

Figure 4.2 shows some (expected) differences between the parking durations of workers and shoppers at off-street car parks in central Croydon; the average length of stay for workers was 6.9 h, whilst for shoppers it was 2.9 h. In this study, 44 per cent of the parkers were journey-to-work parkers who used 65 per cent of the parking space; shoppers, who represented 49 per cent of the parkers, only used 28 per cent. Note, however, the consistently large component of worker-parkers who stayed for durations varying from 1 h to 7 h.

The peak usage data may be further subdivided as shown in Table 4.4 to enable a judgement to be made of the effect which might result from restricting parking to some particular maximum length, say 2 h.

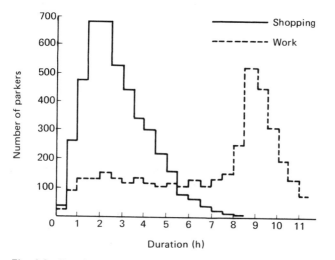

Fig. 4.2   Durations of stay of shopping and journey-to-work parkers at car parks in Croydon[17]

Table 4.4   Some data relating to the peak use of street parking spaces in Central London[8]

Number of times each vehicle was seen	Approximate duration of parking (hours)	Vehicles (total = 39 362)	
		Number	Percentage
1	0.5	2910	7.4
2	1.0	2358	5.7
3–4	1.5–2.0	5651	14.3
5–8	2.5–4.0	6848	17.3
9–12	4.5–6.0	7952	20.1
13–20	6.5–10.0	12560	31.9
21 +	10.0+	1344	3.3

Another way of presenting daily data is shown in Table 4.5. Note that, in this instance, the emphasis is placed on the distribution of the total vehicle-hours of occupation, i.e. the total turnover, thus enabling an estimate to be made of the revenue which might be expected to accrue from a system of paid parking. Alternatively, the average turnover per parking space, i.e. the number of times a space is used by different vehicles during a specified time, may be calculated by dividing the total numbers of vehicle-hours for the period being considered by the numbers of particular types of parking spaces laid out.

Table 4.5   Some data relating to the turnover of parking spaces in Central London[8]

Number of times each vehicle was seen	Approximate duration of parking (hours)	Vehicles (total = 146 624)		Vehicle-hours of occupation (total = 331 048)	
		Number	Percentage	Number	Percentage
1	0.5	50768	34.6	25384	7.7
2	1.0	25872	17.7	25872	7.8
3–4	1.5–2.0	26416	18.0	46228	14.0
5–8	2.5–4.0	19616	13.4	63752	19.3
9–12	4.5–6.0	10048	6.8	52752	15.9
13–20	6.5–10.0	12560	8.6	103620	31.2
21 +	10.0+	1344	0.9	13440	4.1

In relation to turnover, it might be noted that the average turnover rate per parking space decreases as the size of the urban area increases. Furthermore, American experience suggests that kerb parking average turnover rates tend to be three to four times higher than off-street rates, and surface car parks have higher average turnover rates than do garages[18].

Management of on-street parking

Historically, many attempts have been made to control on-street kerb parking. In ancient times, for example, King Sennacherib of Assyria, who ruled about 700 BC, forbade parking on the main road of his capital city; he enforced this decree by directing that any illegal parker on the 24 m wide royal highway should be put to death by being impaled upon a pole in front of his home. Happily, however, while today's urban parking problems are serious, they do not appear to be so severe as to need such stringent penalties.

The legality of modern methods of controlling on-street parking in Britain is fundamentally based on an 1812 ruling by the (then) Lord Chief Justice that 'the King's highway is not to be used as a stable-yard'. It is interesting to note that this same position was taken by the Royal Commission on Transport in 1929, when it reported that 'the streets are intended for the movement of traffic of all kinds, not for the purpose of parking cars while the owners or persons in charge are engaged in business or pleasure'.

Nowadays, the question of when and where to impose some form of kerb parking management is inextricably involved with a town's parking policy. Nevertheless, if one temporarily leaves aside policy as a controlling feature, it is possible to consider objectively the general situations where direct kerbside parking management is desirable, and how it should be carried out.

Prohibited parking

In many small towns, particularly towns which have been developed over the past century, the number of 'natural' on- and off-street parking spaces in and adjacent to the central area is often sufficient to meet the parking demand. In such towns, the only parking management measures which may have to be initiated are as follows.

(1) *Intersections* Cars should never be allowed to park within about 50 m of a major junction. (Desirably, the no-parking zone should extend even further back from the junction.) While this prohibition can be justified on road capacity considerations, even more important is its critical effect on safety, i.e. cars and pedestrians at junctions must have adequate sight distances, while large commercial vehicles must be given sufficient space to negotiate left-hand turns (in Britain).

(2) *Narrow streets* Very often it will be necessary to initiate kerb management measures because of the relative narrowness of streets in relation to the needs of moving vehicles. Guides to when this is necessary are indirectly given in the published figures for the capacities of urban streets. Parking should never be permitted on two-way carriageways in central areas which are less than about 5.75 m wide, and on one-way streets which are less than about 4 m wide; there is just not sufficient room for safe movement *and* parking on these streets.

(3) *Driveways* On no account should parking be permitted in front of driveways from houses and other buildings. Other vehicles must be allowed to enter and leave buildings.

(4) *Pedestrian crossings* For safety reasons, parking should be

prohibited on or adjacent to (within, say, 8 m) pedestrian crossings. Pedestrians should be able to step off the kerb without having their view obstructed by parked cars. Similarly, the driver of a moving vehicle must immediately be able to see any pedestrian leaving the kerb.

(5) *Curvature and grade conditions* Parking should be prohibited if a study shows grade and/or curvature conditions where the removal of parking will improve the safe and efficient movement of traffic.

(6) *Road bridges and tunnels* Although there is no reason why parking should be prohibited on bridges or in tunnels simply because bridges are bridges or tunnels are tunnels, it will be found more often than not that these structures are narrower than the roadway in general and so parking may have to be prohibited.

(7) *Pedestrian concentrations* For safety reasons, it is desirable that parking be prohibited at places of heavy pedestrian concentration, e.g. at exits from schools or hospitals.

(8) *Priority locations* Parking should never be permitted at kerb locations where priority must, of necessity, be given to public services. Thus, for example, parking should be forbidden at or adjacent to fire hydrants and bus stops.

Time-limit parking

In many town centres, the need to impose parking management measures is indicated by the double parking of vehicles, by cruising vehicles awaiting the opportunity to park, or by low turnover and high occupancy of kerb space. In such instances, the measure which suggests itself is that of *rationing* the kerbside spaces so that parking preference is given to the people who are normally the life-blood of the town centre, i.e. the shoppers and visiting business people. This normally means the initiation of a time-limit parking scheme to control the lengths of time that vehicles may spend at given kerbside parking spaces.

There are four types of time-limit control now in general use in parking schemes, as follows:

(1) limited waiting schemes under direct police or warden control,
(2) parking meter control,
(3) parking card control,
(4) parking ticket control.

All of these methods of control can be used in off-street car parks also. However, for comparison purposes, their usage, advantages and disadvantages will be considered in the context of kerbside parking.

Before discussing these different approaches, however, it is useful to consider the basic features which should be built into any time-limit parking scheme.

Scheme design features
Initially, at any rate, the introduction of a time-limit parking scheme into a community is usually viewed with suspicion by both motorists and

business people, principally because of the regulating function of the scheme and inherent doubts as to 'how will it affect me'. To overcome these suspicions, and to justify its introduction to those who suspect its motive and impact, any parking management scheme should contain the following favourable features.

Within the area outlined by the scheme, it should result in a reduction in parking durations and consequent higher parking capacity.

A properly conducted 'before' study will indicate whether there is a real shortage of space for short-time parkers, or whether the problem is simply one of available existing space being occupied by long-time parkers. It will also suggest what time limits might be used in the scheme. An 'after' study should reveal whether the management measure is effective in driving away from the kerb the long-stay, journey-to-work parker; it should also tell whether the time limits are adequate, or whether they should be changed entirely or at certain locations only.

While, ideally—as is suggested above—the time limits prescribed for any given kerb location should be determined in the first instance on the basis of the results from a parking duration study, realistically the decision as to what should be the exact limit is very often an administrative one. The result is that, on the national scene, there is relatively little variability with regard to the time limits actually employed in town centres, e.g. the time limit which is undoubtedly most commonly used is 2 h maximum, with 4 h maximum on the edges of the central area and, in the 'core' part of large towns, a time limit of 40 minutes maximum is often used. Not commonly used in Britain, unfortunately, are 15-minute zones near 'errand-type' establishments such as banks, post offices, chemist shops, public-utility offices, and commercial establishments that prepare telephone orders for immediate collection.

Within the designated area, a motorist should be able to find an empty parking space quickly within a reasonable walking distance of his or her destination.

The optimum utilization of parking spaces is generally considered to be about 85 per cent, leaving about 15 per cent available at any one time during peak periods[19]. The degree of utilization in any instance is, however, regulated by the allowable parking durations and, in the case of parking meters, by the charges imposed. As with the first design feature noted above, the adequacy of particular charges and durations can only be determined from before-and-after studies. Objective methods of evaluating this adequacy are discussed in the literature[20, 21].

The distances from the time-limit zone to principal traffic generators, e.g. large department stores and hospitals, should obviously be taken into account whenever a time limit is being established. In locations where there are no distinctive parking generators, i.e. in a 'well-balanced' central area, the distances that particular types of parkers are willing to walk from particular time-limit zones will need to be taken into account.

The kerbside parking spaces should be arranged so as to make the most efficient use of the road surface with the minimum inconvenience to moving traffic.

In most British towns, this criterion usually means the utilization of parallel parking at kerbs. It does not necessarily mean that the parking spaces are always marked on the carriageway; however, marking with white paint should always be carried out when high turnover is experienced and/or parking meters are used, and/or parking is not parallel to the kerb and/or where non-marking might result in the inefficient use of space.

If parallel parking is utilized (say, in a meter scheme), and if 6.1 m is taken as the space necessary to manoeuvre a car into a kerbside parking position, then two possible marking arrangements are as shown in Fig. 4.3. Of these, the arrangement composed of pairs of, say, 4.87 m bays with a 1.22 m manoeuvring space between each pair is generally to be preferred because it reduces the kerb length required for two vehicles by 10 per cent, still allows each vehicle 6.1 m of manoeuvring space, and yet permits adequate parking space for the rare 'long' car.

Where the carriageway is very wide, consideration may be given to using angle parking, since this provides more spaces for the same length of kerb. If angle parking is utilized, the bays should be marked out with continuous white lines, discontinuous lines, or T-marks. Another advantage of angle parking is that parking/unparking vehicles may cause relatively less delay to moving traffic, since the critical manoeuvre for parallel parking, i.e. entering the space, takes longer than the critical manoeuvre for angle parking, i.e. leaving the space. For example, an American study[22] found that the average driver took 12 seconds to back out of an angle bay and proceed ahead in the traffic stream, whereas for parallel parking the average driver required 32 seconds to back into a bay and be clear of the adjacent traffic lane; however, this study was carried out before cars with power-steering became commonplace, so the time differential may not be as significant nowadays.

Notwithstanding the above, considerable care should be exercised when deciding when or when not to use angle parking. In particular, the accident potential at the location should be queried as cars are particularly liable to accident involvement when backing out of an angled space[23]. In general, therefore, parallel parking is to be preferred, and angle parking should not be permitted unless the street is very wide and traffic is light and slow-moving.

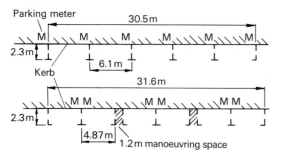

Fig. 4.3 Alternative kerbside parallel-parking arrangements

Within the designated area, the motorist should have no doubt whatsoever as to where, when and for how long a vehicle may be parked. Within the controlled area, the time-limit parking should be properly enforced.

The first of these criteria can be ensured if the controlled area is clearly defined by distinctive road and kerb markings combined with proper signing. The most efficient way of enforcing the parking restrictions in controlled zones is to use patrolling traffic wardens—this applies irrespective of whether parking is free or for payment.

Limited waiting schemes

These are schemes where the authorized time limits are displayed on signs which also show the extent of the permitted (free) parking. Enforcement is carried out by a patrolling police officer or traffic warden—complete with watch and chalk—who marks parked vehicles' tyres and returns to the scene at the end of the authorized time and checks for marked vehicles which have not left.

This form of time-limit parking can be relatively effective at locations where the total number of parking spaces available (on- and off-street) is known to be sufficient to meet the parking demand, so that the parker has little difficulty in leaving his or her car close to the destination for roughly the time desired—which, under the conditions specified, is likely to be less than the posted time limit. In other words, there is little incentive to exceed the specified time limit, and so no great effort is required in relation to enforcement.

Unfortunately, the conditions described above are generally found only in the central areas of quite small towns. In larger towns, this type of scheme requires a very stringent enforcement effort, and will probably fail.

Parking meters

First used in Oklahoma City in 1935, the parking meter is primarily designed to assist in the enforcement of parking regulations, and to increase parking turnover. Meters made their first appearance in Europe in Stockholm in 1953, while their use in Britain began in July 1958 with an experimental installation in London[24]. The parking meter was so effective at meeting its objectives in London that it soon moved out of its experimental stage, and encouraging the use of meters in and about other town centres where time limits needed to be strictly enforced became government policy. The use of parking meters also became big business in Britain; between 1958 and 1978, parking meters had nearly £56m paid into them[25].

There are two main types of parking meter: the 'manual' meter and the 'automatic' meter. The manual meter, which was the first type developed, is operated by the motorist inserting an appropriate coinage and then turning a knob or handle to activate the timing mechanism within the meter. The second type (which is the only meter used in Britain) is automatic in that the insertion of the coinage starts a clockwork timing mechanism[26].

Both types of meter can be set to allow parking for a single fixed period of time for a fixed charge or, dependent upon the coinage inserted, for varying lengths of time up to a predetermined maximum. In either case, as soon as the motorist has used up the allotted parking time, a signal is displayed on the meter which is easily seen by a patrolling traffic warden or police officer. With certain of these meters, a yellow flag bearing the words 'Excess Period' appears in the dial when the time bought has expired. This means that, until a red flag is displayed, the parker owes the local authority a specified but much higher fee for the parking space. The excess-charge meter has the advantages that: (a) if the charge is set sufficiently high, the motorist will be disinclined to use the time, thus making the kerbside space available for other short-time parkers, and (b) it has the practical effect of an 'instant' fine, thus reducing the work of the Courts.

Mention might also be made here of another type of meter which has a built-in facility whereby the time clock always starts from zero when additional money is inserted. This meter, which is in use in some cities on the Continent, ensures that every parker pays for his or her own parking time.

Parking cards
As far as can be determined, the concept of using a parking card located on the car, instead of a parking meter on the kerb, to control the length of time a vehicle may park within a time-limit parking zone was first tried out in Paris in 1958[27]. Following wide usage on the Continent, the first British scheme—which used a fee-free disc card—was initiated in Cheltenham in 1965, and subsequently in Harrogate[28]. A fee-paying system, which used a date-, day- and time-marked rectangular card, was started in Cork, Ireland, in 1975[29].

With the British disc system, the parking cards are obtained free of charge at police stations, garages, newspaper shops, tobacconists, etc. A driver parking at a kerb displays one of these cardboard cards[30] on the car's windscreen. A time-marked circular disc within the card pocket is rotated until the correct time of arrival is displayed in an aperture at the front of the card. The maximum time allowed for parking at the location is shown on kerb signposts. A patrolling traffic warden, by observing the disc and checking a watch, can then determine whether the car is parked in excess of the permitted time.

Unlike the British card, which is reusable at another time, the card displayed in the Cork system can only be used on a given day. The cards are purchased, either singly or in books of ten, at designated retail outlets or from traffic wardens. Upon parking, the motorist indicates the arrival time by punching holes at designated points in the card, to show the month, day of the month, hour and minute (to the nearest five minutes). The perforated card is then displayed on the car window nearest to the kerb and this can be checked by a patrolling warden, as described for the British disc system. Unexpired time on a perforated card may be used at other parking places on the same day.

Parking tickets

A relatively new entry into the parking control equipment field in Britain—but a method widely used overseas for many years—is the usage of ticket machines to sell parking tickets which are displayed for inspection, in the same way as parking cards. The control method is known as pay-and-display parking.

There are a number of machine and ticket variations used with pay-and-display parking control. One type of machine used in surface car parks in Britain[31] is microprocessor controlled, which enables tickets of up to twelve variable prices to be issued; a further twelve prices are held within the machine, and can be brought into operation automatically under the control of an incorporated clock/calendar, thus providing a means for peak/off-peak or weekday/weekend tariff changes.

The machine, which accepts five denominations of coin in any order and in any combination, issues self-adhesive tickets that are marked with the date, parking fee tendered, and the number of the machine from which the ticket was obtained. The expiry time of parking is also printed on the ticket, rather than the time of entry; this helps to avoid any misunderstanding on the part of the motorist and eases the problem of surveillance.

Full audit facilities are available, and a ticket which gives details of the monies entered since the previous occasion on which the coin box was removed, is issued each time the automatically-locked coin box is taken from the machine. Furthermore, the machine has the very valuable capability of printing out on demand an audit trail showing the grand total of cash to date, and the total number of tickets issued against each of the parking tariffs.

Meters versus cards versus tickets

Table 4.6 summarizes some of the main advantages and disadvantages of the various parking control systems.

The particular advantages of parking meters are that they are proven to be effective in regulating on-street parking in both large and small towns, and they are easy to supervise. Disadvantages are that they are fairly easily jammed (sometimes by genuine misuse, at other times by deliberate attempts to obtain free parking time), and subject to theft by the criminal element in society (often juveniles who are difficult to deal with under current legislation, even if caught). Indeed, the entire financial viability of certain schemes can be imperilled by acts of vandalism[32]. In times of inflation, a not insignificant disadvantage is that any attempt to change the tariff requires substantial alteration of the coin acceptance mechanism and recalibration of the timing.

Whilst parking cards can be effective in controlling on-street parking, enforcement requires greater numbers of traffic wardens, as compared with meters. In wet weather also, the cards can be difficult to check. A major problem associated with parking card schemes can be the difficulty experienced by visitors.

Parking tickets are also effective as a parking control measure. Whilst

Table 4.6 Comparison of various parking control systems

System	Advantages	Disadvantages
Parking meters	Simple to use Well proven and accepted by public No additional walking distance Easiest system to supervise	Expensive to install and maintain Aesthetically unsightly Continuous obstacles on the footpath Can attract vandals Fee variations are a function of availability of particular coins
Parking cards	No unsightly installation devices necessary Considerable flexibility Very low cost to local authority No maintenance or vandalism problems Parking bays parallel to kerb do not have to be individually marked Revenue collection (from retail outlets) is easy	Can be complex to use Creates problems for visitors Greater risk of cheating requires higher level of enforcement If cards are free, the cost of administering the scheme is borne by the ratepayers as a whole, and not the motorists
Parking tickets	Simple to use Small number of machines not generally considered unsightly Possible to have variable fee structure Lower maintenance costs than meters	Can attract vandals Imposes short walking distances Surveillance can be difficult

they too attract vandals, the relatively smaller numbers of machines to be serviced means that the cost of collecting the larger sums of money in each machine's coin box at more regular intervals is more easily justified and done. Probably the biggest difficulty associated with a ticketing system is the monitoring problem; it is simply very difficult for the parking wardens to read the ticket through the window of a car, especially if it is poorly displayed (either by accident or deliberately); this problem is of course magnified under inclement weather conditions. (On the whole, therefore, parking ticket control[33] would appear to be more applicable to off-street car parks rather than to kerb parking.)

'Resident' parking

A frequently forgotten part of a time-limit parking scheme is its effects on the on-street parking needs of householders living on residential streets in and about the central area of a large town or adjacent to large traffic generators such as hospitals and universities. It can of course be argued that 'garaging' a car on the public highway is anti-social in that it destroys the amenities and pleasant living conditions of a neighbourhood and, therefore, that it should not be permitted and that everyone owning a car should make an arrangement for it to be parked off the road when it is not in use. As against this, it has to be accepted that in a free society there is likely to be no administrative restraint placed on the *ownership* of a car and, therefore, as car ownership increases so also will the demand for residential on-street parking space—and if provision is not made for residents' cars, then car-owning families may be encouraged to move elsewhere to the detriment of the social balance in the inner residential areas.

The most widespread techniques used in the USA[34] to limit parking on residential streets by non-residents are: (a) non-residents are only allowed to park for a limited amount of time in a regulated area, (b) non-residents are excluded entirely during the day and night, or (c) non-residents are prohibited from parking during certain hours, typically 8 a.m. to 6 p.m., Monday to Friday. Parking permits, where required, are usually sold for a price sufficient to offset administrative costs and are displayed in a car window or stuck on a car bumper.

Legislation initiated in 1967 in Britain made it legally possible to give special consideration to the parking needs of residents, and to allocate for their exclusive use on a prepayment basis a substantial proportion of the on-street space. The first such scheme[35] was introduced into the City of Westminster in January 1968. This *Respark* system has two parts: a residential parking card (on which is printed a monthly calendar) for displaying on the windscreen, and 'stick-on' parking tokens which are prepurchased at a moderate cost for superimposing over the appropriate dates on the card. Each token gives the resident (and family) the right to a full day's use of a parking place anywhere within the resident-priority parking zone. Subsequently, as an improvement on the token system, season tickets for residents parking were made available as an alternative to the parking card.

The Respark system only applies to parking during the day, and no attempt has been made to exercise control at night.

Facilities for commercial vehicles

Commercial vehicle service is vital to the prosperity of a city's central area. Lorries and vans must be able to load/unload merchandise at or close to (say, within 30 m) of business premises. Buses must be able to pick up and drop passengers at convenient locations.

Freight facilities
All loading/unloading of store goods should preferably be handled at off-street facilities such as loading docks or alleyways. Kerb-loading should only be allowed when these facilities are lacking, and even then only to the minimum degree necessary during the normal working day.

Attempts to tackle this aspect of the parking problem have tended to incorporate the following features:

(1) adoption of regulations requiring new or substantially altered commercial buildings to provide adequate off-street parking facilities,
(2) provision of limited kerb-loading zones where justified, each street being studied separately for this purpose,

(3) prohibition of freight vehicle parking on busy streets during peak periods,

(4) prohibition of freight vehicle parking on critical streets during the working day,

(5) continuous review of existing kerb-loading zone needs and uses.

Time-limit parking for lorries and vans is usually difficult to enforce because of opposition by business interests. The easiest restriction to enforce, and the one that is usually of most value in reducing congestion, is the prohibition of parking during the peak traffic hours. If traffic conditions are so chaotic that prohibition must be enforced throughout the day, it may be necessary for deliveries and pick-ups to be made outside the usual commercial hours.

In 1972, the (then) Department of the Environment[36] initiated a programme aimed at the development of a network of high-security lorry parks—50 in the first instance, later reduced to 30—near the edges of towns at locations: (a) where environmental damage would be minimized, (b) within easy reach of motorway, trunk and principal roads, (c) near areas of industrial activity, (d) where there would be adequate access roads, and (e) on level well-drained land. The aim was to encourage heavy freight vehicles to break their loads at these vehicle parks, with the goods then to be carried into town for distribution by a larger number of smaller lorries.

In 1976, however, it was decided to abandon this programme[37] as a national initiative, on the grounds that the heavy capital outlay required by the proposal resulted in correspondingly high charges to users, and this had failed to attract operators. It is still open to local authorities to provide off-street lorry parks, however.

Bus facilities

Kerb stops, i.e. short-stay parking spaces, must be provided for buses. Each stop should have a clearly defined and marked zone and, where appropriate, a layby of sufficient length to contain all buses that can be reasonably expected to stop at a given time. If a bus cannot pull up at a layby inset into the kerb (see also Chapter 6), it will block a lane intended for traffic as it picks up and discharges passengers. If buses do not stop frequently at desirable locations, then passengers will be encouraged to use their own motor cars, thereby increasing traffic congestion—and the demand for road and parking facilities. Undesirable though it may be from a traffic aspect, very many bus stops are located not far from intersections as this is most convenient for passengers.

Preparing the town centre parking plan

It is now becoming increasingly clear that for important social and economic reasons, most medium- and large-sized towns are simply not capable of providing for all the central area demands of the motor car within the foreseeable future. Thus, in order to fulfil their obligations in accordance with the transport and general environmental needs of the community as a whole, most British town councils—because of the immediacy of the traffic congestion problem—feel constrained to turn towards the implementation of stringent traffic control measures when devising their central area parking plans.

The term 'traffic control' as used here refers to the decision by the governing authority to restrict traffic demands to artificially low levels by deliberately imposing stringent parking restrictions in and about town centres—restrictions which result in a limiting of the total number of parking spaces provided, a limiting of the time during which a vehicle may stay in a regulated parking space, and the imposition of a relatively high charge for the use of a parking space. This method of traffic control is becoming the adopted strategy in many towns because there is no practical alternative to it at this time. Its main attributes are as follows: it can be put into operation within the existing legal and administrative framework; although it has a certain unpopularity, it is accepted by the public; most important, if properly implemented it has a good chance of succeeding in its objective of limiting the number of movements made by private cars to levels which are compatible with desired community objectives[38].

How many spaces in the central area?

Any 'true' determination of the potential number of parking spaces required in a town centre—or (perhaps more properly) the number which should be incorporated in its parking plan—presents a most difficult task, even for a city which has the results of a full-scale transportation study to call upon. It is practically an impossible one for a town which has few basic data at hand.

The problem is dominated by the difficulty of calculating the exact effects of the factors which can be expected to influence movement within the town. Some of these factors are as follows:

(1) the future population of the catchment area (including people domiciled outside the town boundaries),
(2) the car-ownership level in the design year,
(3) the number and proportions of person-trips generated by the central area (including, for example, work, business, shopping, educational trips),
(4) the proportions of the daily travel which will take place during the normal daily travel hours and the peak travel times,
(5) the capacity of the road system feeding the town centre,
(6) the availability and quality of the public transport systems,
(7) the relationship between the peak accumulation of parkers and the total number of parkers,
(8) the parking duration times of the various categories of parkers,
(9) efficiency of parking space usage,
(10) the time of year considered for planning purposes,

(11)   anticipated cost of parking,
(12)   increases in the floor area of central area uses under development,
(13)   changes in attraction of the area after redevelopment and in the number of potential customers as a result of new industry and/or overspill schemes in the catchment area,
(14)   the anticipated parking policy, particularly in relation to the modal split and the desired environmental quality.

Examination of the items listed above emphasizes why it is that a comprehensive transport study is normally needed if a realistic appraisal of the future parking needs of a central area of reasonable size is to be obtained (and why it is that the planning of parking facilities cannot be isolated from the planning of the overall transport system for a city). This list further emphasizes that the problem is not simply one of determining the number of parking spaces. What is also required is the correct proportion of on- and off-street parking, of space and time allocated to the long- and short-term parker, and of charges levied for long- and short-term parking.

Nevertheless, as a first step in the preparation of a town centre parking plan, it is useful to obtain a rough estimate of the number of spaces which could be required. Of the methods that have been used to provide estimates of central area parking demand, the following four are probably the most important. It should be appreciated that the methods, all of which are based on the concept of a high degree of motorization, can normally be expected to give quite different answers, so that a choice must then be made as to which is the most applicable to the particular circumstances under consideration.

Method 1
This procedure, an American one, postulates that the 'parking space coefficient' which should be applied to the proportion of the total town centre person-trips made by car, in order to obtain the number of spaces required, is related to the population of the catchment area. The method utilizes curves such as are shown in Fig. 4.4 which are based on the following general formula[39]:

$$P = \frac{drsc}{oe} = \frac{0.70rsc}{1.5 \times 0.85} = 0.55rsc$$

where P = parking space coefficient, d = proportion of daily travel involving the central area which takes place between 7 a.m. and 7 p.m. = 0.70, o = car occupancy = 1.5 person/vehicle, e = efficiency of space usage = 0.85, r = ratio of peak to total daytime parkers (taken as 0.25 in small towns and over 0.40 in large ones), s = seasonal peaking factor, and c = locational 'adjustment' factor to reflect the concentration of demand in the core part of the central area.

The 'desirable' curve in Fig. 4.4 assumes that s and c are each equal to 1.1; the 'tolerable' curve assumes that $s = 1.0$ and $c = 1.1$; the 'minimum' curve assumes that both s and c equal 1.0. The manner in which the figure is used is as follows.

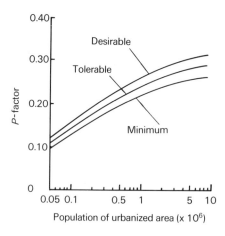

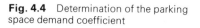

Fig. 4.4   Determination of the parking space demand coefficient

Step 1   Estimate the number of central area person-trip destinations per day.

Step 2   Estimate the percentage of central area person-trips which will be made by car.

Step 3   Calculate the daily number of person-trip destinations by car. (Multiply 1 by 2.)

Step 4   For the urban population in question, read the appropriate *P*-value from the appropriate curve in Fig. 4.4.

Step 5   Calculate the number of central area parking spaces required (Multiply 3 by 4.)

If the 'desirable' curve is used in the above calculation, it means that both the particular needs of the core part of the central area and the seasonal fluctuations will be taken into account. Use of the 'tolerable' curve results in the provision of sufficient space for the typical weekday, while giving some consideration to the concentration of demand in the core area, i.e. it recognizes that certain spaces are beyond acceptable walking distance. Use of the minimum demand level results in the concentration factor not being taken into account. Thus, for example, when the urbanized population is 100 000, the desirable, tolerable and minimum *P*-factors are 0.158, 0.144 and 0.131, respectively. For a population of 1m, the three factors are 0.262, 0.238 and 0.217.

When the trip purposes of the car drivers and passengers are known, it is recommended that Fig. 4.4 not be used but that the parking space requirements should be calculated from a series of adjusted factors. Thus desirable, tolerable and minimum *P*-values for each town centre work trip by car are given as 0.500, 0.454 and 0.412, respectively, while corresponding values for central area non-work trips by car are 0.147, 0.133 and 0.121. For example, if there are 50 000 central-area-oriented work trips and 70 000 other trips by car to the town centre, it would be 'desirable' to provide 25 000 parking spaces for workers, and 10 300 spaces for other parking purposes.

Comment The relationships described above were derived by an internationally known firm of transport engineers from data collected in the course of many parking studies carried out in towns in the USA. They illustrate how parking data collected from many studies can be used to establish patterns which can usefully be utilized by other towns which have not been able to carry out such studies.

Data are now available in the literature[40] relating off-street parking levels in British towns of 75 000–200 000 population to urban population. However, as yet, they do not appear to have been applied in any parking methodology.

Method 2

This method assumes that a relationship exists between the total number of vehicles registered in the city and the number for which parking space within the central area during the peak parking demand period is desired. The number of parking spaces needed is then determined by estimating the number of vehicle registrations for the design year and multiplying it by the appropriate parking proportion value.

It is well known that as urbanized areas increase in size, less use per head is made of cars in order to travel into town centres. This is illustrated by the data in Table 4.7 which are taken from an analysis of parking data collected in the USA[41]. At the time that this study was carried out, there was a strong tendency for the number of vehicles parking in the central areas of the smaller American towns to be approximately constant at 17 per cent; for towns of over 500 000 population, the parking proportion drops to about 10 per cent; when the population is over 1m, the percentage parking drops to 6.

Table 4.7 Percentage of vehicles parked in the central areas of American towns in relation to the number of vehicles registered in those towns[41]

Year	Number of vehicles per 1000 population	Population range (× 10³)	Number of vehicles	Maximum number of vehicles parked in the central area	
				Total	Percentage
1950	380	5–10	3000	490	16.3
1950	380	10–25	6800	1180	17.1
1950	330	25–50	11900	1950	16.5
1950	320	50–100	25600	4450	17.6
1950	320	100–250	52000	5700	10.7
1948	260	250–500	95000	9140	9.6
1947	240	500–1000	132000	12000	9.6
1954	300	>1000	390000	23400	6.0

Comment The apparent simplicity of this method makes it a most attractive way of estimating town centre parking needs. Nevertheless, usage of this technique at this time in Britain can be criticized on many grounds.

It is appropriate to note, for example, that there would appear to be no reason why there should be a firm relationship between the demand for parking spaces in city centres and other variables such as the number of car registrations, the number of people working in the central area, and the size of the area. Town centres, particularly those in Europe, differ so much with regard to type and use of buildings, type of activity, employment, density, peculiarities and size of catchment area, quality of the public transport system, etc., that it is difficult to see why they should be unequivocally described by means of simple formulae. However, this may not necessarily be the case in the USA where more towns have grown up with the motor car and, hence, tend to be better adapted (by basic layout) to the needs of the motor age as compared with the older, more historical, cities of Europe.

Method 3
With this third procedure, the number of parking spaces is calculated on what might be termed a 'standard' basis. In other words, the number of parking spaces is determined by summing estimates of the requirements of the individual parking generators within the central area.

Some parking standards specified by Lancashire County Council in relation to the provision of planning permission for various types of building activities are given in Table 4.8. In this table, the term *operational parking* is used to describe the essential spaces needed for the cars (and other vehicles) that are regularly and necessarily involved in the operation of the business of particular buildings, including commercial vehicles servicing these buildings. *Non-operational parking* refers to the spaces which may or may not be provided for vehicles which do not need to park or wait precisely at the premises in question, i.e. spaces for commuting employees as well as those for shoppers, business callers and tourists. In practice, nowadays, the full space requirement (operational plus non-operational) of a particular land use is only permitted in urban locations well away from the central area; within the central area of larger towns, only operational parking (and, in some circumstances, partial non-operational parking) would normally be allowed.

Comment A major disadvantage of this method of determining the 'ideal' parking needs of a central area is that there is still widespread disagreement over what constitutes a correct relationship between parking space requirements and land use. Another disadvantage is that, in the end, an 'adjustment' factor must be applied to the calculated total value, in order to obtain a reasonable estimate of the parking accommodation actually required during the maximum demand period, i.e. the maximum parking demands of different types of generator do not normally coincide during the day, so that dual use of car parks will be possible.

Table 4.8 Car parking standards for various land uses[42]

Land use	Operational parking	Non-operational parking
1a Family housing	None	Residents—1 garage space per house Visitors—1 space per house
1b Single person housing	None	Residents—3 spaces per 4 dwellings Visitors—2 spaces per 4 dwellings
Retirement housing (owner occupied)	None	Residents and visitors—1 space per 2 dwellings
Sheltered housing (tenanted)	None	Warden (if applicable)—1 garage space and 1 car space Residents and visitors—1 space per 4 dwellings
Homes for the elderly	None	Resident staff—1 garage space Day staff—1 space per 3 staff Visitors—1 space per 6 residents
2 Shops	Minimum space for loading/unloading (m²): *GFA Space ⩽ 500 50 ⩽1000 100 ⩽2000 150 For each extra 500 m² GFA, add 50 m² space	Staff—1 space per 100 m² *or* 1 space per managerial staff + 1 space per 4 other staff Customers—'normal' shops, 1 space per 25 m²; carpet etc. warehouses, 1 space per 20 m²; DIY superstores and garden centres, 1 space per 15 m² (including external sales area); superstores > 2000 m², 1 space per 10 m²
3 Banks	25 m² (minimum)	Staff—1 space per 3 staff Customers—1 space per 10 m² of net public floor space of banking hall
4 Car sale	50 m² (minimum)	Staff—1 space per 2 staff Customers—1 space per 50 m² of display area
5 Car auction rooms	None	Staff—1 space per 2 staff Customers—1 space per 20 m² of display area and vehicle storage area
6 Offices	Minimum space for loading/unloading (m²): GFA Space ⩽ 500 50 ⩽1000 100 ⩽2000 150 ⩽3000 200 For each extra 1000 m² GFA, add 25 m² space	Staff—1 space per 25 m² GFA *or* 1 space per 3 staff Visitors—10% of staff need
7 Industrial premises/ Warehouses	Minimum space for loading/unloading (m²): GFA Space ⩽ 100 70 ⩽ 235 140 ⩽ 500 170 ⩽1000 200 ⩽2000 300 For each extra 1000 m² GFA, add 50 m² space	Staff—1 space per 50 m² GFA (Industrial premises)/1 space per 200 m² GFA (Warehouses) Visitors—10% of staff need (Industrial premises only)

Table 4.8 *(cont.)*

Land use	Operational parking	Non-operational parking
8 Cash-and-carry warehouses (for direct sales to traders only)	Minimum space for loading/unloading (m²): *GFA* *Space* ⩽ 500 50 ⩽1000 100 ⩽2000 150 For each extra 1000 m² GFA, add 50 m² space	Staff—1 space per 200 m² GFA Customers—1 space per 25 m² GFA
9 Garages and service stations	1 space per towing vehicle + 4 spaces per service and/or repair bay (+5 spaces if station has an automatic car wash)	Staff—1 space per 40 m² GFA *or* per 2 employees
10 Petrol stations	5 spaces if station has an automatic car wash	Staff—1 space per 2 employees at busiest time
11 Hotels, motels and public houses	Minimum space for loading/unloading (m²): *GFA* *Space* ⩽ 500 140 ⩽1000 170 ⩽2000 200 For each extra 1000 m² GFA, add 25 m² space	Resident staff—as for (1a) Non-resident staff—1 space per 3 staff at the busiest time Resident guests—1 space per bedroom Bar customers—1 space per 4 m² of net public space in bars plus 1 space per 8 m² in beer garden
12 Restaurants and cafes	Minimum space for loading/unloading (m²): †*DFA* *Space* ⩽100 50 ⩽250 75 ⩽500 100	Resident staff—as for (1a) Non-resident staff—1 space per 3 staff at the busiest time Diners—1 space per 2 seats or 4 m² in the dining area
13 Licensed clubs	50 m² (minimum)	Resident staff—as for (1a) Non-resident staff—1 space per 3 staff at the busiest time Performers—1 space per solo performer and/or groups at the busiest time Patrons—1 space per 2 seats *or* per 4 m² of net public floor space
14 Dance halls and discotheques	50 m² (minimum)	Staff—1 space per 3 staff at the busiest time Performers—3 spaces Patrons—1 space per 10 m² of net public floor space
15 Cinemas	50 m² (minimum) + 2 spaces at main entrance to set down/pick up patrons	Staff—1 space per 3 staff at the busiest time Patrons—1 space per 4 seats
16 Theatres	100 m² (minimum) + 2 spaces at main entrance to set down/pick up patrons	Staff—1 space per 3 staff at the busiest time Performers—1 space per 10 m² of gross dressing room accommodation Patrons—1 space per 3 seats

Table 4.8 *(cont.)*

Land use	Operational parking	Non-operational parking
17 Bingo halls	50 m² (minimum) + 2 spaces at main entrance to set down/pick up patrons	Staff—1 space per 3 staff at the busiest time Patrons—1 space per 10 seats
18 Swimming baths	50 m² (minimum)	Staff—1 space per 2 staff normally present Patrons—1 space per 10 m² of pool area
19 Community centres/ Assembly halls	50 m² + 2 spaces at main entrance to set down/pick up patrons	Staff—1 space per 3 staff normally present Patrons—1 space per 10 m² of floor space
20 Places of worship	50 m² (minimum) + 2 spaces at main entrance to set down/pick up worshippers	Worshippers—1 space per 10 seats
21 Museums/Public art galleries	50 m² (minimum)	Staff—1 space per 2 staff normally present Visitors—1 space per 30 m² of public display space
22 Public libraries	50 m² (minimum) + 50 m² per mobile library van	Staff—1 space per 3 staff normally present Borrowers—3 spaces (minimum) *or* 1 space per 500 adult ticket holders (+1 space per 10 seats, if library has separate reference facilities)
23 Hospitals	Minimum space for loading/unloading (m²): GFA Space ≤1000 200 ≤2000 300 ≤4000 400 ≤6000 500 For each extra 1000 m² GFA, add 50 m² space	Staff—1 space per doctor/ surgeon + 1 space per 3 other staff Outpatients/Visitors—1 space per 3 beds
24 Doctors' surgeries/ Clinics/Health centres	1 space per doctor (minimum)	Other staff—1 space per 2 other staff at the busiest time Patients—2 spaces per consulting room
25 Schools	30 m² (minimum) — Primary/Nursery schools 50 m² (minimum) — Secondary schools/VIth Form Colleges	Staff—3 spaces per 4 staff normally present Visitors—Primary/Nursery schools, 2 spaces; Secondary schools: (a) ≤1000 pupils, 4 spaces, (b) >1000 pupils, 8 spaces; VIth Form and Tertiary Colleges: 1 space per 10 students
26 Colleges of Further Education/ Government Retraining Centres	50 m² (minimum)	Staff—1 space per staff normally present Students/Visitors—1 space per 3 students normally present

* GFA = Gross floor area
† DFA = Dining floor area

Method 4

This procedure relies upon the results of an assessment of the traffic capacities of the arterial roads leading into the central area during the peak inbound traffic period. The number of parking spaces is determined by subtracting the amount of through traffic from this capacity figure and (after allowing a suitable factor for space utilization) adding an appropriate number of extra parking spaces to allow for vehicles entering the town centre after the peak period.

A variation[43] on this method simply suggests that the number of spaces (*P*) required in the central area is given by

$$P = 2CK/100$$

where *C* = capacity of the streets leading into the central area and *K* = percentage of feeder road capacity which is not through traffic.

Comment The basic principle described here is that which is now being implemented in very many European towns. A 'criticism' which could be levelled at this method is that, in this age of high motorization, it generally results in parking 'discrimination' against journey-to-work motorists.

Developing the plan—the map approach

When all the policy decisions and the estimates of needs have been made, the parking plan for a town centre must be devised. It is then that all the information previously gathered must be put together in a manner which will attempt to balance the mobility needs of the community with its social and economic capabilities.

There are many ways in which this can be tackled. Perhaps the most practical—certainly the most detailed—is that which has been devised by the Department of the Environment in Britain as a guide for local authorities. *Parking in Town Centres*[44] describes how a series of maps can be compiled as a convenient and effective way of illustrating the interacting factors which, in most towns, have a bearing on parking problems. A major beneficial quality of this method of preparing the parking plan is that it enables the measure of the parking problem as a whole to be taken very quickly, and with the minimum use of resources. As such its usage is most applicable to towns which have not carried out a transportation survey and, hence, do not have detailed data at their disposal.

The 'map' approach to the problem may be summarized in the following steps.

Step 1 Obtain an appraisal of the effects of parking on the town centre, on the convenience, pleasantness and efficiency with which the central and peripheral areas are used and function, together with the effect on accessibility and environment.

Step 2 Analyse the causes and pressures behind present parking usage and future parking demand.
Step 3 Decide how to make optimum use of the facilities available now and within the future period for which the plan is being devised.

In the following discussion on these steps, use is made of illustrative data from *Parking in Town Centres.* While the maps used in this discussion are shown here separately, in practice they would be prepared on transparencies so that they might be compared as overlays.

Map 1 is prepared in order to show the principal land use generators of parking demand, and the areas where competition for parking space is most intense. In Britain, the area used mainly by shopping parkers is best determined from a parking usage survey carried out on a Saturday afternoon, when offices are closed. The office parking areas are best determined in a similar way on an early closing day. Residential parking is best isolated by means of a usage survey carried out during the night. (These data will suggest where multiple use of public parking facilities might be appropriate, as well as the area where time-limit parking control is most needed.)

Maps 2 and 3 show the 'pinch-points' on the streets, i.e. where parking interferes with the movement of traffic and the operation of essential services. Data for Map 2 are generally available within the city's traffic section or, in the case of a small town, can be easily determined by inspection. Data for Map 3 are obtained in the course of the parking usage survey.

Map 4 is intended to isolate the areas where there is conflict between car parking and the environment.

Map 5 shows how the hinterland of the town centre can be subdivided into catchment areas, and the traffic from each related to the major approach road(s) carrying traffic to the central area. In this case, for example, the southern catchment area is seen to have the highest vehicle-ownership potential; since the two approach roads within this zone converge at the outskirts of the central area, it suggests that the road capacity of the final leg should be examined to see if it is less than the potential parking demand from the catchment area.

Map 6 shows the numbers (if available) of commuters and shoppers using cars which originate from each catchment area outside that outlined in the map. This is most useful in pointing out where peripheral (park-and-ride or park-and-walk) car parks might be located.

Maps 7 and 8 show the existing public transport services and suggest where car travel by the commuter is being encouraged through lack of adequate alternative means of journeying to work. (In so doing, these maps again emphasize how important it is that the town centre parking plan should always be developed in association with a plan for sustaining and improving public transport.)

Map 9 illustrates how the central area can be divided into convenient destination zones (on the basis of figures for employment and floor space) in order to give guidance as to where parking demands are likely to be high.

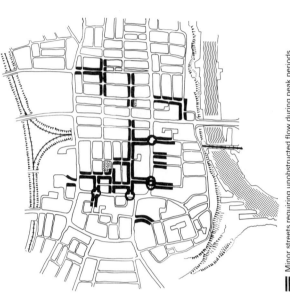

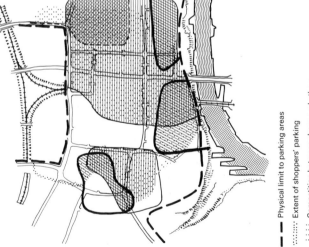

Minor streets requiring unobstructed flow during peak periods

Off-street car parks frequently causing congestion in adjoining streets and junctions

Traffic flow along major roads frequently interrupted by congestion originating in minor streets

Map 2

Physical limit to parking areas

Extent of shoppers' parking

Competition between shops and others

Extent of office parking

Competition between offices and others

Extent of commercial and industrial parking

Competition between commerce/industry and others

Competition between residents, and others

Map 1

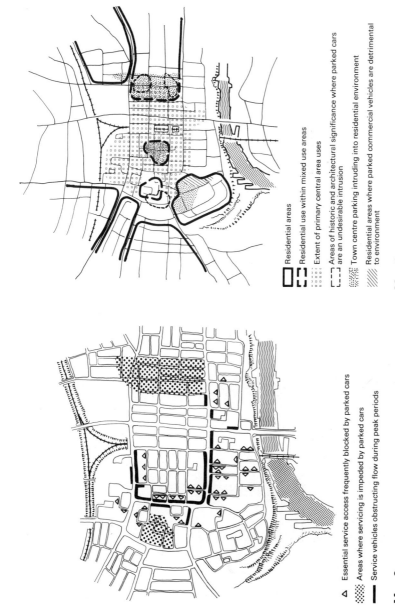

Residential areas

Residential use within mixed use areas

Extent of primary central area uses

Areas of historic and architectural significance where parked cars are an undesirable intrusion

Town centre parking intruding into residential environment

Residential areas where parked commercial vehicles are detrimental to environment

Map 4

△ Essential service access frequently blocked by parked cars

Areas where servicing is impeded by parked cars

Service vehicles obstructing flow during peak periods

Map 3

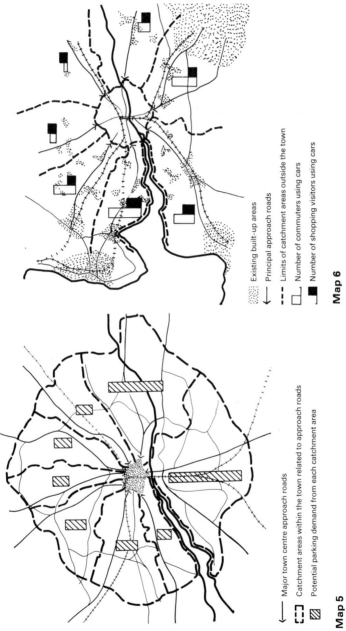

Major town centre approach roads

Catchment areas within the town related to approach roads

Potential parking demand from each catchment area

Map 5

Existing built-up areas

Principal approach roads

Limits of catchment areas outside the town

Number of commuters using cars

Number of shopping visitors using cars

Map 6

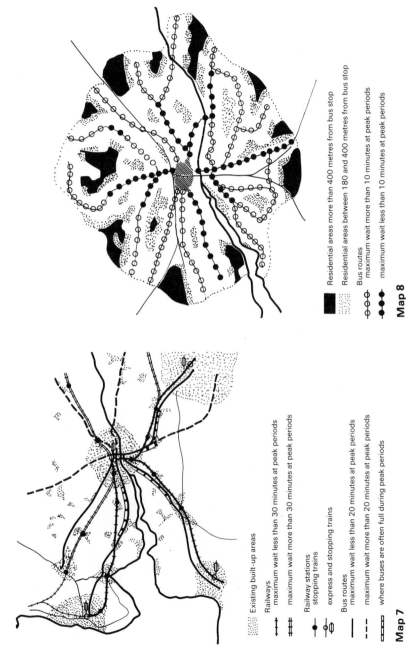

Existing built-up areas

Railways
maximum wait less than 30 minutes at peak periods
maximum wait more than 30 minutes at peak periods

Railway stations
stopping trains
express and stopping trains

Bus routes
maximum wait less than 20 minutes at peak periods
maximum wait more than 20 minutes at peak periods
where buses are often full during peak periods

Map 7

Residential areas more than 400 metres from bus stop
Residential areas between 180 and 400 metres from bus stop

Bus routes
maximum wait more than 10 minutes at peak periods
maximum wait less than 10 minutes at peak periods

Map 8

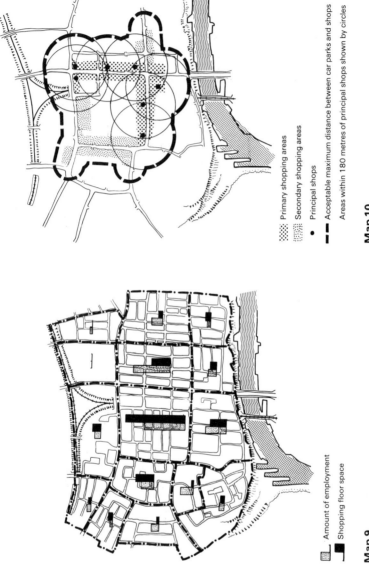

Primary shopping areas

Secondary shopping areas

• Principal shops

Acceptable maximum distance between car parks and shops

Areas within 180 metres of principal shops shown by circles

Map 10

Amount of employment

Shopping floor space

Map 9

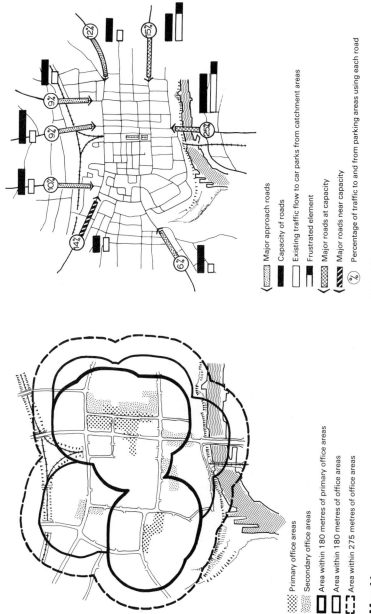

Map 12

- Major approach roads
- Capacity of roads
- Existing traffic flow to car parks from catchment areas
- Frustrated element
- Major roads at capacity
- Major roads near capacity
- (%) Percentage of traffic to and from parking areas using each road

Map 11

- Primary office areas
- Secondary office areas
- Area within 180 metres of primary office areas
- Area within 180 metres of office areas
- Area within 275 metres of office areas

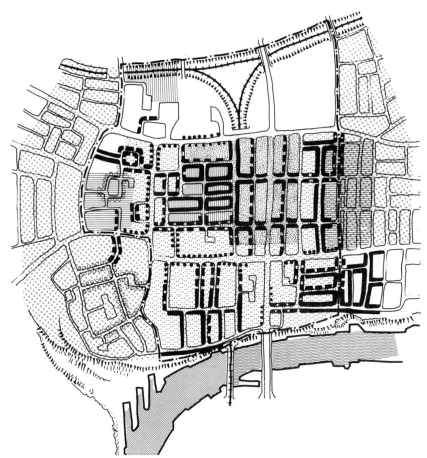

■■■ On-street parking

■□■ On-street parking except from 8.30 to 10 a.m. and from 4 to 6 p.m.

■ ■ On-street parking with special provision for operational vehicles

● ● ● Operational on-street parking only, but not from 8.30 to 10 a.m. or from 4 to 6 p.m.

○ ○ ○ Residents only overnight on-street parking

▦▦▦ Very-short-duration parking areas

⋯⋯ Short-duration parking areas

||||||||| Immediate off-street parking provision

— · — Desirable limit of town centre parking

⋇⋇⋇ Area of priority for residents parking (subject to experiment)

☰ Area requiring special attention to design of parking layout

Map 13

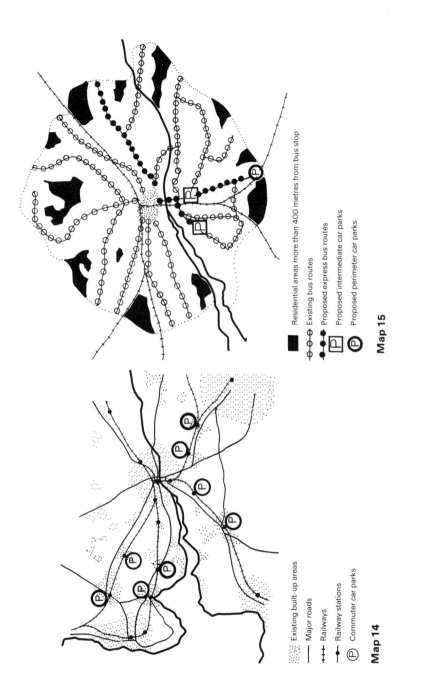

Residential areas more than 400 metres from bus stop

⊖–⊖ Existing bus routes

●–●–● Proposed express bus routes

●–●–● Proposed intermediate car parks

Ⓟ Proposed perimeter car parks

Map 15

Existing built-up areas

—— Major roads

+++ Railways

•+ Railway stations

Ⓟ Commuter car parks

Map 14

Map 10 shows how the shopping parker can be given preference in the development of the parking plan. Note that, in this example, the acceptable walking distance for shoppers is given as 180 m; in fact, the acceptable distance from parking place to destination can vary considerably from town to town. An indication of present practice in any given town can be gained by observation in the central area on a Saturday afternoon.

If insufficient parking spaces can be made available to shoppers within what is judged to be an acceptable walking distance, then it may be in the town's interests to consider the extent to which public transport facilities (e.g. special parker minibuses) might be used to offer a convenient alternative means of reaching the shops.

Map 11 is similar to Map 10 except that it deals with the problem of the journey-to-work parker. Again, ideally, provision should be made for these parkers within an acceptable walking distance. Where, however, this is not possible, consideration will need to be given to the development of park-and-ride services, e.g. by providing parking places at rail and/or bus public transport stations outside the central area.

Map 12 (which is an extension of the analysis initiated in Map 5) shows the existing traffic flows of the major approach roads, and indicates where there is spare capacity. (In this case, to simplify the illustration, the through traffic has been excluded.) Note that the south and south-east approach roads are filled to capacity during peak hours. Obviously, to provide additional parking space in the central area for long-stay cars approaching from the south could simply add to the congestion on the already crowded road (unless, of course, it is intended to carry out major roadworks which will significantly increase its capacity), and so careful consideration will need to be given as to whether or not this added congestion is acceptable.

The town centre parking map

All of the above analysis is designed to culminate in the development of the central area parking plan shown in Map 13. The sequence is to define as follows:

(1) existing off-street car parks, and land to be made available for off-street parking,

(2) where on-street parking is to be prohibited: (a) at access points to buildings, loading bays and service areas, (b) to obtain visibility at road intersections, pedestrian crossings, bus stops and taxi stands, (c) on one side of the street only,

(3) those streets where flow is essential at specific times, such as at peak periods—Map 2,

(4) where only front service access to buildings is possible, and where rear service access is available—Maps 2 and 3,

(5) where priority should be given to service vehicles and other operational parking—Map 3,

(6)   where overnight parking for residents only is desirable—Map 4,

(7)   those areas of historic and architectural significance where special attention needs to be given to the detailed design and layout of any parking provision—Map 4,

(8)   the areas where very-short-term car parks should be given preference, e.g. near post offices and banks,

(9)   the areas where shoppers and visitors to offices, etc., should be given preference—Maps 1, 9, 10 and 11,

(10)   the extent of the displacement of long-duration parkers into the adjoining areas, related to acceptable walking distances—Maps 10 and 11.

Essential developments in the hinterland of the town cannot be isolated from central area proposals. Thus Map 14 is also developed in order to show where special parking facilities are needed to serve commuters at railway stations, while Map 15 (which is derived from Maps 7 and 8) shows where express and improved bus services should be provided. Note that both outlying and intermediate car parks are proposed on the southern approach routes, and intermediate car parks only on the south-western approach route; these are intended both to alleviate congestion on the southern approach road (see Maps 5, 6 and 12) and to limit the overspill parking into the residential areas which adjoin the central area.

Note also (Map 15) that an express bus service which does not have a terminal car park is proposed for the eastern approach road; this would probably be best considered, initially at any rate, as an experimental service to see how many commuters will voluntarily use this facility in preference to their private cars. Similarly, the parking restrictions in the residential areas adjacent to the town centre (Map 13) should be considered as experimental, and subject to changes which will ensure the best solution in both traffic and environmental terms. If the experiments are successful, then similar provisions can be made at a later date in other parts of the town which would relate to the land use and traffic proposals both for the particular areas and for the town as a whole.

To conclude this discussion, it cannot be too strongly emphasized that *this town centre parking plan is in no sense static; modifications can be made from time to time to meet changes in traffic demand and land use, and to accord with other changes in the comprehensive town centre plan of which it forms a part.*

Locating town centre off-street parking facilities

The correct siting of a central area car park is critical to its success in attracting motorists. *If the parking facility is to thrive, then it must be sited conveniently for the customers it is intended to serve.*

The above principle is the governing one influencing the location of a large parking facility. It holds true irrespective of whether the car park is offered as an inducement (i.e. if it is free or only a nominal fee is charged) in an area where adequate parking space already exists, or whether its function is to meet a demand at economic charges in an area deficient in parking supply. It cannot be overemphasized that there is little point in taking a wedge of land, simply because it is lying there 'wasted', and putting a car park on it, if that piece of land happens to be in the wrong place.

Obviously there are other factors also which influence the decision as to where a public car park should be located, viz. the locations of the parking generators, the origins of the travellers, the adequacy of the access street(s), the nature of the topography, and the intended usage of the car park. Another important factor—albeit one which may not necessarily be taken into account on all occasions—is the effect of car park siting upon individual businesses.

Location and business

An indication of the importance of location in relation to an individual business may be gathered from the results of a study[45] carried out in Leeds. This showed that a department store (or complex of shops) which is situated on a main pedestrian route between a car park and the main shopping area of a town will be the first destination of a considerably greater proportion of shoppers from the parking facility than if the store (or complex) is at an equal or greater distance, but in an alternative direction, from the car park. In other words, a shopping facility which is close to a car park, and is between it and the centre of attraction of a central area, is most likely to benefit from the car park. It is equally obvious that shopping facilities which are not in such favourable positions may lose a certain amount of trade.

Nevertheless, it is not always practical for the planner/designer of a public parking facility to attempt to take its effect upon individual businesses into account. Indeed, to do so could be considered wrong (if he or she is a government employee), since the planner must think primarily in terms of its effects in relation to the community as a whole rather than upon any one particular member.

Parking generators

The location of the parking generators (i.e. the destinations of the travellers after parking) is probably the major criterion influencing the siting of a car park. Cost factors also prevail here in that the part of the town where land is low in cost is very often in run-down or slum areas at a distance from the main generators. The result is that, given a choice, parkers may not use the parking facility if they have to walk large distances to their destinations, particularly if their walk is through distressed areas.

No definitive statement can be made regarding how far people are willing to walk from an off-street car park. The controlling factors are, primarily, the extent to which there are other closer parking spaces, the size of the town, the degree of parking enforcement, the attractiveness of the destination(s), the cost of parking in the facility in question as compared with alternative ones, and the purpose of the trip. Table 4.9 summarizes American experience in respect of walking distances from parking place to destinations, classified according to trip purpose. On the basis of these data, the general principle can be stated that *the larger the town the further parkers are willing to walk, and worker-parkers will accept greater walking distances than will shopping parkers.*

Table 4.9 Average distances walked from parking place to destinations (m)

Population group of urban area (× 10³)	Trip purpose			
	Shopping	Personal business	Work	Other
10–25	61	61	82	58
25–50	85	73	122	64
50–100	107	88	125	79
100–250	143	119	152	104
250–500	174	137	204	116
500–1000	171	180	198	152

Origins of travellers

In many urban areas, it is usually found that great numbers of particular types of parkers, e.g. shoppers or workers, come from certain areas of the town. In such instances, the car park(s) intended for these users should be situated on the side of the central area which is towards their origins. Not only will these locations be more attractive to their potential patrons, but they will also have the added effect of causing less traffic congestion within the town centre, i.e. congestion caused by car trips through the central area and a walk back to the destination.

Access streets

A car park should always be considered as forming an integral part of the traffic plan for a town. This means, amongst other things, ensuring that the location is selected with a mind to the capability of the street system to handle safely and efficiently the entering and leaving vehicles. For example, the sudden disgorging of a car park at peak times may well create chaos either on the immediate access street(s) or at important nearby junctions as a result of, say, heavy movements.

The road access problem is one of the main reasons why it is often recommended (see, for example, reference 8) that off-street parking facilities should be limited to capacities no greater than, say, 400–500 cars, especially if they are to be used by commuters. Whilst some locations in large towns can, and do, justify larger car parks, it is generally held to be more desirable to have several car parks of lower capacity strategically located throughout the area, rather than a few car parks of high capacity. Alternatively, larger car parks may be used in certain areas if they are located so that their entrances and exits are dispersed onto lightly-trafficked (preferably one-way) streets leading to high-capacity routes.

Car parks should preferably be located at mid-block locations as compared with corner ones. No car park exit or entrance should be within at least 50 m of a street junction, due to the possibility of interference with or from the intersection control. Indeed, if the intersection is an important one, it may well be that the car park entrance or exit will be blocked by back-up traffic. Alternatively, if the car park is small, and the demand for parking is high, queueing cars may extend back onto the access road and into the junction.

Very many towns have, or are developing, ring roads about their town centres, the aims being: (a) to allow through traffic to bypass the central area, and (b) to minimize the intrusion by vehicles which must enter the town centre. In such instances, radial route car parks may logically be placed on the left-hand inbound side—this minimizes hold-ups in the morning—connecting the ring road to the core of the town centre. Indeed, if the ring road and/or radial road is a high-quality route, it may be possible to incorporate the car park in its design, e.g. under an elevated intersection structure.

Topography

The nature of the topography can influence the location of a car park. For example, a site on a steep hill has the obvious disadvantage of requiring extra walking effort from its customers—and thus it may be shunned by, in particular, older and/or heavier people who will not like walking uphill, especially from a shopping area when carrying packages, etc. However—and this applies particularly to towns whose residents are used to walking along grades—a sloping site may be very appropriate for a parking garage since it may allow direct access at different levels without the need for ramps.

Intended usage

A car park which is convenient to, for example, the shopping part of a central area may enjoy an excellent turnover during the daytime hours, but be practically empty in the evenings. It may be that, in certain circumstances, a little foresight could result in it being used on a day *and* evening basis. For example, if it is located so as to be convenient not only to the shopping area, but also to other diverse forms of traffic generators, such as cinemas, theatres, large hotels and restaurants, then there is every likelihood that its usage—and its income—will be increased because of its appeal to evening visitors to the central area.

Peripheral car parks

A particular example of an off-street parking facility is the peripheral car park. Nowadays, it is possible to define a number of types of peripheral parking developments. These may be loosely differentiated as: park-and-walk, park-and-ride, and kiss-and-ride. Before discussing these, however, it is useful to consider the entire concept of peripheral car parks.

Peripheral parking is not at all a modern-day phenomenon. Its practical origins may be associated with the development of railway travel in the mid-late 19th century, when it was not uncommon for outlying-living travellers to leave their horse-drawn carriages at a suburban railway station and complete their journey into town by train; nowadays, of course, it is common practice for motorists to leave their cars at railway stations and complete their trips by train. What is often forgotten, however, is that there are hundreds of bus stops in the public transport network in any large town, and most of these are located in areas where there is ample opportunity for all-day kerb parking; hence, it is now regular practice for many drivers, of their own accord, to drive to convenient bus stops and take the regular bus service for the remainder of the trip into the central area.

The idea of *deliberately* attempting to persuade motorists to park their cars at preselected locations and then to use another form of transport to travel to the central area (and back again) is, however, of much more recent origin. Its practical beginnings can be traced to the USA in the mid-1940s, and since then its usage has spread throughout North America and Europe[34].

It is appropriate to comment at this time upon what is meant by a 'successful' peripheral parking operation, for this is something which historically has tended to distort its usefulness. From the point of view of the car park operator, a successful system is generally one which not only pays its way but, if possible, also makes a financial profit. In contrast, the highway planner and engineer may well consider the operation successful if it takes significant numbers of vehicles off the roads leading into the town centre and helps to reduce traffic congestion and parking problems—even though it may not be self-supporting in the process. *If there is one lesson to be learned from previous experiments in this area, it is that any authority which is considering a peripheral parking project should have a clear view of what it means by a 'successful' system before embarking on any experimentation.*

Park-and-walk

This relates to the concept whereby the motorist completes the greater part of the journey to the city centre by car, parks it in a facility just outside the central area, and then walks (or perhaps takes a bus) to a destination which is a relatively short distance away.

Park-and-walk is primarily aimed at keeping the journey-to-work motor car out of the city centre. This concept, which may be applied to any size of town, is now in use in very many British urban areas and, properly implemented, is likely to prove very successful in all but the largest towns. Proper implementation in any given urban area requires: (a) that the local authority be able to exercise strong control over parking both *within* the central area and for a reasonable distance *outside* its periphery, and (b) that the scheme be initiated as part of an overall transport plan for the urban area.

The key to successful operation is undoubtedly the first of the above two criteria. If the local authority does not have a firm grip on parking within the central area, then motorists will not use the designated outlying facilities. If, however, firm control is exercised, say through a pricing mechanism which makes it completely uneconomic for the commuter to leave his or her car within the town centre, then this type of parker will be forced outside its periphery. If parking control is exercised over the roads immediately outside the periphery, then the commuters will have no option but to use the peripheral parks—in effect, they become captive users of the car parks.

There is no technical reason why the park-and-walk motorist should not be charged for using peripheral car parks—provided that the fee imposed is significantly less than that charged within the town centre. It is generally found that motorists reasonably willingly accept such a charge. The income derived in this way may be used to pay for the operation of the car parks (which will usually be owned by the municipality).

Park-and-ride

This refers to the concept whereby motorists drive their cars to long-stay car parks which are located well away from the town centre, usually in suburbia, and then travel by public transport to their destinations.

Park-and-ride differs from park-and-walk not only in the fact that the car park is much further away from the town centre, but also in that its success is much more dependent upon the *voluntary* cooperation of the motoring public. While generally aimed at the commuter, park-and-ride schemes can also (in very big cities) appeal to long-stay shoppers or business visitors.

In theory, park-and-ride schemes are excellent traffic planning measures. In practice, the degree of excellence achieved is dependent upon how they are used, and upon the extent to which the commitments implicit in their usage are understood. Proper usage generally infers that a considerable advantage is to be gained by the user of the scheme, in the form of, say, substantial savings in cost and/or travel time. Proper understanding of the commitments involved means, for example, realizing that a 'successful' park-and-ride scheme may not necessarily be a financially viable one, and thus its operator(s) may have to be subsidized to ensure its continued service.

Criteria for success

The successful operation of a park-and-ride scheme is dependent upon many, if not all, of the following features.

The central area being serviced should normally be surrounded by an urbanized area of large population.

From a design aspect, the car park should be as far as possible from the town centre so as to remove the cars from the road; however, if the town is not a large one, then the interchange located well away from the central area may not be able to generate sufficient traffic to justify itself. For obvious reasons, therefore, the likelihood of a park-and-ride system being successful increases the greater the population of the catchment area of the town.

An early investigation of park-and-ride practices in the USA[46] found that of 19 cities that had established park-and-ride (bus) schemes in the 1940s, nine had abandoned them. Only two of the cities with abandoned schemes had populations above 500 000; of the cities retaining park-and-ride schemes, only one had a population of under 500 000.

Another compilation of results by a prominent firm of international consulting engineers[39] suggests that park-and-ride schemes are most beneficial in urban regions which exceed 2m population. Its recommendations regarding the extent to which these schemes can usefully substitute for central area parking are given in Table 4.10.

Table 4.10 Estimated outlying parking requirements along motorways or rapid transport routes into urban areas in the USA

Population of urbanized area ($\times 10^6$)	Range of outlying parking spaces as a percentage of central area spaces
0.5	10–20
1.0	15–25
2.0	20–30
5.0	25–35

The relationship between the peripheral car park and the town centre must be such that the total travel time by car and public transport is not significantly more than that by car alone.

It must always be remembered that the traveller's decision to park-and-ride is primarily determined by the 'trade-off' relative to the inconvenience and possible lost time associated with fringe parking, as compared with the high parking costs and congested traffic conditions experienced in and on the way to the central area. From a travel time aspect, the trip will always be slower if the motorist transfers to a bus, even if it be an express bus—unless (as may happen) the bus travels on a separate lane directly to the town centre. If, as is common practice, the buses become part of the general traffic stream then, *at best* (i.e. if the buses are non-stop), only the time required to park and transfer is lost. Furthermore, the car park should be easily accessible by car and located within, say, 100 metres of, and have direct access to, a direct route into the central area; within the town centre, the main bus stop(s) should be located close to the main destination points, so that the total door-to-door travel time is minimized.

It is appropriate to comment here upon the desirable frequency of service of the public transport undertaking, as this is also a major factor influencing the decision to park-and-ride, i.e. people are willing to 'waste' time on a bus but not at a bus stop. (One widely used rule-of-thumb is that the effect of a waiting time of one minute at a bus stop is roughly equivalent—in terms of influence on public transport usage—to three minutes on a bus.) Good service frequency may mean, for example, that the park-and-ride scheme may require bus headways of not more than about five minutes during peak demand periods. If the service is discontinued during off-peak hours, early returning users will have to utilize either the local bus services or else a special bus service laid on, say, every hour. Rail park-and-ride services probably can be operated at lower frequencies because of their generally faster travel times and greater arrival time reliability, as well as convenience.

The parking fee plus the two-way public transport fee must be appreciably less than the cost of parking in and about the town centre.

For obvious reasons, the cost of the round trip discernible to the park-and-rider must be significantly less than the cost of parking in or adjacent to the town centre. American experience[47] suggests that free parking is a necessity for the successful operation of a park-and-ride bus scheme, as otherwise motorists will: (a) park in the surrounding streets and walk into the car park, or (b) bypass the special car park entirely and drive to a regular bus stop further in town where not only is free kerbside parking permitted, but a lower fare is charged on the bus. If, however, the park-and-ride operation involves a public transport vehicle on a reserved right-of-way (rail or bus), then there is a greater likelihood of a small parking fee being acceptable.

The car park must be properly sited.

To attract users, the park-and-ride car park should be well sited with regard to the origins of the potential parkers at which it is aimed, the highway network, and the public transport route. Care should be taken in this respect, however, to ensure that the car park is not too close to large local daytime generators of parking demand. The danger here is that the spaces designated for the park-and-ride patrons may be utilized by the local users (if there is a local shortage of parking) or vice versa (if the demand for park-and-ride is in excess of that provided). Nevertheless, experience has shown that outlying car parks used in bus schemes are best located on paved land which is already used for public parking purposes. Ideal in this respect are parking areas at stadia, auditoria, shopping centres, churches, etc.; not only are the peak activities at these locations likely to occur during evenings or weekends, but they are also generally well serviced by good access roads.

The car park must be properly designed and operated. Continuing and ample publicity must also be given to the scheme.

Proper design implies that the car park should be clearly arranged, and sufficient entry and exit capacities provided to cater for the peak demands. The walking distance from the parked car to the bus boarding area should be short and, preferably, covered; this could mean that very large peripheral car parks might have to be constructed as multistorey garages.

Proper operation implies that the parking facility be a safe place in which to leave the car all day—and in certain neighbourhoods this may well require the appointment of a car park supervisor. Good operation also requires an assurance to the motorist of a place to park, since excessive space hunting will discourage potential users. Substantial cover should be provided at the public transport waiting area so that patrons are protected from inclement weather while waiting for, and boarding, the public transport vehicles.

The formula for the success of a park-and-ride service requires one further major ingredient. It is that the system be given continuing and ample publicity so that the commuter, in particular, is made fully aware of the advantages associated with the service.

Kiss-and-ride

Kiss-and-ride (which refers to the practice whereby, for example, a wife drives her husband to/from a bus stop) is the most frequently forgotten part of peripheral parking programmes—even though it can be a major feature of a properly designed *park-and-ride* scheme. It poses its most severe problems where an express bus (or rail) service is the mode of transport, and buses (or trains) operate with very short headways. The dropping of passengers in the morning need not be a serious problem; this arises in the evening when waiting cars parked in the surrounding streets or on the internal roadways of the car park become a cause of serious congestion.

Experience has shown that it is practically impossible to regulate this form of congestion by legal means alone. (What police officer or traffic warden is going to risk continually giving tickets to bevies of harassed housewives at the wheels of cars full of starry-eyed or rampaging children (as the case may be) while they wait for the breadwinner to return from work?) A much better solution is to provide adequate numbers of short-term parking spaces for these vehicles within the park-and-ride car park.

How many spaces?

Inevitably, the time comes when the decision has to be taken as to how many places are provided at any given peripheral car park.

The total number of *park-and-walk* places provided will be to a large extent determined by the policy adopted by the city council with regard to parking within the central area—and this, in turn, will be primarily affected by the size and composition of the urban area, the availability of existing parking, and the capacity of the road system serving the central area. The number of spaces provided at any given car park will then be dependent upon the total number of parking facilities, the demand created close to each facility, and the interaction between the car park and the traffic on the road(s) immediately servicing it.

There is no simple way of determining the number of spaces at any given outlying *park-and-ride* terminal. Obviously this depends upon the success of the project, and upon the nature and size of the area serviced by the facility. However, an indication of what might be required can be gained by looking at the needs created by a particular service interval.

Let it be assumed, for example, that a commuter park-and-ride service is to be serviced by double-decker express buses (each capable of carrying 76 passengers) operating at a service interval of five minutes during the peak hour. It is anticipated that the demand for this service will be such that every seat will be occupied when each bus leaves the car park. Then the maximum number of spaces needed is

$$76 \times \frac{60}{5} \times \frac{1}{1.15} = 793$$

assuming that the average car occupancy is (typically) 1.15. If, as is likely, significant percentages of the patrons are kiss-and-ride passengers or walk-in passengers, then, of course, the number of long-term spaces would be considerably reduced.

To cater for the *kiss-and-ride* parkers at car parks, it will usually be necessary to provide special separate short-term spaces where cars can drop off passengers or await their arrival. The number of pick-up spaces will normally be much greater than the number of drop-off spaces, simply because passengers can be disembarked and the car driven away in approximately one minute, whereas an awaiting vehicle may stay between five and fifteen minutes before the arrival of the public transport vehicle.

Functional design of off-street car parks

Off-street parking facilities can be either surface parks or garages. Whichever is used in a particular situation usually depends upon the value of the land occupied by the facility and the intensity of usage anticipated. High-cost land frequently justifies a multistorey car park, i.e. the land cost per vehicle may be less than for a surface park. Conversely, expensive garage construction is rarely justified (except for environmental reasons) on low-cost land where equal capacity can be obtained on a greater land area at a lower cost per space.

Design car

The first step in the functional design process is to determine the car or cars to be used for design purposes. However, it should also be appreciated that there is no obvious vehicle in the car fleet that can be selected for this purpose. Car sizes and capabilities vary tremendously from country to country, and from car manufacturer to car manufacturer within a country. Furthermore, consumer preferences can change significantly for reasons as varied as vehicle shape or changes in family size or reactions to the 'energy crisis'. Once on the road, cars last for a long while, i.e. the average age of the British car fleet is just over six years, whilst the average longevity is just short of twelve years[48].

To cope with these tremendous variations, it is usual practice to design for at least one hypothetical standard design car which reflects the space requirements for the majority of car models. This design car need not have the dimensions of any particular 'real' car model. In fact, good practice suggests that it is desirable to design particular parking facilities for at least two design cars, i.e. the standard car and (for occasional use) the large car. Table 4.11 and Fig. 4.5 show some critical dimensions for new cars, published in Britain in 1971. Since then, car dimensions have changed; however, investigations from time to time have indicated that these are insufficient to justify altering the design standards of car parks[50]. It might also be noted that different countries use different design cars, e.g. the American design car has a length, width, and turning (wall) radius of 5.72, 2.03, and 7.47 m, respectively.

Note that this table does not include vehicle mass, as this is not a critical criterion in functional car park design. Included are vehicle

Table 4.11   Some design-car sizes[49]

Dimension	Symbol (as Fig. 4.5)	Value of dimension (m)		
		Small car	Standard car	Large car
Length	*a*	4.100	4.750	6.100
Width	*d*	1.600	1.800	2.000
Height	—	1.500	1.700	2.200*
Door opening clearance				
normal	*p*	0.500	0.500	†
minimum	*p*	0.400	0.400	†
Wheelbase (worst cases)	*m*	†	0.900	†
	c	†	2.900	3.700
	f	†	1.100	†
Nearside clearance				
normal	*n*	0.200	0.200	†
minimum (piers, etc.)	*n*	0.150	0.150	†
Turning radius				
kerb	*r*	5.500	6.500	7.500
wall	r_1	†.	7.000	†
Ground clearance	—	0.130	0.100	0.050*
New registrations falling within these limits (approximate percentage)		50	95	100

* Specialized vehicles
† Generalized dimensions not available

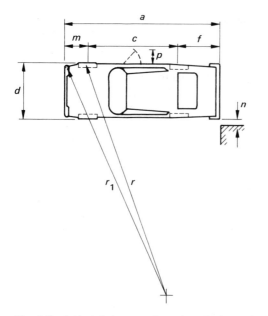

Fig. 4.5   Critical design-car dimensions (to be read with Table 4.11)

turning capabilities, which influence both aisle width and corner radii; to minimize these turning radii, it is usual to assume that cars travel at very slow speeds in the parking facility. Ground clearance and end clearance can be very important in relation to ramp design and entrances and exits, if slope (grade) changes which cause a vehicle's ends or centre to 'ground' are to be avoided.

Surface parks

Surface car parks are the most commonly used type of general-purpose parking facility, primarily because of the ease with which they can be quickly and economically constructed when suitable sites become available. The following is a brief description of the basic physical features underlying the functional design of a surface car park. Since this type of car park can be similar to one floor of a parking garage, many of the considerations discussed here will also apply to the design of a driver-parking garage facility.

The effective operation of an off-street car park is greatly influenced by its functional design. It is useful therefore to discuss the design of the facility in a step-by-step operational context.

Entry from the street system

In most instances, off-street car parks require driveways which cross the footpath, in order to link the car park entrances (and exits) with the street system. Thus cars entering (and leaving) the car park have the potential for direct conflict with pedestrians. The severity of this problem at entrances depends upon the volume and approach speed of the entering traffic, and upon the numbers of pedestrians on the footpath during the same period.

Key elements of driveway design include the following[51].

(1) *Width and radii*—in general, the greater the driveway width (measured at the throat limit when kerbs are constructed) and radii, the easier it is for motorists to access the car park quickly and efficiently. Particularly generous design is required at shopping centres and factory car parks, as these are generally susceptible to peak access (and egress) problems; in contrast, driveways in areas with high pedestrian volumes should probably be designed for lower vehicle entry (and exit) speeds.

(2) *Angle of entry*—by this is meant the minimum acute angle, measured from the edge of the carriageway, through which the entering vehicle has to turn. In line with what has been said at (1) above, an angle of about 75 degrees is appropriate for entrance driveways at locations of high pedestrian activity, and 45 degrees at other locations.

(3) *Directional flow*—the design will obviously be affected by whether the driveway is to handle one- or two-way traffic movements.

(4) *The spacing from the nearest junction and between adjacent driveways*—see pp. 263–267.

The capacity design of the car park entrance is discussed later in the context of the exit design.

Search for a parking bay

Figure 4.6 illustrates some of the alternative traffic circulation movements used in car parks. As a general principle, it should be kept in mind that one-way circulation with vehicles driving directly into the parking bays (also known as 'stalls') will ensure a more efficient traffic flow with the minimum of conflict points. Moreover, the layout of the car park should seek to discourage vehicle speeding by having shorter rather than longer aisle lengths.

The most desirable internal circulation is one in which each potentially vacant parking bay within a small car park, or segment of a larger facility, must be passed once by an incoming patron seeking a space. Thus, for example, the aisle circulation pattern most commonly sought in conjunction with angle parking is a continuous system of alternating-direction one-way aisles. In practice, however, this ideal circulating system is not always possible and most car parks are therefore arranged so that a motorist has to circulate on a random basis until a vacant space is found.

Parking bays and aisles

Whilst developing the layout of a car park it is usual for the designer to work with bays, aisles and combinations of bays and aisles called 'bins' or 'modules'. From Fig. 4.7, it can be seen that modules may be composed of either one- or two-way aisles, with parking bays on either or both sides

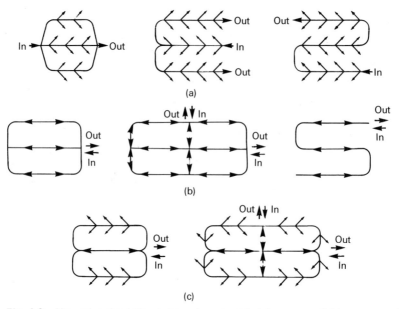

Fig. 4.6 Alternative circulation patterns in a car park: (a) one-way, (b) two-way and (c) combined one- and two-way

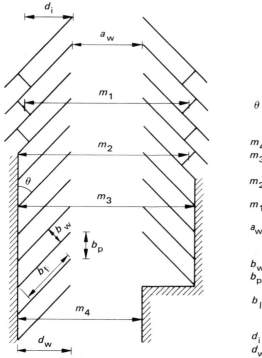

θ Parking angle

Wall-to-wall module
m_4 single-loaded aisle
m_3 double-loaded aisle

m_2 Wall-to-interlock module

m_1 Interlock-to-interlock module

a_w Aisle width

Bay width
b_w normal
b_p parallel to aisle

b_l Bay length

Bay depth
d_i to interlock
d_w to wall, perpendicular to aisle

Fig. 4.7 Basic dimensional elements of a functional parking design

of each aisle. A major factor influencing the width of a module is the boundary condition, i.e. whether there are walls on each side, parking bays on each side, or a wall on one side and a parking bay on the other. Aisles having parking bays on both sides are called 'double-loaded' aisles; those with bays on one side only are termed 'single-loaded' aisles. Single-loaded aisles are very inefficient in respect of space usage per vehicle and should be avoided wherever possible.

The access aisle width required to allow a car to park/unpark in one manoeuvre varies mainly with the angle of parking, and then with the bay width; it is obviously also related to bay length. Equations have been developed[52] which enable the aisle width to be theoretically derived for any particular parking arrangement; however, these should be used with care, as practical experience has shown that the assumptions used in the derivations do not always hold for actual parking manoeuvres.

All other factors being equal, 90-degree parking bays with aisles parallel to the long dimensions of the site, and 60-degree interlocking bays with one-way aisles, are generally considered to require the least amount of space per parking bay. It used to be considered that some groups of drivers (e.g. women) had difficulty in manoeuvring into 90-degree bays; however, experience would suggest that this concern has been exaggerated. Angle parking is often used where site dimensions do not allow an integral number of 90-degree bays. It is reported[53] that experiments carried out by the Transport and Road Research Laboratory have proved that bays placed at an angle of 70–75 degrees, with one-way traffic flow, make parking easier and quicker. A very flat angle layout is significantly less efficient in rectangular car parks as it tends to result in considerable space wastage.

Figure 4.8 shows some possible layouts associated with interlocking parking bays. The drive-through arrangement, whereby parkers are asked to drive into the opposing bay in the adjacent row, and to exit into the aisle next to the one used for entering (Fig. 4.8(b)) would normally be only used in large industrial commuter car parks, where the major concern is to expedite the peak period flow of cars into and out of the parking facility, and to prevent the unnecessary wastage of space, i.e. a more conventional arrangement could require a large reservoir space for the accumulation of cars delayed at exit points. The herringbone arrangement (Fig. 4.8(d))—which is suitable only for 45-degree parking— is often recommended on the grounds of being economical in space; in practice, it should be avoided as it is likely to result in significant damage to adjacent vehicles. Kerbs/wheel-stops are normally provided at the 'butt' end of the spaces shown in Figs 4.8(a) and 4.8(c) to ensure that vehicles do not encroach onto adjacent parking bays.

It is logical that as the parking angle becomes flatter, parking manoeuvres should require less aisle width. In practice, however, minimum aisle widths associated with particular parking angles vary slightly according to their source. Table 4.12 shows some aisle and bin (module) width recommendations in current use in Britain[50].

The bay width is affected by the clearance needed for motorists to get

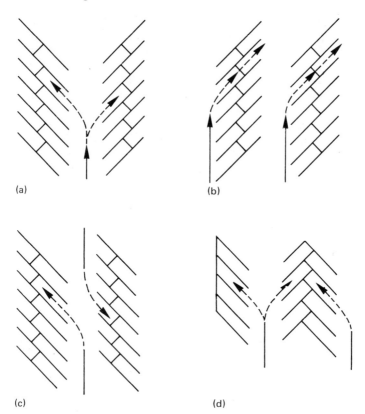

Fig. 4.8 Some interlocking parking arrangements: (a) one-way aisle, butt bays, 60° angle, (b) one-way aisle, drive-through bays, 60° angle, (c) two-way aisle, butt bays, 60° angle, and (d) one-way aisle, herringbone bay arrangement, 45° angle

in and out of their cars. This clearance is, in turn, dependent upon such factors as the trip purpose, e.g. whether or not shopping bundles will be carried, and the age and sex of the motorists. Widths used vary, with 2.300, 2.400, and 2.500 m being used for long-stay, general, and short-stay car parks, respectively. Narrower bays may be used in commuter car parks located on high-cost land or in attendant-parking facilities. Wider bays may be used for disabled persons; 3.600 m is the preferred minimum for wheel-chair users.

Table 4.12 Some aisle and bin widths (preferred minimum) used with 2.30 m wide by 4.75 m long parking bays

Parking angle (degrees)	Aisle width (m)	Bin width (m)
45 (one-way)	3.600	13.570
60	4.200	14.727
90	6.000	15.500
90 (two-way)	6.950	16.450

A parking bay should be designed to accommodate the overall length of nearly all cars expected to use the space. A bay length of 4.75 m is most commonly used to cater for the standard design car, although 5.0 m has also been used.

Pedestrian egress and access

Wherever possible, pedestrian entrances, exits and footpaths should be separated from the travelways used by cars, and located so as to lead naturally to such goals as shopping areas, factory entrances, ticket offices, etc. In practice, however, this is often not possible and pedestrians must share the access aisles with cars.

If formal pedestrian ways are possible, it is best if they are raised above the general parking area, as this allows people to scan the entire car park and get their bearings more easily. If walkways cannot be separated from the circulation ways, then they should be well defined by a change of surface material and/or pavement markings. Widths vary, but walkways about 2.4 m wide enable two people to walk abreast, and allow for car bumper overhang.

All well-designed car parks are well lit, well drained, kept clean and attractively landscaped. Good lighting is essential if night-time use is expected, e.g. it reduces loitering, vandalism and pilferage as well as assuring patrons regarding their personal safety. Similarly, good drainage, cleanliness and landscaping make the car park more attractive to use (as well as helping to maintain the values of adjacent properties).

Clear signposting and, in large car parks, the coding of aisles and bays by numbers and colours will help the pedestrian to find the way to/from the pedestrian exit. In large car parks also, this can be aided by dividing the mass of the 'carscape' into smaller units, and by providing identifiable landmark features such as trees, bushes and mounds which also improve the appearance of the facility.

Earth mounds planted with shrubs are a most effective form of subdivision, but satisfactory aesthetic results can also be obtained with planting boxes about 1 m high and constructed of brick, stone or concrete, whichever is most appropriate to the site and its surroundings; these should be stocked with plants which are in proportion to the size of the box and complement it in colour, tone and texture.

All trees and shrubs planted in urban car parks should be of a pollution-free variety. Whilst expert advice should always be obtained from the Parks Department of the local authority, one useful set of recommendations re trees and shrubs suitable for use in such car parks is available in the literature[53].

Car egress

As a basic principle, the vehicle pathways to the nearest exits (in large car parks) should be well signposted so that the motorist is in no doubt as to what route(s) to use to leave the facility.

Ideally, exits should be located away from entrances so that the potential for vehicle and pedestrian conflict—both inside and outside the car park—is minimized; this, however, is not always possible. As with entrances, exit design should take into account driveway width, turning radii, angle of approach in relation to the egress street, the direction of the traffic flow, and spacing from the next intersection on the street. Particular attention should be paid at exits towards ensuring that unobstructed sight lines are provided for exiting drivers and pedestrians (outside, on the footpath). Signs and carriageway markings should also be used to alert drivers and pedestrians to potential crossing conflicts. In certain instances, mirrors may need to be provided to allow pedestrians and motorists to observe conflict-point conditions which are obscured from straight-line vision. Although opposed by some designers, step-down kerbs at the footpath–driveway interface do have the merit of clearly warning the pedestrian that he/she is about to enter a conflict area—and thus may prevent a serious accident from occurring as a vehicle exits (or enters) the car park.

The interaction between the car park and its immediate access road has been examined in some detail[54], so it is possible to design the exit (and entrance) points so that the car park traffic can be made as compatible as possible with the road traffic. In relation to this, it should be kept in mind that there are three basic ways of collecting parking fees, viz.: pay-on-entry/free-exit, free-entry/pay-before-exit, and free-entry/pay-on-exit operation.

Pay-on-entry/free-exit operation requires the levying of a flat charge for parking, and is therefore only used at parking facilities which lend themselves to this method of charging, e.g. at special events, all-day parking, or car parks which may be unattended when the driver returns. A major disadvantage of this type of fee collection is that it requires large reservoir areas at the entrances to handle peak surges in car arrivals, as otherwise large queues will develop on adjacent streets.

Free-entry/pay-before-exit operation usually requires the driver to collect a ticket from a ticket dispenser when driving into the car park, and to return to the car via a central cashier location where the parking fee is paid for the time parked; the driver then enters the car and leaves the facility after (momentarily) presenting some form of receipt at the exit point. As no cash transactions take place at the exit, this form of fee collection results in the most efficient exit/entrance operations. Thus it is reported[52] that, compared with pay-at-exit cashiering, the exit lane capacity can be doubled or trebled by the use of central cashiering.

Free-entry/pay-on-exit operation is the most usual way of collecting fees at surface car parks (and multistorey parking garages) in and about central areas. With this type of operation, the driver collects a ticket from a dispenser upon entering, presents it to an attendant from the car at the time of departing from the car park, and then pays the requested fee. Whilst exit cashiering significantly reduces the reservoir requirement at the entrance, as compared with pay-on-entry cashiering, it normally requires reservoir space within the car park at the exit point where cars are most likely to queue during peak exit periods.

The following example of car park exit (and entrance) capacity design is concerned with a free-entry/pay-on-exit type of operation.

Example of car park dynamic capacity calculation
Assume that a 900-space multistorey car park is to be located so that its exits discharge into a 7 m wide one-way street carrying a peak flow of 200 vehicles per hour. The car park, which will stand on a rectangular site about 30 m wide by 120 m long, will be primarily used by commuters. The peak hour departure demand, during the evening, is estimated to be 800 cars per hour. An early decision was that exit gates of the lifting-barrier and pay-from-car type would be used. Determine the compatibility of the entrance and exit system with the service roads.

Solution
Step 1 Calculate the number of exit gates required.
Exit gates of the type specified can handle about 200 vehicles per hour. Thus four gates will be required to deal with the evening flow. If four exits are provided, no reservoir space should be needed at the end of the internal aisle; apart, of course, from the space needed for the fan-out from the aisle or ramp to the four gates, assuming that all four gates are placed together. If fewer than four gates are provided, occasions could arise when a queue might extend back into the aisle, thus preventing vehicles from unparking.
Let it be assumed that each exit lane is 3 m wide, and that a GIVE WAY sign will be provided at the junction with the main road.
Step 2 Determine whether the 'junction' between the access (main) street and each car park exit (minor) road is capable of handling the peak flows.
The query here is whether the maximum flow from the car park exit can enter the traffic stream on the main road without causing congestion. Assuming the junction to be a T-type, the problem resolves into determining the maximum flow (along the stem of the T) that can enter the main stream during the heaviest flow conditions on the main road, i.e. when the commuters are on their way home. This determination can be made with the aid of the following formula:

$$q_{max} = \frac{Q(1 - Q/S)}{e^{Q(0.0015 - 1/S)}[1 - e^{-Q/s}]}$$

where q_{max} = maximum possible flow on the exit road (car/h), Q = flow ($\neq 0$) in the nearside lane of the main road (veh/h), S = saturation flow of the nearside lane on the main road (veh/h), and s = saturation flow ($\geqslant 667$) on the exit road (car/h).
The saturation flow, S, on the nearside part of the main road is obtained from

$$S = 525w_m/N$$

where w_m = carriageway width (metres) and N = number of traffic lanes.

Actual exit road saturation-flow values measured along a 2.75 m wide exit into a 6.1 m road (kerb radius $= 1.22$ m, angle of entry $= 90$ degrees, gradient $= 0$, and estimated path radius $= 6.1$ m) are given in Table 4.13.

Table 4.13 Actual saturation-flow values measured at a car park exit[54]

Condition	Saturation flow (s) (car/h)	Reducing factor (k)
1 Ideal, exiting cars turning left, no other traffic	1250	1.00
2 Limited visibility (requires GIVE WAY sign), cars turning left, possibility of traffic on main road	760	0.60
3 Blind exit, cars turning left, possibility of traffic on main road	710	0.55
4 Ideal (as 1 above) but with WAIT sign and stop line	620	0.50

The saturation flow of a stream of vehicles following a curved path is less than for a straight section of similar width. If the exit lane from the car park is curved, an approximate representation of the saturation flow, s, can be obtained from

$$s = \frac{1850k}{1 + 100/r_m^3}$$

where $r_m =$ estimated radius of curvature of the path of cars (metres) and $k =$ reducing factor (see Table 4.13).

All of the above formulae are applicable to the situation where a single stream of left-turning cars enters a single lane of traffic on a two-way street, or right-turning cars enter a single lane of traffic on a one-way street. (Additional traffic in other lanes of the main road are ignored.)

For the problem considered here, the most critical situation occurs at the fourth exit, where the traffic flow for the main road will be $200 + (3 \times 200) = 800$ veh/h, and 200 veh/h on the exit lane. Thus

$$S = \frac{525 \times 7}{2} \simeq 1840 \text{ veh/h}$$

and

$$s = \frac{1850 \times 0.6}{1 + 100/6.1^3} \simeq 770 \text{ car/h}$$

Therefore

$$q_{max} = \frac{400(1 - 400/1840)}{e^{400(0.0015 - 1/1840)}[1 - e^{-400/770}]} \simeq 545 \text{ car/h}$$

This value of q_{max} considerably exceeds the gate capacity (200 car/h), so it appears (from a local congestion aspect) that there is no reason why the car park exits should not be located as intended.

Step 3 Check the size of the exit reservoir between the exit gate and the main road.

Since there will be fairly heavy traffic on the main road, storage space will be needed on the exit road, after the exit barriers, where cars can wait until suitable gaps appear in the main road traffic stream. If sufficient space is not provided, the exit gate may not be able to achieve its maximum output of 200 car/h. However, the calculated value for q_{max} is so much greater than 200 car/h that experience would suggest that a reservoir of 2–3 car lengths will be adequate at each gate. (A detailed procedure for estimating the reservoir sizes required at car park exits is readily available in the literature[55].)

Step 4 Determine the number of entrance gates.

In this instance, let it be assumed that cars will enter the car park via entrance gates of the lifting-barrier take-ticket type which will be installed on slightly curved approaches (to suit the site). Table 4.14 gives the maximum flows measured at different entrances in a simulation carried out by the Transport and Road Research Laboratory. In this instance, therefore, a capacity value of 450 car/h is appropriate, which suggests that at least two entrances are needed. (These entrances should be located such that they are upstream of the exits on a one-way street.)

Table 4.14 Suggested capacities for car park entrances[54]

Condition	Maximum flow (car/h)
Tight left-hand turn and take ticket	350–450+
Straight-approach and take ticket	650–670
Tight left-hand turn only (no ticket taken)	575–970

Car park capacity

Prior to the 1970s, car park capacity was simply defined as the number of parking bays within a system, i.e. *static capacity*. Then research[54] determined that a parking bin had *dynamic capacities* as well as a static capacity. Thus *dynamic inflow capacity* was defined as the maximum rate (cars/h) that a system can park cars, and *dynamic outflow capacity* as the maximum rate that cars can unpark from it. The capacity design procedure at a car park with tidal-flow traffic conditions then became one of estimating the traffic demands and the dynamic capacities of the critical parts of the system (e.g. entrances, exits, aisles and bays, and ramps), and then bringing the two into balance by adjusting design details or the layout.

Recent work[56] examined the dynamic capacity concept further, and defined *turnover capacity* as the maximum rate at which cars can be unparked when others are parking at the same rate in a bin. This capacity can be explained by considering a 300-bay facility that is operating such that each vacated bay is almost immediately re-occupied by a new car. If the average car duration is 50 minutes, the flow is $300 \times 60/50 = 360$ cars/h; the turnover in this instance is a function of static capacity and duration. If the average duration is reduced to 40 minutes, it might be expected that the flow would rise to 450 cars/h; however, due to some physical characteristic (e.g. angle of parking), the flow only rises to 400 cars/h. In this instance, the turnover capacity would be defined as 400 cars/h; this is not a function of static capacity or duration, but rather of the parking angle.

Reference 56 describes how turnover capacity can be determined and used in car park design.

Parking garages

There are so many types of multistorey car parks that it is difficult to classify them. Basically, however, any parking garage can be described by a combination of three groupings, viz.:

(1)   by vertical location, e.g. whether the parking garage is underground or above ground,
(2)   by method of interfloor vehicle travel, e.g. whether the cars move up and down via ramps/sloping floors or mechanical lifts,
(3)   by the person who parks the car, e.g. whether the cars are attendant-parked or patron-parked.

In addition, there are the 'non-permanent' demountable steel or concrete panel (on steel frame) garages, some of which are operationally most efficient. These lend themselves to interim usage on land intended for other longer-term purposes, leased land, in air rights space, and over existing high-volume surface car parks. Most of the demountables are generally regarded as being somewhat lacking in terms of their contribution to the aesthetic environment. Furthermore, they tend to have a life span beyond their original purpose, i.e. once constructed they are 'politically' difficult to remove and relocate.

The major justification for all parking garages relates to the high cost of land—which is why most facilities of this nature are to be found in and about the central areas of towns, i.e. in the case of most town centre off-street surface car parks there comes a time when the purchase of land to provide for additional parking becomes uneconomical, and it is cheaper to provide parking vertically. At other non-central-area locations, however, the geography of the land, e.g. hillside sites, may make it more suitable to construct vertical parking structures rather than horizontal ones. Alternatively, the need for integration with other non-parking facilities, e.g. shopping centres, in the course of their construction, may make it appropriate to consider the building of a vertical car park earlier than might otherwise be the case.

Underground garages
The principal benefits associated with underground car parks are the conservation of surface land, the preservation of visual amenities, and the capability of greater integration—and the sharing of construction costs—

with non-parking (above-ground) usages such as shopping centres or office buildings. The major disadvantage is that they present special construction problems that typically make them as much as twice as expensive as parking garages that are built above ground[51].

Major contributors to the extra cost of underground car parks include extensive excavation (magnified if rock is encountered), the need for a load-bearing 'roof', and the relocation of public utilities. The greatest single design problem that has to be overcome is that of water; the design must ensure that water cannot enter from above by flooding or seepage, or from below or through the walls as a result of excess hydrostatic pressure. Other significant cost contributors, as compared with above-ground garages, are the need for daytime illumination, special fire-fighting provisions, and artificial ventilation; these latter two requirements alone can add 20–30 per cent onto basic construction costs.

Underground car parks have been built with a variety of vehicle access/egress arrangements. Obviously, also, they may operate under attendant-parking or patron-parking arrangements.

Ramped/sloping-floor garages

A variety of interfloor ramp systems can be used to allow vehicles to travel between adjacent floors in multistorey underground and/or above-ground parking garages. Figure 4.9 shows (schematically) examples of some well-known types.

Clearway ramp systems provide interfloor travel paths that are entirely separated from potentially conflicting parking/unparking movements. The ramps can be straight (Fig. 4.9(a)) or curved and lead to normal or split-level floors, internal (Fig. 4.9(b)) or external (Fig. 4.9(c)) to the building, and one-way (Fig. 4.9(a)) or two-way (Figs 4.9(b) and 4.9(c)) operated. Concentrically-constructed spiral ramps are most commonly used to provide a one-way clearway exit for 'down' traffic from the various parking floors. Vehicles travelling on these one-way clearway ramp systems can move either clockwise or anticlockwise; a clockwise design is usually preferred in Britain (and in other countries where drivers normally sit on the right-hand side of the vehicle) as it places the driver on the inside of turns, thereby enabling better and safer car handling. When two-way traffic is handled on a single spiral, the outer lane is normally used for up-movements, since it has a larger radius and smaller grade; these up-movements are usually clockwise, and down-movements anticlockwise.

Clearway ramped systems are generally considered to provide the safest movement with the least delay and (except for sloping-floor designs) are preferred for patron-parking designs[52]. However, these systems are seldom feasible for small garage sites.

Adjacent-parking ramped garages are those in which the aisles between parked cars are used as links between the ramps. The number of parking bays adjacent to the ramp may range from the full capacity ('sloping-floor' garages) to only a small number. The ramps used with these garages may be one-way (Fig. 4.9(d)) or two-way (Fig. 4.9(e)) operated, and lead to full-size (normal) parking floors. These designs usually have the ramp

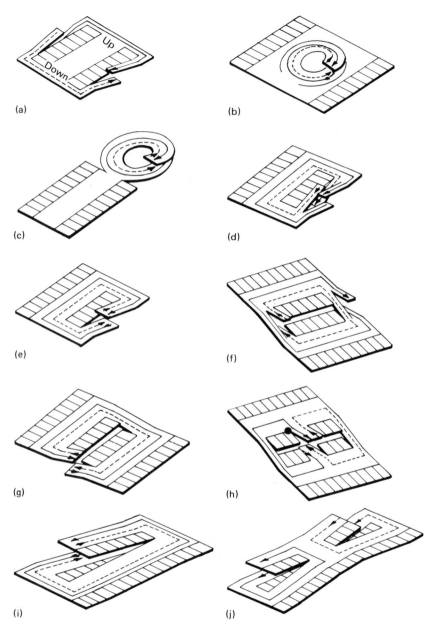

Fig. 4.9 Diagrammatic examples of ramped/sloping-floor garage systems

well(s) along the garage's longer-side dimension, as this makes it easier to satisfy ramp grade criteria.

Alternatively, the ramps may lead to split-level floors. With this type of design, the garage is essentially constructed in two sections, with the floors in one section staggered vertically by a half-storey from those in the other section. Short, straight ramps connect the half-floors; these are sloped in opposite directions and separated sufficiently to allow a 180-degree turn between parallel ramps. Again the ramp operation may be either one-way with two-way aisle movements (Fig. 4.9(f)), or two-way with two-way aisle movements (Fig. 4.9(g)) or one-way aisle movements (Fig. 4.9(h)). The division between the split-level garage halves may be at right-angles to the street, or parallel to it; in the latter case, either the front or the back half may be the elevated section.

Split-level garages can also be designed so that the floors overlap, i.e. so that the bonnets of the cars below are tucked in under the tails of the vehicles above. Whilst this arrangement does increase the space efficiency and renders usable otherwise unusable narrow sites, it has the potential disadvantage of making the design of the garage inflexible should future car shapes change to a 'forward drive' or van-type silhouette.

A particularly interesting example of an adjacent-parking ramped garage is the sloping-floor car park. There are no separate areas set aside for ramps with this type of design; rather the floors on which the cars are parked are themselves sloped (usually not more than 4 per cent) so that the aisle between two rows of parked cars forms a gradual ramp leading onto the next floor. Whilst the cross-aisles may be sloped or level, it is generally considered that they should be level in the larger parking garages. Figure 4.9(i) is an example of a basic sloping-floor system with level cross-aisles, under two-way circulation. If two sloping-floor units are placed end-to-end, it is possible for the garage to have one-way traffic flow operation (Fig. 4.9(j)); this has the advantage of allowing angle parking.

Since there are no interconnecting ramps, a sloping-floor garage provides more parking bays than are provided by a comparable split-level design[57]. However, as the cars are parked across the slope of the floor, the floor gradient must be restricted, and hence these garages tend to suit larger sites which permit long 'ramp' lengths. The natural slope of the land is unlikely to take up all the waste space at ground level created by the long, shallow ramp, and so it is usually necessary to have a partially sunk basement section of parking to get the full use of the site. Very often, also, the floors of these garages are linked with clearway spiral exit ramps in order to minimize traffic congestion at the lower levels.

A disadvantage of a sloping-floor garage can be its relative unacceptability on aesthetic grounds, i.e. the long, sloping floors give an undesirable effect in external elevation. To overcome this, a garage construction has been developed which uses warped (hyperbolic paraboloid) floors to give a horizontal external line to the building, while containing the internal circulation features of a split-level facility in combination with the efficiency of a sloping-floor parking system.

Mechanical garages

With mechanical garages, lifts take the place of ramps, the headroom required is for cars rather than people—which enables more cars to be parked per unit height of building—and automation is employed either wholly or partly. Since mechanical garages make the most effective use of available space, they have therefore tended to be developed in areas of high-cost land and limited space.

One of the classic examples of the earlier mechanical garages is the 'Park-O-Mat' constructed in Washington DC. This 17-storey building was constructed to hold 68 cars on a frontage width of only 7.6 m. The design concept was that the parking of cars would be carried out by a single attendant who never moved from the office; thus the garage was designed with two lifts, each serving parking bays at the back and front of the lift shaft on each floor. When a customer arrives at the garage, the car is left locked in front of the lift, with the brakes released. The attendant, by push-button control, causes a 'dolly' to extend from the lift platform which moves the car onto the lift platform, and upon arrival at the selected floor, the lift dolly moves the car backwards or forwards into an empty bay.

Electronic controls and faster, more reliable lifts have enabled significant improvements in mechanical garages in the past decade. Lift systems are now used that can move laterally as well as vertically, so that each tier of parking bays can be reached by more than one lift and the problems of lift breakdown are therefore minimized. The newer garages are also faster than their predecessors and a single vehicle can now be stored or delivered in less than one minute; when one car is stored and another collected on the same trip, the round trip can be completed in 90 seconds.

Notwithstanding the very significant developments in mechanical parking garage design and operation, relatively few of these facilities have been constructed in recent years, in comparison with ramp garages. The high initial costs, teething problems with new equipment, aesthetic considerations, environmental problems resulting from vehicles queueing in adjacent streets, high mechanical maintenance costs, and the need for attendant-operation have all tended to limit the construction of mechanical garages. Nevertheless, it must be pointed out that significant numbers have been built abroad and, as far as can be determined, they are operating both efficiently and profitably.

Attendant-parking ramped garages

With attendant-parking surface and garage car parks, the motorist delivers the car to the entrance (or reservoir adjacent to the entrance) of the parking facility, leaves the key in the ignition, and an attendant then drives the car to a parking bay within the facility. The reverse operation takes place when the motorist returns to retrieve the vehicle.

The major advantage of an attendant-parking operation is the reduced space required, as compared with the space needs of a patron-parking facility, i.e. where the parking bay, aisle widths, ramp slopes, and turning radii must all be designed to cater for patrons with widely-varying driving skills. The design requirements of an attendant-parking facility can be significantly lower—American experience is that attendants are able to park up to 25 per cent more cars in a given space, as compared with driver-parkers[52].

In the case of ramp garages, the headroom can also be significantly lower with attendant-parking. Consequently, the initial capital cost of the garage can be significantly reduced. In contrast, of course, the operating costs of the car park are high, due to the need to hire the attendants.

From an operating viewpoint, the most important area in any attendant-parking facility is the reservoir at the entrance. The entrance reservoir is intended to absorb peak inbound flows, when cars arrive at a rate greater than that the attendants can handle.

If the storage rate is given by

$$A = 60N/t$$

where A = average storage rate (cars/h), N = number of attendants on duty, and t = time for attendant to make the round trip (minutes), then it can be seen that the number of attendants, the time required to move each car to a parking bay, and the size of the reservoir space are all closely interrelated. An inadequate reservoir will need to be compensated for by employing extra attendants, whilst a large reservoir should permit fewer attendants.

Good attendant-parking operations encourage the parking activity to be spread throughout the garage to reduce the possibility of car conflicts. Cars to be parked for long periods of time are normally placed in rear rows of double-parking, or in bays adjacent to garage ramps, leaving the most accessible bays for short-term, high-turnover parking.

Patron-parking ramped garages: some design considerations
Many of the design considerations discussed in the context of off-street surface car parks also apply to parking garages, and so these will not be repeated here. The following are additional aspects that are of particular interest to the design of both underground and above-ground patron-parking ramped garages.

There is significant debate in the technical literature as to whether *long- or short-span construction* should be used in multistorey parking garages. Currently, long-span (also known as 'clear-span') construction is favoured on the grounds that it gives greater parking capacity and efficiency, is more adaptable to future changes in motor car design, provides the column-free parking that is preferred by motorists, and gives better visibility on each floor. Major advantages of short-span construction are that it more easily permits the erection of an office, flats or shop structure above the top parking floor, as well as permitting shallower beam depth and a reduced floor-to-floor dimension within the garage.

Notwithstanding the many *varieties of ramped garages* (only a few of which have been described here), there is no single system that can be said to be superior to another in all circumstances. The choice under any given circumstances should therefore be based on site shape and dimensions, user characteristics, and available funds. At all times, sites located on sloping ground should be examined to see whether an operational system can be devised which will allow direct access to/from the street system.

Maximum ramp grades are mainly limited by safety considerations and the psychological effect on drivers, with the braking and hill-climbing abilities of cars being of secondary importance. Steep ramps hinder traffic movement and can be dangerous when wet. For patron-parking garages, therefore, the maximum grades for straight ramps should be skid-free, and not exceed 10 per cent normally or 14 per cent for short lengths; on curved ramps, the gradient along the centreline should not exceed about 8.5 per cent.

Where a ramp joins a parking floor, a change in surface planes occurs. The change of 'breakover' angle should not be too sudden, as otherwise the ends or centres of cars may scrape the floor. A good rule-of-thumb used to overcome this problem is to apply a transition grade of half the ramp grade, at the top and bottom of the ramp, over a minimum blending distance of at least 3 m at each location. The intersection points should be rounded also.

The preferred *minimum width* of a straight ramp lane is 3.0 m with at least a 0.3 m wide by 0.1 m high kerb on either side. In the case of two-way ramps, a median kerb is essential to prevent car drivers from using the full width of the ramp as a single lane. End sections, where a car will start or complete a turn, are normally flared (e.g. by about 0.6 m over the length of the vertical transition gradient), to allow the rear wheels of the car to turn to a straight line.

In the case of curved one-way ramps, the minimum width at minimum radius should be greater, e.g. 3.65 m between kerbs. The preferred minimum radius of the outside kerb of a curved ramp is 9 m. The side clearance (between the outside kerb and the structure) for curved ramps should be at least 0.6 m, compared with 0.3 m for straight ramps.

Superelevation up to 0.10 should be incorporated to make turning easier on curved ramps.

The headroom under ramps and beams should be not less than 2.05 m[58]; this permits a greater ceiling height (which is also less oppressive). The minimum clearance should allow for any projections below structure, e.g. signs, lighting conduits, sprinkler systems, and drainage pipes. To allow for caravan-type vehicles, which require greater clearances, the usual practice is to give the entrance floor of the car park a minimum headroom of, say, 2.3 m. All low points should be clearly marked with the relevant dimensions, making allowance for the 'bridging effect' of cars at ramp ends.

Ramp structures should be as open as possible to provide good sight distances and *reduce closed-in impressions*. The optical trick of obscuring horizontal and vertical lines of reference can be used to reduce the apparent steepness of ramp grades, improve driver confidence, and reduce the adverse psychological effects associated with driving in restricted spaces. For example, ramp walls can be painted with stripes that are parallel to the ramp surface or at steeper angles, and the normal angles between vertical columns and the travelway can be obscured by paint markings.

Assuming that building regulations are not a controlling factor, the *maximum height of a parking garage* is generally governed by the number of full 360-degree rotations a car requires to circulate through the garage in order to get to the furthest parking bay. General experience would suggest that 5–6 rotations are the maximum tolerated by driver-parkers[52]. Another rule-of-thumb is that 400 parking bays are the maximum that a driver can be expected to pass before reaching a parking space.

Perimeter walls along ramps and parking floors should be designed to limit the driver's view of the surroundings outside the car park, as this may detract from the driving task. This is particularly important when travelling on parking levels that extend higher than the adjacent buildings.

Perimeter walls on parking floors are normally designed to withstand the impact from a 1500 kg car rolling free at 16 km/h (the statutory requirement). The adequacy of this design criterion has been criticized severely, however[59], and it would appear that consideration should be given to the usage of stronger perimeters than is specified by the statutory requirement.

As will have been gathered from the previous discussion re surface car parks, the determination of *reservoir needs* at exits and entrances requires a thorough analysis involving the arrival rates of parking/unparking cars as well as the capacities of cashier-control-points and street lanes. American experience would generally suggest, however, that reservoir requirements at entrances can be satisfactorily met by the provisions recommended in Table 4.15.

Table 4.15 Typical inbound reservoir requirements at parking garages[52]

Method of operation	Reservoir requirement
Free-flow entry	1 space per entry lane
Ticket-dispenser entry	2 spaces per entry lane
Manual-ticket dispensing	8 spaces per entry lane
Attendant-parking	10 per cent of that portion of the car park capacity served by the entry lane

Mechanical and electrical services The service elements of importance in patron-parking garages are lighting, electrical installation, road and ramp heating, lightning protection, ventilation, fire prevention and fighting, drainage, and lifts. The following summary is primarily based on an excellent review that is available in the literature[60].

Careful design of the *lighting layout* should ensure an installation that is economic to install and operate, and yet will provide a level of illumination that is adequate for safety and security purposes. Desirable standards of illumination in parking garages are given in Table 4.16. Note that higher lighting levels are required at entrances and exits to compensate the driver for the sudden changes in lighting levels between the parking garage and the outside. Areas subject to security problems, e.g. stairways and lifts, should also be well lit.

Table 4.16   Lighting levels suggested for patron-parking multistorey parking garages[50]

Area	Illumination level (lux)
Parking stalls	30 (at floor level)
Access aisles	50
Ramps	50 (at vertical walls)
Entrances and exits	150 (at floor level)

One further point that may be noted in relation to the implementation of the standards of lighting prescribed above is that the output from a fluorescent lighting tube—the most common lighting fixture in a car park—is normally measured at 25 °C; if the air temperature drops to the freezing point, then the output from the tube will be reduced by about 30 per cent. Also, the value of good maintenance procedures relative to lighting must be emphasized. For example, it is not uncommon for an old fluorescent tube on a cold winter's night, and in a dirty fitting, to emit only 25–30 per cent of its designed output—which explains why so many car parks appear gloomy and dangerous at night.

There are many variables involved in relation to the design of the *electrical installation*, and so the advice of a qualified electrical engineer should be sought at the planning stage, in order to ensure that an installation is provided that will be safe and efficient to use, as well as being adequate for its intended purpose throughout the life of the parking structure. (See also reference 50.)

Many parking garages are open to the elements—of these, *ice and snow* can create hazardous driving conditions on exposed circulation routes, particularly on steep, sloping ramps. This can be overcome, where circumstances justify it, by warming driveways subject to rain or condensation, with embedded electric cables. Heating can be automatically activated by control probes set in the carriageway surface, which energize the cables when they sense an appropriate combination of surface temperature and moisture.

Whether or not a parking building requires major protection against the effects of *lightning* depends upon: (a) the numbers of people likely to congregate in or adjacent to it, (b) whether it is in an area where thunderstorms are prevalent, (c) if it is a very tall or isolated structure, and (d) whether it is either self-protecting or needs only minor expenditure to render it safe, viz. structures composed entirely of metal and adequately earthed are self-protected, while structures with frames of steel or reinforced concrete require only minor improvements. When it is considered that the consequential effects of a lightning strike will be restricted to slight damage to the building fabric, then it may be more economical to accept the risk rather than incur the extra cost involved in protection.

Ventilation is essential in a parking garage in order: (a) to avoid the risk of fire and explosion arising from the presence of inflammable or explosive vapours, and (b) to minimize any damage to health which might arise from the presence of excessive concentrations of exhaust gases. As many multistorey parking garages are open-sided, this is not as great a problem as it is often considered; generally, it is safe to assume that a ventilation problem does not exist if an open space equivalent to 2.5 per cent of the total floor area is provided on each long elevation of the garage. The problem is perhaps most critical in underground car parks, and here the designer is well advised to refer to a qualified ventilation engineer. Generally, it will be found desirable to install appropriate alarm equipment, particularly at tunnel locations where cars have to queue, which will start up the ventilation plant when, for example, the level of carbon monoxide reaches a predetermined level.

The *fire problem* in parking garages is not primarily related to any danger of the building burning down (because it is normally very hard for the structure to catch fire), but rather to the hazard of vehicles in the garage going on fire as a result of, say, a lighted cigarette being left on a seat[61]. Here also the problem is most serious in underground car parks; these may, for example, require the installation of sophisticated detector devices which will 'sniff' smoke, or flames, or register a significant rise in temperature and sound the alarm.

Fire-fighting equipment in use in parking buildings includes wet rising mains and hose reels, dry rising mains, automatic sprinkler and drencher systems, and simple fire extinguishers. It is probable that portable extinguishers are the most effective equipment for fighting fires likely to be encountered in parking buildings, as they are most easily used by the public and they can be complemented by dry rising mains for use by the fire department. To minimize the opportunities for theft and vandalism, it is better not to locate small fire extinguishers throughout the parking garage; larger extinguishers mounted on handcars should be sited at central locations in the car park (preferably where they can be readily monitored). The benefits to be derived from hose reels can also be nullified by vandalism and they, as well as automatic sprinkler systems, can be subject to frost hazards. More important is the fact that water is not suitable for fighting petrol fires, and improper usage (as could occur with hose reels and sprinklers) could result in the fire being spread indiscriminately and made worse.

Special attention should be paid to the *drainage* of parking garages because they are so open to the elements, as well as subject to water carried in by cars. This generally involves, for example, the deliberate sloping of parking floors (a gradient of about 1.67 per cent is appropriate) so as to provide falls in all directions to gulley outlets.

The design of the *lift system* to serve a parking garage is another frequently forgotten part of car park design. Lifts should be used in all multistorey car parks that have three or more floors above ground level.

They are normally installed in banks (the number of lifts located together) of two or more, as close as possible to the principal pedestrian generators serviced by the parking garage.

Lifts can be raised and lowered either hydraulically or electrically. Hydraulic lifts are generally limited to heights of about 15 m; above 5–7 floors electrically-operated lifts (which are also much faster) should be considered.

As a general rule-of-thumb, two lifts should be adequate for parking structures with up to 600 bays, plus one extra lift for every additional 500 bays or substantial parts thereof. While the design elements for passenger lifts are now well established, it is still advisable to leave the decision as to the number and type of lifts to the lift consultants. However, an approach to a satisfactory design may be gathered from the following calculations for a six-storey parking garage with space for 100 cars on each floor, which assume that during the peak hour there is a complete turnover of vehicles.

From the above,

total number of cars above ground level $= 500$

Assuming average car occupancy of 1.5 person/car,

total number of persons $= 750$

If, say, 75 per cent of these use the lift facilities, then

number of people to move in 5 minutes $= (750 \times \tfrac{75}{100})/12 = 47$

The number of stops made by the lift is 5 and the approximate length of travel is 12.2 m.

Table 4.17 shows that in this instance two eight-person lifts should be able to provide a fairly reasonable service (provided that there is not a

Table 4.17 Handling capacities of an eight-person, 544 kg lift

Number of floors served	Speed			
	45.75 m/minute		61 m/minute	
	Capacity (persons/5 minutes)	Round trip travel time (seconds)	Capacity (persons/5 minutes)	Round trip travel time (seconds)
5	24	100	26	94
6	21	114	23	104
7	19	127	21	115

sudden heavy surge of travellers, e.g. as might occur with journey-to-work parkers, or that the passengers are not continually laden with shopping, perambulators, etc.). Obviously, the provision of lifts with speeds lower than 45.75 m/minute would also result in an unsatisfactory service during periods of peak demand.

Fee-collection equipment

As will have been gathered from the previous discussions on surface and multistorey car parks, one of the first items to be considered when designing an off-street parking facility is the method of fee collection to be used, and the check system to be employed to ensure that the proper fees are passed on to the car park owner/operator. Decisions taken at this stage can have a major effect both on the design of the facility, e.g. on the location or the number of entrances/exits, and on its entire financial viability.

The simplest and most obvious method of fee collection is to employ an attendant to stand at an entrance and collect a fixed fee from every driver entering the car park. This method, which can result in hold-ups to traffic at the entrance, depends very much for its financial success upon the honesty and integrity of the attendant. Nowadays, its usage is not recommended for other than minor parking facilities.

There is ample evidence in the technical literature, and in the Courts, to the effect that parking thefts are a common occurrence if proper control procedures are not initiated and maintained. Thefts can happen via a variety of methods, ranging from the simple expedient of an attendant not issuing or disposing of a ticket and pocketing the money, to resetting time clocks, switching or substituting tickets, vandalizing equipment, or altering records. Another form of theft is that carried out by patrons of the car park, i.e. some parkers have been known to mutilate, destroy, throw away, swap, and alter tickets in order to escape paying the proper fees; others have been detected in the act of vandalizing equipment, exiting through unauthorized points, and working in collusion with employees and/or other parkers in order to evade paying the parking fee.

To help to overcome these illegal activities, the following elementary steps have been recommended[62] as being essential to the successful financial operation of a car park.

(1) There should be an accurate count made of all vehicles entering and leaving the car park. There should be no way whereby the car park's patrons, executives or floor staff are able to enter or leave the parking facility without being counted.

(2) No vehicles should be allowed to enter the parking area without the driver taking a ticket that can be surrendered when leaving. A hardcopy record of the entry should always be left in the cashier booth, or a positive vehicle registration count made with some recording device.

(3) All time clocks used in the control system should be operated accurately and in conjunction with each other. As well as being reliable (i.e. time differentials from one system component to another can mean revenue losses, inaccurate records, and conflicts with patrons), all clocks should be of a type designed to ensure that no unauthorized person can reset them.

(4) Some kind of validating device, cash register or fee calculator should

be used to record all exit transactions. Tickets should always be stamped with all appropriate information so that each cashier-attendant can be held responsible for his/her transactions.

(5) Ticket-dispensing machines should be operated so that it is impossible for any vehicle to be given more than one ticket. Reliable and well-proven machines should be purchased. Additional supplies of tickets should be kept in a secure place.

(6) Parking gates or similar preventative devices should be used to stop cars from entering through an exit or exiting through an entrance.

(7) A display of the fee charged should be visible to all car parkers.

Control equipment

The equipment used in normal car park operations can be expected to vary according to the size and type of car park being operated, as well as according to the individual functions that the operator wishes to see performed. Commonly used pieces of basic equipment are detectors and counters, ticket-dispensing machines, gates, clocks and fee indicators.

Vehicle detectors are most commonly used in car parks to register the presence of a vehicle in order to count it and/or send an impulse to another machine such as a ticket-dispensing machine or an entrance or exit gate. The most widely used sensing devices are inductive loop detectors and contact strip (treadle) detectors; as these have been described in Chapter 3, their method of operation need not be further discussed here.

Differential counters are also a normal requirement for large car parks. Usually located at entrances and exits, they enable a running score to be kept of the number of vehicles in the car park at any given time, and thus knowledge of the number of vacant spaces is always available. Hence, when the car park is nearly full, it can be closed for a period until sufficient vacancies are available to allow for re-opening, i.e. it is better to close a car park at capacity rather than allow vehicles to enter, circulate and cause congestion—and frustration.

Ticket-dispensing machines replace the function carried out in the past by the parking attendant. These machines, which can typically hold 1000–2500 tickets, are placed on the driver's side of the entrance driveway(s) into the car park so that, after actuation by a vehicle detector, a parking ticket is automatically issued to the driver who then proceeds into the parking facility. Tickets issued by these machines vary considerably in configuration and use. Most commonly, they are made of paper and stamped with the time and date of entry and, in the case of a large car park, the entrance of issue. Some ticket machines are designed to issue plastic tickets or cards instead of paper ones.

In the case of computer-operated control systems, the time and date of entry are recorded on the ticket in a coded form. Most commonly, magnetic tickets are used for this purpose. Some types of magnetic ticket have the entire back covered in a magnetic substance which is encoded with the entry information; others have a narrow strip of magnetic tape attached to the back on which the information is placed. Encoding onto the magnetic-sensitive ticket is done by bar codes, binary and alpha-numeric, depending upon the type of equipment used.

Some tickets are punched with holes which must be sensed by a card reader and the code transmitted to calculate the elapsed time; others are embossed with a bar code to indicate the time of arrival. The difficulty with punched-hole tickets is that they can be easily mutilated or changed, whilst bar-coded tickets have the problem that they are easily mashed. Whatever the type of ticket used, the subsequent action is the same, i.e. the computer eventually reads the time of entry on the ticket, as well as the time of departure, and then calculates the intervening time and fee to be charged.

The purpose of a *parking gate* is to stop cars from making unauthorized movements into or out of the car park. Basically, the 'gate' consists of a metal cabinet which contains an electric motor, various electrical connections, a belt or drive system, and a gate arm. The arms used at outside locations are usually made of wood and 2–3 m long; in parking garages with limited headroom at the entrance or exit, gate arms must be used that are capable of folding when raised.

The removal of a ticket from the ticket-dispensing machine at the entrance is the activating action that causes the gate arm to rise so that vehicles can enter the car park. As the car passes over a detector placed beyond the arm another signal is sent back to the gate mechanism which causes the gate arm to close.

In the case of a fixed-charge, fully-automatic car park, the entrance to (or exit from) the parking facility is controlled by a gate arm which is automatically raised when an appropriate fee is paid into a payment machine. Machines can be made to accept coins and/or currency, and are easily adjusted to charge any fee desired. A change-giving unit can also be incorporated.

In the case of a variable-charge, fully-automatic car park, the exiting driver inserts the ticket in a ticket-and-coin acceptance unit; this causes the fee due to be calculated and automatically illuminated on a digital screen. Upon insertion of the correct money (units able to handle change are also available), the exit arm is raised and the vehicle can be driven away. If the ticket-and-coin unit is located away from the exit, the same procedure applies except that the motorist receives, in return for the fee, a 'short-life' coded token/ticket. If the token/ticket is placed in a small acceptance unit, located at the exit, within a short period of time, the barrier arm is raised and the car can be driven away with the minimum delay at the exit.

As a vehicle leaves a non-automated variable-charge car park, the most common practice is for the cashier-attendant to collect the ticket from the motorist and place it in a *time clock* which stamps the exit time and date on the ticket, after which the elapsed time and fee are calculated. Since exit-clocks, ticket-dispensing machines, and auditing equipment are all based on time, it is essential that the clocks used within the system are controlled by a single master clock.

The *fee indicator* is usually an illuminated device at the exit booth which informs the driver of the parking charge. Its purpose is to minimize overcharging and to keep the cashier honest. The fee indicator is normally connected to some device such as the cash register, so that whatever data are punched into the cash register (or read by the computer) are automatically visible to the motorist as a confirmation of the charge being levied, e.g. if the fee charged is 75p, but only 50p is entered in the register by the cashier, the error (or theft) would be evident to the driver.

A British Standard[63] has been developed in respect of specifications for the following items of basic equipment: rising arm/kerb gate barriers, collapsible plates, multiple-coin units to control the operation of barriers, ticket-dispensing machines, and the marking and usage of tickets.

'Season ticket' holder control Many off-street car parks in central areas also have to cater for long-stay parkers who pay for their parking spaces on a monthly, quarterly or annual basis. Ideally, these parkers should be provided with separate access and egress gates, but this may not always be possible and so the normal entry and exit points must be designed to cater also for this type of parker.

Particular attention has to be paid in this instance towards ensuring that a patron is not able to switch tickets or that a cashier cannot classify a fee-paying vehicle as non-fee-paying.

One way of ensuring proper control is to require season ticket holders to insert a special plastic card into a reader located in front of the normal ticket dispenser at the entrance. This causes a special ticket—one without a date- or time-stamp—to be issued to the motorist. When exiting, the driver must present this ticket together with the plastic card for identification to the exit cashier, who then validates the non-date/time-stamped ticket in the cash register and removes it from circulation.

Alternatively, the season ticket holder may be issued with a 'cycling card' which is also inserted in a reader at the entrance. In this case, however, no ticket is issued but instead the cycling card is encoded to indicate that it has been used at the entrance. The card must then be inserted into an exit reader to neutralize the entry code before it can be used at an entrance again. This method has the advantage that there is no need to provide tickets at the entrance and, obviously, the cashier has only to deal with cash-paying patrons at the exit. It is also a good way of controlling access to single-purpose (e.g. employee) parking facilities. In this latter case, an additional visual check may be provided by issuing all legal parkers with special stickers which can be attached to their car windows.

Special event parking control Whereas the previous data have been concerned with parking control at multipurpose parking facilities, the following discussion is concerned with control and audit procedures at special event parking areas, e.g. at racecourses, sports grounds and convention centres.

At these latter locations, vehicles typically enter the car parking areas throughout the hour preceding the start of the event, but attempt to leave after the event within a much shorter time span. Any attempt to collect fees from these drivers as they leave is impractical at the least, and could result in traffic chaos. In this instance, the usual approach adopted is to charge a flat fee at the point of entry, and upon conclusion of the event, all lanes are opened to exiting traffic.

The most important auditing control with this type of operation is the recording of each vehicle entry, as each such entry should represent one fee collected. The total number of vehicles is most easily determined with the aid of lane detection equipment of the type described previously. These automatic counts can then be used to match the manual counts and monies collected by the attendants at the fee-collection points.

An additional advantage of this type of operation is that many of the parking attendants can be released from duty shortly after the formal start of the special event, as only a few access lanes need to be kept open for latecomers to the car parks.

Special access and parking arrangements normally have to be made for buses, taxis and special vehicles for which parking fees are not normally levied. These vehicles should be directed to particular parking areas, preferably via special lanes. If this is not possible, transactions need to be initiated at the collection points and refund tickets/passes issued which can be collected as the vehicles leave.

Audit procedures

It is a basic requirement of successful car park operation that the cashier-attendants be held completely responsible for their own operations. Thus when, say, an exit cashier goes on duty at a conventional car park, it should be necessary for that attendant's (coded) personal key to be inserted into the booth equipment before it can be operated. The booth's audit recorder then notes the time the equipment was switched on, as well as the date, cashier number, and other information required for auditing purposes.

As the cashier-attendant begins operation, data are fed into the recorder (or computer) and held in storage for a given length of time, typically one hour. At the end of this time, a summary of all activities is normally printed out so that it is possible for a supervisor to check, for example, the total number of vehicles going through each exit, the total number of transactions the cashier has recorded on the register or other device, and an accumulative total of the fees collected. At all times, the number of vehicles counted as going through the exit should correspond with the total number of cashier transactions registered.

In the case of a small car park, the simplest and most effective way of auditing the honesty of the system is to check the time of arrival and departure of each vehicle throughout the entire day, using the car licence number as a check. Total receipts can then be calculated and checked against the reported receipts for that day, and against the pattern of receipts for comparable days.

Emergencies often arise at car parks, e.g. drivers arrive at the exit without cash or tickets are lost, or mutilated, which must be handled in an ad hoc way. When these incidents occur, the usual practice is for the cashier to use designated blank ticket stock to record whatever transaction then takes place.

Functional design of off-street lorry parks[64]

As with off-street car parks, the functional design of a lorry park is very much dependent upon the selection of the design vehicle. Unlike car parks, however, it is not practicable to design lorry parks on the basis of a single design vehicle for all or the majority of lorry spaces, i.e. lorry length dimensions vary so significantly that this would be very uneconomical in terms of land utilization.

The maximum vehicle width permitted by law in Britain is 2.5 m. In practice, many heavy goods vehicles have a width of at least 2.0 m, so it is good practice to work with a 2.5 m vehicle width. On this basis, the recommended *minimum* bay width is 3.3 m; this allows for wing mirrors which may protrude beyond the vehicle's body and for the opening of the cab door.

Heavy goods vehicle lengths, and thus parking bay lengths, vary according to whether the lorry is rigid or articulated, as well as according to manufacturer and intended function. Typically, overall lengths range from about 8 to 11 m for rigid vehicles and from 12 to 15 m (the maximum permitted) for articulated vehicles. To minimize the opportunity for the heavy vehicles to protrude into an aisle and to allow for minor vehicle maintenance, the *minimum* bay length used in any design therefore should be the vehicle length plus 0.5 m.

As with car parks, the aisle width in a lorry park also varies according to the angle of parking and the design vehicle's turning capabilities. Experience would suggest that single-row, 45-degree parking with one-way aisles and a drive-in/drive-out method of operation (to avoid the need for reversing) is appropriate for lorry parks. (With 45-degree parking, geometry dictates that the minimum bay length then becomes the lorry length plus the vehicle width (2.5 m) plus 0.5 m.)

Figure 4.10 shows the interrelationships established between minimum bay width and minimum aisle width for heavy goods vehicles of different type and length. From this figure it can be seen that a wide range of combinations are possible, so that a combination can be chosen which best suits the size and shape of the land available for the lorry park, e.g. a long narrow site may preclude the use of wide approach aisles. Analysis has shown, however, that the most efficient use of land is generally gained by using a 3.3 m wide bay in combination with the corresponding aisle width. On this basis, the recommended dimensions to use for the most economical land use layout are as given in Table 4.18. Note that the aisle widths used in this table include 0.5 m additional to that given by Fig. 4.10, to allow for possible driver error.

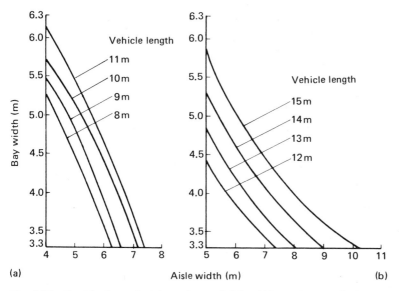

Fig. 4.10 Combinations of minimum bay and aisle width requirements for heavy goods vehicles of different lengths: (a) rigid vehicles and (b) articulated vehicles

Table 4.18 Recommended dimensions for use in a one-way, 45-degree lorry park (metres)

Vehicle length	Bay length	Bay width	Aisle width
8	11	3.3	6.8
9	12	3.3	7.1
10	13	3.3	7.7
11	14	3.3	7.9
12	15	3.3	7.9
13	16	3.3	8.6
14	17	3.3	9.5
15	18	3.3	10.8

Selected bibliography

(1) Khisty CJ, Some views on traffic management strategies—with emphasis on parking and energy, *Traffic Quarterly*, 1980, **34**, No. 4, pp. 511–522.

(2) Department of Transport, *Transport Act 1978: Control of Off-street Parking*, Circular Roads No. 3/79, London, HMSO, January 1979.

(3) Buchanan CD et al., *Traffic in Towns*, London, HMSO, 1963.

(4) Bendsten PH, *Town and Traffic*, Copenhagen, Danish Technical Press, 1961.

(5) Jonassen CT, *The Shopping Center vs Downtown*, Columbus, Ohio State University Bureau of Business Research, 1955.

(6) Regional Planning Research Ltd, *Parking: Who Pays?*, London, The Automobile Association, **13**, No. 3, 1967.

(7) Department of the Environment, *Policy for the Inner Cities*, Cmnd 6845, London, HMSO, June 1977.

(8) Ministry of Transport, *Urban Traffic Engineering Techniques*, London, HMSO, 1965.

(9) Smith WS, Influence of parking on accidents, *Traffic Quarterly*, April 1947, pp. 162-178.

(10) Department of Transport, *Road Accidents in Great Britain 1983*, London, HMSO, 1984.

(11) Heyman JH, Parking trends and recommendations, *Traffic Quarterly*, 1968, **22**, No. 2, pp. 245-258.

(12) Seburn TJ, Relationship between curb uses and traffic accidents, *Traffic Engineering*, 1967, **37**, No. 8, pp. 42-47.

(13) Cleveland DE, Accuracy of the periodic parking survey, *Traffic Engineering*, September 1963, pp. 14-17.

(14) Benn GK, Time-lapse photography applied to a parking duration survey, *Traffic Engineering & Control*, 1971, **12**, No. 11, p. 581.

(15) Alexander LA, The downtown parking system, *Traffic Quarterly*, 1969, **23**, No. 2, pp. 197-206.

(16) The City Engineer et al., *City of Edinburgh Traffic and Transport Plan 1970-1974*, Edinburgh, The Corporation, September 1970.

(17) Pearce H and Jackson P, A question of parking, *Traffic Engineering & Control*, 1979, **20**, No. 8/9, pp. 403-405.

(18) Stout RW, Trends in CBD parking characteristics, 1958 to 1968, *Highway Research Record*, 1970, No. 317, pp. 40-47.

(19) Ministry of Transport, *Parking—The Next Stage*, London, HMSO, 1963.

(20) Mohr EA, A performance index for parking facilities, *Traffic Engineering*, 1954, **28**, No. 7, pp. 11-14 and 38.

(21) Knoflacher HA, A method of verifying the effectiveness of limited parking zones, *Traffic Engineering & Control*, 1978, **19**, No. 2, pp. 82-85.

(22) Johnson BK, *Angle vs Parallel Parking—Time and Street Width Required for Manoeuvering*, Berkeley, California, The Institute of Transportation and Traffic Engineering, 1960.

(23) Box PC, Curb parking effect, *Traffic Digest and Review*, 1970, **18**, No. 2, pp. 6-10.

(24) Inwood J and Green H, Parking meters in London's West End, *Traffic Engineering & Control*, 1962, **4**, No. 7, pp. 402-403 and 405.

(25) Parking: meters ingest £56m in 20 years, *Traffic Engineering & Control*, 1978, **19**, No. 8/9, p. 419.

(26) BS 4684: *Clockwork Parking Meters*, London, The British Standards Institution, 1971.

(27) Bourlon P, Parking discs in the Blue Zone of Paris, *Proceedings of the Fourth International Study Week in Traffic Engineering*, Copenhagen, 1958.

(28) Morgan EJ and Lewis CG, A before-and-after study of the effect of the introduction of disc parking in Harrogate, *Journal of the Institution of Municipal Engineers*, 1971, **98**, No. 2, pp. 35-41.

(29) O'Cinneide D, Parking control by the Park System in Cork, *Traffic Engineering & Control*, 1977, **18**, No. 4, pp. 171-174.

(30) BS 4631: *Parking Discs*, London, The British Standards Institution, 1970.

(31) Pay and display parking, *Chartered Municipal Engineer*, 1980, **107**, No. 7/8, p. 193.

(32) Benn GK and Hildick B, Parking meters—their future, *Traffic Engineering & Control*, 1976, **17**, No. 5, pp. 218-219.

(33) Ball DJ, The merits and problems of pay and display, in *Paying for Parking—Theory and Practice*, London, The British Parking Association, November 1977.

(34) Research Group UT5, *Methods of Evaluating Urban Parking Systems*, RR/UT5/80.3, Paris, OECD, June 1980.

(35) Cave FJ, Resident parking problems and schemes, in *Making Parking Pay—The Economics of Car Parking*, London, The British Parking Association, 1972.

(36) Department of the Environment, *Lorry Parking*, London HMSO, 1972.

(37) Parking: lorry-park plan is scuttled, *Traffic Engineering & Control*, 1976, **17**, No. 5, p. 219.

(38) O'Flaherty CA, Town, transport and parking, in *Off-street Parking*, London, The British Parking Association, August, 1970, pp. 4–6.

(39) Wilbur Smith and Associates, *Parking in the City Center*, Washington DC, The Automobile Manufacturers Association, 1965.

(40) Sheppard BR, Off-street car parking—a comparative urban study, *Municipal Engineer*, 1984, **111**, No. 3, pp. 90–94.

(41) Bureau of Public Roads, *Parking Guide for Cities*, Washington DC, US Government Printing Office, 1956.

(42) Lancashire County Planning Department, *Car Parking Standards*, Lancashire County Council, 1986.

(43) Bentfeld G, Effect of parking policies on traffic volumes, *Proceedings of the Tenth International Study Week in Traffic Safety and Engineering*, 1971, **10**, pp. 51–58.

(44) Ministry of Housing and Local Government, *Parking in Town Centres*, Planning Bulletin No. 7, London, HMSO, 1965.

(45) O'Flaherty CA, This parking business, *Journal of the Institution of Municipal Engineers*, 1968, **95**, No. 12, pp. 368–379.

(46) Lovejoy FW, Resume of fringe parking practice, *Highway Research Board Bulletin*, 1949, **19**, pp. 57–60.

(47) Deen TB, A study of transit fringe parking usage, *Highway Research Record*, 1966, No. 130, pp. 1–19.

(48) Bennett TH, The physical characteristics of the British motor vehicle fleet, *The Highway Engineer*, 1976, **23**, No. 10, pp. 24–29.

(49) Department of the Environment, *Cars and Housing/2*, London, HMSO, 1971 (metric edition).

(50) Joint Committee of the Institution of Structural Engineers and the Institution of Highways and Transportation, *Design Recommendations for Multi-storey and Underground Car Parks*, London, The Institution of Structural Engineers, 1984 (2nd edition).

(51) *Parking Principles*, Special Report 125, Washington DC, The Highway Research Board, 1971.

(52) Weant RA, *Parking Garage Planning and Operation*, Westport, Conn., The Eno Foundation for Transportation, 1978.

(53) Lancashire County Planning Department, *Design Notes for Car Parks*, Lancashire County Council, 1976.

(54) Ellson PB, *Parking: Dynamic Capacities of Car Parks*, RRL Report LR221, Crowthorne, Berks., The Road Research Laboratory, 1969.

(55) Layfield RE, *Parking: Reservoir Sizes at Car Park Exits*, TRRL Report 39UC, Crowthorne, Berks., The Transport and Road Research Laboratory, 1974.

(56) Ellson PB, *Parking: Turnover Capacities of Car Parks*, TRRL Report 1126, Crowthorne, Berks., The Transport and Road Research Laboratory, 1984.

(57) Glanville J, Development in multi-storey car parking in Great Britain, Paper presented at the Fifth World Meeting of the International Road Federation held at London, September 1966.

(58) Sharp R, Design recommendations for multi-storey and underground car parks, *Highways and Transportation*, 1984, **31,** No. 2, pp. 11-13.

(59) Hill J and Cannon J, Multi-storey car park design, *Concrete*, December 1975, pp. 33-37.

(60) Glanville J and Richards DL, The design of mechanical and electrical services for car parking buildings, in *Off-street Parking*, pp. 25-35, London, The British Parking Association, August 1970.

(61) Butcher EG, Fire and car park buildings, *Traffic Engineering & Control*, 1970, **12,** No. 7, pp. 377-388.

(62) *Parking Revenue Control*, Transportation Research Circular No. 184, Washington DC, The Transportation Research Board, June 1977.

(63) BS 4469: *Car Parking Control Equipment*, London, The British Standards Institution, 1973.

(64) Brannam M, Design criteria for the layout of lorry parks with 45° parking bays, *Traffic Engineering & Control*, 1974, **15,** No. 20, pp. 912-914.

5
The assessment of road improvements

Possibly the most difficult and controversial step in the highway transport/ traffic planning process is that of proposal evaluation. From the 1930s onwards, very many alternative assessment methods have been considered for this purpose, but it was not until the 1960s, when the number of planning studies being conducted was at its height, that the highway administrator was confronted with the real need to indicate scheme priority in expenditure of the very considerable funds then being provided by the Government to improve and replace portions of the highway system. At that time, there was general acceptance of an evaluation approach based on cost–benefit analysis, which involved searching for a plan which maximized economic efficiency.

Soon, however, controversy again began to flare regarding the adequacy of this method of assessment. It was argued that whilst the economic cost–benefit approach had many admirable virtues, its narrowness resulted inevitably in the determination of decision-making values which could be considerably in error from the *true* values. Particular criticisms mooted were that there were many effects of a highway scheme that could not be quantified in monetary terms, e.g. traffic noise, visual intrusion and severance effects, and the effect upon historical values, and thus these were invariably omitted from the evaluation. Furthermore, the economic analysis approach made no attempt to distinguish the effects upon the different community groups affected by the highway decisions, and this was claimed to result in a built-in bias in the methodology. In addition, the methodology was not understood by the general public, even though it was the main plan-selection criterion.

These and other concerns were voiced at many public inquiries in the late 1960s and early 1970s and, as a result, the economic approach began to fall into increasing disrepute. Eventually, the Government sought an independent assessment of the cost–benefit method then in use to evaluate trunk road schemes. This investigation[1] concluded that there was a need for a wider assessment framework for evaluating trunk road proposals, and that whilst the cost–benefit approach was still valid, it should be considered as simply one, albeit an important one, of many criteria.

The following discussion on a wider framework for assessing trunk road schemes is primarily based on the report of this governmental investigation.

A framework for assessing trunk road schemes

Criteria for assessment

In the view of the Leitch Committee[1], the assessment of any highway proposal should seek as far as possible to meet the following criteria.

(1) It should be generally comprehensible to members of the public and command their respect.
(2) The public should be able to identify how different groups of individuals are affected by the scheme.
(3) It should be comprehensive in terms of evaluating the different kinds of road scheme effects.
(4) It should allow for the effective control of decentralized minor decisions.
(5) It should not be expensive to use.
(6) It should balance costs and benefits (however described) in a rational manner.

Public understanding

If long, acrimonious and expensive public inquiries are to be avoided, then logically any procedure that is developed for highway assessment purposes must be comprehensible to the public and considered by them to be fair, and to be in their and society's interest.

The groups affected

The conventional cost–benefit analysis (as will be discussed later) emphasizes the effects of a highway proposal upon the road user and the financing authority. Other groupings which should be taken into account include non-road-users directly affected by the scheme, those concerned with the intrinsic value of the area, and those indirectly affected.

Road users affected by the scheme are spread throughout the network. They are concerned with reducing accidents, and with saving travel time and vehicle operating costs, all of which can be quantified. Also of interest to them are any increases in their general comfort, and the attractiveness of the view from the road; these considerations are intangible.

Non-road-users directly affected by the scheme include the owners/occupiers/users of buildings and land adjacent to the route. Their objective is normally to ensure that any environmental disadvantages which might arise from the proposal are minimized whilst any associated benefits are maximized. This group should also include those whose properties are bypassed as a result of a route diversion.

Those concerned with intrinsic values normally have as their objective the preservation of the landscape and ecology of the area. If historical sites are located within the zone of influence of the highway proposal, their interest is in ensuring that these should be disturbed as little as possible. In some instances, e.g. in respect of an area of industrial dereliction, the objective is to see how the proposal can be used to enhance the area.

Those directly affected by a highway scheme include local authorities who are concerned with its general land use and job opportunity implications, governmental agencies with interests in resource consumption, and the operators and users of other (non-highway) transport modes who are interested in its implications for their systems.

The *financing authority*—in the case of trunk roads, the central government—is concerned with ensuring that the best possible scheme is completed at the least net cost to public funds.

Comprehensiveness of the assessment

This aspect of the assessment is essentially concerned with the means by which the concerns of the groups noted above are taken into account. Some of these can be measured and quantified in monetary terms; the means by which this is done (discussed later) are well established. Other effects, primarily those which are thought of in social and environmental terms, e.g. noise, air pollution (in urban areas), visual intrusion, severance effects, and construction impacts, cannot be easily valued in financial terms, even though they may be capable of being measured.

It is often suggested that a highway development promotes the dispersal of homes and that this has an unfortunate effect upon land use. In practice, however, there is scant factual information available to show how to dissociate the effects of road construction from the general effects of increasing wealth or how to make valid adjustments for land use planning policies. Consequently, the treatment of this consideration is largely subjective and reflected in the views put forward by local planning authorities.

It is also often advocated that the indirect beneficial impacts of the highway proposal upon the economic development of a region should be taken into account in any comprehensive assessment, i.e. the assumption is that highway development is an efficient way of attracting economic growth to a region and that the more easily quantified road user benefits substantially underestimate the total benefit to the economy. In practice, however, the evidence available suggests that in only relatively few cases is the quality of the highway network a predominant factor influencing decisions on industrial location. Whilst the industrialist obviously welcomes any investment which improves the road system, it is only when more immediate requirements (e.g. the availability of a skilled labour force with a good industrial record) have been taken into account that marginal changes to the transport infrastructure are normally allowed to influence the location decision. As a result, the inclusion of this consideration in evaluation is not normally recommended.

Control of decentralized decisions

This assessment criterion is primarily concerned with evaluating and influencing the numerous design decisions of apparently a minor nature that are made in the course of developing a road scheme. In practice,

Table 5.1 Example of a framework for the assessment of trunk road schemes[1]

Group	Interest	Units	Red	Green	Blue	'Do-nothing'
Road users directly affected						
All road users	Reduction in casualties	Number				
	fatal		3	4	2	—
	serious		15	20	11	—
	slight		30	26	15	—
	Value of accident savings	£m (*NPV*)	0.80	0.59	0.59	—
	Comfort/convenience	Rank	2	1	4	3
	Attractiveness of view from road	Rank	3	4	1	2
Car drivers/passengers (working time)	Time savings	£m (*NPV*)	1.94	1.70	1.70	—
	Vehicle operating costs		−0.21	−0.14	+0.13	—
Car drivers/passengers (to/from work: leisure time)	Time savings		1.83	1.66	1.67	—
	Vehicle operating costs		−0.21	−0.08	+0.19	—
Heavy goods vehicle operators	Time savings		0.91	0.43	0.09	—
	Vehicle operating costs		−0.08	−0.03	+0.01	—
Light goods vehicle operators	Time savings		1.00	0.93	0.88	—
	Vehicle operating costs		−0.08	−0.03	+0.06	—
Bus operators and users	Time savings		0.59	0.58	0.58	—
	Vehicle operating costs		−0.02	−0.07	+0.07	—
Pedestrians	Time savings		1.03	0.41	0.50	—
	Amenity	Rank	4	1	2	3
Non-road-users directly affected						
Residential property owners/occupiers	Demolition	Number of properties demolished	—	—	—	—
	Noise increase	Number of properties with L_{10} increases of				
		10–20 dB(A)	96	64	9	—
		5–10 dB(A)	180	116	102	—
	Noise decrease	Number of properties with an L_{10} decrease of >5 dB(A)	1450	1682	1276	—
	Visual intrusion	Number of properties with				
		severe	30	12	3	—
		significant	183	139	9	—
		slight	399	351	17	—
	Construction disruption	Rank	4	2	3	1

Table 5.1 (*cont.*)

Group	Interest	Units	Scheme alternatives			
			Red	Green	Blue	'Do-nothing'
Shops and businesses' owners/occupiers	Demolition	Number of properties demolished	—	—	—	—
	Noise increase	Number of properties with an L_{10} increase of 5 dB(A)	1	1	—	—
	Noise decrease	Number of properties with an L_{10} decrease of 5 dB(A)	70	96	85	—
	Visual intrusion	Number of properties with				
		severe	—	—	—	—
		significant	—	—	—	—
		slight	1	1	1	1
	Construction disruption	Rank	3	4	2	1
Industrial and commercial properties' owners/occupiers	Demolition	Number of properties demolished	—	—	—	—
	Noise increase	Number of properties with an L_{10} increase of 5 dB(A)	—	—	—	—
	Noise decrease	Number of properties with an L_{10} decrease of 5 dB(A)	—	—	—	—
	Visual intrusion	Number of properties with				
		severe	—	—	—	—
		significant	—	—	—	—
		slight	1	—	—	—
	Construction disruption	Rank	4	2	3	1
Public buildings' occupiers/users						
1 Schools	Demolition	Number of properties demolished	—	—	—	—
	Noise increase	Number of pupils subject to an L_{10} increase of <5 dB(A)	150	—	—	—
	Noise decrease	Number of pupils subject to an L_{10} decrease of >5 dB(A)	97	97	52	—
	Visual intrusion	Number of pupils affected	150	—	—	—
	Construction disruption	Rank	4	2	3	1
2 Churches	Demolition	Number of properties demolished	—	—	—	—
	Noise increase	Number of properties with an L_{10} increase of <5 dB(A)	—	—	—	—

Table 5.1 (cont.)

Group	Interest	Units	Red	Green	Blue	'Do-nothing'
Public open space users	Noise decrease	Number of properties with an L_{10} decrease of >5 dB(A)	1	1	—	—
	Visual intrusion	Number of properties affected	1	1	—	—
	Construction disruption	Rank	3	4	2	1
	Landtake	Ha taken	—	—	—	—
	Noise increase	Number of people with an L_{10} increase of <5 dB(A)	350	—	—	.
	Noise decrease	Number of people with an L_{10} decrease of >5 dB(A)	—	50	185	—
	Visual intrusion	Number of people affected	350	—	—	—
	Construction disruption	Rank	4	3	2	1
Farmers	Landtake	Ha taken — Grade II	18	26	24	—
		Grade III	3	3	17	—
	Severance	Number of farms affected	—	—	8	—
	Noise increase	Number of farms with an L_{10} increase of <5 dB(A)	—	1	3	—
	Noise decrease	Number of farms with an L_{10} decrease of >5 dB(A)	1	—	—	—
	Visual intrusion	Number of farms affected	—	1	3	—
	Construction disruption	Rank	2	3	4	1
Those concerned with intrinsic value of area Landscape/townscape value	General assessment	Rank	2	1	3	4
	Items specific to scheme National parks Areas of outstanding natural beauty Heritage coasts Country parks National Trust land Conservation areas Other items	Effect description				

Table 5.1 (cont.)

Group	Interest	Units	Scheme alternatives			
			Red	Green	Blue	'Do-nothing'

Historic buildings value	Ancient monuments
	Listed buildings
	  Grade 1
	  Grade 2
	Other structures of character
		Effect description	Part of castle moat to be used as carriageway			

Ecological value	General assessment	Rank	3	1	4	2
	Special scientific interest sites Nature reserves	Effect description				
Archeological/historic value	General assessment	Rank	1	3	4	2
	Specific sites interest	Effect description				
Those indirectly affected Resource effects	Mineral deposit sterilization	Effect description	—	—	—	—
Land use planning effects	Job opportunities	Number of jobs gained	3	—	—	4
		Number of jobs lost	1	3	2	3
	County council's view	Rank	2	1	4	3
	District councils' views		2	1	4	3
	Parish councils' views		1	3	2	4
	Statutory objectors' views					
Other transport operators/users	Rail Air Waterways	Effect description				
Other factors specific to the scheme		Effect description				
Financing authority	Construction cost	£m (*NPV*)	6.2	4.9	4.1	—
	Land cost		1.0	1.0	1.0	—
	Compensation cost		0.5	—	—	—
	Maintenance cost		0.0284	0.0285	0.0435	—
	Total cost		7.7284	5.9285	5.1435	—
	Total gross benefits		7.50	5.95	6.83	—
	Differences between discounted costs and benefits		−0.2284	+0.0215	+1.6865	—

some of these decisions can have major costs associated with them. For example, whilst all-purpose dual carriageways are normally designed to a 4 per cent gradient, it may be appropriate to use steeper gradients (up to, say, 7 per cent) in hilly terrain so that both the capital cost and the environmental impact of the scheme are significantly reduced. There are many other examples which could be quoted where the use of a sensible, more flexible, approach to design criteria and standards could result in significant benefits of a financial and/or environmental nature.

Since these variations in standard are physical in nature, they are most easily evaluated by means of conventional cost–benefit analyses which balance road user benefits against construction costs within a range of permitted limits.

Cost of assessment

Any method of appraisal costs money, and normally the greater the degree of scrutiny the more costly becomes the exercise. This was most noticeable throughout the 1960s and 1970s as public consultation and involvement in the decision-making process was increased. Logically, there must come a time when improving the appraisal procedures becomes counter-productive, and the cost of the assessment becomes too great in comparison with the benefits to be gained.

In practice, however, this cost has some way to go before it is likely to be considered excessive. Typically, design and appraisal account for some 3 per cent of the total cost of a highway scheme, and it would be unfortunate to seek small savings in assessment costs if by so doing it resulted in substantial increases in total cost.

Use of the framework

Table 5.1 is an example of an assessment framework which embodies the criteria discussed above. Whilst the scheme evaluated in this table is hypothetical, the data used in the development of the four alternative routes were taken from actual road proposals.

One major use of this method of assessment is to decide between options within a scheme, *on a basis of judgement*, comprehending all factors whether valued in monetary terms or not. The simplest way of doing this is to compare the options available two at a time until a preference is reached; the best of these options can then be compared with the 'do-nothing' option. In so doing, the advantages and disadvantages of each alternative are traded-off, and the eventual decision as to which alternative to build is based on a subjective comparison of quantifiable and intangible benefits and losses.

Note that in the examples given in Table 5.1 the total cost of the red route is considerably more (by £2.585m) than that of the blue route. Whilst the red route offers an extra £0.670m in road user benefits, its net present value (*NPV*) is seen to be £1.915m lower than that of the blue when the discounted costs and benefits are compared. The red option also affords noise relief to slightly more houses than does the blue, but has a significantly greater landtake. In favour of the blue route is that it is preferred in planning terms and for its effects upon the intrinsic value of the area. The blue route also causes an increase in noise and visual intrusion for only one-third as many properties as does the red route. Thus it can be seen that those responsible for the decision between the red and blue alternatives must judge explicitly upon whether the additional benefits to the road users are sufficient to outweigh the losses elsewhere.

The Leitch Committee also considered that this type of framework could be used to provide a sound basis for settling scheme priorities at the national level. Thus it advocated that each scheme put forward for inclusion in the national programme should be given a merit rating of 1, 2 or 3. The date at which construction of a scheme would begin would then be based on its merit rating as determined from the assessment framework, as well as on such practical factors as the overall availability of funds and the stage achieved in design.

Cost–benefit methods

Economic appraisals of road improvement proposals may be carried out for a number of reasons[2], as follows:

(1)   to determine whether a scheme is economically justified,
(2)   to aid in the choice of engineering features of design,
(3)   to provide one way of determining priority of one scheme as related to others,
(4)   to assist in tax or cost allocation studies or decisions,
(5)   to develop information which will aid in evaluating a specific improvement as against other proposals in public works or community projects.

As may have been gathered from the previous discussion, objectives (1), (2) and (3) underly most British highway economic studies.

The field of economics abounds with concepts and procedures by which highway cost-benefit studies may be (theoretically, if not practically) carried out. It is not possible within the limited scope of this text to attempt to describe these in any detail, and the reader is referred to the literature (see, for example, references 2, 3, 4, 5 and 6) for many excellent treatises and further references. The following discussion is primarily concerned with those appraisal methods which may be of particular interest to highway and traffic engineers in Britain, i.e. the net present value, single year rate-of-return, internal rate-of-return, and benefit–cost ratio procedures.

Net present value (*NPV*) method

Also known as the *present worth* method, this procedure assesses whether a scheme is worthwhile by determining the base year market value of the proposed project; this is the present value of the benefits, *PVB*, less the present value of the costs, *PVC*. In formula terms, this is expressed as

$$NPV = PVB - PVC$$

With this method, the costs and benefits which arise in different future years are transformed to their present value by a process known as discounting. This discounting concept is most easily explained by considering the concept of compound interest. If, say, £1 is invested at an interest rate of *r* per cent then, expressing *r* as a fraction and neglecting inflation, at the end of one year it will be worth £$(1+r)$, at the end of two years £$(1+r)^2$, and so on. Conversely, £1 that is received in *n* years time is worth only £$1/(1+r)^n$ in the base (present value) year, again neglecting inflation.

Formally, any sum, *S*, received in any future year, *n*, may be reduced to its present value, *PV*, by means of the formula:

$$PV = \frac{S}{(1+r)^n}$$

r is known as the discount rate because this process involves the notion of charging interest against a project rather than paying interest to an investor. The discount rate used in road improvement appraisals in Britain is 7 per cent; this figure was set in 1978 when the White Paper on nationalized industries (Cmnd 7131) was published.

The present value method of cost–benefit analysis was originally put forward by the Transport and Road Research Laboratory[7] as being the appropriate way 'of bringing scientific technique to bear as an aid to administrative judgement'. The steps involved in the procedure were outlined as follows.

Step 1   For each year for which the benefits are being determined, calculate: (a) the difference in travel costs (i.e. vehicle operating costs and time costs of vehicle occupants) in conditions with and without the improvement, on all roads or sectors of roads affected by the improvement, (b) the reduction in accident costs as a result of the improvement, and (c) the gains to generated traffic. Sum (a), (b) and (c) to obtain the total road user benefits in each year. Discount the benefits in each year to their present value. Sum the discounted benefits over the life of the scheme.

Step 2   Calculate the following costs for each of the years in which they occur: (a) capital cost and (b) maintenance costs and costs of delay to traffic during construction. Sum these to give the total costs in each year. Discount these total costs to present values using the same base year as for the benefits.

Step 3   Subtract the discounted costs from the discounted benefits in order to obtain the net present value of the investment. (If the difference is positive at the going discount rate, then the scheme concerned is economically acceptable.)

Step 4   Express the net present value as a proportion of the discounted capital cost; this gives the present value to cost ratio. (If there is a budgetary restraint on the amount of highway construction capital available—which is normally the case—then this step can be used to arrange different road proposals in order of priority.)

Step 5 For the first year of full operation, express the net benefit, i.e. the total benefits less maintenance costs, for that year as a percentage of all the capital costs incurred by that date. This is the annual rate of return for the first year. (If the rate of return is less than the discount rate used in the calculation, the project should not be initiated as planned as the net present value will be increased by postponement. Rather the project should be started during the year when the rate of return equals the discount rate as this is when the maximum net present value will be achieved, i.e. postponement beyond this date will mean that the costs saved—the discount rate applied to the capital cost—will be less than the benefits foregone.)

The COBA program

The net present value method of cost-benefit analysis used by the Department of Transport to evaluate the economics of new road proposals is of the form described above. However, rather than requiring the engineer to attempt to carry out all the calculations manually, use is made in the analysis of a complex computer program known as COBA 9[8]. Figure 5.1 sets out the means by which the COBA program calculates the changes in user costs and net present value. The program measures costs and benefits over the entire road network affected by a scheme, but assumes that the pattern of trip-making is unaffected by the routes used, i.e. it assumes that there is a fixed trip matrix.

With COBA 9, the road user benefits are calculated for each year of operation of the new roadway over a period of 30 years from the scheme's opening year. Although a highway can be expected to have a useful life in excess of this time, the program assumes that the current value of its residual life is zero. This assumption is justified by the effects of discounting at 7 per cent over 30 years—reducing the benefit in year 31 to 12 per cent of its undiscounted value—and by the uncertainty associated with such a distant time.

The 30-year analysis is obtained by applying high and low forecast rates of traffic growth (see Table 5.2) to the original assignments input to COBA 9. (The economic assumptions underlying these forecasts are given in Table 5.3.) These are then associated with a consistent set of economic parameters to give 'high' and 'low' 30-year profiles of user benefits which encompass a range of likely outcomes.

Since the present value year in COBA 9 is 1979, the program calculates the present value of a stream of benefits according to the formula:

$$PVB = B(79) + \frac{B(80)}{(1+r)} + \frac{B(81)}{(1+r)^2} + \cdots + \frac{B(n)}{(1+r)^{n-79}}$$

where $B(n)$ is the benefit occurring in the particular year indicated by each bracket. The present value of the stream of costs is determined by a similar formula, so that the net present value (1979 prices) is then given by

$$NPV = PVB - PVC$$

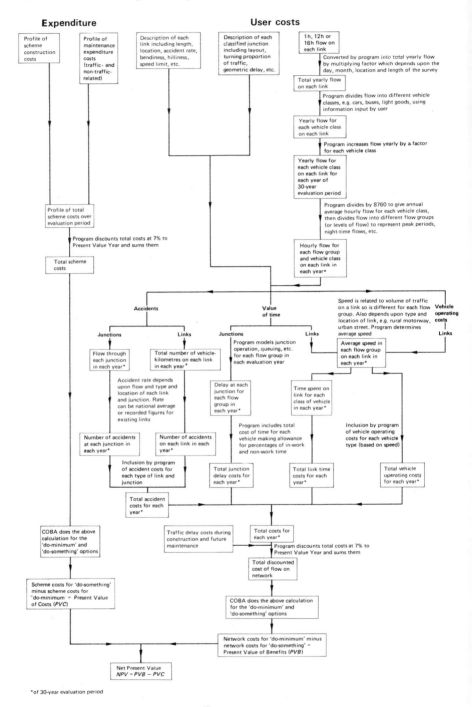

Fig. 5.1 The COBA evaluation scheme[8]

Table 5.2 The Department of Transport's (December 1984) long-term national traffic growth forecasts of vehicle-kilometres: (a) low growth forecasts and (b) high growth forecasts (1982 = 100)

Year	Cars and taxis	Heavy goods vehicles	Light vans	Buses and coaches	All (excluding 2-wheelers)
1982	100	100	100	100	100
1990	116	99	107	100	114
1995	122	99	111	100	119
2000	126	99	115	100	123
2005	131	99	120	100	127
2010	135	99	125	100	131
2015	140	99	130	100	136

(a)

Year	Cars and taxis	Heavy goods vehicles	Light vans	Buses and coaches	All (excluding 2-wheelers)
1982	100	100	100	100	100
1990	129	107	113	100	126
1995	143	110	124	100	139
2000	154	114	135	100	149
2005	165	117	147	100	159
2010	175	119	159	100	167
2015	182	121	173	100	176

(b)

Table 5.3 Economic assumptions underlying the road traffic forecasts

Parameter	Low traffic growth	High traffic growth
GDP growth * (% per annum)		
1983–88	1.5	3.0
1988–93	1.0	2.5
1993+	1.0	2.0
Petrol price increases * (% per annum)		
1980–2010	2.6	1.0
2010+	0.0	0.0

*In real prices

In practice, COBA is used by the Department of Transport to contribute to the following decisions:

(1)   the assessment of the need for a corridor improvement (either on-line or new route) prior to entry to the Department's preparation pool,
(2)   the priority to be accorded to individual schemes by comparison of economic returns with those from other schemes in the region or in the country,

(3) the optimal timing of the scheme, including consideration of the merits of staged construction and timing in relation to complementary or competing schemes in the network,
(4) the selection of a short list of potentially attractive solutions to present at public consultations,
(5) the selection of the preferred option to recommend to the Minister after public consultation,
(6) the optimal link design standards for the options under consideration by comparing the economic returns from the feasible alternatives,
(7) the initial assessment of the optimal junction designs for options under consideration.

COBA is applicable to the great majority of schemes in the trunk road programme, although there are circumstances in which additional or alternative techniques are more appropriate. Examples of the latter are: (a) schemes costing less than £1m where a less sophisticated approach can be adopted, (b) larger schemes where the effects of redistribution are assessed in the traffic modelling process, and (c) schemes where the formulae used in the COBA program are considered inappropriate, e.g. some conurbation schemes.

Single year rate-of-return method

Whilst there is general agreement that the net present value method is the correct one to apply in most highway economic appraisals, there are occasions when the complex data required by the computational procedure are not available. In addition, the situation may arise where there is a need to provide a simple rate for the development of priorities, which is readily understood by people who are not well trained in methods of economic evaluation. In such instances, it may be appropriate to use the single year rate-of-return method of appraisal.

With this relatively simple appraisal method, the rate of return, i.e. the ratio of the road user benefits to the total capital cost of the scheme (expressed as a percentage), is calculated for a specific year in the future. This year may be the year when the improvement is open to traffic (most commonly), the design year, or indeed any other intermediate year which may be chosen for some particular purpose, e.g. comparability with some other proposals. Whatever the year selected, the single year rate of return tends to be known (deceptively) as the first year rate of return.

In equation form, the single year rate of return may be expressed as

$$R = \frac{O + A - M}{C} \times 100$$

where R = rate of return (%), O = savings in vehicle occupants' costs and vehicle operating costs, A = savings in costs of accidents, M = additional annual maintenance costs, and C = total capital cost of the improvement.

This rate-of-return method is most appropriately used when the following three assumptions apply[6]: (a) that the rate of growth of benefits is the same for all schemes, (b) that this rate is always positive and sufficient to ensure a net present value at whatever is the going discount rate, and (c) that the capital costs of the schemes are spent quickly or have the same profile of expenditure over time.

Internal rate-of-return method

This procedure involves determining a discount rate which brings all future benefits into equality with the capital costs. It attempts to allow for some of the limitations in the single year rate-of-return method by recognizing that all future years within the appraisal period must be taken into account in the analysis. In its simplest form, the method may be described as follows.

The total capital cost of the improvement is first determined. Next the benefits occurring in each year of the future are estimated separately, and from these are taken the maintenance costs likely to be experienced in the same years. The values obtained are then substituted in the following formula and the rate of return calculated:

$$C = \frac{(B_1 - M_1)}{(1+r)} + \frac{(B_2 - M_2)}{(1+r)^2} + \ldots + \frac{(B_n - M_n)}{(1+r)^n}$$

where C = value (i.e. capital cost) of the investment in year zero, B = value of the benefits which occur in any particular year, M = maintenance costs incurred in the same year, n = number of years into the future for which the rate of return is to be calculated, and r = rate of return per year expressed as a fraction.

If the rate of return calculated in this way is greater than the discount rate, then the road improvement is considered to be worthwhile.

Supporters of the internal rate-of-return method of appraisal have tended to stress its practical aspects, particularly its ready intelligibility. In practice, however, it has rarely been applied to highway economic evaluations. Whilst it has the advantage of being easily understood, the way in which it uses the discounting principle is generally considered by economists to be contrary to the intentions underlying the use of the discount rate in the public sector.

Benefit–cost ratio method

This method of appraisal attempts to assess the merit of a particular scheme by comparing the annual benefits from the scheme with the increase in annual costs. The increase in annual highway costs used in this analysis is equal to the annual maintenance costs on the improvement plus its estimated annual amortized cost (irrespective of whether it is publicly financed or not) less the annual cost associated with the existing highway. The comparison is made by the following simple formula in which the estimated annual road user benefits are divided by the increased annual highway costs.

$$B/C \text{ ratio} = \frac{\text{annual benefits from scheme}}{\text{annual costs of scheme}} = \frac{R - R_1}{H_1 - H}$$

where R = total annual road user cost for the basic condition (usually the existing road), R_1 = total annual road user cost for the proposed scheme (this includes travel on existing facilities affected by the scheme), H = total annual highway cost for the basic condition, and H_1 = total annual highway cost for the proposed scheme.

A benefit–cost ratio of less than unity is normally interpreted as indicating a poor investment, i.e. it implies that the benefits derived by the road users are less than the funds to be invested in the highway. In reality, because of the general paucity of highway construction funds, a project is usually required to have a B/C ratio considerably in excess of unity before it is considered suitable for development purposes.

When the appraisal method is used in the selection of an alternative location or design, each alternative is normally compared with the basic condition with the same volume of traffic being used in all cases; the preferred alternative indicated by the analysis is then the one for which the B/C ratio is the highest. Where several major alternatives are under study, however, a second analysis may be made using the preferred one as the base, to determine if an added increment of investment might yield a proportionately larger increase in road user benefits on another alternative.

Components of an economic appraisal

As will have been gathered from the above, the traffic planner/engineer involved in an economic evaluation of a road improvement proposal must carry out an analysis involving such factors as motor vehicle operating costs, time value of individuals, safety of individuals and property, the economic value of land, buildings, businesses and resources, the economic cost of capital investments, and the maintenance and operating costs of physical property. Educational, social, environmental and general community values are not included in economic analyses since they cannot be reduced to supportable and realistic monetary values. The following discussion therefore is primarily concerned with those factors, and their consequences, which can be readily quantified and converted to monetary values—in particular, those which are of special interest in British highway economic studies.

Cost of highway improvements

The cost of a highway improvement really consists of both the total initial cost and the subsequent maintenance costs.

Initial-cost items
The items making up the total initial cost of the road improvement can be summarized as follows. (Typical construction costs are given in Fig. 6.3, whilst representative maintenance costs are included in Table 6.10.)

(1) *Land* This includes the costs of acquiring land, less the proceeds from the sale of surplus land, plus other costs such as payments for severance, injurious affection, disturbance and rehousing and compensation for depreciation. Where land has been purchased in advance of its use for a scheme, its value may have changed in the interval, i.e. reflecting a change in the opportunity cost of the land. Notwithstanding what has been paid for the land, it is its current value in relation to its most probable alternative use that is used in the economic analysis.

(2) *Main works contract* Components of this initial-cost item normally include preliminary work, e.g. site clearance and preparation, drainage, earthworks, main carriageway, interchanges, side roads, road signs, lighting and road furniture, structures (bridges, subways, etc.), and accommodation works. Accommodation works are those carried out on land and property adjoining the road, which remain the property of somebody other than the highway authority; typical examples are fencing and walling. Earthworks is an important component of the major works contract; its cost normally includes expenditures on retaining walls since they are closely allied to this construction activity. Carriageway and side road costs include those for kerbs and channels, shoulders and verges, footpaths, cycle tracks, central reservations and, of course, the construction costs for surfacing, roadbase and subbase.

(3) *Ancillary works contracts* These typically include the costs of maintenance compounds, lighting, motorway communications, landscaping, and noise insulation.

(4) *Work by other authorities* Normally, these involve the expenses incurred by bodies such as British Rail, local authorities, and statutory undertakers. The costs of moving and relaying gas and water mains, electric and telecommunication cables, etc. are typical statutory undertaking costs.

(5) *Administration and preparation* These include consulting engineer and/or agent authority fees, the actual costs of pursuing alternative routes in the early stages of a scheme, public consultation and public inquiry costs, and the costs of any surveys carried out during a scheme's preparation. Typically, in Britain, these costs vary from 12 per cent of scheme construction and land costs at the roads' programme entry stage, to 2 per cent at the works commitment stage.

(6) *On-site supervision and testing* These relate to the costs of site staff, and include expenditures on the on-site testing of materials. Typically, these on-site costs total about 3 per cent of scheme construction and land costs.

Construction costs used in highway economic assessments are normally derived using item rates from comparable schemes previously quoted by contractors. It is usually necessary for the original estimates to be updated at different stages in the scheme development, and the Road Construction Price Index (RCPI), published by the Department of the Environment, is used for this purpose.

The construction costs need also to be allocated to each year during which construction takes place. Table 5.4 provides a guide to the distribution of construction costs (excluding land costs) over varying construction periods.

Table 5.4   Time distribution of construction work costs for a typical highway scheme

Contract period (years)	Percentage of total cost				First scheme year
	Years before opening				
	4	3	2	1	
1.5	—	—	29	68	3
2.0	—	—	47	50	3
2.5	—	16	42	39	3
3.0	—	30	34	33	3
3.5	11	29	30	27	3
4.0	22	25	25	25	3

Maintenance costs

In road improvement economic studies, maintenance costs are normally regarded as an equivalent stream of constant costs arising in each year of the evaluation period, even though some expenditures may actually be incurred at infrequent intervals. Costs taken into account are divided into two broad categories, viz. non-traffic-related costs and traffic-related costs.

As the description suggests, *non-traffic-related costs* comprise expenditures which are not mainly due to traffic flow. Typically, they include the costs incurred in the maintenance of: drainage systems; footpaths and cycle tracks; safety barriers and fences; boundary fences; culverts and subways; remedial earthworks; verges; road sweeping and gulley emptying; traffic signals and signs; street lighting; roadstuds and markings; salting, snow ploughing and snow fencing; motorway compounds. These maintenance costs are generally described as an annual cost per unit length of road.

The *traffic-related costs* comprise maintenance expenditures that result from traffic flow. Typical examples of this type of cost are: overlay strengthening of the road structure; reconstruction; surface treatments to restore loss of resistance to skidding; resealing of joints in concrete pavements; resurfacing of hard shoulders, raised kerbs, etc.; local patching. Also included in this category are the time, accident and vehicle costs associated with the carrying out of traffic-related maintenance operations.

Evaluation of road user benefits

Highway improvements can bring benefits to the community in many ways, the more obvious ones being related to the road user. These are

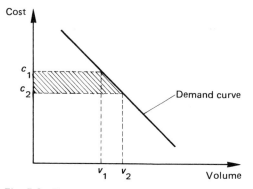

Fig. 5.2 The road user benefit concept

reflected by reduction in the cost of vehicle operation, savings in time and accidents, and increased comfort and convenience, with accompanying increases in the volumes of traffic that can be handled. The benefits derived are shown schematically in relation to the demand (or 'indifference') curve in Fig. 5.2.

Let it be assumed that an economic appraisal is being carried out on a proposal to replace a heavily-travelled winding country road with a new well-aligned single carriageway highway. On the old road, the volume of traffic is v_1 and the road user cost is c_1; as the cost of the trip decreases via the new road, the volume of traffic will increase to v_2 until equilibrium is achieved at cost c_2. The road user benefit derived from the highway change is then the hatched area in Fig. 5.2; this comprises the rectangle of cost savings to the existing v_1 of users, plus a triangle of benefits to new users as the traffic volume increases from v_1 to v_2.

The demand curve, established by joining the points of equilibrium, represents the consumer's choice for the highway in question. If the slope of the curve is towards the horizontal, the demand is said to be elastic, i.e. small cost changes result in fairly large changes in consumption. However, if the curve tends towards the vertical, the demand is said to be inelastic, i.e. large cost changes are needed to produce small changes in consumption.

From traffic planning studies, it is possible to estimate the changes in traffic volumes (and speeds, delays and accidents) that may be associated with a highway scheme. Once these have been determined in terms of person- and vehicle-hours, vehicle-kilometres, and accidents saved, they are translated into monetary terms so that the benefits can be calculated for use in the appraisal.

The benefits associated with road proposals will obviously vary from project to project. Typical savings, divided according to benefit type, are shown in Table 5.5. The non-working-time savings referred to in this table are comprised as follows: journeys to work and education establishments, 9 per cent; shopping, 4 per cent; personal business, 2 per cent; social activities, 14 per cent; holidays, 8 per cent.

Table 5.5   Average benefits from a road improvement scheme[9]

Component	Percentage
Accident savings	15
Vehicle operating costs	−11
Working-time savings	
cars	37
light goods vehicles	9
heavy goods vehicles	12
buses	1
Non-working-time savings	
cars	34
buses	3

Road user operating costs
The main factors affecting road user operating costs are the vehicle operating costs (which are mainly influenced by the type of vehicle, the type of locale in which the vehicle is operated (e.g. urban or rural), speed of vehicle operation, the type of highway and the quality of the carriageway surface, and the horizontal and vertical alignment of the highway) and the value placed upon the vehicle occupants' time.

Vehicle operating costs   The changes in total vehicle operating costs in relation to a new highway or highway system depend upon the distances travelled by vehicles and upon changes in average link speeds. British practice is to include the following six cost items when determining formulae for vehicle operating costs: fuel, oil, tyres, maintenance, and depreciation and size of vehicle fleets. Only resource items which vary with the use of the vehicle are measured; thus indirect taxes such as fuel tax (VAT) and licence fees are excluded, i.e. the revenue raised by these taxes would probably have been raised anyway[10]. Insurance premiums are also excluded from the calculations, since to include these and to value savings in accidents would be double-counting.
The resource cost of *fuel consumption* is determined by a formula of the general form:

$$C = (a + b/V + cV^2)(1 + mH + nH^2)$$

where C = cost (pence/km), V = average link speed (km/h), H = average link hilliness (either the total rise and fall per unit distance for two-way links, or the total rise per unit distance for one-way links), and a, b, c, m and n are parameters defined for each category. This function results in high vehicle operating costs at low speeds, reflecting the effects of stop–start motoring in congested traffic conditions. The adjustment for gradient is to take account of increased fuel consumption at steeper gradients.
The marginal resource costs of *oil* and *tyres* are assumed to be fixed costs per kilometre. *Vehicle maintenance* is assumed to vary with speed in the same way as fuel consumption, but without the adjustment for gradient. *Depreciation* for goods and public service vehicles is assumed to be entirely related to the kilometres travelled. Only the kilometre-related

element of car depreciation is considered to be a marginal resource cost; the passage-of-time-related component is excluded. Increased speeds yield savings in the *service fleets* of working cars, commercial vehicles, and public service vehicles, and an estimate of this effect is also included in vehicle operating costs.

The non-fuel elements of the marginal resource costs are combined in a formula similar to that used for fuel consumption, but without a gradient effect, viz:

$$C = a + b/V + cV^2$$

The economic parameter values used in vehicle operating cost determinations in Britain are included in Table 5.6.

Value of time Also shown in Table 5.6 are the values of time for different types of vehicle occupant used in economic appraisals (1979 prices). As this table reflects, these normally contribute the greatest benefits in the economic analysis, and so it is appropriate to comment upon them.

There is widespread acceptance of the premise that a saving in commercial vehicle time has value corresponding to a reduction in those operating costs, such as hire of drivers and the hourly rental of equipment, which are directly related to time. However, not all passenger vehicle trips are for the purpose of business; they may be shopping trips, pleasure trips or journey-to-work trips, and controversy regularly arises over the values to be attached to these non-working trips.

In relation to non-working trips, it might be noted that, notwithstanding the many traffic planning and highway economic analyses that have been carried out, there is still only limited tangible evidence[11] available as to how the marginal value of time savings via a specific mode is affected by: (a) the size of the time saving, (b) the overall length of journey (or of the particular activity engaged in), (c) the direction of travel, i.e. whether to or from the home, (d) the income level of the respondent, and (e) the effect of the weather upon the marginal value of walking and waiting times. Various studies have given different answers, and so it is probably true to say that the only point about which there is good agreement is that it is better that some price, rather than no price, should be attached to non-working-time savings.

An important consideration in relation to any pricing of time savings is the relative importance of *comfort and convenience*. That most motorists place a value on comfort and convenience is well demonstrated by the significant diversions to new and modern highway facilities, even though greater distances and times can be required. Certainly there is comfort value, over and above the saving in operating cost, in being able to drive without frequent brake applications, stops or starts, or unexpected interference to travel. There is further value in the conservation of health through driving in a relaxed manner without the tensions necessary where roadside interference is imminent. These facts must be drawn upon to explain *in part* instances on modern highway facilities, such as motorways

Table 5.6 Summary of parameter values used in highway economic appraisals (1979 figures)[8]

Item description	Cars	Light goods vehicles	Heavy goods vehicles		Public service vehicles
			<25 t	>25 t	
National average veh-km proportions (%)					
in work mode	16.7	100.0	100.0	100.0	100.0
in non-work mode	83.3	0	0	0	0
Vehicle occupancy					
in work mode	1.21	1.30	1.20	1.20	1.25 crew, 0.15 passengers
in non-work mode	1.87	0	0	0	14.17 passengers
Value of time (pence/h)					
driver in work mode	411.2	258.4	300.0	300.0	276.8 driver, 270.6 conductor
driver in non-work mode	55.4	—	—	—	—
passenger in work mode	328.8	258.4	300.0	300.0	272.9
passenger in non-work mode	55.4	—	—	—	55.4
Resource vehicle operating cost formulae parameter values					
a fuel*	0.51	0.70	1.87	3.54	2.76
non-fuel	2.35	5.33	10.86	15.21	18.81
b fuel*	26.57	29.98	26.42	49.98	44.53
non-fuel	14.72	27.69	39.00	72.99	92.40
c fuel*	0.000065	0.000077	0.000116	0.000219	0.000131
non-fuel	0.000029	0.000050	0.000081	0.000101	0.000093
m fuel	−0.002030	−0.001250	0.003460	0.003460	0.003460
n fuel	0.000102	0.000067	0.000048	0.000048	0.000048
Speed-flow category	Light	Light	Heavy	Heavy	Heavy

*a, b and c for cars and light goods vehicles for fuel only are 15% lower on motorways

into urban areas, where the traffic volumes are substantially greater than those predicted from the origin and destination surveys.

Positive identification of values for assignment to various degrees of comfort and convenience are not yet possible. Present practice in highway economic studies in Britain is to omit this factor altogether, except of course inasmuch as it is already taken into account in the operating costs.

One further feature which might be noted is that values of time are related to levels of personal income, and these are expected to grow over the long term at the same rate of growth as the Gross Domestic Product (GDP) per head. The manner in which the gross domestic product is expected to vary in Britain over the course of the next decade is summarized in Table 5.3, for both low and high traffic growths.

Accident costs

Any increase in safety of vehicle operation is of benefit to the road user. In most countries, the approach taken in respect of determining the expected cost to society of a fatal, serious or slight injury accident assumes that the cost has three components, as follows:

(1) loss of output due to death or injury,
(2) medical and ambulance, police and administration, and property damage costs,
(3) costs of pain, grief and suffering to the casualty, relatives and friends.

In Britain, the loss-of-output calculation involves estimating the present value of the expected loss of earnings plus any non-wage payments (national insurance contributions, etc.) paid by the employer. Note that this definition does not attempt to make any deduction for what otherwise might have been consumed had the person(s) in the accident, for example, not been killed. The reasoning employed is that the costs determined should be those which relate to the benefits which arise when accidents are *prevented*, i.e. the accidents costed are those which do not occur but which, without the introduction of the road improvement, would have occurred; thus, since the persons concerned are still alive, their consumption should not be deducted when assessing the benefits of preventing accidents.

The medical and ambulance costs include the costs of hospital care. Obviously, these and the police, administrative and damage costs are taken into account as they represent a real expense to the community as a result of accidents.

The most difficult determination of all is that of the costs to be used for pain, grief and suffering. These are very real costs to society; however, by their very nature they are not directly quantifiable in monetary terms. In recognition of the relevance of these costs, a notional allowance has always been included in the accident cost calculation. It used to be that this allowance was a minimal one; however, following a detailed examination of the governmental approach to the assessment of road

Table 5.7   Components of accident costs in 1979, in 1979 prices[8]

Component	£ per accident
Direct financial costs (medical, police, damage to property, etc.)	1000
Lost output	2150
Pain, grief and suffering	1640
Total	4790

improvements[1], which suggested that this was not in line with general principles of cost–benefit analysis, the (then) minimal value per accident was raised by 50 per cent. (However, it should be kept in mind that a survey of studies which have attempted to evaluate how individuals value risk revealed 'sorrow' figures for value of life which were between 2.5 and 10.0 times the *raised* value.)

Table 5.7 lists the component costs of an average personal-injury accident, as determined for 1979, at 1979 prices. In respect of this table, it should be appreciated that the relative size of each component item varies according to the severity of the accident under consideration.

Overall, the costs of personal injury and death form the main element of any valuation of accidents, i.e. in 1979, the average cost of a damage-only accident was £350 (1979 prices). Table 5.8 lists the average cost per casualty figures for 1979; these are used by the Department of Transport in economic assessments involving accidents.

The severity of an accident varies according to the locale in which it took place. Table 5.9 summarizes the injury and damage-only accident costs for 1979, divided according to highway type. Note, for example, that the costs for rural general-purpose highways are higher than those for urban roads; this is because a rural accident involves more casualties and more seriously injured casualties than does an urban one, together with a greater amount of vehicle damage.

Comparing Tables 5.8 and 5.9, it can also be seen that the casualty costs are lower than the corresponding accident costs. There are two reasons for this. Firstly, accident costs include certain costs that are not normally included in casualty costs, e.g. costs of damage to vehicles and property, costs of police, and administrative costs of accident insurance. Secondly, although accidents are classified according to the most seriously injured casualty, they usually involve more than one casualty (see Table 5.10).

Table 5.8   The average cost per casualty in 1979, in 1979 prices[8]

Type of casualty	Cost (£)
Fatal	101900
Serious	4310
Slight	100
Average for all casualties	3040

Table 5.9 The average costs of accidents on different types of road in 1979, in 1979 prices[8]

Highway type	Cost of accident (£)					Average accident cost per injury accident
	Fatal	Serious	Slight	Average injury	Average damage-only	
All-purpose roads						
urban	108000	5510	720	3600	340	5750
rural	121300	7230	1260	8680	410	10570
Motorway	128900	7410	1380	8300	490	10480
All roads	114200	6060	830	4790	350	6890

In any given analysis, the most appropriate cost figures used will depend upon the amount of data available at the time.

If the numbers of fatal, serious and slight accidents are known, then the costs are normally estimated separately for each type. Where only the total number of injury accidents is known, the average injury accident cost may be used. If there are no prior data available about the number of damage-only accidents, average cost figures may be applied on estimates derived from the number of injury accidents, i.e. on average, the ratios of damage-only accidents to injury accidents on urban roads, rural roads, motorways, and 'all roads' are 6.4 to 1, 4.6 to 1, 4.5 to 1, and 6.0 to 1, respectively.

If injury accidents cannot be differentiated by severity, then instead of making separate calculations for the cost of damage-only accidents and the cost of injury accidents, it is not uncommon to use a cost figure which values all injury accidents by incorporating an appropriate allowance for damage-only accidents. This cost is called the average accident cost per injury accident (see Table 5.9).

For consistency purposes, all of the accident costs quoted in this chapter

Table 5.10 Average number of casualties per accident[12]

Casualty, classified according to road type	Number of casualties per accident, classified according to severity		
	Fatal	Serious	Slight
Urban roads			
Fatal	1.03	—	—
Serious	0.23	1.11	—
Slight	0.25	0.25	1.18
Rural roads			
Fatal	1.16	—	—
Serious	0.71	1.36	—
Slight	0.50	0.51	1.42
Motorways			
Fatal	1.35	—	—
Serious	0.85	1.50	—
Slight	0.69	0.66	1.52

Table 5.11 Methods of determining annual average daily flow[(13)]

Number	Method description	Accuracy classification*	Possible variations	Comments on analysis
I	*Using manual counters only*			
(i)	Count for 1 hour, on a weekday, between 9 a.m. and 6 p.m.	D	Count could be lengthened by any convenient amount	
(ii)	Count on 1 weekday from 6 a.m. to 10 p.m.	C or D		
(iii)	Count from 6 a.m. to 10 p.m. on a successive Friday, Saturday and Sunday	C	Could be extended to 4 days by including Monday	Estimate week's total as (5 × Friday) + Saturday + Sunday
(iv)	Count from 6 a.m. to 10 p.m. on 7 consecutive days	C		
(v)	As (i) to (iv) but carried out on 4 occasions at 3-monthly intervals (for (i) and (ii) use different hours and days)	C	The number of occasions could be 2, 3 or 6 instead of 4, with appropriate alteration in spacing	As (i) to (iv) to estimate weekly totals, then average the four weekly totals
(vi)		C		
(vii)		B		
(viii)		B		
(ix)	Count from 6 a.m. to 10 p.m. every 52nd day for a year (7 counts in all)	B		
(x)	Count from 6 a.m. to 10 p.m. every 26th day for a year (14 counts in all)	A or B		
(xi)	Count from 6 a.m. to 10 p.m. every 13th day for a year (28 counts in all)	A		
(xii)	As (ix) but divide the part of the day of interest into 7 equal parts (e.g. of 2 hours each); on each of the 7 days count successively parts 1, 4, 7, 3, 6, 2, 5	C	Other similar arrangements of parts are equally suitable (e.g. 4, 7, 3, 6, 2, 5, 1)	
(xiii)	As (x) but divide the day into 14 equal parts and count successively parts 1, 4, 7, 10, 13, 2, 5, 8, 11, 14, 3, 6, 9, 12	C		

Table 5.11 (cont.)

Number	Method description	Accuracy classification*	Possible variations	Comments on analysis
(xiv)	As (xi) but again divide the day into 14 equal parts and count successively parts 1, 6, 11, 2, 7, 12, 3, 8, 13, 4, 9, 14, 5, 10 and then repeat the cycle	B		
II	*Using automatic counters only*			
(i)	Continuous count for 1 week	C		
(ii)	4 continuous counts for 1 week at 3-monthly intervals	B	Replace 4 counts by 2, 3 or 6 at appropriate intervals	
(iii)	Continuous count for 1 year	A		
III	*Manual and automatic counts*			
(i) to (xiv)	As I but with continuous automatic count for whole year	A	The continuous count could be reduced to 1, 2, 3, 4 or 6 equally spaced 1-week counts. Accuracy would then be C, C, B, B. B	The manual counts should be analysed in the same way as in I(i)–(xiv), but the results of this should be solely to give the average percentage composition of the traffic. The actual flows of each type of vehicle should be obtained by applying these percentages to the total flow obtained from the automatic count

Notes: The total number of vehicles counted divided by the number of hours of counting gives the average hourly flow for the times of day covered. The hours 6 a.m. to 10 p.m. may be extended if desired, but usually it will be sufficient to add 10 per cent for the night hours 10 p.m. to 6 a.m. Methods I(i)–(iv) and II(i) are the only ones available if an answer is required at short notice. Methods I(ix)–(xiv) are especially useful when counts have to be made at a number of points in the same area.
*See Table 5.12

have been given in June 1979 prices. For evaluation purposes, however, it is necessary to adjust these as they vary in real terms in future years (for example, see the 1983 values given on p. 452). As the main elements of accident costs are proportional to national income, it is usual practice to assume that the values change in line with the gross domestic product. This is comparable with the practice in respect of the future value of time.

Traffic and accident data required

Before economic studies can be carried out, it is necessary to obtain information about the traffic flows, speeds, delays, accidents, etc., on the roads concerned. Present conditions should be obtained by direct observation, and these, when taken in combination with knowledge based on previous research, can be used to predict changes in the future. The following discussion assumes that an analysis is to be carried out on a single link of a route.

Traffic flows

The first and most basic step is to measure the existing traffic flow, and its composition, in each direction.

The annual traffic flows on existing facilities may be conveniently estimated by means of sample counts. These counts can be of two types: firstly, the counts that are made so as to be sufficiently representative of the whole year (i.e. so that all times of the day, all days of the week, and all the seasons occur in appropriate proportions) and, secondly, the counts that are not made during representative periods, but in which this is allowed for in the analysis by applying correction factors. Both methods provide suitable results, but for practical reasons the second is the one most usually carried out.

Methods of determining the current annual average daily traffic flows have been tabulated by the Transport and Road Research Laboratory, and are shown in Table 5.11. Each of these methods can be, under certain circumstances, the 'best' one to use, and the decision in any instance is made on the basis of local knowledge and availability of facilities.

The accuracy classification used in Table 5.11 is explained in Table 5.12. It should be emphasized that these measures of accuracy are not universally applicable; there may be, for instance, circumstances in which a method classified as C may be relied upon to give perfectly satisfactory results. Again, each case should be considered individually in the light of local conditions.

In practice, traffic census data are most usually gathered on a weekday (Monday–Friday for 12 or 16 hours, starting at 07.00 hours or 06.00 hours, respectively) and the count is then grossed up by applying a multiplier, M, to obtain the annual traffic flow (i.e. all motor vehicles for 365 24-hour days). Thus, for the 16-hour count,

annual traffic flow $= M \times$ 16-hour flow

Table 5.12 Accuracy classifications to be read in association with Table 5.11[13]

Category	Error exceeded with probability of 1 in 10	Interpretation
A	Up to 5 per cent	Very satisfactory
B	5 to 10 per cent	Satisfactory for all normal purposes
C	10 to 25 per cent	Good enough for a rough guide
D	25 to 50 per cent	Unsatisfactory
E	Over 50 per cent	Useless

The value of the M-factor depends upon the month of the year and the type of site at which the count is taken. Appropriate monthly 16-hour M-factors for different types of highway are given in Table 5.13.

Site-to-site variations in the sample count are minimized by ensuring that it is taken at 'neutral' times of the year, e.g. May and September. Day-to-day variations within a given month can be minimized by not collecting data during weeks containing a Bank Holiday and Fridays preceding a Bank Holiday Monday. These day-to-day variations are particularly significant in recreational areas, and additional steps suggested[15] for these locales include: (a) avoiding market days and half-day-closing days, (b) avoiding days with exceptionally bad weather or exceptionally good weather, (c) avoiding counts during weeks immediately prior to and immediately after Bank Holidays, and (d) repeating the count in the same month.

Once the current annual traffic volume determinations have been made, it is necessary to project them ahead into future years, starting with the year when the highway is opened to traffic. One way of doing this—and this method is most applicable to non-urban roads—is simply to expand the data using expansion factors such as are given in Table 5.2.

Table 5.13 M-factors used to expand 16-hour daily traffic flows to annual traffic flows[14]

Month	Route			
	Urban/ commuter	Non-recreational, low-flow	Rural, long-distance	Recreational
January	393	408	478	637
February	387	404	456	596
March	391	398	445	560
April	373	377	414	495
May	363	367	400	432
June	367	354	391	323
July	375	359	356	271
August	371	369	338	227
September	363	369	372	330
October	364	382	399	477
November	368	384	438	583
December	362	391	441	603

Speed and delay

The next step is the estimation of present speeds or journey times over the roads and/or intersections to be improved or affected by the improvement. It is important that the speed and time determinations should refer to all stretches of road likely to be affected. In the case of a new highway or a bypass, all existing roads should be studied from which substantial amounts of traffic are likely to be diverted; it is probable that speeds on these roads will rise if traffic is diverted from them.

Speed   Methods of obtaining speeds of traffic on existing facilities are described in Chapter 3 and so will not be discussed here. Ideally, the most up-to-date speed data at the site being evaluated should be obtained whenever possible.

Measurements taken at various locations have resulted in the establishment of speed–flow relationships[16, 17] which can be used as the basis of 'future' speed data input to analyses of suburban and rural road improvements; these relationships—which include the effects of traffic composition, and those of the hilliness and bendiness of the highway alignment—are described in Chapter 6.

Delay   The relationships given above, and those described in Chapter 6, relate only to speeds of vehicles when travelling between major intersections. Obviously, however, any economic analysis must take into account changes in the amount of time during which vehicles are forced to stop or slow down at major junctions.

Current delays are best determined by field surveys (see Chapter 3). In the case of future traffic delays, the delay calculation for *traffic-signal-controlled intersections* can be based on the Transport and Road Research Laboratory's traffic signal formula[18, 19]. This involves determining the optimal cycle time for the signals at given times and then using the values determined in a delay formula. The determination of the total delay represents a fairly considerable calculation in that it involves reworking the optimal cycle time and delay calculation for each hourly flow group and each new year considered in the appraisal.

The basic process involved in estimating the delay at a signal-controlled intersection is described in Chapter 7.

The delay which an individual vehicle experiences at a *priority-controlled conventional roundabout* can be regarded as consisting of two main components, viz.: (a) the waiting time due to other vehicles, i.e. the traffic-dependent element of delay, and (b) the layout effect due to the extra distance travelled plus the deceleration and acceleration times.

The waiting or 'congestion' delay is often assumed to be equal to twice that which occurs in the critical 'weaving' section (i.e. the section with the greatest flow–capacity ratio), on the assumption that the average vehicle passes through two sections of a four-arm normal roundabout. This delay can be estimated using the following well-known function:

$$D = e^{3.15k}$$

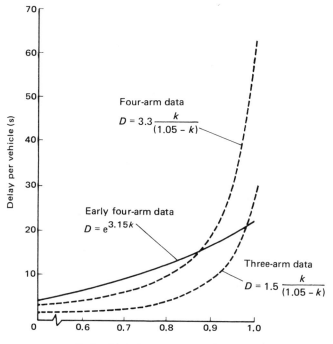

Ratio of exit flow to capacity (k)

Fig. 5.3 Observed delay–flow relationships at three- and four-arm roundabouts

where D = average congestion delay per vehicle (s) and k = flow-capacity ratio in the critical section.

Figure 5.3 shows the average delay as given by this equation compared with more recent comparable data at three- and four-arm roundabouts. Note that there is a considerably lower delay associated with the three-arm roundabouts at the same level of flow. The formula used to describe the capacity of the critical section is given in reference 20.

The 'fixed' or geometric delay per vehicle—which results from a vehicle slowing down, travelling the extra distance about a normal roundabout (the average journey distance is taken as passing through two 'weaving' sections), and then accelerating—can be estimated from Table 5.14.

Table 5.14 Calculated average fixed delay per vehicle (in seconds) at a priority-controlled conventional roundabout due to layout effects

Normal speed for road (km/h)	Diameter of roundabout (m)		
	75	75–150	150
50	5	6	8
65	6	8	10
80	10	12	15

Total daily delays to vehicles at *mini roundabouts* can be determined using formulae that are described in Chapter 6 (see p. 437).

Delays to traffic at *priority intersections* are generally considered to be due to two main factors: (a) the waiting time due to main road traffic flow, and (b) the deceleration and acceleration effect associated with turning and crossing vehicles. In this context, it is normally assumed that through traffic on the major road is not delayed at all, and that delay is caused to major and minor road vehicles as they wait for suitable gaps to carry out cutting or merging manoeuvres.

It is now possible (see p. 437) to make an empirical calculation of the level of delay experienced at side roads of priority junctions as a result of traffic flow on the main road. An estimate of the average 'waiting' delay per vehicle can also be made using a complex theoretical mathematical model that is described in the literature[21]. An indication of the scale of these delays can also be gained from Table 5.15; the calculations on which these theoretical data are based assume that traffic along the minor road divides at the (four-way) junction in the proportions 50 per cent straight ahead, 25 per cent turning left, and 25 per cent turning right.

Typically, 5–10 seconds need to be added to the theoretical delay determined for the average side road vehicle to allow for the acceleration and deceleration of vehicles.

Delays to vehicles due to *gate and swing bridge closures* can be estimated from the following formula:

$$D = Qnt^2/48$$

where D = total daily delay (s), Q = total flow, assumed random (number of vehicles), n = number of closures per 24-hour day, and t = duration of each closure (minutes).

Table 5.15 Congestion delays at priority junctions to vehicles on the minor road (in seconds per vehicle)

Flow on minor road (veh/day)	Flow on major road (veh/day)					
	5000	10000	15000	20000	25000	30000
100	10	15	20	30	49	295
500	12	18	24	35	75	395
2000	12	18	28	67	370	990
5000	12	23	138	395	985	>1800
10000	17	190	985	>1800	>1800	—

Accidents

Information on accidents is most necessary in any economic evaluation of the effects of highway improvements. The accidents which should be considered can be divided into two categories, i.e. personal-injury and non-injury (damage-only) accidents. The former, however, are generally more serious and important, while the reporting of the latter is sometimes unreliable; hence a common practice is to concentrate attention on the personal-injury accidents and to make some allowance for associated non-injury accidents.

The first step in forecasting possible changes in personal-injury accidents is to establish from accident records the average annual occurrence of such accidents over several years on the stretches of road to be improved and which will be affected by the improvement. For each existing section, the accident rate is normally determined from the following formula:

$$ PIA = \frac{(A_1 + A_2 + A_3) \times 10^6}{(F_1 + F_2 + F_3) \times L} $$

where PIA = personal-injury accident rate per 10^6 veh-km, A_1, A_2, A_3 = annual totals of all personal-injury accidents (including fatals) for each of the most recent three consecutive years for which data are available, F_1, F_2, F_3 = annual total traffic flows for the same three years (number of vehicles), and L = length of the section (km).

This number can then be corrected for future traffic increases by assuming that personal-injury accidents will increase at the same rate as the total traffic. The result is an estimate of the number of accidents before the improvement is opened at the future traffic flow.

The change in accidents resulting from the improvement may then be estimated by either or both of two different ways. The first method uses data from appropriate before-and-after studies, which give an indication of the percentage change in accidents which can be expected. The second way of estimating changes in accidents due to road improvements is by using known data regarding accident rates on similar types of road. The use of known accident rates is preferable when dealing with future bypasses, entirely new facilities, or radical road improvements, since the improvement can then be regarded as the transfer of traffic from conditions bearing one accident rate to conditions bearing an entirely different rate. Some accident rate data which can be used to predict personal-injury accidents on different types of new or improved highway are given in Table 5.16.

Table 5.16 Personal-injury accident rates on various types of highway

Type of highway	PIAs per 10^6 veh-km
Rural motorways	0.4
All-purpose rural roads	
single 2- and 3-lane	1.2
widening to dual on existing alignment	1.0
dual carriageway on new alignment	
(with and without grade-separation)	0.6
All-purpose dual carriageways	
with urban characteristics	3.1
with all junctions grade-separated	1.2

Selected bibliography

(1) Leitch Sir G et al., *Report of the Advisory Committee on Trunk Road Assessment*, London, HMSO, October 1977.

(2) Winfrey R, *Concepts and Applications of Engineering Economy in the Highway Field*, Highway Research Board Special Report 56, 1959, pp. 19–33.

(3) *A Manual on User Benefit Analysis of Highway and Bus-transit Improvements*, Washington DC, The American Association of State Highway and Transportation Officials, 1977.

(4) Winfrey R, *Economic Analysis for Highways*, Scranton, Pa., International Textbook, 1969.

(5) *An Introduction to Engineering Economics*, London, The Institution of Civil Engineers, 1962.

(6) Harrison A J, *The Economics of Transport Appraisal*, London, Croom Helm, 1974.

(7) Dawson R F F, *The Economic Assessment of Road Improvement Schemes*, Road Research Technical Paper No. 75, London, HMSO, 1968.

(8) *COBA: A Method of Economic Appraisal of Highway Schemes*, London, The Department of Transport, May 1981 (updated to May 1983).

(9) Searle G A C, Cost benefit analysis: felicific calculus or black art?, *Proceedings of the Seminar on Highway Design*, PTRC Report P169, pp. 257–266, London, Planning and Transport Research and Computation (International) Company Ltd, July 1978.

(10) *The Treatment of Taxation in Cost Benefit Analysis*, Highway Economics Note No. 3, London, The Department of Transport, December 1977.

(11) Heggie I G, The value of time, modal choice and the justification for road improvements, *Proceedings of the Seminar on Highway Planning*, PTRC Report P107, pp. 14–40, London, Planning and Transport Research and Computation (International) Company Ltd, July 1974.

(12) Dawson R F F, *Current Costs of Road Accidents in Great Britain*, RRL Report LR396, Crowthorne, Berks., The Road Research Laboratory, 1971.

(13) Road Research Laboratory, *Research on Road Traffic*, London, HMSO, 1965.

(14) *Local Speed Limits*, Circular Roads No. 1/80, London, The Department of Transport, February 1980.

(15) Davies P W and Hunjan J S, Some comments on the COBA method, *The Highway Engineer*, 1981, **28**, No. 1, pp. 14–18.

(16) Duncan N C, *Rural Speed/Flow Relations*, TRRL Report LR651, Crowthorne, Berks., The Transport and Road Research Laboratory, 1974.

(17) *Speed/Flow Relationships on Suburban Main Roads*, London, Freeman Fox and Associates, January 1972. (A report of a study carried out for the Road Research Laboratory.)

(18) Webster F V, *Traffic Signal Settings*, Road Research Technical Paper No. 39, London, HMSO, 1958.

(19) Webster F V and Cobbe B M, *Traffic Signals*, Road Research Technical Paper No. 56, London, HMSO, 1966.

(20) *Junctions and Accesses: Determination of Size of Roundabouts and Major/Minor Junctions*, Advice Note TA23/81, London, The Department of Transport, August 1984.

(21) Tanner J C, A theoretical analysis of delays at an uncontrolled intersection, *Biometrika*, 1962, **49**, Nos 1/2, pp. 163–170.

6
Geometric design of highways

Geometric design is primarily concerned with relating the physical elements of the highway to the requirements of the driver and the vehicle. It is mainly concerned with those elements which make up the visible features of the roadway, and it does not include the structural design of the facility.

Features which have to be considered in geometric design are, primarily, horizontal and vertical curvature, the cross-section elements (including noise barriers), highway grades, and the layout of intersections. The design of these features is considerably influenced by driver behaviour and psychology, vehicle characteristics and trends, and traffic speeds and volumes.

Proper geometric design will inevitably reduce the number and severity of highway accidents while ensuring high traffic flow with the minimum of delay to vehicles. While these are the main factors to be considered, care must also be taken that the highway presents an aesthetically satisfying picture to both the driver and the onlooker. The aim should be to design a facility that blends harmoniously with the topography, and not one that leaves an ugly scar on an otherwise pleasant landscape. When geometric design is improperly carried out, it may result in early obsolescence of the new highway, with consequent economic loss to the community.

Design speed

A most important consideration in geometric design is the design speed of the highway. Notwithstanding the apparent simplicity of the term 'design speed', there is still some confusion over what exactly it means. Loose terminology and imprecise definitions have contributed to this confusion, and as a result various definitions are currently in use.

The original design speed concept represented a driver-behavioural approach to highway alignment design, with design speed being regarded as an upper estimate of a relatively uniform travel speed[1]. As such it is generally known today as the highest continuous speed at which individual vehicles can travel with safety on the highway when weather conditions are favourable, traffic volumes are low, and the design features of the highway are the factors governing safety. In practice, however, driver behaviour does not always conform to the assumptions implicit in this description, so that the most valid definition is simply that the design speed of a highway is a 'design criterion' that is aimed at providing a consistent and coordinated alignment.

The main factors that should affect the choice of design speed are the type of highway and the environmental terrain through which it will run. Other qualifying factors are traffic volumes and characteristics, cost of land, speed capabilities of vehicles, and aesthetic features. The choice of design speed has a considerable effect upon the design of such geometric features as horizontal and vertical curvature, safe stopping and overtaking sight distances, and acceptable highway grades. In addition, therefore, it has a significant effect upon highway cost.

Generally, the design speeds of highways are chosen by administrative decision. As might be expected, therefore, the design speed (and geometric design standards) for a particular type of highway will vary from country to country[2].

Rural roads

The Department of Transport's recommendations with regard to design speed recognize that vehicle speeds vary according to the impression of constraint that the mainline road alignment and layout (including access points) impart to the driver. Consequently, the Department utilizes a graphical procedure (see reference 3 for details) which is based on measurements of this constraint when selecting mainline design speeds for rural roads. Table 6.1 shows typical design speeds for both mainline and connector carriageways. The following discussion relates only to mainline design speeds, i.e. to the design speeds of carriageways carrying the main flows of traffic and from which/to which connector roads diverge/merge. (Connector road design speeds are discussed elsewhere in this chapter.)

There is still debate as to whether speeds greater than 120 km/h should

Table 6.1 Typical control speeds for rural roads in Britain (km/h): (a) mainline carriageways and (b) connector roads to motorways and dual carriageways

Road type	Design speed	Speed limit
Motorways and dual carriageways	120	112
Single carriageways		
derestricted	100	96
others	85,70	—

(a)

Connector type	Design speed	
	Desirable minimum	Absolute minimum
Link roads	85	70
Slip roads	70	60

(b)

be used for design purposes on mainline carriageways of rural motorways. It is often pointed out that a higher design speed not only safeguards against early obsolescence of the highway, but also provides an increased margin of safety for those driving at high speeds. That there is some validity in this statement is reflected by the fact that the design speed of high-speed-type roads is now 120 km/h as compared with 56 km/h in 1927. These increases are probably due primarily to continued increases in vehicle performance, e.g. nearly all small family cars today can exceed 130 km/h, whereas similar cars of the late 1940s had average top speeds of 95 km/h.

In practice, the great majority of vehicle speeds on 120 km/h dual and single carriageway highways are generally less than the design speed, and vehicle operating behaviour is generally in line with the conditions assumed in the formulation of the design speed concept.

Design speeds lower than 120 km/h are often applied to single carriageway roads in order to keep construction costs within certain limits. There is danger in this philosophy since, although drivers will obviously accept lower values in what are clearly difficult locations, repeated studies have shown that they do not adjust their speeds to the importance of the facility. Instead, they endeavour to operate at speeds consistent with the traffic on the facility and their view of its physical limitations.

For a highway that is designed in the range 80–120 km/h, operating speeds tend to vary according to the actual speed standard of individual alignment features; thus they may often be in excess of the design speed at particular locations. Typically, however, operating speeds rarely exceed the speed standard for individual elements of horizontal alignment, although they may be greater for vertical alignment elements.

When the highway has a design speed less than about 80 km/h, actual speeds again vary considerably according to the alignment conditions; they are also generally greater than the design speed. In this situation, it is quite invalid to consider design speeds as reflective in any real way of driver speed behaviour.

Lower design speeds, e.g. 40 or 50 km/h, are rarely used nowadays as a basis for the design of any significant length of rural highway; rather, they may be used to design isolated curves under severely constrained conditions. Their application generally results from an attempt to fit the most liberal curves permitted by the terrain constraints.

In relation to the above, it might be noted that observations in rural areas have shown that the speeds exceeded by the faster car drivers are closely related to the mean speed of cars. Studies have shown that the 85th and 95th percentile speeds are approximately 1.19 and 1.32 times the average speeds of cars[4].

When designing a substantial length of highway, it is obviously desirable that a constant design speed be used. However, this may not always be feasible because of topographical or other physical limitations. Where these occur, a change in design speed should not be introduced abruptly, but instead a transition stretch of adequate length should be inserted. Within this stretch, drivers should be encouraged to reduce speed gradually by means of adequate signing notifications.

Urban roads

The design speed of mainline carriageways of an urban motorway should ideally be as high as possible. However, the basic consideration in choosing the design speed of the higher-quality urban roads should be that it is at least greater than the running speed desired for maximum flow during the peak demand periods. Safety considerations, influenced by the greater frequency of intersections on urban roads, combined with the greater (and more expensive) landtake required by high-speed highways, strongly suggest that major highways in urban areas should be designed to lower speed standards than equivalent highways in rural areas. From a practical aspect, it should also be appreciated that speed limits are more readily imposed and accepted in urban areas, so that motorists who might wish to travel in excess of the speed standard for the road during off-peak hours are more easily regulated.

Design speeds for lower-quality urban roads are generally limited by the place of these roads in the hierarchical highway system which in turn is influenced by safety, cost and environmental concerns.

Primary distributors form the primary network for the town as a whole; longer-distance traffic movements to, from and within the town are normally canalized onto these routes. Such distributors are very often constructed as motorways or dual carriageways. *District distributors*, which are mostly single carriageways, distribute traffic within the residential, industrial and principal business districts of the town; they form the link between the primary network and the roads within environmental areas (i.e. areas free from extraneous traffic in which considerations of environment predominate over the use of vehicles). *Local distributors* distribute traffic within environmental areas; they form

Table 6.2 Typical control speeds for urban roads in Britain (km/h): (a) mainline carriageways and (b) connector roads to motorways and dual carriageways

Design speed	Speed limit
60	48
70	64
85	80
100	96

(a)

Connector type	Design speed	
	Desirable minimum	Absolute minimum
Link roads	70	50
Slip roads	60	50

(b)

the link between district distributors and access roads. *Access roads* give direct access to buildings and land within environmental areas.

In British practice, the design speeds for urban roads are normally selected[3] with reference to the speed limits envisaged for the road, so as to permit a small margin for speeds in excess of the speed limit (see, for example, Table 6.2). The minimum design speed used on primary distributors is 70 km/h.

Traffic flow and design capacity

Basic considerations

The terms 'design flow' and 'design capacity' are used here to define the ability of a road to accommodate traffic under given circumstances. Factors which must be taken into account in determining the governing circumstances are the physical features of the road itself and the prevailing traffic conditions. In addition, it must be considered whether designs for uninterrupted or interrupted traffic flow are being evaluated. This is important since uninterrupted-flow conditions apply to mainline road sections along which intersecting flows do not interfere with continuous movement, whereas interrupted-flow conditions refer to highways with junctions at-grade where intersecting flows interfere with each other.

Prevailing road conditions

Design flow figures for uninterrupted flow on highways should be modified if certain minimum physical design features are not adhered to. Poor physical features which tend to cause reductions include the following.

(1) *Narrow traffic lanes*—lane widths of about 3.65 m are now accepted as being necessary for *heavy* volumes of mixed traffic.

(2) *Inadequate shoulders*—too narrow, or the lack of, shoulders alongside a road causes vehicles to travel closer to the centre of the carriageway, thereby increasing the medial traffic friction. In addition, vehicles making emergency stops must of necessity park on the carriageway. This inadequacy causes a reduction in the effective width of the road and thereby reduces capacity.

(3) *Side obstructions*—vertical obstructions such as poles, bridge abutments, retaining walls, or parked cars that are located within about 2 m of the edge of the carriageway contribute towards a reduction in the effective width of the outside lane.

(4) *Imperfect horizontal or vertical curvature*—long and/or steep hills and sharp bends result in inadequate sight distances. Drivers are then restricted in opportunities to pass and hence the capacity of the facility will be reduced.

In addition to the above physical features, the capacities of certain rural roads and the great majority of urban roads are controlled by the layouts of intersections on these roads. Some physical features having considerable influence are the type of intersection, e.g. whether plain, channelized, roundabout or signalized, the number of intersecting traffic lanes, and the adequacy of speed change lanes.

Prevailing traffic conditions

Unlike the physical features of the highway, which are literally fixed in position and have definite measurable effects upon uninterrupted and interrupted traffic flows, the prevailing traffic conditions are not fixed but vary from hour to hour throughout the day. Hence the flows at any particular time are a function of the speeds of vehicles, the composition of the traffic streams, and the manner in which the vehicles interact with each other, as well as the physical features of the roadway itself. A general understanding of how capacity is affected by traffic conditions can be obtained by examining Fig. 6.1.

The term 'concentration' used in this figure is the same as the term 'density' used in American traffic literature, and is defined as the number of vehicles occupying a unit length of a traffic lane at a given instant. Concentration is usually expressed in vehicles per kilometre.

A not too inaccurate picture of the relationship between spot speed and concentration is given by the curves in Fig. 6.1(a). One of the curves is based on an early study conducted in the USA on speed–flow relationships[5]; the other curve is based on speed–flow and headway–speed data collected by the Transport and Road Research Laboratory[6]. Both of these curves indicate quite clearly that as the concentration per

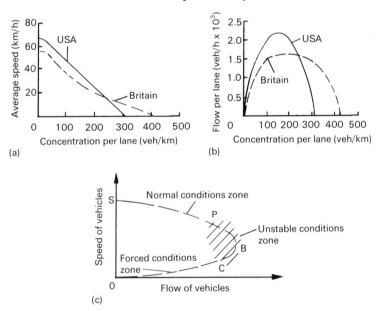

Fig. 6.1   Fundamental relationships between traffic concentration, speed and flow on a highway

lane increases, the average spot speed of traffic decreases. The kink introduced at the top of each curve reflects the manner in which vehicle speeds flatten out for each road at an independently determined free speed; a flat portion must be expected on any speed–concentration curve, since the mean speed of traffic will be unaffected by concentrations below a certain limiting value. In addition, the average rate of speed decrease can be expected to vary with the physical features of the road: for instance, the slope will be steeper for narrow roads, but more gradual for wider roads.

The relationships between flow and concentration for the same data are shown in Fig. 6.1(b). The term 'flow' as used in this context refers to the number of vehicles that can pass a given point in unit time; it is usually expressed in vehicles per hour. The basic features of these curves are most important. As the concentration of vehicles on the road tends towards zero, obviously the flow must also tend towards zero. However, when the concentration is very high, vehicles travelling in a given direction are in a saturated condition of road occupancy, each vehicle being nearly bumper to bumper with the vehicle in front. This represents an almost impossible operating condition so that the flow of traffic again tends towards zero. Practically, some headway must be allowed between vehicles for flow to take place. The headway distribution at any particular time on any given road is dependent upon the traffic composition, driver perception and reaction times, brake application time, braking distance, and a suitable factor of safety. Finally, at some concentration value between the two extremes, the flow is a maximum value which can be described as the *ultimate* capacity of the road.

From these two figures, and using the relationship

$$\text{concentration (veh/km)} = \frac{\text{flow (veh/h)}}{\text{speed (km/h)}}$$

it is possible to derive a theoretical relationship between speed and flow for given road situations. (For useful brief discussions with examples on the practical validity of such a relationship, see references 7 and 8. For an excellent publication on traffic flow theory as a whole, see reference 9.) On the assumption that Fig. 6.1(c) indicates the form that such a relationship might take, it illustrates the following very basic features relating to traffic movement:

(1)   the zone of normal conditions, i.e. portion SP on the curve,
(2)   the zone of forced conditions, i.e. portion 0C on the curve,
(3)   the zone of unstable conditions, i.e. portion PBC on the curve.

The portion SP represents the situation where free driving occurs on the highway. Within this zone, mean speeds are higher and more variable, and traffic flow increases as speed decreases. This arm marks a relationship between spot speed and flow that reflects primarily on the prevailing highway geometric standards. The higher the standards of highway design, the freer are the driving conditions and the flatter the curve SP; the lower the standards, the steeper is the slope of the curve SP.

In practice, therefore, the designer seeks to insert geometric features into the highway which ensure that the curve SP is as flat as is reasonable for the circumstances under consideration, so that there is only a relatively small drop in speed as flow increases within the design limits.

Table 6.3 contains examples of the highest recorded whole-hour flows on two-lane single carriageway roads in various countries. Automatic counters on the A4 Colnbrook Bypass (before it was bypassed by the M4 motorway) regularly recorded 3500 veh/h on a three-lane, 10 m wide single carriageway, with the highest totals going above 4000 veh/h (1450 veh/h per standard lane). Whole-hour flows per carriageway of more than 4000 veh/h have been regularly recorded by automatic counters on the dual two-lane M4 Slough Bypass (subsequently widened to dual three-lane); the highest manual counts on this highway, in September 1968, gave a whole-hour count of 4630 veh/h (at an average running speed of about 70 km/h) and a peak 6-minute rate of 5020 veh/h, with traffic containing about 4 per cent heavy commercial vehicles.

Within zone 0C, mean speeds are much lower and vehicles move under conditions of forced driving. A decrease in speed under these saturated conditions is associated with a decrease in traffic flow. The shape of the arm 0C depends primarily upon the interaction between vehicles, i.e. the concentration is very high and hence the control exercised on each vehicle by the one in front has a most important influence on the flow. To all intents, therefore, the curve 0C is relatively independent of the standard of geometrical design of the road.

Within zone PBC, flows can be very high but driving conditions are very unstable, i.e. it is possible for vehicles to be moving along under uncongested driving conditions and then, under the influence of some restricting bottleneck factor (which may or may not be determinable), forced conditions of driving may supervene abruptly, even though the flow may be the same as before. In fact, within this zone, traffic conditions seldom remain sufficiently stable to enable reliable survey figures during operational measurement of traffic flows to be obtained—which is why the theoretical ultimate capacity (at point B) is never sought for design purposes, but rather some point about P.

British speed–flow studies

The general procedure adopted by the Transport and Road Research Laboratory in Britain has been to study capacity in terms of the relationship between the average journey speed of vehicles and the traffic flow (see, for example, references 10–14). The decision to use average journey speed for speed–flow studies is based on the acceptance of loss of time as the factor with which motorists are most concerned.

Rural roads
Speed–flow formulae derived for rural roads are shown in Table 6.4. Relationships established for representative average traffic composition (15 per cent heavy commercial vehicles), average 'hilliness' (15 m/km all

Table 6.3 Highest whole-hour flows observed on two-lane single carriageway roads in various countries[10]

Country	Situation	Carriageway width		Flow (vehicles per hour)		Heavy vehicle content (%)	Directional split
		Metres	Standard lanes	Actual	Per standard lane		
USA	Rural	6.1	1.67	1777	1070	n.a.	69–31
	Urban	6.1	1.67	2024	1210	n.a.	53–47
	Tunnel	6.7	1.84	2595	1410	n.a.	50–50
Denmark	Bridge	5.6	1.53	2718	1770	0	n.a.
Germany	Rural	7.5	2.05	2415	1180	28	n.a.
Japan	Rural	7.5	2.05	3060	1490	20	n.a.
Norway	Rural	7.5	2.05	2494	1210	2	n.a.
Spain	Rural	7.0	1.92	2310	1200	n.a.	n.a.
Switzerland	Rural	7.5*	2.05	2605	1270	n.a.	n.a.
	Urban	7.5*	2.05	2964	1440	n.a.	n.a.

*Assumed width

Table 6.4 Speed–flow formulae for rural roads[10]

Type of carriageway	Free speed (V_0)	Speed–flow slope (S) in observed flow range	Values for the average road					
			In observed flow range			Suggested extensions		
			S	V (Q)	V (Q)	S	V_c	Q_c
Single								
two-lane	$82.5^* - \dfrac{(P-15)}{10} - \dfrac{H}{7.5}\left(\dfrac{185+P}{200}\right) - \dfrac{B}{7.5}\left(\dfrac{215-P}{200}\right)$	$-0.5(V_0-62.5^*)-\left(\dfrac{P}{5}\right)\left(\dfrac{H}{7.5}\right)$	-10	70.5^* (300)	63.5^* (1000)	-10	62.5^*	1100
three-lane	$85.5^* - \dfrac{(P-15)}{8} - \dfrac{H}{7.5}\left(\dfrac{185+P}{200}\right) - \dfrac{B}{7.5}\left(\dfrac{215-P}{200}\right)$	$-0.5(V_0-62.5^*)-\left(\dfrac{P}{7.5}\right)\left(\dfrac{H}{7.5}\right)$	-11.5	77.5^* (300)	69.5^* (1000)	-11.5	64.5^*	1400
Dual two- or three-lane	$92.5^* - \dfrac{(P-15)}{5} - \dfrac{H}{2.5}$	Zero	Zero	86.5^* (300)	86.5^* (1200)	-15	77.5^*	1800
Motorway								
two-lane	$97.5^* - \dfrac{(P-15)}{5}$	Zero ($Q \leqslant 1400$)	Zero	97.5^* (300)	97.5^* (1400)	—	—	—
		-24 ($Q>1400$)	-24	97.5^* (1400)	78.5^* (2200)	(-24)	83.5^*	2000
three-lane	$98.5^* - \dfrac{(P-15)}{5}$	Zero	Zero	98.5^* (300)	98.5^* (1400)	-24	84.5^*	2000

Definitions

V = average journey speed, all vehicles (km/h).

Q = total flow, all vehicles, per standard lane of 3.65 m width (veh/h).

V_0 = free speed (km/h): the value of V when $Q=300$ veh/h.

S = speed–flow slope: the change in speed when Q increases by 1000 veh/h (km/h). Basic speed–flow relation: $V = V_0 + S(Q-300)/1000$.

Q_c = suggested typical absolute capacity per standard lane (veh/h).

V_c = value of V when $Q=Q_c$ (km/h).

P = percentage of heavy vehicles (those with more than four tyres). The average value for rural main roads is $P=15$ per cent.

H = hilliness: the total rise-and-fall per unit distance (m/km). Values of 0, 15 and 30 may be used for roads judged to be flat, average and hilly, respectively, if no measurements are available.

B = bendiness: the total change of direction per unit distance (degrees/km). Values of 0, 75 and 150 (on two-lane single carriageways), and 0, 45 and 90 (on three-lane single carriageways), may be used for roads judged to be straight, average and bendy, respectively, if no measurements are available.

Note: All values marked * must be adjusted to allow for the tendency for speeds to increase over the years.

roads), average 'bendiness' (75 and 45 degrees/km for two- and three-lane single carriageway roads, respectively, and 25 degrees/km for dual carriageway roads) and, in the case of single carriageway roads only, equal flows in both directions, are shown graphically in Fig. 6.2. Note that these data were gathered over road lengths up to 10 km which included minor junctions, but excluded roundabouts or other major junctions where the highway being studied did not have priority.

In relation to these data, it should be noted that most drivers experience a fair degree of freedom in choosing their speed at low traffic volumes; the average speed under these conditions is called the free speed. Since low-flow regression analysis showed that speeds are normally lower on busier roads, it was necessary for inter-road-type comparison purposes to define the level of 'busyness' associated with free speed for each type of highway. The value eventually chosen for the practical definition of free speed was 300 veh/h per *standard lane* of width 3.65 m, which is a level of flow that occurs at some time on most main rural highways. This meant, for example, that a one-way flow of 550 veh/h on a 6.7 m (1.835 standard lane) wide carriageway of a dual carriageway highway could be expressed as 550/1.835 = 300 veh/h per standard lane. Similarly, a two-way flow of 550 veh/h on a 6.7 m single carriageway road could be expressed as 300 veh/h per standard lane, i.e. the directional split of flow was regarded as unimportant for rural two-lane highways.

Examination of these data showed the following.

(1) The free speed was different for each highway type, with the higher speeds being associated with the higher-quality highway layouts.

(2) Carriageway width was not an important determinant of free speed at low traffic flows. Hilliness (m/km), i.e. the sum of all height changes (rises and falls) divided by the section length, and bendiness (degrees/km), i.e. the sum of all changes in direction divided by the section length, were strong determinants of free speed on both single and dual carriageway roads. (Subsequent work[15] has indicated that speeds on motorways are influenced by hilliness in a similar way to those on all-purpose dual carriageway roads, and that the hilliness factor should therefore also be included in motorway equations.)

(3) On single carriageway roads, there was generally a marked straight-line reduction in speed from the free speed as flow increased. Even at the highest flow levels, there was no suggestion of curvature in the speed–flow relationship. On dual carriageways, speeds were constant over the range of flows observed. On motorways, speeds did not fall until the flow exceeded about 1400 veh/h per lane.

(4) For all highway types, speeds at both low and high flows were more affected by heavy vehicles than by light vehicles, i.e. the greater the heavy vehicle content, the lower was the average speed.

(5) On flat, 'bendy', two-lane single carriageway roads, free speeds were restrained by highway layout and there was little further reduction in speed as traffic flow increased, i.e. the speed–flow slope tended to be flat. On 'hilly', straight, two-lane single carriageway roads, light vehicles were

able to maintain fairly high speeds at both low and high traffic flows, unless they were obstructed by slow heavy vehicles which could not be passed; the combination of hilliness and a high proportion of heavy vehicles produced a steep speed–flow slope.

(6) On all-purpose dual two-lane highways, no speed reduction was observed within the measured flow range; however, a slope of minus 15 km/h per 1000 veh/h per standard lane was taken to apply at flows above 1200 veh/h per standard lane. This slope is roughly midway between the single carriageway and motorway values.

(7) On three-lane motorway carriageways, the two-lane motorway slope was assumed to apply to traffic flows in excess of 1400 veh/h per lane.

(8) The highest flows observed on the different types of highway, either in this study or in long-period counts elsewhere (see Table 6.3), were taken as the 'absolute capacities'. The absolute capacity as defined here is the level of flow that is not normally exceeded on a given carriageway type; it is a much higher level than is used for highway design purposes, where considerations of comfort and safety apply. The absolute capacities determined in this study for flows of standard composition (15 per cent heavy commercials) are shown in Fig. 6.2 and tabulated in the last column of Table 6.4. (The values for the single carriageway highways are considered applicable regardless of directional split, although the value for three-lane roads may be difficult to achieve with equal flows in two directions.)

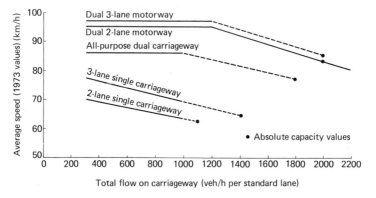

Fig. 6.2 Standard adjusted speed–flow relations for rural highways[10]

(9) Only at traffic flows at the absolute capacity level was it considered appropriate to apply a passenger car unit (p.c.u.) adjustment for heavy vehicles. From this and other studies, it was accepted that the p.c.u.-rating for a heavy commercial vehicle at high traffic flows on rural highways (between major intersections) is about 1.5. Hence absolute capacities expressed in vehicles per hour at standard composition may be corrected to the level appropriate to P per cent heavy commercial vehicles by multiplying the standard values by $215/(200+P)$. For example, an absolute capacity of 1800 veh/h per standard lane at standard composition

becomes 1890 and 1720 with 5 and 25 per cent heavy commercial vehicles, respectively.

Suburban roads

As with rural highways (see Table 6.4), the basic speed–flow relationship determined for suburban roads[11] was of the form:

$$V = V_0 + \frac{S(Q-300)}{1000}$$

In this instance, however, measurements on principal, trunk and 'A' class routes (including all intersections) on the outskirts of large conurbations found that

$$V_0 = (50+0.1d) - 10(i-0.8) - 0.15(a-27.5)$$

and $$S = -25 - \tfrac{4}{3}[V_0 - (50+0.1d)] - 30(i-0.8) - 0.4(b-65)$$

where the terms are as follows. V is the average journey speed of all vehicles in the traffic stream (km/h). V_0 is the free speed (km/h) when $Q = 300$ veh/h per standard lane at a traffic composition of 85 per cent light vehicles (i.e. vehicles with three or four tyres), 13.5 per cent heavy vehicles (i.e. vehicles with more than four tyres), and 1.5 per cent buses and coaches on scheduled local stopping services. Q is the one-way flow of all vehicles (excluding solo motor cycles and pedal cycles), measured in veh/h per standard lane width of 3.65 m. (It should be noted that for single carriageway suburban roads, Q was found by dividing the one-way flow by *half* of the carriageway width expressed in standard lanes; for dual carriageway roads, Q was obtained by dividing the one-way flow by the *full* width (in standard lanes) of the carriageway on which the flow was travelling.) S is the speed–flow slope. i is the density of major intersections (i.e. roundabouts or traffic signals) per kilometre. a is the number of accessways per kilometre, an accessway being defined as either a minor road junction (i.e. where the road in question had priority) or a driveway entrance/exit; where counting driveways, the total for both sides of the road was used (even for dual carriageways). Typical accessway values were: high = 70 per km, medium = 27.5 per km, and low = 10.0 per km. High values were typical of areas of dense housing or commercial activity, e.g. shops and service stations, where the development was continuous along the length of the roadway; medium values applied to roads through areas of mixed development, including some open space; low values related to partially developed areas, or areas with restricted direct access and served by service roads. d is the proportion of dual carriageway (%). Finally, b is the proportion of roadside that was developed (average of both sides of the road) (%).

When the traffic composition was other than the standard 85 : 13.5 : 1.5, it was found that the 'non-standard' V_0 could be determined from the following formula:

$$\text{non-standard } V_0 = \frac{\text{standard } V_0(102 - 0.12P_H - 0.23P_B)}{100}$$

where P_H = percentage of heavy vehicles in the stream observed, varying between 5 and 40 per cent, and P_B = percentage of buses and coaches in the stream observed, varying between 0 and 7 per cent.

All of the variables noted above are subject to upper and lower limits, as given in Table 6.5.

Table 6.5   Upper and lower limits of the variables used in the speed–flow relationships derived for suburban roads

Variable	Lower limit	Upper limit
V_0	38	71
Q	300	1700
S	−75	0
i	0	2
a	5	75
d	0	100
b	20	100

A free speed correction factor for weather conditions was also determined in this study. However, this correction was only of significance under extreme conditions of weather (e.g. fog), visibility (e.g. twilight), and road surface (e.g. icy) and is therefore not discussed here further.

It might be noted that this study also showed that the speeds of heavy vehicles and buses or coaches could be deduced from those of light vehicles by means of the following formulae.

$$V_H = 1.92 + 0.845 V_L$$

and $$V_B = 1.71 + 0.738 V_L$$

where V_L = unadjusted average speed of light vehicles (km/h), V_H = average speed of heavy vehicles (km/h) for $V_L \geqslant 12.4$ km/h, and V_B = average speed of buses (km/h) for $V_L \geqslant 6.5$ km/h.

Department of Transport design flow recommendations

Before discussing the Department of Transport's recommendations in relation to the design capacity of rural and urban highways, it is useful to appreciate that there have been substantial changes in the recommended flows over the past thirty years, resulting mainly from the changing character of traffic, improvements to roads and in driver performance, and a better technical understanding of how traffic behaves. For example, Table 6.6 gives a summary of the major changes in the recommendations for rural roads between 1950 and 1968.

In this table, the recommendations are given in terms of total two-way 16-hour (i.e. 6 a.m. to 10 p.m.) flows, and are defined as the maximum traffic volumes that will ensure comfortable free-flow conditions for the movement of traffic. In the cases of the 1961 and 1968 recommendations, the 16-hour capacity values were expressed in equivalent passenger car units to allow for mixed traffic conditions, where one pedal

Table 6.6 Department of Transport capacity recommendations for rural roads at various times

Road type	1950 Memorandum 653 (vehicles per day)	1961 Memorandum 780 (p.c.u. per day)	1968 *Layout of Roads in Rural Areas* (p.c.u. per day)
Single			
two-lane	3000	6000	9000
three-lane	6000	11000	15000
Dual two-lane	15000	25000	33000

cycle = 0.5 p.c.u. and one medium/heavy goods vehicle, bus or coach = 3 p.c.u. (The peak hour capacities were taken as 10 per cent of the daily values for all rural highways.)

Rural highway design flows

Table 6.7 shows the traffic flow values now used in the design of rural roads in Britain. There are a number of points about this table that are worth noting, as follows.

(1) The concept of a 'design flow' is used instead of the previously used 'capacity'.

(2) Design flows are expressed in vehicles, not equivalent passenger car units. Reasons for not using p.c.us are[10]: (a) the p.c.u.-value of a heavy vehicle is different under different highway and traffic conditions, so that the main attraction of its usage—its simplicity—is lost, (b) there is very little experimental backing for the conventional all-purpose rural p.c.u.-values, and (c) even if traffic flows are expressed in p.c.us, the average traffic speed is still a function of the proportion of heavy vehicles, so that the use of p.c.u.-factors does not fully express the effect of traffic composition.

(3) The flow ranges in Table 6.7 are 24 h $AADT$ flows for highway links in the 15th year after opening. They have no significance other than providing convenient starting points for assessment, and are not intended to provide any indication of the ultimate flow which a road can carry[21]. Rather, they result from economic and operational analyses which reflect the costs and benefits of the different carriageway widths both during normal operating conditions and during maintenance. As such they are intended to assist designers in providing a starting point for scheme assessment by helping decide which widths are most likely to be economically and operationally acceptable in normal circumstances for any given traffic flows. Local assessment is then carried out for those widths within whose range either or both of the high/low traffic forecasts fall.

(4) The lower bound of a flow range is determined only from the economic assessment; it indicates the lowest flow at which a given width is likely to be economically preferred to a lesser width.

The upper limits of the flow ranges are based on the principle of an

Table 6.7 Design flow levels recommended for rural road assessment purposes[16]

Road type	24 h AADT flow (vehicles)	Edge treatment	Access treatment	Junction options relating to flow	
				Minor road junction	Major road junction
Normal single carriageway, S2	Up to 13 000*	1 m hardstrips	Restriction of access. Turning movements concentrated. Clearway at top of the flow range	Simple junctions or ghost islands (reference 17)	Ghost islands, single lane dualling or roundabouts (references 17–20)
Wide single carriageway, WS2	10 000 to 18 000	As S2	As S2	Ghost islands, single lane dualling or roundabouts (references 17–20)	As S2
Dual two-lane all-purpose carriageways, D2AP	11 000 to 30 000	As S2	As S2	Priority junctions. No other gaps in the central reserve	Generally at-grade roundabouts
	30 000 to 46 000†	As S2	Restriction of access severely enforced and left-turns only. Clearway	No gaps in the central reserve	Generally grade-separation
Dual three-lane all-purpose carriageways, D3AP	40 000 and above	As S2	Restriction of access severely enforced and left-turns only. Clearway	No gaps in the central reserve	Generally grade-separation
Dual two-lane motorway, D2M	28 000 to 54 000†	Motorway standards	Motorway regulations	None	Grade-separation
Dual three-lane motorway, D3M	50 000 to 79 000†	As D2M	As D2M	As D2M	As D2M
Dual four-lane motorway, D4M	77 000 and above	As D2M	As D2M	As D2M	As D2M

* Upper limit of flow range assumes maximum diverting flow of about 2000 veh/day during maintenance works
† Upper limit of flow range assumes maximum diverting flow of about 10 000 veh/day during maintenance works

acceptable diversion flow (see Table 6.8) to other roads during maintenance. Table 6.8(a) relates the two-way *AADT* (i.e. the carriageway flow) in the 15th year after opening to the maximum permitted diverting flow during maintenance. This table should not be regarded as applying directly to individual schemes but rather as a guide to the maximum daily two-way diverting flow permitted at any time in the 30-year (evaluation) life of the road, e.g. if the maximum diversion to be permitted from an S2 highway at any time in the 30-year period is 2000 veh/day, then the carriageway flow should be not greater than 13 000 *AADT* in the 15th year after opening. Table 6.8(b) shows the assumed traffic management arrangements during maintenance; in this table the 'primary carriageway' is the one being maintained, whilst the other carriageway is the 'secondary carriageway'.

Table 6.8 Maintenance arrangements relating to the derivation of the upper limits of recommended traffic flows[21]: (a) permitted diverting traffic and (b) traffic management measures

Road type	Diverting flow (veh/day)				
	500	2000	5000	10 000	12 000
	Two-way carriageway flow (*AADT*)				
S2	10 000	13 000	18 000	23 000	25 000
WS2	26 000	32 000	38 000	—	
D2AP	27 000	32 000	38 000	46 000	49 000
D3AP	51 000	58 000	67 000	78 000	81 000
D2M	27 000	33 000	42 000	54 000	59 000
D3M	52 000	60 000	67 000	79 000	82 000
D4M	65 000	74 000	86 000	101 000	104 000

(a)

Road type	Width available without works (m)	Traffic management during maintenance
S2	9.3	Shuttle working
WS2	12.0	Two-way working
D2AP	2 × 9.3	1 + 1 contraflow on secondary carriageway
D3AP	2 × 13.0	2 + 2 contraflow on secondary carriageway
D2M	2 × 11.3	2 + 1 contraflow on secondary carriageway
D3M	2 × 14.3	2 + 2 contraflow on secondary carriageway
D4M	2 × 17.9	2 lanes open on primary carriageway
		3 + 2 contraflow on secondary carriageway

(b)

(5) The 15th year is chosen as the reference year as flows expressed at this level are commonly understood by designers and are normally produced for other purposes, e.g. noise calculations and geometric design of junctions.
(6) The unit of traffic flow now used is the 24 h annual average daily traffic (*AADT*); this is the total annual traffic on the highway divided by 365. Appropriate peak hour flows for the main types of road can be

estimated by multiplying the *AADT* for the road by the appropriate factor in Table 6.9 and dividing by 24.

Table 6.9   Peak hour flow factors[16]

Highest hour	Main urban*	Inter-urban*	Recreational inter-urban*
10th	2.837	3.231	4.400
30th	2.703	3.017	3.974
50th	2.649	2.891	3.742
100th	2.549	2.711	3.381
200th	2.424	2.501	3.024

*These factors need to be divided by 24 to give an hourly flow

(7)   A motorway is only considered as an option if: (a) the high-growth traffic forecast exceeds 28 000 *AADT* in the 15th year after opening, and (b) it is intended to be connected with the national motorway network *or* it will form part of a new motorway at least 20 km long.

(8)   General road layout characteristics are also included in Table 6.7 to establish an indication of suitable provision related to carriageway width and *AADT*.

Procedure for the assessment of carriageway widths   The procedure involved in selecting the appropriate highway layout is most easily described on a step-by-step basis.

   Step 1   Forecast the high- and low-growth *AADT* traffic flows for the 15th year after opening.

   Step 2   Select for local assessment those carriageway widths within whose Table 6.7 flow range either or both of the forecasts fall.

   For example, if the forecast flow range is 16 000–20 000 *AADT*, then Table 6.7 shows WS2 and D2AP highways for 16 000 *AADT* and D2AP highway for 20 000 *AADT*. Hence the widths to be used for local assessment are WS2 and D2AP.

   Step 3   Consider whether there are any local circumstances which suggest that different widths outside the Table 6.7 flow ranges should be assessed.

   Typical examples of such circumstances would be usually high or low costs, severe environmental effects, operational considerations, and major network changes during the evaluation period.

   Step 4   Determine the net present values (*NPV*) for each width.

   Figure 6.3 provides typical construction costs for the different types of highway, whilst Table 6.10 suggests typical maintenance works profiles, durations and costs.

   Step 5   Enter *NPV*s and all other relevant factors in an assessment framework (see Chapter 5) and select an optimal width.

   Overall, it can be seen that the above approach to selecting a highway layout to be used for design purposes is a significant move away from the 'standards' approach of the past, whereby a given layout with a specified

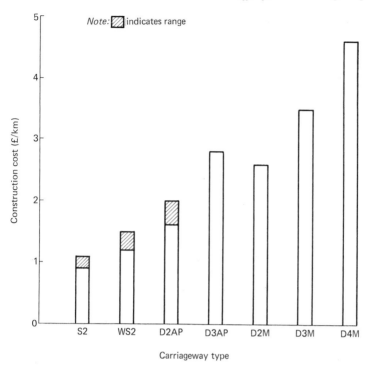

Fig. 6.3 Typical construction costs for different types of highway (August 1983 prices)

traffic capacity was matched with the predicted traffic flows for the design year. The Department's current approach of using a range of traffic forecasts for the 15th year after opening to identify the approximate carriageway requirements, and then to use framework criteria to test alternatives consistent with this range to determine the optimal design over the (assumed) 30-year life of the scheme, is in fact more similar to that utilized in selecting a preferred route.

Reference 22 provides a comprehensive example of the way in which the framework approach can be used in analysis to arrive at an optimal decision regarding the selection of a rural highway layout within a specified route option.

Urban highway design flows
Department of Transport flow recommendations for peak hours on new two- and one-way urban roadways are given in Tables 6.11 and 6.12, respectively. All the design flows are for traffic compositions including up to 15 per cent heavy (>1.52 t unladen mass) commercial vehicles. In the case of heavy vehicle contents between the ranges 15–20 and 20–25 per cent, the motorway and dual carriageway all-purpose road flows should be reduced by 100 and 150 veh/h per lane, respectively; for 10 m wide and above single carriageway roads, the recommended design flow reductions are 150 and 225 veh/h per carriageway, respectively; for single carriageway

Table 6.10   Indicative maintenance works profiles, durations and costs (August 1983 prices) [21]

Road type	Year after opening	Duration (working days per km)	Cost (£/km)
S2	10	3	7 000
	15	3	7 000
	20	10	70 000
	28	3	7 000
WS2	10	4	9 000
	15	4	9 000
	20	13	90 000
	28	4	9 000
D2AP	10	6	15 000
	15	6	15 000
	20	16	410 000
	30	10	60 000
D3AP	10	8	20 000
	15	8	20 000
	20	22	580 000
	30	10	60 000
D2M	9	7	40 000
	17	16	550 000
	26	7	40 000
D3M	8	7	40 000
	11	7	40 000
	17	22	620 000
	25	7	40 000
	28	7	40 000
D4M	8	7	40 000
	11	7	40 000
	17	28	850 000
	25	7	40 000
	28	7	40 000

roads that are less than 10 m wide, the recommended reductions in flow level are 100 and 150 veh/h per carriageway, respectively.

It should be appreciated that for an all-purpose road in an urban area, capacity is most usually controlled by the main intersections on the route. Thus the design flows given in Tables 6.11 and 6.12 are for roads that are intended to act solely as traffic links and are independent of junction capacities. Where such roads serve other purposes also, it is for the highway designer to assess their effects and to arrive at a judgement of the maximum traffic flow to be considered suitable for the conditions pertaining. In practice, the design flows quoted in Tables 6.11 and 6.12 are more likely to be used for the capacity analysis of a network or the balancing of junction capacity with link capacities, rather than the design of new links with particular characteristics.

In the particular case of wide carriageways that are used as traffic links but where parking is not prohibited during the peak periods, an assessment of the design flow for the link can be obtained from Table 6.11 using the effective carriageway width derived following application of data such as are given in Table 4.2.

Table 6.11 Recommended design flows for two-way urban roads[16]

Road type	Description	Single carriageway									Dual carriageway		
		2-lane*					4-lane†			6-lane†	2-lane†		3-lane†
		6.1	6.75	7.3	9	10	12.3	13.5	14.6	18	6.75	7.3	11
A	Urban motorway, no frontage access, grade-separation at junctions	—	—	—	—	—	—	—	—	—	—	3600	5700
B	All-purpose road, no frontage access, no standing vehicles, negligible cross-traffic	—	—	2000	—	3000	2550	2800	3050	—	2950‡	3200‡	4800‡
C	All-purpose road, frontage development, side roads, pedestrian crossings, bus stops, waiting restrictions throughout the day, loading restrictions at peak hours	1100	1400	1700	2200	2500	1700	1900	2100	2700	—	—	—

*Total for both directions of flow
†For one direction of flow
‡Includes division by line of refuges as well as central reservation; effective carriageway width excluding refuge width is used

Table 6.12 Recommended design flows for one-way urban roads[16]

Road type	Description	Peak flows (veh/h) for carriageways by width (m)					
		6.1	6.75	7.3	9	10	11
B1	All-purpose road, no frontage access, no standing vehicles, negligible cross-traffic	—	2950	3200	—	—	4800
C1	All-purpose road, frontage development, side roads, pedestrian crossings, bus stops, waiting restrictions throughout the day, loading restrictions at peak hours	1800	2000	2200	2850	3250	3550

Design flows are not now given for roads of lesser traffic-handling calibre than all-purpose type C roads, as it is obvious that their physical characteristics would be more prohibitive to the flow of traffic. Furthermore, on some of these lower order roads, environmental considerations should predominate over traffic considerations in determining acceptable traffic flow levels.

The data in Tables 6.11 and 6.12 are intended to provide adequate capacity to meet the peak hour flow demand in the design year, i.e. the highest flow for any specific hour of the week averaged over any consecutive 13 weeks during the busiest period of that year. The current urban peak hour demand is normally determined from measurements taken during the busiest three-month period of the year (June–August); the weekday peak hour is normally 5–6 p.m. Friday. In practice, it is not essential to make observations for the full number of 13 weeks; measurements over 5–7 weeks are normally sufficient, not least due to the difficulties inherent in projecting forward the current flows. It is not now expected that design year flows should necessarily be determined for unrestrained-flow conditions; rather they are expected to reflect the influence of restraint measures adopted by the highway authority as part of its overall transport policy.

Intersection capacity

The traffic performances of many rural and suburban roads, and of the great majority of intra-urban streets, are influenced significantly by the restrictions imposed by closely-spaced at-grade intersections, parking and unparking manoeuvres, public transport vehicles, bicycle and pedestrian movements, and generally low geometric design standards. Of these, intersections normally form the major deterrent to free traffic flow, and are the main cause of interrupted-flow conditions in urban areas.

Generally, it is the major intersections—whether controlled or not— which determine the overall capacity and performance of the road network in an urban area.

For capacity purposes, at-grade intersections can be divided into the following main types:

(1) uncontrolled and priority intersections where one road takes precedence over another—at either type of intersection, little or no delay is caused to the traffic on the major road,
(2) space-sharing intersections where weaving of traffic may take place— roundabout intersections are the most typical examples,
(3) time-sharing intersections where the right-of-way is transferred from one traffic stream to another in sequence—while police-controlled intersections can be considered as belonging to this category, the most important from a capacity point of view are those controlled by traffic signals.

Priority intersections

At these intersections, the traffic flow from the minor road gives way to traffic on the main highway, and only enters the major road traffic stream when 'spare time' gaps occur. These intersections are normally controlled by STOP or GIVE WAY signs and markings on the minor road; at less important junctions, only GIVE WAY markings may be provided.

The majority of existing intersections in Britain are priority-controlled. As well as being a simple way of resolving potential conflicts, this form of intersection control is economical in that it ensures that the heavier-volume traffic on the main highway is not delayed. In the case of intersections on new highways, it is now usual practice to use another form of control, e.g. roundabout, when the minor to major flow rate exceeds about 0.5.

Traffic may enter a priority intersection by carrying out a number of manoeuvres. A *merging manoeuvre* consists of joining the major road traffic stream from either the nearside or the offside of the traffic stream. A *cutting manoeuvre* involves a simple straightover or right-turn cut by one stream of traffic across one or more other streams. On single carriageway highways, the other streams travel in opposite directions, and on dual carriageway highways, the other stream occupies more than one lane. On single carriageway highways, a *cutting and merging manoeuvre* requires the turning vehicle to cut the near traffic stream, and then accelerate to merge with the far traffic stream in one movement without pausing.

The capacity of a priority intersection is theoretically dependent upon the ratio of the flows on the major and minor roads, the minimum gap in the major road traffic stream acceptable to the merging or crossing traffic, and the maximum delay that is acceptable to the minor road traffic.

As is to be expected when traffic builds up on the major road, fewer gaps become available and delays on the minor road increase accordingly, theoretically to infinity. Early theoretical research derived the following formula to describe the maximum flow that can enter the intersection from the minor road:

$$Q_m = \frac{Q_M(1 - \beta_M Q_M)}{e^{Q_M(\alpha - \beta_M)}(1 - e^{-\beta_m Q_M})}$$

where Q_m = maximum flow from the minor road, Q_M = flow on the major road, β_M = minimum headway in the major road traffic stream, β_m = minimum headway in the minor road traffic stream, and α = minimum acceptable gap in the major road traffic stream.

The sum of Q_m and Q_M is the 'ultimate' capacity of the intersection for the particular flow ratio.

As can be seen, this model is based on a gap-acceptance mechanism whereby minor road vehicles either accept or reject gaps between vehicles in the major road stream. Field measurements suggested that the median values of the observed gaps accepted by side road motorists were 4.9–5.6 seconds for left-turners and 6.0–6.3 seconds for right-turners.

Capacity calculations, particularly those involving high flows, based on formulae such as the above, are of limited practical value, however, for a number of reasons. Firstly, they are based on the assumption that at intersection-capacity the vehicle arrival rate equals throughput, queueing is continuous, and delays become infinite in length—all of which clearly do not happen in practice. Secondly, the effects of significant platooning on the major road (from, say, traffic signals upstream) are not taken into account. Thirdly, no account is taken of the manner in which gap-acceptance parameters depend upon junction geometry.

Department of Transport recommendations

More recent research work[23] has resulted in the development of an empirical method for predicting the stream capacities of major/minor junctions directly from the traffic flows themselves and from certain broad features of junction layout. This methodology has now been adopted by the Department of Transport in major/minor junction selection and layout design[17, 18].

The equations used to predict the possible minor road entry flows into a major/minor junction, as a function of the flow/geometry of the junction, are summarized in Table 6.13. These equations are applicable to all types of major/minor T-junctions, including staggered junctions, when the major road traffic is travelling at approach speeds of up to 85 km/h.

Table 6.13 Predictive capacity equations for T-junctions[23]

$$q_{B-A}^s = X_1 \{627 + 14W_{CR} - Y[0.364q_{A-C} + 0.144q_{A-B} + 0.229q_{C-A} + 0.520q_{C-B}]\}$$

$$q_{B-C}^s = X_2 \{745 - Y[0.364q_{A-C} + 0.144q_{A-B}]\}$$

$$q_{C-B}^s = X_2' \{745 - 0.364Y[q_{A-C} + q_{A-B}]\}$$

where

$$Y = [1 - 0.0345W]$$

$$X_1 = [1 + 0.094(w_{B-A} - 3.65)][1 + 0.0009(V_{rB-A} - 120)][1 + 0.0006(V_{lB-A} - 150)]$$

$$X_2 = [1 + 0.094(w_{B-C} - 3.65)][1 + 0.0009(V_{rB-C} - 120)]$$

$$X_2' = [1 + 0.094(w_{C-B} - 3.65)][1 + 0.0009(V_{rC-B} - 120)]$$

Note: The equations for q_{B-A}^s and q_{B-C}^s assume separate lanes for right- and left-turning minor road traffic and no simultaneous queueing for the two movements.

The notation $q_{C-A}(l)$ denotes the flow of light vehicles for the stream C–A (see also Fig. 6.4(a)), $q_{C-A}(h)$ denotes the flow of heavy vehicles for the stream C–A, and so on. The superscript s (e.g. q_{B-A}^s) denotes the flow from a saturated stream, i.e. one in which there is stable queueing.

In each of the equations the geometric parameters represented by X_1, X_2 and X_2' are stream specific. Thus w_{B-A} and w_{B-C} denote the average widths of each of the minor road approach lanes for waiting vehicles in streams B–A and B–C, respectively, measured over a distance of 20 m upstream from the GIVE WAY line; w_{C-B} is the average width of the right-turn central reserve lane on the major road, or 2.1 m if there is no explicit provision for right-turners, in stream C–B. V_{rB-A} and V_{lB-A} denote the right and left visibility distances, respectively, measured at a height of 1.05 m above the carriageway surface from a point 10 m upstream of the minor road GIVE WAY line to a line bisecting the relevant side (farside for

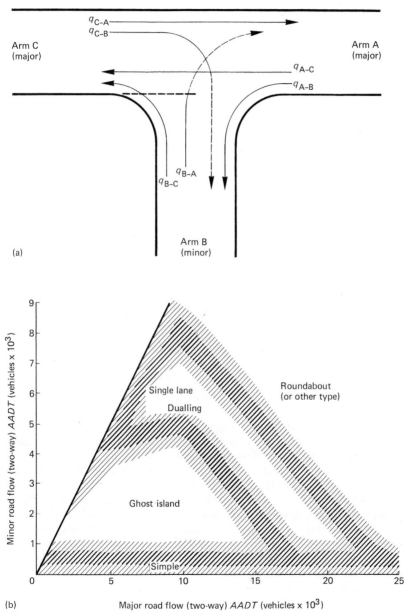

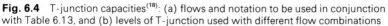

Fig. 6.4 T-junction capacities[18]: (a) flows and notation to be used in conjunction with Table 6.13, and (b) levels of T-junction used with different flow combinations.

V_1, nearside for V_r) of the major road carriageway; V_{C-B} is the visibility at a height of 1.05 m measured from the position assumed by vehicles waiting to turn right, or from the midpoint of the central reserve right-turning lane (in the case of a dual carriageway), towards the oncoming major road traffic.

W is the average major road carriageway width at the junction; in the case of dual carriageways and single carriageways with ghost islands, W excludes the width of the central reserve lane. W_{CR} is the average width of the central reserve lane on a dual carriageway major road, at the junction.

For all the equations in Table 6.13, the capacities and flows are in passenger car units per hour (p.c.u./h) and the distances are in metres. Capacities are positive or zero; if the right-hand side of any equation is negative, the capacity is zero. One heavy goods vehicle (HGV) is considered equivalent to 2 p.c.u., for calculation purposes. The maximum values of visibility and W_{CR} used in computation are 250 m and 10 m, respectively, even though greater values may be provided. The ranges within which the geometric data were considered valid in the base data are: $w = 2.05-4.70$ m, $V_r = 17.0-250.0$ m, $V_1 = 22.0-250.0$ m, $W_{CR} = 1.2-9.0$ m (dual carriageway sites only), and $W = 6.4-20.0$ m.

Design Reference Flows (*DRFs*) are the hourly traffic flow rates against which the capacity values derived from the equations in Table 6.13 are compared, when undertaking the detailed design of alternative practical layouts for a site. The range of *DRF*-values used for the generation of options at a site is normally wide enough to embrace junction designs which will be optimal economically or operationally, but not so wide as to produce junction designs which will clearly be an under- or over-provision.

When choosing a peak hourly flow to represent the Design Reference Flow, the function of the road within the highway network (i.e. whether it is recreational, urban or inter-urban) should always be taken into account. For example, it is most unlikely that it would be economically viable to design a junction to carry the very highest peak hourly traffic flow in a future year, i.e. such a design hour would be many times greater than the annual average hourly traffic flow ($AAHT = AADT/24$). Consequently, a lower hourly flow is normally selected instead.

The Department of Transport considers that the highest hourly flow that would typically generate viable junction options on recreational roads is the 200th highest hour, whilst in urban areas—where there is relatively little seasonal variation—designs for the 30th highest hour would probably be acceptable. Designs which cater for the 50th highest hour are normally acceptable for inter-urban roads. *DRF*-values used for design purposes are also expected to encompass the practical range of turning movements that can be expected at a site.

Figure 6.4(b) shows for single carriageway roads the various levels of T-junction that are likely to be considered for different combinations of major and minor road traffic flows. This diagram takes into account geometric and traffic delays, entry and turning stream capacities, and accident costs. In practice, new simple major/minor T-junctions—these are basic junctions without any ghost or channelizing islands on either the major or minor roads—are only used when the flow on the minor road (or access) is not expected to exceed about 300 vehicles (two-way total) *AADT*. Even though the cost of upgrading a simple T-junction to the next highest junction in the hierarchy (i.e. a ghost island junction) can appear to be quite expensive, it is usually considered for existing rural and urban T-junctions when the minor road flow exceeds 500 vehicles (two-way) *AADT* or where a right-turning accident problem is known to exist.

It might be noted here that Fig. 6.4(b) makes no mention of cross-roads. In practice, staggered T-junctions (with right–left staggers being preferred) are normally used instead of cross-roads. For safety reasons, cross-roads are appropriate only at very low minor road flows; they are never used on new dual carriageway roads or at single lane dualling. Where existing rural cross-roads are experiencing an accident problem, they are normally improved by siting channelizing islands in the minor road approaches.

Ghost island T-junctions use painted hatched islands in the middle of single carriageway roads to provide a diverging lane and waiting space for vehicles turning right from the major road onto the minor road. Cars and light goods vehicles turning right from the minor road onto the main road are also helped by ghost islands, i.e. they are able to complete the turning manoeuvre in two stages when necessary. Ghost islands are highly effective in improving safety and they are relatively cheap—especially on wide, two-lane single carriageway roads, where very little extra construction cost is involved.

Single lane dualling refers to the introduction of a physical island, at a junction, in the middle of the carriageway of a two-lane major road. This provides an offside diverging lane for vehicles turning right from the main road, and a safe central waiting area for these vehicles and the ones turning right from the minor road. Most importantly, with single lane dualling, only one through lane is provided in each direction on the major road, thereby preventing excessive speeds through the conflict zones. Single lane dualling is best suited to rural single carriageway roads which have good overtaking opportunities between intersections; it is normally provided when the volume of turning traffic is greater than that recommended for a ghost island junction and when long-vehicle turning is a problem, i.e. it allows vehicles of nearly all lengths to carry out the right-turn manoeuvre in two stages.

Major/minor T-junctions are normally unlikely to be cost effective on continuous, all-purpose, two-lane dual carriageway roads when the minor road flow is expected to exceed about 3000 vehicles *AADT* (two-way). They are never used on all-purpose, three-lane dual carriageway roads.

The Department of Transport has documented[17, 18] twenty-nine T-junction layouts which give guidance on the types most commonly encountered in practice. These layouts are presented as targets for the designer to aim at, especially for new construction; they may be modified or reduced in scale to suit the needs of any individual scheme.

Roundabouts

There are three main types of space-sharing intersections that can be classified under the generic title of roundabouts, as follows.

(1)   A *normal roundabout* has a one-way circulatory carriageway around a kerbed central island. Depending upon its mode of entry operation, the circulatory carriageway may or may not be composed of weaving sections and the entries may or may not have flared approaches. In Britain, the central island is normally 4 m or more in diameter and the entries are flared.

(2)   A *mini roundabout* has a one-way circulatory carriageway around a flush or slightly raised circular marking less than 4 m in diameter (in Britain) and with or without flared approaches.

(3)   A *double roundabout* has two normal (with flared approaches) or mini roundabouts either contiguous or connected by a central link road or kerbed island.

Variants of these main types include: a *ring junction* where the usual clockwise one-way circulation of vehicles around a large island is replaced by two-way circulation with three-arm mini roundabouts and/or traffic signals at the junction of each approach arm with the circulatory carriageway; a *signalized roundabout* which has traffic signals installed on one or more of the approach arms; a *grade-separated* roundabout which has at least one entry road via an interconnecting slip road from a road at a different level, e.g. underpasses, flyovers or multiple level intersections.

Mini roundabouts can be particularly effective in improving existing urban junctions that experience capacity and safety problems. They are only used in Britain when all the approaches are subject to a 48 km/h speed limit.

Double roundabouts have a number of special applications, for example:

(1)   at an awkward site such as a scissors junction where the installation of a single island roundabout would require extensive realignment of the approaches or extensive landtake,

(2)   at an existing staggered junction where their usage avoids the need to realign one of the approach roads,

(3)   to join two parallel routes separated by a feature such as a river or a railway line,

(4)   at overloaded single roundabouts where, by reducing the circulating flow past critical entries, the junction's capacity is increased,

(5)   at junctions with more than four entries, to achieve better capacity with acceptable safety characteristics in conjunction with a more efficient use of space,

(6)   at existing cross-roads where a double-roundabout junction separates opposing right-turning movements allowing them to pass nearside to nearside.

For capacity determination purposes, roundabouts can be divided into two groups: conventional weaving roundabouts without offside-priority operation, and roundabouts operating under offside-priority control.

Conventional weaving roundabouts

The capacity of a *conventional weaving roundabout* (without offside-priority control) is related to the capacity of each weaving section incorporated within the intersection. If any of the weaving sections is overloaded, then locking of the roundabout may occur and it can be said that the capacity of the roundabout is exceeded. The true weaving roundabout (without offside-priority control) is no longer constructed in Britain (see later discussion for current British practice); however, it is discussed here for comparative purposes and to illustrate its role in the development of today's offside-priority roundabout.

Within a particular weaving section, true non-stop weaving will only occur when the headways between the vehicles are of sufficient lengths and frequencies that safe merging and diverging movements can take place. Discontinuous flow, due to stop–go movements of the weaving vehicles, occurs when these headways are not available, or when the weaving section length is so short that the paths of the weaving vehicles cross at large intersecting angles.

The main factors controlling the capacity of a conventional weaving section are the geometric layout, including entrances and exits, and the volume and composition of the weaving traffic. Research by the Transport and Road Research Laboratory has shown that the flow through a weaving section at speeds between 14.5 and 24.0 km/h is given by

$$Q = \frac{282(1 + e/w)(1 - p/3)}{1 + w/l}$$

where Q = practical capacity (p.c.u./h), w = width of weaving section in metres (within range 6–18 m), e = average entry width in metres (within e/w range 0.4–1.0), l = length of weaving section in metres varying between 18 and 90 m (within w/l range 0.12–0.40), and p = ratio of weaving traffic to total traffic in weaving section (within range 0.4–1.0).

Figure 6.5(a) illustrates these dimensions at a conventional weaving roundabout. Since, in practice, the weaving sections may differ from this idealized layout, it may be necessary to develop 'effective' dimensions for use in the formula, as shown in Fig. 6.5(b).

The equivalent passenger car unit values recommended for use with the above formula are shown in Table 6.14.

The practical capacity derived from the above formula is about 80 per cent of the maximum dry weather flow found in experiments on isolated weaving sections. The 80 per cent value was chosen to provide a margin of safety which would cater for the effects of wet weather, possible interaction between weaving sections, variations in flow over the hour, and possible interference from pedestrians crossing the road. The formula is valid within the ranges given, provided that there are no parked vehicles on the

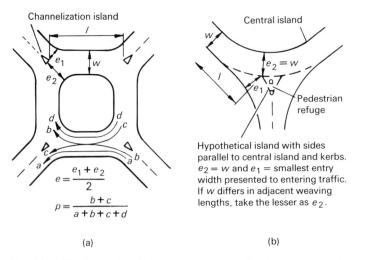

Channelization island

Central island

$$e = \frac{e_1 + e_2}{2}$$

$$p = \frac{b+c}{a+b+c+d}$$

Hypothetical island with sides parallel to central island and kerbs. $e_2 = w$ and $e_1 =$ smallest entry width presented to entering traffic. If w differs in adjacent weaving lengths, take the lesser as e_2.

Pedestrian refuge

(a)   (b)

Fig. 6.5   Weaving section dimensions at a conventional weaving roundabout, with and without channelization island

Table 6.14   Passenger-car-unit values used in the capacity formula for conventional weaving roundabouts

Vehicle type	Passenger-car-unit value
Cars, taxis, light goods	1
Motor cycles/scooters, mopeds	0.75
Pedal cycles	0.50
Heavy (>1.52 t) goods, horse-drawn, buses/coaches, trolley buses, trams	2.80

approaches to the roundabout and the site is level, with approach gradients not exceeding 4 per cent.

Where the layout is not conducive to uniform speeds, the arbitrary adjustments given in Table 6.15 should be made to the computed practical capacity of the weaving section. The angles in Table 6.15 are defined as follows.

(1)   The entry angle is the acute angle between the centreline of the approach carriageway and the axis of the weaving section under consideration.

(2)   The exit angle is the acute angle between the axis of the weaving section under consideration and the centreline of the exit carriageway.

(3)   The internal angle is the angle through which a vehicle turns when proceeding from one weaving section into another, i.e. into that section which is under consideration.

In addition, where the pedestrian flow across the road at an exit from a weaving section exceeds 300 pedestrians per hour, an arbitrary reduction of one-sixth is applied to the practical capacity of the section.

Table 6.15 Adjustments to the practical capacity of a conventional weaving roundabout, depending upon the angles through which vehicles must turn[24]

Angle (degrees)	Reduction (percentage)
0 < entry angle < 15	5.0
15 < entry angle < 30	2.5
60 < exit angle < 75	2.5
Exit angle > 75	5.0
Internal angle > 95	5.0

Table 6.16 summarizes the results of an analysis of over 400 weaving sections at conventional weaving roundabouts in Britain which was carried out in the early 1960s. Weaving widths in rural areas were normally about 10 m and speeds were higher than in urban areas. A detailed example of the calculations involved in conventional weaving roundabout design is available in the literature[26].

Table 6.16 Average length to width ratios at sections on conventional weaving roundabouts[25]

Site location	Road classification	
	Trunk	Non-trunk
Rural	4.8	4.4
Suburban		
semi-rural	3.8	3.9
built-up	3.8	3.9
Urban	3.4	

Roundabouts under offside-priority control

The introduction in 1963 of the 'give way to the right' rule at test roundabouts in Britain initiated significant changes in the method of operation of a conventional roundabout[27]. One result was that each entry began to function as a priority intersection, with the circulating vehicles in the roundabout taking precedence and entering traffic only joining as gaps occurred; the priority rule caused the actual amount of weaving within the roundabout to be very significantly reduced, and long weaving sections therefore became of less benefit in attaining high capacity. Further study[28] showed that the capacity of a junction as a whole could be considerably improved by widening the entry approach, providing a smaller central island, and installing such traffic management devices as bollards and carriageway delineators to deflect traffic well to the left upon entry. A consequent result was that the above capacity formula became obsolete as a design tool in Britain.

More recent research has shown that the traffic flow entering any type of roundabout operating under priority control from a saturated approach is linearly dependent upon the circulating flow crossing the entry. It has been determined that the most important factors influencing the capacity of any given roundabout are the width and flare of each entry; the entry angle and radius also have small but significant effects.

The predictive equation now recommended[17, 18] for use in estimating the capacity of any *at-grade roundabout* (i.e. other than those that are part of junctions with grade-separation) is

$$Q_e = K(F - f_c Q_c) \qquad \text{when } f_c Q_c \leqslant F$$
$$\quad = 0 \qquad\qquad\quad \text{when } f_c Q_c > F$$

where Q_e = entry flow (p.c.u./h) and Q_c = circulating flow across the entry (p.c.u./h), and K, F and f_c are as follows. $K = 1 - 0.003\,47(\phi - 30) - 0.987[(1/r) - 0.05]$, $F = 303x_2$, where $x_2 = v + (e - v)/(1 + 2S)$, and $S = 1.6(e - v)/l'$, and $f_c = 0.210t_D(1 + 0.2x_2)$, where $t_D = 1 + 0.5/(1 + M)$, and $M = \exp[(D - 60)/10]$. The symbols e, v, l', S, D, ϕ and r are as defined in Table 6.17 and Fig. 6.6.

Table 6.17   The limits of the geometric parameters used in the predictive roundabout equation

Geometric parameter	Symbol	Unit	Database limits	Recommended limits
Entry width	e	m	3.6–16.5	4.0–15.0
Approach half-width	v	m	1.9–12.5	2.0–7.3
Average effective flare length	l'	m	1.0–∞	1.0–100.0
Sharpness of flare	S	—	0.0–2.9	0.0–2.9
Inscribed circle diameter	D	m	13.5–171.6	15–100
Entry angle	ϕ	Degrees	0.0–77.0	10–60
Entry radius	r	m	3.4–∞	6.0–100.0

Figure 6.6(a) describes the entry and circulating flows, Q_e and Q_c, respectively, and shows the entry width e and the approach half-width v. In relation to Q_e and Q_c, it should be noted that one heavy goods vehicle is considered to be equivalent to two passenger car units for computation purposes.

The inscribed circle diameter D is the diameter of the largest circle that can be inscribed within the junction outline; this is also shown in Fig. 6.6(a). In cases where the outline is asymmetric, the local value in the region of the entry is taken. Figure 6.6(c) shows the determination of D for a double roundabout at a scissors intersection.

The average effective flare length is found as shown in Fig. 6.6(b).

The sharpness of flare S is a measure of the rate at which extra width

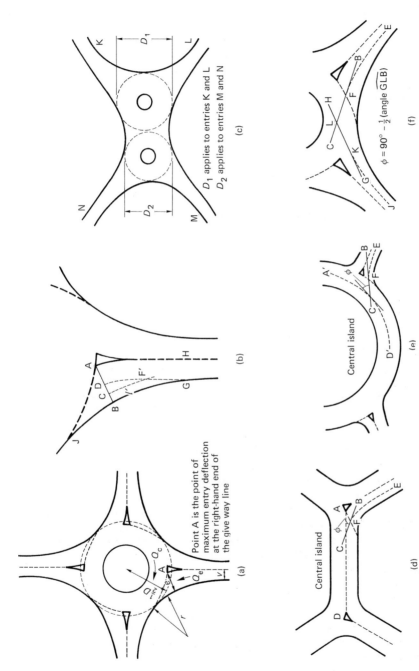

Point A is the point of
maximum entry deflection
at the right-hand end of
the give way line

(a)

(b)

(c)

D_1 applies to entries K and L
D_2 applies to entries M and N

Central island

(d)

Central island

(e)

$\phi = 90° - \frac{1}{2}$ (angle $\widehat{GLB}$)

(f)

Fig. 6.6 Parameters used in the predictive equation for roundabout entry capacity[17]

is developed in the entry flare. Large values of S correspond to short, severe flares, and small values to long, gradual flares.

The entry angle ϕ acts as a geometric proxy for the conflict angle between the entering and circulating streams. For roundabouts with a distance of more than 30 m between the offsides of an entry and the next exit, the construction is as shown in Fig. 6.6(d) for roundabouts where the circulatory carriageway between the entry and the next exit is approximately straight, and in Fig. 6.6(e) for curved circulatory carriageways. For all other roundabouts, ϕ is defined by the equation shown in Fig. 6.6(f). For this equation, the angle $\widehat{\text{GLB}}$ is measured on the side facing away from the central island; when the right-hand side of the equation is zero or negative, $\phi = 0$.

The entry radius r is measured as the minimum radius of curvature of the nearside kerbline at entry (see Fig. 6.6(a)).

As noted previously, the above predictive formula is applicable to all types of at-grade roundabout, whether normal or mini. For *grade-separated roundabouts*, the predictive capacity is

$$Q_e = 1.11F - 1.40f_c Q_c$$

where Q_e, F, f_c and Q_c are as defined before.

As with priority junctions, the roundabout capacity design process involves the derivation of a range of design reference flows, and the predictive equation is then used to produce trial designs for assessment. For each trial design, the reference flow to capacity (RFC) ratio is normally also determined. This ratio is an indicator of the likely performance of a junction under a future year traffic loading.

It has been determined that, as a result of site to site variation, the standard error of prediction of entry capacity given by the above equation is plus or minus 15 per cent. Thus, if an entry RFC-ratio of about 85 per cent occurs, queueing will theoretically be avoided in the chosen design year peak hour in 5 out of 6 cases (schemes); in the case of an entry RFC-ratio of 70 per cent, queueing will theoretically be avoided in 39 out of 40 cases (schemes). In practice, the use of designs with an RFC-ratio of about 85 per cent is generally considered as likely to result in a level of provision which will be economically justified.

A detailed example of the use of the predictive roundabout equation in generating alternative junction designs, and the manner in which the most appropriate roundabout design is selected, is readily available in the literature[17, 18].

Signalized intersections

The main factors affecting the capacity of an intersection controlled by traffic signals are the characteristics of the junction layout, the traffic composition and needs, and the setting of the signal control.

The two types of signals in general use are fixed-time signals and vehicle-actuated signals. With the fixed-time signals, the green, amber and red periods are of fixed duration and are repeated in equal recurring cycles. With the vehicle-actuated signals, however, there are maximum and minimum limits on the amount of green time in each cycle, and the length of the green period utilized in any particular cycle is dependent upon the demands of traffic at that time.

Details of signal types and systems are given in Chapter 7. The following discussion relates only to the capacities of fixed-time signals and vehicle-actuated signals which act as fixed-time signals, i.e. the maximum lengths of green period are utilized.

Field studies indicate that the capacities of the approach roads to a signalized intersection control the capacity of the intersection. The possible capacity of an individual approach road is in turn proportional to the approach width at the intersection. The approach width is taken as the distance from the edge of the carriageway to the centreline, or to the edge of the pedestrian refuge, or to the edge of any barrier dividing opposing traffic streams, whichever is the minimum.

Lane marking the carriageway generally has little effect upon capacity at the intersection itself. When the approach road is not marked or, if marked, has narrow lanes, vehicles of different sizes take up irregular but closely-packed arrangements behind the stop line. When the lane widths are about 3.5 m, similar capacity values are obtained, since the vehicles stay in their proper lanes but assume closely-packed arrangements with small headway values.

At an intersection, the maximum number of vehicles that can enter from an approach road is affected by the amount of green time available to that approach. When the green period commences, vehicles in the traffic queue take some time to accelerate to a constant running speed. When this speed has been reached, the queue discharges into the intersection at a more or less constant rate, as illustrated in Fig. 6.7. At the end of the green period, some of the vehicles in the queue make use of the amber period to enter the intersection, and the discharge rate falls away to zero during this period.

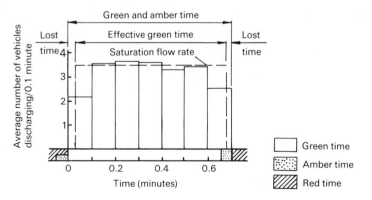

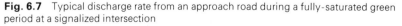

Fig. 6.7 Typical discharge rate from an approach road during a fully-saturated green period at a signalized intersection

The term *saturation flow* is used to describe the constant maximum rate of discharge from the approach road, and it is usually expressed in veh/h of green time for existing intersections and p.c.u./h for new intersections. That part of the actual green plus amber time which, when multiplied by the saturation-flow rate, gives the maximum number of vehicles that can enter the intersection is known as the effective green time. The difference between the actual green plus amber time and the effective green time is termed lost time.

Observations at intersections throughout Britain have resulted in the saturation-flow values given in Table 6.18 for approach widths between 3.0 and 5.2 m. For greater approach widths (at least up to 18.0 m), the saturation flow can be estimated from the equation:

$$S = 525w$$

where S = saturation flow (p.c.u./h) and w = approach width (m).

Table 6.18   Saturation-flow values for signalized intersections at average sites[24]

Approach width (m)	Saturation-flow value (p.c.u./h)
3.0	1850
3.5	1875
4.0	1975
4.5	2175
5.0	2550
5.5	2900

The p.c.u.-values used in the above saturation-flow determinations are 1.00 for cars and light vans, 1.75 for medium or heavy goods vehicles (> 1.52 t) and horse-drawn vehicles, and 2.25 for buses, coaches, trolley buses and trams. It is also assumed that no parked, right-turning or two-wheeled vehicles are present at the intersection approach under consideration. If right-turning vehicles are present, but they are so few that a separate right-turn lane is not necessary, each right-turning vehicle should be given a weighting of 1.75 straight-ahead vehicles. If, however, the right-turning vehicle content is sufficiently great to justify a separate right-turn lane, but none is provided, then the saturation flow should be determined for a reduced approach width, e.g. a reduction of 2.10 m may be adequate if the right-turning vehicles are mainly cars and light vehicles, but 2.75 m may be necessary if the heavy vehicle content is greater than about 20 per cent. If a right-turn lane is provided, then the saturation-flow determination relates only to the remaining approach width and the flow of straight-through and left-turning vehicles.

The saturation flows also need to be adjusted for site location according to the values given in Table 6.19.

If the site location is on a slope, then the saturation-flow values are reduced by 3 per cent for each 1 per cent uphill grade averaged over the

Table 6.19   Effect of approach road location upon saturation flow

Site designation	Description	Percentage of average saturation flow
Good	Dual carriageway, no noticeable interference from pedestrians or right-turning vehicles, good visibility and adequate turning radii, exit of adequate width and alignment	120
Average	Average site, some conditions good and some poor	100
Poor	Average low speed, some interference from standing vehicles and right-turning traffic, poor visibility and/or alignment, busy shopping street	85

61 m section of approach road just prior to the stop line. Similarly, the saturation flows are increased by 3 per cent for each 1 per cent of downhill grade. These corrections apply to gradients ranging from -5 to $+10$ per cent.

If passenger cars are regularly parked near the intersection, the effective approach width is determined by applying the following formula:

$$\text{loss of width (m)} = 1.7 - \frac{0.9(Z - 7.6)}{g}$$

where Z = clear distance between the nearest parked car and the stop line, in metres (if $Z < 7.6$ m, it is taken as equal to 7.6 m), and g = green time, in seconds.

If the parked vehicle is a commercial one, then the loss of width is taken as 1.5 times that obtained by the equation.

Example of traffic signal capacity determination
Determine the possible capacity, in vehicles per hour, of an approach road to an intersection that is being reconstructed just outside the central area of a town. The total carriageway width is 14.6 m and it is level just prior to the intersection. Commercial vehicles park regularly outside an adjacent builder's yard at a distance of 38 m back from the stop line. During the peak traffic period, the traffic is composed of 75 per cent passenger cars, 22 per cent commercial vehicles, and 3 per cent buses. One-third of the passenger cars and one-half of the commercial vehicles turn right at the intersection at this time. The number of buses turning right is negligible. The effective green time on the approach road will be 41 s and the total cycle length will be 90 s.

Solution
This intersection can be considered to be an average one, so that no adjustment for site location is necessary. Since commercial vehicles stop regularly on the carriageway, the loss of width is estimated as follows:

$$\text{loss of width} = 1.5 \left(1.7 - \frac{0.9(38 - 7.6)}{40} \right) = 1.5\,\text{m}$$

$$\text{effective approach width} = 7.3 - 1.5 = 5.8\,\text{m}$$

$$\text{saturation flow} = 525 \times 5.8 = 3045 \text{ p.c.u./h of green time}$$

However, saturation flow exists for only 41/90 of the cycle time. Therefore

$$\text{possible capacity} = 3045 \times 41/90 = 1387 \text{ p.c.u./h}$$

To obtain the possible capacity in vehicles per hour, assume that the possible flow $= Q$ veh/h. The following adjustments need to be made:

straight-through and left-turning passenger cars $= \frac{50}{100}Q$ at 1.00 p.c.u./h $= 0.5000Q$

right-turning passenger cars $= \frac{25}{100}Q$ at 1.75 p.c.u./h $= 0.4375Q$

straight-through and left-turning commercial vehicles $= \frac{11}{100}Q$ at 1.75 p.c.u./h $= 0.1925Q$

right-turning commercial vehicles $= \frac{11}{100}Q$ at $(1.75)^2$ p.c.u./h $= 0.3369Q$

straight-through buses $= \frac{3}{100}Q$ at 2.25 p.c.u./h $= 0.0675Q$

Therefore

$$\text{possible capacity} = 1.5344Q = 1387 \text{ p.c.u./h}$$

Therefore

$$Q = 1387/1.5344 = 904 \text{ veh/h}$$

Traffic signals versus roundabouts

The problem often arises as to whether an intersection should be signalized or whether a roundabout should be constructed. While each situation has its own characteristics, the choice of form at many junctions may be based on the following[29, 30].

(1) The land requirement of a conventional roundabout is usually greater than that for traffic signals, especially if the flows on one pair of arms (e.g. the minor road) are relatively low. (Even so, it may be easier to acquire corner plots for a roundabout than to take up the same area in the long narrow multiplot strips required for the parallel widening of a signalized junction.)
(2) If traffic entering from one arm greatly exceeds traffic leaving by it, there will often be a shortage of gaps in the circulating stream in both conventional and non-conventional roundabouts, which can lead to excessive delays to side road vehicles.
(3) When the proportion of offside-turning vehicles (right-turning in Britain) is large, say greater than 30 to 40 per cent, either a special

turning phase is required at a signalized junction or the vehicles must queue in the approach—in which case a roundabout may be preferred.

(4) Three- or five-way junctions are not really satisfactorily treated by traffic lights, especially when the flows are balanced. Mini and small roundabouts are very successful at three-way junctions, especially if they are Y-shaped.

(5) If a signalized junction has a high accident rate, its safety performance normally will be improved by replacement with a roundabout design. (Figure 6.31 shows that on well-trafficked dual carriageways, for similar flows on both roads, a roundabout will generally have fewer accidents than a signalized junction; no definitive data are yet available, however, on comparative accident rates for single carriageways.)

Merging/diverging on high-speed roads

Department of Transport recommendations

It is obviously of critical importance that grade-separated intersections and interchanges on high-quality highways are designed to cope with high traffic flows. If they are not, then of course the entire philosophy of a special (expensive) high-speed highway system can be challenged.

Before discussing the methodology currently used in the capacity design of merging and diverging junctions on motorways and all-purpose dual carriageways, it should be noted that the maximum hourly flows per lane that are used for design purposes in Britain are as follows:

(1) *all-purpose roads:* 1600 veh/h per lane for mainline carriageways, two-lane links and slip roads ≥ 6 m; 1200 veh/h per lane for single lane links and slip roads ≤ 5 m,

(2) *motorways:* 1800 veh/h per lane for mainline carriageways, two-lane links and slip roads ≥ 6 m; 1350 veh/h per lane for single lane links and slip roads ≤ 5 m.

British practice in respect of the design of junctions with grade-separation[31] is based on the 30th, 50th or 200th highest hourly flow in the 15th year after opening for main urban, inter-urban and recreational highways, respectively. The hourly flow is normally derived from the 24 h *AADT* flow, using the peak hour factors in Table 6.9. (It should be

Table 6.20 Percentage corrections to be applied to the mainline and connector design hour volumes, for varying gradients and heavy goods vehicles[31]

Heavy goods vehicle content (%)	Mainline gradient (%)		Connector gradient (%)		
	<2	>2	<2	>2 <4	>4
5	—	+10	—	+15	+30
10	—	+15	—	+20	+35
15	—	+20	+5	+25	+40
20	+5	+25	+10	+30	+45

appreciated, however, that these hourly flows can have a high degree of uncertainty attached to them and therefore should not be used inflexibly in design.)

Depending upon the siting of the junction, the predicted hourly flows may have to be adjusted for uphill gradients, as specified in Table 6.20. In this table the average mainline gradient is measured over a 1 km section

Table 6.21   Merging and diverging lane types to be used with design flow regions[31]: (a) merging layouts—see Fig. 6.8(a)—and (b) diverging layouts—see Fig. 6.8(b)

	Merging lane type										
	1	2	3	4	5	6	7	8	9*	10*	11*
Number of lanes											
upstream main carriageway	2	2	2	2	3	3	3	3	2	2	3
link	1	1	2	2	1	1	2	2	1	2	1
entry	1	2	2	2	1	2	2	2	1	2	1
downstream main carriageway	2	2	2	3	3	3	3	4	3	4	4
Flow region											
A		✓	✓	✓	✓	✓	✓	✓	✓	✓	✓
B			✓	✓	✓	✓	✓	✓	✓	✓	✓
C				✓	✓	✓	✓	✓	✓	✓	✓
D				✓		✓	✓	✓	✓	✓	✓
E			✓	✓			✓	✓	✓	✓	
F				✓			✓		✓	✓	
G				✓					✓	✓	
H									✓	✓	
I										✓	✓
J				✓	✓	✓	✓		✓		
K						✓	✓	✓			✓
L							✓				✓
M									✓		

(a)

	Diverging lane type								
	1	2	3	4	5	6	7*	8*	9*
Number of lanes									
upstream main carriageway	2	2	3	3	3	4	3	4	4
link	1	2	1	2	2	2	1	1	2
downstream main carriageway	2	2	3	2	3	3	2	3	2
Flow region									
P		✓	✓	✓	✓	✓	✓	✓	✓
Q			✓	✓	✓	✓	✓	✓	✓
R				✓	✓	✓	✓	✓	✓
S			✓		✓	✓		✓	
T					✓	✓			
U				✓	✓	✓			
V						✓			✓
W						✓			✓
X							✓	✓	

(b)

*Figures for merging and diverging lanes marked thus are not shown in Fig. 6.8

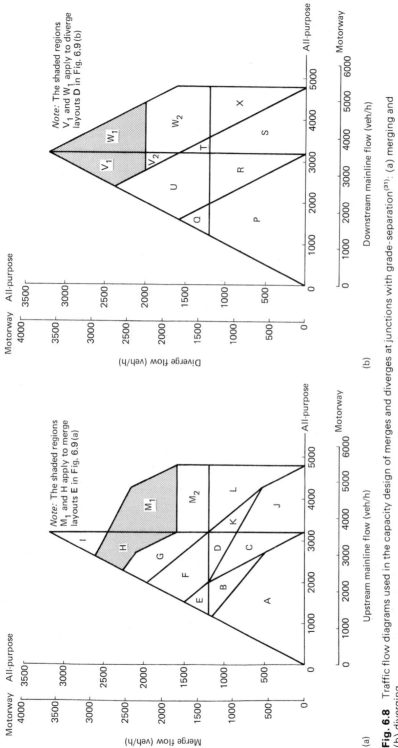

Fig. 6.8 Traffic flow diagrams used in the capacity design of merges and diverges at junctions with grade-separation[31]: (a) merging and (b) diverging

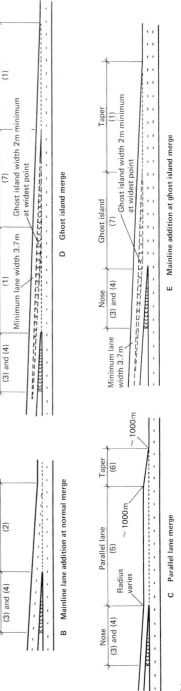

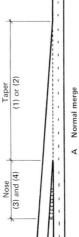

A Normal merge

B Mainline lane addition at normal merge

C Parallel lane merge

D Ghost island merge

E Mainline addition at ghost island merge

Nose
(3) and (4)

Taper
(1) or (2)

Taper
(2)

Parallel lane
(5)

Radius
varies

~1000m

Taper
(6)

~1000m

Taper
(slip road right-hand lane)
(1)

Minimum lane width 3.7 m

Ghost island
(7)

Ghost island width 2m minimum
at widest point

Taper
(1)

Ghost island
(7)

Ghost island width 2m minimum
at widest point

Minimum lane
width 3.7 m

(a)

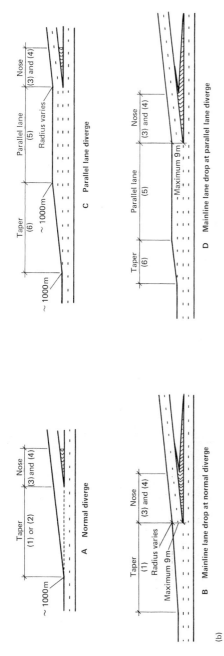

Fig. 6.9 Alternative connector layouts used at junctions with grade-separation[31]: (a) merge lane layouts for use with Table 6.22(a), and (b) diverge lane layouts for use with Table 6.22(b). (Note that the taper, nose, parallel lane, and ghost island dimensions described in brackets are given in reference 31.)

composed of 0.5 km on either side of the merge or diverge nose in question; the connector road gradient is the average of the 0.5 km section prior to the nose.

From a traffic capacity aspect, the *merging* situation where a slip or link road joins a main carriageway is not dissimilar to the situation at a major/minor junction or an offside-priority roundabout entry, in that all are designed to operate on a gap-acceptance basis. Thus slip or link road entry capacity can also be considered in the context of major carriageway flows and junction layout factors.

In practice, the capacity design methodology requires the worst combination of corrected predicted hourly flows, in the 15th year after opening, for the merge traffic and the mainline upstream traffic to be inserted in Fig. 6.8(a). In this context, the mainline upstream traffic is defined as the traffic on the section of the major carriageway prior to where the mainline and connector flows combine (i.e. prior to the merge nose). The flow region defined by Fig. 6.8(a) is then checked with Table 6.21(a) to identify a merging lane type that is applicable to the selected flow region. Table 6.22(a) is then used to relate the merging lane type to a typical layout in Fig. 6.9(a).

In the case of diverge intersection design, the procedure is as described above except that the worst combination of corrected merge traffic and mainline downstream traffic is inserted in Fig. 6.8(b) to select a flow region. In this case, the mainline downstream traffic is the traffic on the section of the major carriageway after where the mainline and connector flows have separated (i.e. after the diverge nose). Tables 6.21(b) and 6.22(b) are then used to select a typical layout from Fig. 6.9(b).

Table 6.22   Lane types used in the layout of junctions with grade-separation[31]: (a) merging lanes, related to Fig. 6.9(a), and (b) diverging lanes, related to Fig. 6.9(b)

Merging type	Diagram	Layout
1, 3, 5	A	Direct taper
	C	Parallel lane
2, 3, 6, 7	D	Direct taper with ghost island
4, 8	B	Direct taper
	E	Direct taper with ghost island

(a)

Diverging type	Diagram	Layout
1, 2, 3, 5	A	Direct taper
	C	Parallel lane
4, 6	B	Direct taper
	D	Parallel lane

(b)

Sight distances

The ability to see ahead is probably the most important feature in the safe and efficient operation of a highway. Ideally, the geometric design should ensure that at no time is one vehicle invisible to another within normal eyesight distance. This is not always economically possible because of topographical and other considerations, and so it is necessary to design roads on the basis of sight distances that are the minimum necessary for safety.

Sight distance, which can be defined as the length of carriageway that is visible to the driver, is a basic element of geometric highway design and the standards set for minimum acceptable values of sight distance have a direct bearing on the cost of a highway. To the highway engineer engaged in designing a road, there are two forms of sight distance that are of particular interest. If safety is to be built into the highway, then sufficient sight distance must always be available to drivers to enable them to stop their vehicles prior to striking an unexpected object on the carriageway; this sight distance is known as the safe stopping sight distance. If efficiency is to be built into the highway, then sufficient sight distance must be provided for drivers to overtake and pass slower vehicles in complete safety; this sight distance is called the safe passing sight distance.

Before discussing the factors which compose the different sight distances, it might be noted that whilst there is now a general body of well-informed technical knowledge available regarding these, there is relatively little agreement internationally regarding the actual values that should be used for design purposes. This is most simply illustrated by comparing the 100 km/h standards for some different countries, as shown in Table 6.23.

Table 6.23 International rural road sight distance standards (in 1977) for design speeds of 100 km/h[32]

	Sight distance (m)	
Country	Stopping	Passing
Britain	210	450
Czechoslovakia	150	900
France	140	745
USSR	140	280

A second point which might be noted is the variation in the driver eye height used for geometric design purposes. For example, in 1962, the British design eye height was 1143 mm; in 1969, this value was lowered to 1050 mm. In 1962, the actual mean driver eye height above the carriageway was 1239.5 mm, with a measured standard deviation of 96.5 mm; in 1976, the actual mean value was 1141.2 mm, with a standard deviation of 49.8 mm. The current design eye-height criterion of 1050 mm corresponds to the 4th percentile of the 1976 actual eye height; in 1962, the 1143 mm criterion was the 15th percentile value[33].

As can be seen, therefore, the significant changes that have occurred in car styling since 1962 have resulted in the average height of drivers' eyes above the carriageway being decreased by 98 mm. However, the smaller standard deviation suggests that there is a styling trend towards a lower profile, even for larger cars, and could imply that a limiting height is now being approached.

Safe stopping sight distance

Whether on a two-, three- or multi-lane roadway, the most important consideration is that at all times the driver travelling at the design speed of the highway must have sufficient carriageway distance within his or her line of vision to allow the vehicle to be stopped before colliding with a slowly-moving or stationary object appearing suddenly on the carriageway. Unfortunately, there is ample evidence in the accident records and in the literature to indicate that the available sight distance does not have a great effect upon operating speed, and that drivers will not adjust their speeds sufficiently to accommodate inadequate sight distances on vertical curves in particular.

Calculation of the minimum distance required to stop the vehicle involves establishing values for speed, perception–reaction time, braking distance, eye height, and object height.

Unlike the situation with respect to overtaking sight distance, the object with which the designer is concerned when determining safe stopping sight distance may be a person lying on the carriageway, small objects, water on a causeway, or a pothole in the road surface. As a result, object eye heights used for design purposes in different countries range from a low of zero to various heights. In Britain, stopping sight distance is measured from a minimum driver's eye height of between 1.05 and 2.00 m to an object height of between 0.26 and 2.00 m above the carriageway surface[3]. It is applied in both the horizontal and vertical plane, between any two points in the centre of the lane on the inside of a curve (for each carriageway in the case of dual carriageways).

Perception–reaction time
Perception time is the time which elapses between the instant the driver sees the object on the carriageway and the instant of realization that brake action is required. The length of perception time varies considerably since it depends upon the distance to the object, the natural rapidity with which the driver reacts, the optical ability of the driver, atmospheric visibility, the type, condition and location of the roadway, and the type, colour and condition of the hazardous object.

Reliable data on perception time are meagre due to the difficulty of relating laboratory test conditions and results to actual situations. Most studies indicate, however, that the perception time at high speeds is usually less than at low speeds, since fast drivers are usually more alert.

Also, the perception time in urban areas tends to be lower than in rural areas, due to the varied conditions present in built-up locations which cause drivers to be continually alert.

Brake reaction time is the time taken by the driver to actuate the brake pedal after perceiving the object. Since it begins after the driver has been alerted, more reliable data are available concerning this time. Tests indicate that the average driver's brake reaction time is about 0.5 s. In order to provide a safety factor for drivers whose reaction times are above average, a time of 0.75 s is often suggested as appropriate for design purposes.

In practice, driver perception time is usually combined with brake reaction time in order to arrive at a total perception–reaction time that is suitable for highway design purposes. Measurements of this combined time showed that many drivers required as much as 1.7 s under normal roadway conditions. Since there is little control over the calibre of most motorists on the road today, it is clear that a value in excess of this must be used for design purposes. In line with this concept, a perception–reaction time of 2.5 s has been proposed as being desirable for rural design purposes[34]. A value proposed for use in urban areas is 1.5 s.

Expressed in formula form, the distance travelled by a vehicle during the perception–reaction time is as follows:

$$\text{perception–reaction distance (m)} = tv = 0.278tV$$

where v = initial speed (m/s), V = initial speed (km/h), and t = perception–reaction time (s).

Braking distance

The distance needed by a vehicle to stop on a level road, after the brakes have been applied, can be visualized as depending primarily upon the initial speed of the vehicle and the friction developed between the wheels and the carriageway surface. This distance can be estimated by utilizing the principle that the change in kinetic energy is equal to force multiplied by distance. Thus

$$\tfrac{1}{2}Mv^2 = Mgf \times d$$

Therefore

$$d = \frac{v^2}{2fg} = \frac{V^2}{254f}$$

where g = acceleration of free fall = 9.81 m/s^2, v = initial speed (m/s), V = initial speed (km/h), d = braking distance (m), and f = coefficient of friction developed between the tyre and the surface of the carriageway.

There is considerable variation in the value of f used for design purposes. If the comfort of the motorist is considered to be the sole criterion, then deceleration rates greater than 4.9 m/s^2 should not be used, as it is at about this rate that passengers are caused to slide from their seats. This deceleration rate is equivalent to a developed coefficient of friction of nearly 0.5 which, as will now be discussed, tends to be too high a value from a safety point of view.

Friction factor   The choice of the value of f is a most complicated one because of the many variables involved. Some physical elements which cause considerable variation in test results are the resilience and hardness of the tread rubber, tread design, amount of tread, efficiency of the brakes, and the condition and type of carriageway surface. Other important considerations are whether the friction value should be selected on the basis of dry, wet, icy or muddy conditions, and whether or not the wheel is rotating after the driver slams on the brakes.

The quality of the tyre tread is something over which the highway engineer has little control, so most investigators assume that the tyre is relatively smooth.

The choice of whether the friction factor should be measured under conditions of incipient sliding or after the wheel is locked and skidding is taking place is usually resolved in favour of the locked wheel concept. When an unexpected object appears on the road, most motorists automatically step on the brakes, thus causing the brakes to lock on very many vehicles. Until such time, therefore, that brake systems are widely used that will prevent the wheels of cars from locking after the sudden application of brakes, design calculations will tend to be based on the lower and locked wheel values.

The choice of road surface conditions is also a straightforward one. Ideally, design should be based on icy conditions, since repeated studies have shown that braking distances on icy surfaces are from three to twelve times as great as those for dry surfaces. In Britain, however, this is not a practical design concept because of the relatively mild climate; more important are the friction values which may be developed on wet carriageway surfaces. Values measured on wet surfaces are substantially lower than those measured on dry surfaces, since water tends to act as a lubricant between the road and the tyre. In this respect also, it should be noted that the value of the friction factor decreases as the vehicle speed increases. This can be related to the variation in contact time between the tyre tread and the road surfacing. At low speeds, the contact time is relatively long and there is ample time for water film to be expelled from between the rubber and the surfacing. This is facilitated by a well-patterned tyre with effective drainage characteristics. However, when the vehicle speed is increased, the relative contact time is smaller, the expulsion of the moisture becomes less complete, and the value of the friction coefficient becomes smaller.

Friction values measured at given speeds on particular carriageway surfaces also show considerable variations. For instance, friction measurements on a concrete road at one location may result in very different values from those taken on a concrete road at a different site under similar conditions. Attempting to take into account the many variables involved, as well as the prime consideration of safety, friction values of 0.36, 0.33, 0.31, 0.30 and 0.29 at 48, 64, 80, 96.5 and 112 km/h, respectively, have been proposed for use in designing safe stopping sight distances on most highways when using the above equations.

The question also arises as to whether it is realistic to assume that vehicles travel at the full design speed of the highway when the carriageway is wet, i.e. the conditions chosen for measuring the friction factor. American experience in this matter suggests that the speed for wet conditions is approximately 85 to 95 per cent of the design speed. Studies on motorways in Britain, however, indicate that drivers pay relatively little attention to wet weather conditions. For this reason, it is probably safer to design on the assumption that vehicles will be travelling at the design speed of the highway.

British practice
Table 6.24 shows the safe stopping sight distances used in the design of single and dual carriageway urban and rural roads in Britain. These are based on passenger car operations only.

Table 6.24 Minimum sight distances currently used on highway design in Britain[3]

Design speed (km/h)	Stopping distance (m)		Minimum passing distance (m)
	Desirable	Minimum	
120	295	215	*
100	215	160	580
85	160	120	490
70	120	90	410
60	90	70	345
50	70	50	290

*Not used with single carriageways

Commercial vehicles
Commercial vehicles have longer braking distances (often as much as twice the distances) as compared with passenger cars for the following reasons: (a) commercial vehicle tyres do not grip the carriageway as well as do car tyres, (b) the brakes on commercial vehicles are often not as powerful, in relation to the laden weight of such vehicles, as are those of cars, and (c) being air-operated, the brakes on heavy goods vehicles take an appreciable time (up to 1 s) to become fully applied after the driver pushes the brake pedal. In fact, whilst most cars can achieve an overall average deceleration through an emergency braking stop of 0.8g, heavily laden goods vehicles often cannot reach much above 0.4g[35].

In practice, however, it is usually not considered necessary to give special consideration to commercial vehicle stopping sight distance requirements. There are two main reasons for this. Firstly, commercial vehicles travel at slower speeds than do passenger cars at a given location, either because of regulation or by operator choice. Secondly, by occupying a higher position in the vehicle, the commercial vehicle driver is usually able to see a vertical obstruction on the roadway sooner than it would be seen by the passenger car driver. Thus, in the great majority of situations, these two additional considerations have the effect of compensating for the additional braking distances that might otherwise be needed.

The main exception where additional stopping distance should be provided for commercial vehicles is in the case of a horizontal sight restriction at the end of a steep hill. In locations where this occurs, the commercial vehicle speeds may approach those of passenger cars, while the additional eye height of the commercial vehicle operator is of little value because of the sight restriction.

Passing sight distance

On single carriageway roads, a primary driver comfort requirement is the ability to overtake slower vehicles without undue danger or delay. Furthermore, the frequency of the overtaking opportunities affects the capacity of a heavily-travelled highway, i.e. the fewer the overtaking sites the less is the capacity of the road, as the faster motorists are constrained by the speeds of the slower ones from filling gaps in the traffic stream.

The minimum sight distance required by a vehicle to overtake safely on two- or three-lane roads is the visibility distance which will enable the overtaking driver to pass a slower vehicle without interfering with the speed of an oncoming vehicle travelling at the design speed of the highway.

Two-lane roads
As indicated in Fig. 6.10, there are four components of the minimum distance required for safe overtaking on two-lane roads.

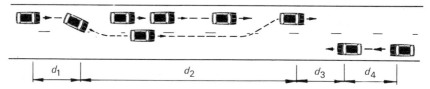

Fig. 6.10   Elements of total passing sight distance requirements on two-lane roads

The dimension d_1 represents the time taken or distance travelled by a vehicle while its driver decides whether or not it is safe to pass the vehicle in front. This time period has been described as the hesitation time[36] and is about 3.5 s for comfortable overtaking conditions.

The dimension d_2 represents the time taken or distance travelled by the overtaking vehicle in carrying out the actual passing manoeuvre. Thus it begins the instant the overtaking driver turns the wheel and ends when the vehicle is returned to its own lane. Measurements under controlled conditions regarding the length of this dimension are shown as solid lines in Fig. 6.11. The dotted lines shown in this graph provide estimates of the requirements at higher speeds.

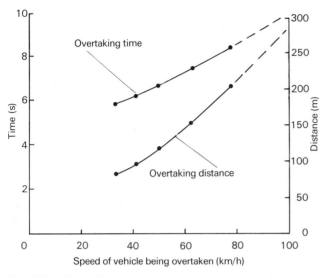

Fig. 6.11   Overtaking time and distance requirements

The dimension d_3 has been called the safety dimension and is the time or distance between the overtaking vehicle and the oncoming vehicle at the instant the overtaking vehicle has returned to its own lane. From a safety aspect, it is of course desirable that this distance should be as large as possible, whereas practical economic requirements necessitate that it should be as small as possible. One value that has been suggested is 1.5 s. This means that if the combined relative speed is 160 km/h, then a safety margin of 67 m is available between the two vehicles.

The dimension d_4 represents the time taken or distance travelled by the opposing vehicle at the design speed of the road while the actual overtaking manoeuvre is taking place. Conservatively, it should be the total distance travelled by the opposing vehicle during the time $d_1 + d_2$, but in practice this may be questioned as being too long. Examination of the overtaking vehicle's track in Fig. 6.10 shows that it can return freely to its own lane at any instant prior to drawing alongside the overtaken vehicle. If this initial time is not taken into account, then the opposing vehicle dimension d_4 can be taken as approximately equal to $\frac{2}{3}d_2$.

The recommended safe passing sight distances for use on single carriageway roads in Britain are given in Table 6.24. These are based on a visibility envelope encompassing a driver eye height of 1.05 to 2.00 m and an object height of 1.05 to 2.00 m, above the carriageway centre.

Three-lane roads
Adequate data are meagre relative to the distances required for safe

overtaking on three-lane roads. Ideally, the minimum passing sight distance requirements for three-lane roads should be the same as for two-lane roads. From a practical aspect, however, this requirement can be considered rather stringent, since the centre lane makes it feasible for very many overtakings to be accomplished in the face of oncoming traffic, *provided* that the traffic on the highway does not exceed its capacity. The centre lane also makes the overtaking manoeuvre more flexible since either or both vehicles in the centre lane can more easily return to the proper lane if danger arises. For these reasons, therefore, the minimum passing sight distance requirements can be assumed to be composed of the same elements as for two-lane roads, except that the dimension d_4 can be omitted.

As with two-lane roads, it must be emphasized that values determined in this manner represent the minimum for safe overtaking. Where economically possible, much longer sight distances should be designed into the highway.

Multilane roads

Most multilane roads in Britain, and especially those in rural areas, are dual carriageways. With these roads, there is no need to provide sections for one vehicle to overtake another in the face of oncoming traffic. Unless the capacity of a dual carriageway is severely overtaxed, opportunities for safe overtaking will normally present themselves at frequent intervals on each carriageway.

Commercial vehicles

The values discussed previously with respect to safe passing sight distances on two- and three-lane roads relate to passenger car requirements only. The evidence available suggests that overtaking vehicles are relatively little affected by the type of oncoming vehicle[36] or by the type of overtaken vehicle[37]. Other things being equal, however, motorists tend to require a larger passing opportunity and hence tend to take fewer risks when the oncoming vehicles are larger ones.

The greatest difference between the various types of vehicle is reflected in the overtaking times required by the heavier vehicles. On average, heavy commercial vehicles take about 4 s longer than cars to complete the overtaking manoeuvre. Nevertheless, it is doubtful whether overtaking sight distances should be based on commercial vehicle needs, unless exceptional circumstances apply. Apart from the extra expense that this would involve, the commercial vehicle driver is able to see further due to the greater eye height associated with this type of vehicle, thereby compensating for any additional overtaking length that might be required.

Horizontal curvature

Horizontal curvature design is one of the most important features influencing the efficiency and safety of a highway. Improper design will result in lower speeds and a lowering of highway capacity. The importance of curve design upon safety is reflected by the seminal accident statistics given in Table 6.25, which show that the sharper the curve the greater is the tendency for accidents to occur.

Table 6.25 Accident rates per 1.6 million veh-km for curves of various radii in Lancashire, 1946–47

Curve radius (m)	Total accident rate	
	Urban	Rural
Straight to 610	4.70	1.47
610 to 305	3.42	2.46
305 to 150	4.20	4.01
150 to 60	7.20	3.72
<60	20.00	16.70

The maximum comfortable speed on a horizontal curve is primarily dependent upon the radius of the curve and the superelevation of the carriageway. In addition, vehicle speeds and safety on high-speed roads are aided by the presence of such features as extra carriageway widths at the curves themselves and the insertion of transition curves between straights and curves.

Properties of the circular curve

A circular curve joining two road tangents can be described either by its radius or by its degree of curvature. In Britain, the radius is usually utilized, whereas in certain other countries, notably the USA, the degree of curve concept is preferred. The (metric) degree of curve is defined as the central angle which subtends a 100 m arc of the curve.

Figure 6.12 shows the principal properties of a simple highway curve. PI, PC and PT are the points of intersection, curvature and tangency, respectively. Some of the more important formulae for such a curve are

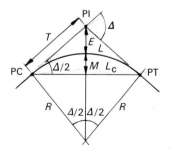

Fig. 6.12   A simple circular curve

given below without proof—their derivation can be found in any good textbook on surveying.

$$T = R\tan(\varDelta/2)$$

$$L_c = 2R\sin(\varDelta/2)$$

$$L = 100\varDelta/D$$

$$E = R[\sec(\varDelta/2) - 1] = T\tan(\varDelta/4)$$

$$M = R[1 - \cos(\varDelta/2)]$$

$$D = 5729.6/R$$

$$R = 50/\sin(D/2)$$

The symbols in these formulae are defined as follows: $\varDelta$ = external angle, D = degree of curvature, E = external distance, M = chord-to-curve length, L = length of the curve, L_c = long chord length, T = length of the tangent, and R = radius of the curve.

Curvature and centrifugal force

Any body moving rapidly along a curved path is subject to an outward reactive force called the centrifugal force. On highway curves, this force tends to cause vehicles to overturn or to slide outwards from the centre of road curvature. Figure 6.13(a) illustrates the forces acting on a vehicle as it moves about a horizontal curve. Since the carriageway surface is flat and the forces are in equilibrium, then

$$P = Mv^2/R$$

where P = lateral frictional force resisting the centrifugal force (newtons),

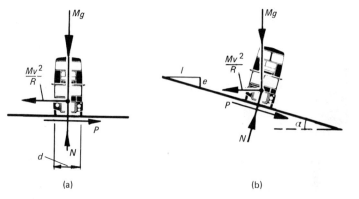

Fig. 6.13 Forces acting on a motor vehicle moving about a horizontal curve: (a) non-superelevated and (b) superelevated

M = mass of the vehicle (kg), v = speed of the vehicle (m/s) and R = radius of the curve (m). In addition,

$$P = N\mu = Mg\mu$$

where N = reaction (newtons). Therefore

$$\mu = v^2/gR = P/Mg$$

where μ = developed lateral coefficient of friction and g = acceleration of free fall (m/s²). Also, v^2/gR is known as the centrifugal ratio.

When the velocity is expressed in V km/h and g is taken as 9.81 m/s², then

$$\mu = V^2/127R \quad \text{(approx.)}$$

Thus, if the limiting value of μ is known, the minimum curve radius can be calculated for any given design speed.

The centrifugal force also tends to cause a vehicle to overturn. This force, acting through the centre of mass of the vehicle, causes an overturning moment about the points of contact between the outer wheels of the vehicle and the carriageway surface. This overturning moment is resisted by a righting moment caused by the mass of the vehicle acting through its centre of mass. Thus, for equilibrium conditions,

$$(Mv^2/R) \times h = Mg \times d/2$$

and $h = d \times 1/(2v^2/gR) = d/2\mu$

where d = lateral width between the wheels (m) and h = height of the centre of mass above the carriageway (m).

The above equation illustrates, for instance, that if $\mu = 0.5$, then the height to the centre of mass must be greater than the lateral distance between the wheels before overturning will take place. Modern vehicles, however, and especially passenger cars, have low centres of mass so that relatively high values of μ would have to be developed before overturning could take place. In practice, the frictional value is usually sufficiently low for sliding to take place long before overturning. It is only with certain commercial vehicles having high centres of mass that the problem of overturning may arise.

Superelevation

In order to resist the outward-acting centrifugal force, it is customary to superelevate or slope the carriageway cross-section of curved sections of a modern highway in the manner shown in Fig. 6.13(b). For every combination of radius of curvature and highway design speed, there is a particular rate of superelevation that exactly balances the centrifugal force. Where the superelevation is insufficient to balance the outward force, it is necessary for some frictional force to be developed between the tyres and the road surface in order to keep the vehicle from sliding laterally.

When the carriageway is superelevated, the forces acting on the vehicle are as shown in Fig. 6.13(b). Since the forces are in equilibrium,

$$(Mv^2/R)\cos \alpha = Mg \sin \alpha + P$$
$$= Mg \sin \alpha + \mu[Mg \cos \alpha + (Mv^2/R)\sin \alpha]$$

Therefore

$$v^2/gR = \tan \alpha + \mu + (\mu v^2/gR)\tan \alpha$$

where α = angle of superelevation.

The quantity $(\mu v^2/gR)\tan \alpha$ is so small that it can be neglected. If $\tan \alpha$ is expressed in terms of the slope e, then

$$v^2/gR = e + \mu$$

This indicates clearly that the centrifugal force is resisted partly by the superelevation and partly by the lateral friction. If v m/s is replaced by V km/h, and if $g = 9.81$ m/s^2, then the *minimum* radius equation becomes

$$V^2/127R = e + \mu$$

If $\mu = 0$ and the forces acting on the vehicle are in equilibrium, then the situation occurs where the centrifugal force is entirely counteracted by the superelevation. In this ideal situation, a moving vehicle will steer itself about the superelevation curve once its steering wheels have been set in the required track. When this occurs, it must be remembered that, although no lateral friction is brought into use, sufficient friction must be developed in the longitudinal direction of travel to provide for traction of the wheels.

In practice, high design speeds are only utilized in car-racing tracks, and cannot be used alone on normal highways because of the danger to slow-moving vehicles. For instance, a carriageway surface may be relatively steeply superelevated in order to allow one vehicle travelling at high speed to traverse a curve in safety without making use of the friction component. Another vehicle travelling at a slow speed about the same curve will develop a much smaller outward centrifugal force and the result may well be that the vehicle will tend to slide inwards. In the extreme case of a vehicle at rest on the curved section, no centrifugal force at all is developed by the vehicle. Therefore, in order to minimize the danger of sliding, the superelevation slope must never be greater than the minimum lateral coefficient of friction that can be developed between the tyre and the carriageway under the design weather conditions.

Proper curve design does not normally take full advantage of the obtainable lateral coefficients of friction since, obviously, design should not be based on a condition of incipient sliding. Investigations have shown that the main factor controlling vehicle speed on a curve is the feeling of discomfort felt by the motorist whilst negotiating the curve at a given speed. This sensation of the driver is related to the centrifugal ratio v^2/gR which, in turn, is related to the sliding resistance required to carry out the manoeuvres. Various observers have studied this problem and it appears that motorists feel uncomfortable, and the vehicle becomes harder to hold on the road, when the centrifugal ratio exceeds about 0.30; a value of 0.15 is a safe basis for curve design as it will eliminate most feelings of unease. This lateral ratio is equivalent to a lateral coefficient of friction of 0.15, which is considerably below friction values that can be developed on the carriageway before slipping occurs. Since, however, highway curves should be designed to avoid slipping conditions, driver comfort represents a rational method of selecting a lateral coefficient of friction that can be used for design purposes.

Department of Transport recommendations General practice in Britain with respect to superelevation design is not to think in terms of a self-steering vehicle travelling at the design speed of the road but rather of one negotiating a curve at about the average speed of the road; this will result in a more gentle superelevation, and hence there is less chance of slow-moving vehicles slipping sideways under icy conditions or, indeed, overturning when carrying exceptionally high loads. The 'average' speed tacitly accepted for normal usage is 67 per cent of the road design speed; in practice, this means that the self-steering equation becomes

$$(0.67V)^2/127R = e = V^2/283R$$

When the superelevation is expressed as a percentage (the usual practice), the above formula becomes the familiar

$$e \text{ (per cent)} = 0.353V^2/R$$

Generally, the superelevation should never exceed 7 per cent, with 5 per cent being the maximum in urban areas.

Table 6.26 summarizes current British practice[3] in respect of highway radii and superelevation. In this table the desirable values represent the

Table 6.26 Radii and superelevation criteria currently used in highway design

Horizontal curvature condition	Radius (m) for design speeds (km/h) of					
	120	100	85	70	60	50
Minimum R^* without elimination of adverse camber and transitions	2880	2040	1440	1020	720	510
Minimum R^* with $e = 2.5\%$	2040	1440	1020	720	510	360
Minimum R^* with $e = 3.5\%$	1440	1020	720	510	360	255
Desirable minimum R with $e = 5\%$	1020	720	510	360	255	180
Absolute minimum R with $e = 7\%$	720	510	360	255	180	127
Limiting radius at sites of special difficulty	510	360	255	180	127	90

*Not used with single carriageways

comfortable values dictated by the design speed; normal design would seek to achieve these at least. Note that the absolute minimum values, which are acceptable using minimum dynamic parameters, are identical to the desirable values for one design speed below.

Sight distances at horizontal curves

In certain instances, the radius of curvature determined by using limiting values of e and μ may not be adequate to ensure that, at the very least, the minimum safe stopping sight distance requirements are met. To provide the necessary horizontal sight distances, it may be necessary to set back slopes of cuttings, fences, buildings or other such obstructions adjacent to the carriageway. If these obstructions are immovable, it may be necessary to redesign the road alignment in order to meet the safety requirements.

The most usual methods of calculating the offset distance necessary to secure the required horizontal sight distance are illustrated in Fig. 6.14. In both cases, the driver's eye and the dangerous object are assumed to be at the centre of the nearside lane. The chord AB is the sight line and the curve ACB is the distance required to meet the sight distance criterion.

Figure 6.14(a) illustrates the situation where the required sight distance lies wholly within the length of the curved road section and ACB is equal to the required sight distance S. The minimum offset distance M from the centreline to the obstruction can be estimated most simply by considering the track of the vehicle to be along the chords AC and CB rather than along the arc of the curve. By geometry,

$$R^2 = x^2 + (R-M)^2$$

and $x^2 = (S/2)^2 - M^2$

Therefore

$$R^2 = S^2/4 - M^2 + R^2 - 2RM + M^2$$

Thus $M = S^2/8R$

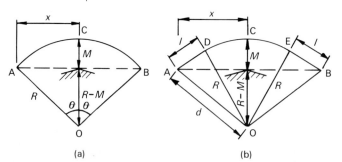

Fig. 6.14   Sight distance considerations on a horizontal curve

Figure 6.14(b) illustrates the situation where the required sight distance is greater than the available length of curve. Thus the sight distance overlaps the highway curve onto the tangents for a distance l on either side. If the length of the curve is equal to L and the stopping distance is equal to S, then

$$S = L + 2l$$

Therefore

$$l = (S - L)/2$$

By geometry,

$$(S/2)^2 = x^2 + M^2$$
$$x^2 = d^2 - (R - M)^2$$

and $d^2 = [(S - L)/2]^2 + R^2$

Therefore

$$S^2/4 = \tfrac{1}{4}(S^2 - 2LS + L^2) + R^2 - (R^2 - 2RM + M^2) + M^2$$

Therefore

$$M = L(2S - L)/8R$$

It should be noted that, when $S = L$, the above equation reduces to $M = S^2/8R$.

Widening of circular curves

Curve widening refers to the extra width of carriageway that is required on a curved section of a highway over and above that required on a tangent section.

Most heavy goods vehicles are steered by their front wheels or, in the case of an articulated trailer, by the hinge point, whilst the rear wheels are fixed in the straight-ahead position. The steering geometrics of heavy commercials are based on the Acherman principle whereby a vehicle turns about a point approximately in line with its rear axle, so that when the vehicle is turning the rear wheels follow a path of shorter radius than that followed by the front wheels. This has the practical effect of increasing the effective carriageway width required if the same clearance is to be maintained between opposing goods vehicles on curved sections as on tangent sections of single carriageway roads, or between adjacent vehicles travelling in the same direction on dual carriageway roads.

The difference between the arcs of the front and rear wheels on the same side of the vehicle, defined as the *cut-in*[38], is the basic amount of 'extra' carriageway width required on the curve as compared with the tangent section of highway. The clearance *cut-out* is determined by the amount of overhang forward of the front wheels or, in the case of trailers, the hinge point; it is the difference between the radii of turn of the outer front wheel and the foremost outer corner of the vehicle. The extra carriageway and clearance needs are accentuated on low-speed, tight radius curves/corners.

A second practical reason for providing an extra carriageway width on curves/corners is the natural tendency of drivers to shy away from the edges of the carriageway as they carry out turning manoeuvres. This further reduces the clearance between vehicles and increases the accident potential at these locations.

Figure 6.15 shows in diagrammatic form the cut-in, cut-out and basic

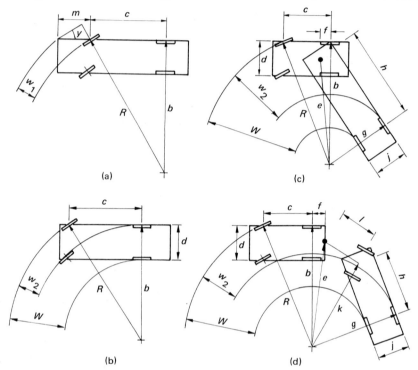

Fig. 6.15 Extra width requirements at circular curves: (a) cut-out requirements, and (b), (c) and (d) cut-in requirements for rigid vehicle, articulated vehicle and draw-bar trailer combination, respectively

road width requirements of various types of vehicle. Mathematically, these requirements can be expressed as follows.

Cut-out

$$w_1 = m \sin y = mc/R$$

Cut-in: rigid vehicle

$$w_2 = R - b$$

and $$W = R - b + d = R + d - (R^2 - c^2)^{1/2}$$

Cut-in: articulated vehicle

$$w_2 = R - g$$

and $$W = R - g + j = R + j - [(e^2 - h^2)^{1/2} + 0.5j]$$
$$= R + 0.5j - (e^2 - h^2)^{1/2}$$

Cut-in: draw-bar trailer combination

$$w_2 = R - g$$

and
$$W = R - g + j = R + j - [(k^2 - h^2)^{1/2} + 0.5j]$$
$$= R + 0.5j - (k^2 - h^2)^{1/2}$$

In the above formulae, the symbols are defined as follows. w_1 is the maximum cut-out. w_2 is the maximum cut-in. W is the basic turning width. R is the outside turning radius of the first axle. m is the overhang at the front of the vehicle. b is the outside turning radius of the second axle $= (R^2 - c^2)^{1/2}$. c is the wheelbase, i.e. the length between the first and second axles. d is the width of the rigid or tractor vehicle. e is the turning radius of the king pin or coupling $= [(b - 0.5d)^2 + f^2]^{1/2}$. f is the distance from the second tractor axle to the king pin or coupling. g is the outside turning radius of the trailer rear axle. h is the wheelbase of the trailer. j is the width of the trailer. k is the turning radius of the trailer steering axle $= (e^2 - l^2)^{1/2}$. l is the draw-bar length. Finally, y is the steered angle of the outside front wheel.

In relation to the above, it might also be noted that the vehicle width plus cut-in plus cut-out gives the total theoretical minimum turning width required by a vehicle for a given curve/corner radius and angle of turn. In determining actual widths required in addition to cut-out, however, allowance has to be made for body side overhang and driver tolerance, as well as for the additional tyre slip which is likely to occur at higher speeds and on wet surfaces[38].

A graphical method of predicting the swept paths of both rigid and articulated vehicles through any series of turns is available in the technical literature[39].

To determine the extra carriageway width required in any particular situation, it is necessary to choose an appropriate design vehicle for the curve/corner radius in question. It is generally considered good practice to provide the same total extra width to each carriageway of a dual carriageway highway as is applied to a two-lane, two-way single carriageway road; this ensures the high standard of geometric design which is compatible with dual carriageway facilities. On three-lane single carriageways, the need for extra widening depends upon whether the highway is to be operated as a two- or three-lane facility at the sections in question, i.e. if operated as a three-lane road, then widening is usually needed, whereas if operation is to be on a two-lane, two-way basis, then extra widening is normally not necessary.

British practice re widening on highway curves[3] and at corners of intersections[18] is given in Table 6.27. In this table, the data given for intersections are considered to be satisfactory for use by the longest permitted articulated vehicle. Widening on curves is required for substandard conditions, and for low radius curves below 70 km/h design speed. In the case of two-lane roads wider than 7.9 m, no additional widening is required.

Table 6.27 British practice re extra widths at curves and junctions: (a) widenings on highway curves and (b) maximum widths of carriageways in junctions

Radius (m)	Increase per lane width of carriageway (m)	Application
<150	0.30	To standard width carriageways, i.e. 7.3 m (2 lanes), 11.0 m (3 lanes) and 14.6 m (4 lanes)
	0.60	To carriageways of less than the standard widths, subject to maximum widths of 7.9 m (2 lanes), 11.9 m (3 lanes) and 15.8 m (4 lanes)
150–300	0.50 ⎫	To carriageways of less than the standard widths, subject to the maximum widths not being greater than the standard widths
300–400	0.30 ⎭	

(a)

Inside corner radius (m)	Single lane width (m)		Two-lane width (without hardstrip) for one- or two-way traffic (m)	
	With hardstrip	Without hardstrip	Inside lane	Outside lane
10	8.4	10.9	8.4	6.5
15	7.1	9.6	7.1	6.0
20	6.2	8.7	6.2	5.6
25	5.7	8.2	5.7	5.2
30	5.3	7.8	5.3	5.0
40	4.7	7.2	4.7	4.6
50	4.4	6.9	4.4	4.3
75	4.0	6.5	4.0	4.0
100	3.8	6.3	3.8	3.8
150	3.65	6.1	3.65	3.65

(b)

Widening procedure

When a carriageway is to be widened at a curve, the extra width should be attained gradually over the approaches to the curve so that the 'natural' inclinations of vehicles entering or leaving the curved section are facilitated. The four main considerations to be taken into account are as follows.

(1) On simple circular curves, the total extra width should be applied to the inside of the carriageway, while the outside edge and the centreline are both kept as concentric circular arcs. The result is much the same as providing a transition curve on the centreline. The main drawback is that the full width of the carriageway may not always be in use, since a fast-moving vehicle will tend to curve in from the outer edge at the beginning and end of the curved section as it makes its own transition.

(2) Where transition curves are provided before and after a simple curve, the widening may be equally divided between the inside and outside of the

curve, or it may be wholly applied to the inside edge of the carriageway. In either case, the final centreline marking, if it is a two-lane road, should be placed halfway between the edges of the widened carriageway.

(3) The extra width should always be attained gradually, and never abruptly, to ensure that the entire carriageway is usable. When the curved sections utilize transition curves, they should be attained over the entire transition length so that there is zero extra width at the beginning of transition and the full extra width at the end. Where transition lengths do not occur, smooth alignment will be obtained if one-half to two-thirds of the extra width is distributed along the tangent just prior to the curve, and the balance along the curve itself.

(4) From an aesthetic point of view, the edges of the carriageway should at all times form smooth and graceful curves—not only is this pleasing to the eye but it also facilitates construction and provides reassurance for the motorist that the highway curve is properly laid out.

Transition curves

As a vehicle enters or leaves a circular curve, the driver will, for comfort and safety, gradually turn the steering wheel from the normal position to that of the necessary deflection in order to combat the developing effect of the centrifugal force. By thus gradually changing direction, the motorist will cause the vehicle to trace out its own transition curve from the tangent to the curve.

The transition path traced out in this way will vary depending upon the speed of the vehicle, the radius of the curve, the superelevation of the carriageway, and the steering action of the motorist. At moderate values of speed and curvature, the average driver is usually able to effect this path within the limits of the normal lane width. With higher speeds and sharper curves, vehicles in the inner lane tend to trace their transitions by crowding onto the adjoining lane, while those in the outer lane tend to cut the corner. This reduces the effective width of the roadway and can be a cause of accidents.

The primary purpose of transition curves is to enable vehicles moving at high speeds to make the change from the tangent section to the curved section to the tangent section of a road in a safe and comfortable fashion. The proper introduction of transition curves will provide a natural easy-to-follow path for motorists so that the centrifugal force increases and decreases gradually as the vehicle enters and leaves the circular curve. This minimizes the intrusion of vehicles onto the wrong lanes, tends to encourage uniformity in speed, and increases vehicle safety at the curve.

Spiral transition curves

The essential requirement of any transition curve is that its radius of curvature should decrease gradually from infinity at the tangent–spiral intersection to the radius of the circular curve at the spiral–circular-curve intersection. Various forms of curves are used for this purpose, each having its own special advantages. The most common of these are the lemniscate, the spiral (or clothoid), and the cubic parabola; they are considered the 'natural' transition curves in contrast to others which constrain the motorist into a more definite path. In practice, there is very little difference between the results obtained by all three procedures.

Of the various forms, the spiral transition curve is probably the most widely accepted by highway engineers, so it will be briefly discussed here. Its acceptance is primarily based on the ease with which it can be set out in the field compared with the other two curves, and not because of any particular superiority as a transition curve.

Spiral properties A circular curve joined to two tangents by spiral transition curves is shown in Fig. 6.16. The dotted lines illustrate the circular curve as it would appear if the transition curves were omitted. TS is the tangent–spiral intersection. SC is the junction of the spiral and circular curves. CS is the intersection of the circular arc and the second spiral. Finally, ST is the second tangent point.

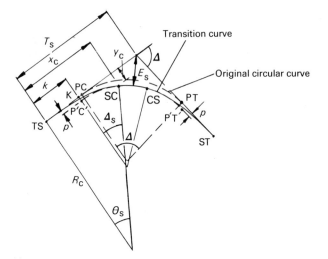

Fig. 6.16   A simple spiral transition curve

In the figure, the amount by which either end of the circular curve is shifted inwards from the tangent is indicated by the dimension K to P′C′. This distance p is called the *shift*. It is convenient to remember that the offset from point K on the tangent to the transition spiral is very nearly $P/2$, and that the line from K to P′C′ approximately bisects the spiral.

Some of the more important formulae affecting the use of the spiral curve in conjunction with the circular curve are given below without proof—detailed derivation of these can be found in surveying textbooks.

$$\theta_s = L_s/2R_c \text{ radians} = 57.3L_s/2R_c \text{ degrees}$$

$$T_s = (R_c + p)\tan(\varDelta/2) + k = T + p\tan(\varDelta/2) + k$$

$$E_s = (R_c + p)\sec(\varDelta/2) - R_c = E + p\sec(\varDelta/2)$$

$$p = y_c - R_c(1 - \cos\varDelta_s) = L_s^2/24R_c \quad \text{(approx.)}$$

$$k = x_c - R_c\sin\varDelta_s = L_s/2 \quad \text{(approx.)}$$

$$x_c = L_s(1 - \theta_s^2/10)$$

$$y_c = L_s(\theta_s/3 - \theta_s^3/42)$$

The symbols in these formulae are defined as follows. θ_s is the spiral angle, i.e. the total angle subtended by the spiral. $\varDelta_s$ and $\varDelta$ are the shifted and unshifted external angles, respectively. E_s and E are the external distances to the shifted and unshifted curves, respectively. L_s is the length of the spiral. R_c is the radius of the circular arc. T is the tangent distance for the unshifted circular curve. T_s is the tangent distance from the point of intersection to the TS. k is the abscissa of P'C', the shifted PC. Finally, x_c and y_c are the coordinates of the SC.

Length of transition The three major factors governing transition curve design are the radius of curvature R_c, the external angle $\varDelta$, and the length of transition L_s. Of these, R_c and $\varDelta$ are usually selected on the basis of conditions existing in the field, whereas L_s is selected on the basis of factors affecting the comfort and safety of the motorist.

As a vehicle passes along the transition curve, its centrifugal acceleration changes from zero at TS to v^2/R_c at SC. The transition length over which this change takes place is equal to the vehicle velocity v multiplied by the travel time t. Thus

$$t\,(\text{s}) = \frac{L_s\,(\text{m})}{v\,(\text{m/s})}$$

If the rate of gain of radial acceleration is denoted by C, then

$$C = \frac{v^2/R_c}{L_s/v} = \frac{v^3}{R_cL_s}\,\text{m/s}^3$$

If C is controlled and becomes the allowable rate of change of radial acceleration, then

$$L_s = \frac{v^3}{CR_c} = \frac{V^3}{3.6^3CR_c}$$

where L_s = transition length (m), V = design speed (km/h), and R_c = radius of circular curve (m).

There is considerable difference of opinion as to what constitutes a proper value for C. In railroad practice, a value of $C = 0.3\,\text{m/s}^3$ has tended to be accepted as the value at which the increase in acceleration will be unnoticed by a passenger sitting in a railway coach over a bogie. Conditions are not the same for drivers of motor vehicles, however, since the motorist can much more readily make travel path corrections than can the railway passenger. Hence the tendency in highway design in many countries has been to utilize higher values of C as these result in shorter transition lengths; for instance, the C-values historically used in Britain[40] vary between 0.3 and $0.6 \, \text{m/s}^3$, with values in excess of $0.3 \, \text{m/s}^3$ only being used in difficult cases[3].

Use of transition curves

Transition curves should be used on all highways whenever there is a significant change in road curvature. Obviously, however, there is less need for transitions as curves become flatter. At some point, the difference between road curves with and without transition lengths is so small that the transition curve has little significance, except perhaps as a graceful method of changing from a cambered to a superelevated section. This point of no significance is usually reached when the shift is approximately 0.3 m.

British practice is for transitions to be applied to all curves the radii of which are less than those in the first row of Table 6.26. On such bends (substandard curves for the appropriate design speed) the transition length is normally limited to $(24R)^{0.5}$.

Application of superelevation   As discussed previously, superelevation should be introduced on curves with other than very large radii. In these cases, it is most important that the superelevation of the carriageway be built up uniformly over as long a distance as possible, so that the full superelevation rate is achieved at the circular arc part of the curve. This is not only for reasons of comfort and safety but also for aesthetic reasons. To avoid the carriageway edges having an unpleasing, distorted appearance, the change in cross-section from a camber-section to a fully-superelevated one should be carried out, ideally, such that the slope of the outside edge of the carriageway does not exceed 1 in 200 when compared with the centreline. In other words, the difference in grade between the centreline profile and the edge of a two-lane road should not exceed 0.5 per cent.

When a spiral transition curve is used in conjunction with a circular curve, it is usual for the superelevation run-off to be carried out over the total length of the spiral. The actual attainment of superelevation requires that the carriageway cross-section be tilted, the amount of movement varying with its location on the transition or circular curves. This tilting of the carriageway can be carried out by any one of the following three ways:

(1)   rotation about the centreline of the carriageway,
(2)   rotation about the inside edge of the carriageway,
(3)   rotation about the outside edge of the carriageway.

In practice, there is little to choose between the three methods, so that the choice in any particular situation can be that which provides the most pleasing and functional results. Regardless of which method is used, however, care should always be taken that the drainage requirements at the inside of the roadway are not detrimentally affected by the rotation procedure.

The method which this author prefers is that of rotation about the centreline. In the normal course of events, the longitudinal profile of the carriageway will have been decided upon before any superelevation calculations take place. More often than not, therefore, it is more convenient to keep the centreline levels unchanged and to move both edges of the carriageway instead.

One method of attaining superelevation by rotating about the longitudinal centreline is illustrated in Fig. 6.17. With this method, the

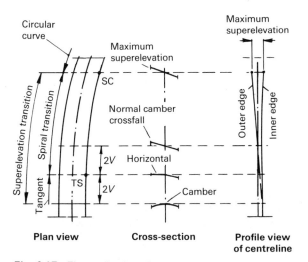

Fig. 6.17 The application of superelevation to transitional and circular curves

superelevation transition is started on the tangent at a distance of $1.4V$–$2.0V$ metres back from the beginning of the horizontal transition curve. In this case, V is the design speed of the highway expressed in *metres per second*. By the time the tangent-to-the-spiral point is reached, the outer half of the carriageway is warped upwards so that it is horizontal, while the inner half is still cambered. Warping of the outer half is continued for a further distance of $1.4V$–$2.0V$ metres along the spiral transition until both lanes are sloping inwards at a cross-slope equal to the normal camber crossfall rate on the tangents. Between this point and the spiral–circular-curve intersection, the full superelevation is developed at a uniform rate. The circular curve is then entirely superelevated at the maximum rate for its entire length. From the circular–spiral-curve intersection, the superelevation is reduced in reverse order to the manner just described.

Vertical alignment

Vertical alignment design refers to the design of the tangents and curves along the profile of the road. The primary aim of this profile design is to ensure that a continuously unfolding ribbon of road is presented to the motorist so that his or her anticipation of directional change and future action is instantaneous and correct.

When designing the vertical alignment of a road, it must be ensured that, at the very least, minimum stopping sight distance requirements are met. Furthermore, on two- and three-lane roads, the design must provide adequate stretches of highway to enable faster vehicles to overtake comfortably and safely. In addition, long stretches of steep gradients must be avoided because of their detrimental effect upon speeds, capacity and safety.

Gradients

Long, steep uphill grades can have a considerable effect upon vehicle speeds[41]. This is of the utmost importance on highways where commercial vehicles form a significant portion of the traffic flow. Since restrictive sight distances are usually associated with steep uphill gradients, the speed of traffic is often controlled by the speeds of the slower commercial vehicles. As a result, not only are the operating costs of vehicles increased but, in addition, the capacity of the road will have to be reduced in order to maintain the required level of service.

Uphill gradients can be a cause of accidents between vehicles in opposing traffic streams, since the drivers of faster vehicles are tempted to overtake where normally they might not do so. Safety can also be a very important consideration on downhill gradients due to the possibility of relatively great increases in vehicle speeds. As a result, the ability of motorists to stop or take other emergency measures can be seriously curtailed during inclement weather conditions. In particular, if vehicles have to be braked in order to traverse a curve at the bottom of a gradient, and the road is icy or wet, serious accidents can be caused by vehicles skidding out of control.

Maximum grades

Grades of up to about 7 per cent have relatively little effect upon the speeds of passenger cars. However, the speeds of commercial vehicles are considerably reduced when long gradients, with grades in excess of about 2 per cent, are features of the highway profile. In certain instances, when the uphill sections are short, grades of 5 to 6 per cent have little detrimental effect; vehicles usually accelerate upon entering an uphill section and this extra momentum can be sufficient to overcome the effect of the extra increase in slope over these shorter distances.

The desirable maximum gradients currently used in British design practice[3] are 3, 4 and 6 per cent for motorways, dual carriageways and single carriageways, respectively.

Climbing lanes

The maximum grade is not in itself a complete design control. It is also necessary to consider the length of the gradient and its effect upon desirable vehicle operation, particularly on single carriageway roads. (For a detailed explanation of the economic implications inherent in using steep gradients, see reference 42.)

To ensure the safe and economic use of the highway network, and to sustain flow levels consistent with those recommended for design purposes, extra 'climbing' lanes therefore are normally provided on long uphill climbs. These allow the slower-moving vehicles to be removed from the main uphill traffic stream—so that the delays to the faster vehicles are reduced—and avoid the platooning which results from impedance by heavy vehicles, thereby also increasing safety.

It must be emphasized that the extra lane is not intended to convert a two-lane highway into a three-lane one. Rather it is a traffic management aid to vehicle movement, just as acceleration and deceleration lanes facilitate vehicle movement at intersections.

It is British practice[3] to provide an additional uphill climbing lane, where it can be economically or environmentally justified, on hills with gradients (= 100 height/length) greater than 2 per cent and longer than 500 m. An indication of the scale of the economic justification required can be gained from the data in Table 6.28 which show, for various hill heights risen, the design year traffic flows (with various heavy goods vehicle contents) which must normally be forecast for a two-lane road before an extra climbing lane is justified. Table 6.29 shows similar justification data for non-motorway dual carriageways. In either case, the data assume a standard cost provision for a climbing lane in relatively easy terrain. (Climbing lanes on motorways in Britain are extremely rare occurrences.)

Where a hill occurs which has varying grades, then consideration is given as to whether the extra lane is needed on all sections, e.g. if an easy grade follows one justifying a climbing lane, then it too may require an extra lane as the speeds of the heavy commercial vehicles will be very slow at the start of the second section.

In the case of a three-lane single carriageway requiring an extra climbing

Table 6.28 Traffic flow criteria for the provision of uphill climbing lanes on two-lane roads

| Height (m) | Design year traffic flows, two-way *AADT*, with HGV contents (%) of | | | |
	5	10	15	30
20	—	—	12 500	10 500
40	—	10 000	8 500	7 500
60	9 700	7 100	6 100	5 300
80	7 400	5 400	4 500	3 900
100	5 800	4 300	—	—
140	4 000	—	—	—

Table 6.29 Traffic flow criteria for the provision of uphill climbing lanes on all-purpose dual carriageways

	Road type			
	D2AP		D3AP	
	Design year traffic flows, two-way *AADT*, with HGV contents (%) of			
Grade (%)	30	10	30	10
2	30 000	35 200	46;900	52 200
4	27 500	32 000	44 500	50 500
6	23 200	25 000	42 700	47 500
8	19 000	19 000	40 500	44 200

lane, local dualling is preferred. In some instances, however, it may be desirable to convert a single three-lane road to priority uphill working at steep or prolonged gradients without widening; in such cases, the carriageway is marked in three lanes with the climbing and overtaking lanes given priority by means of offset double white lines.

Logically, the point at which a climbing lane should be initiated depends upon the speeds at which the commercial vehicles begin to ascend the hill. Where there are no restrictive sight distances or other features that might result in low vehicle approach speeds, British practice is for the full width of the extra lane to be provided at a point 100 m uphill from the 2 per cent instantaneous grade point of the sag curve, preceded by a transition taper of 1:30 to 1:40 for single carriageway roads and 1:45 for dual carriageway roads.

Where restrictions cause lower approach speeds, then the climbing lane should obviously be initiated closer to the bottom of the hill. The extra lane should end well beyond the crest of the hill so that the slower vehicles, especially the heavy commercials, can return to their normal lane with safety. In addition, it should not be ended abruptly but should be tapered so that the vehicles can make the transition efficiently as well as safely. British practice in this respect is to continue the full width of the climbing lane to a point 220 m beyond the 2 per cent instantaneous grade point of the crest curve, followed by a taper of 1:30 to 1:40 for single carriageway roads and 1:45 for dual carriageway roads.

Minimum grades
These are of interest only at locations where surface drainage is of particular importance. Normally, the camber of the road is sufficient to take care of the lateral carriageway surface drainage. In cut sections, however, it may be necessary to introduce a slight longitudinal gradient into the road surface in order to achieve longitudinal drainage in the side-ditches; a similar situation arises when the road is kerbed. A minimum grade of 0.5 per cent (1 in 200) is desirable in both of these instances. Satisfactory drainage has, however, been obtained with grades of the order of 0.3 per cent.

Vertical curves

Just as a circular curve is used to connect horizontal straight stretches of road, a parabolic curve is usually used to connect gradients in the profile alignment. These curves are convex when two grades meet at a 'summit' and concave when they meet at a 'sag'.

While other curve forms can be used with satisfactory results, the tendency has been to utilize the parabolic curve in profile alignment design. This is primarily because of the ease with which it can be laid out as well as enabling the comfortable transition from one grade to another. Normally, vertical curves of this type are not considered necessary when the total grade change from one tangent to the other does not exceed 0.5 per cent.

Parabolic curve properties

The form of parabolic curve most often used is the simple parabola, an

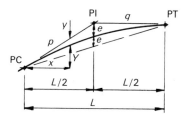

Fig. 6.18 A simple symmetrical parabolic curve

example of which is shown in Fig. 6.18. If Y in this figure is considered to be the elevation of the curve at a point x along the curve, then

$$\frac{\mathrm{d}^2 Y}{\mathrm{d}x^2} = k \quad \text{(a constant)}$$

since the rate of change of slope along a parabola is a constant. Integrating,

$$\frac{\mathrm{d}Y}{\mathrm{d}x} = kx + C$$

When $x = 0$,

$$\frac{\mathrm{d}Y}{\mathrm{d}x} = p \quad \text{(the slope of the first tangent)}$$

When $x = L$,

$$\frac{\mathrm{d}Y}{\mathrm{d}x} = q \quad \text{(the slope of the second tangent)}$$

Therefore

$$C = p \quad \text{and} \quad k = \frac{q-p}{L}$$

and $\quad \dfrac{dY}{dx} = \left(\dfrac{q-p}{L}\right)x + p$

Integrating,

$$Y = \left(\frac{q-p}{L}\right)\frac{x^2}{2} + px + C_1$$

When $x = 0$,

$$Y = 0 \quad \text{and} \quad C_1 = 0$$

Therefore

$$Y = \left(\frac{q-p}{L}\right)\frac{x^2}{2} + px$$

From geometry,

$$\frac{y+Y}{x} = \frac{p}{1}$$

Therefore

$$y = -\left(\frac{q-p}{2L}\right)x^2$$

y, which is measured downwards from the tangent, gives the vertical offset at any point along this curve. These vertical offsets from the tangents are used to lay out the curve.

It is often necessary to calculate the highest (or lowest) point on the curve to ensure that minimum sight distance (or drainage) requirements are met. The location of this point is given by

$$x = \frac{Lp}{p-q} \quad \text{and} \quad y = \frac{Lp^2}{2(p-q)}$$

As these equations illustrate, the high or low point of a symmetrical curve is not necessarily directly below or above the point of intersection of the tangents, but may in fact be located on either side of this point.

Figure 6.19 illustrates in exaggeration some varying forms of highway vertical curves. The algebraic signs given to the slopes are quite important; the plus sign is used for ascending slopes from the point of curvature, while the minus sign is used for descending ones. Proper use of these algebraic signs indicates automatically whether the elevations along the vertical curve are obtained by addition or subtraction of offsets from the tangent elevations. Negative answers are measured downwards from the tangents for all summit curves, while positive answers are measured upwards for sag curves.

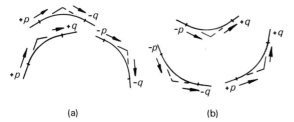

(a)   (b)

Fig. 6.19 Typical vertical curve forms: (a) summit curves and (b) sag curves

Example of vertical offset calculation

A -2 per cent grade is being joined to a -4 per cent grade by means of a parabolic curve of length 1000 m. Calculate the vertical offset at the point of intersection of the tangents.

Solution

Assume that the length of the curve is equal to its horizontal projection. The actual difference between the two can be neglected for all practical purposes. Thus $L = 1000$ m.

The vertical offset y at any point x along the curve is given by

$$y = \left(\frac{q-p}{2L} \right) x^2$$

When $x = L/2$, $y = e$, the vertical offset at the PI. Therefore

$$e = \left(\frac{q-p}{2L} \right) \frac{L^2}{4} = \frac{L}{8}(q-p)$$

$$= \frac{1000}{8}[(-4/100)-(-2/100)] = -2.5 \, \text{m}$$

Sight distance requirements

In determining the lengths of vertical curves, the controlling factors are the security and comfort of the motorists, and the appearance of the profile alignment. Of these, the sight distance requirements for safety are by far the most important on summit curves. With sag curves, safety is of less importance and more consideration can be given to the other factors.

Summit curves When deriving formulae for the *minimum* lengths of summit curves, there are two design conditions that have to be considered.

Figure 6.20(a) illustrates the first condition, where the required sight distance is contained entirely within the length of the vertical curve, i.e. where $S < L$. In this figure, h_1 and h_2 are the height of the driver's eye and the height of the dangerous object on the roadway, respectively. Since the curve is a parabola, the offsets from the line of sight are proportional to

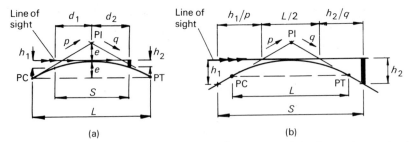

Fig. 6.20   Sight distances over summit curves: (a) required sight distance is contained entirely within the length of the vertical curve and (b) required sight distance is greater than the length of the vertical curve

the square of the distance from the point where the curve is tangential to the line of sight. Thus

$$h_1 = kd_1^2 \quad \text{and} \quad h_2 = kd_2^2$$

However, $e = k(L/2)^2$, therefore

$$\frac{h_1 + h_2}{e} = \frac{4d_1^2 + 4d_2^2}{L^2}$$

and $\quad d_1 + d_2 = \left(\dfrac{h_1 L^2}{4e}\right)^{1/2} + \left(\dfrac{h_2 L^2}{4e}\right)^{1/2}$

Also $e = LA/8$, where A = algebraic difference in slopes expressed in absolute values; it is always given in decimal form. Substituting,

$$d_1 = \left(\frac{2h_1 L}{A}\right)^{1/2} \quad \text{and} \quad d_2 = \left(\frac{2h_2 L}{A}\right)^{1/2}$$

Therefore

$$L = \frac{A(d_1 + d_2)^2}{[\sqrt{(2h_1)} + \sqrt{(2h_2)}]^2}$$

However, $d_1 + d_2 = S$, the sight distance, therefore

$$L = \frac{AS^2}{[\sqrt{(2h_1)} + \sqrt{(2h_2)}]^2}$$

If the dangerous object is assumed to be at carriageway level, then

$$L = AS^2/2h_1$$

If $h_1 = h_2$, then

$$L = AS^2/8h_1$$

Figure 6.20(b) illustrates the second condition, where the required sight distance overlaps onto the tangent sections on either side of the parabolic curve. In this figure, let g represent the difference between the slope of the sight line and the slope p of the rising gradient. Then $(A - g)$

is the difference between the slope of the sight line and the gradient with slope q. Therefore

$$S = \frac{L}{2} + \frac{h_1}{g} + \frac{h_2}{A-g}$$

For the sight distance S to be a minimum, $dS/dg = 0$. Therefore

$$\frac{dS}{dg} = -\frac{h_1}{g^2} + \frac{h_2}{(A-g)^2} = 0$$

Solving,

$$g = \frac{A\sqrt{(h_1 h_2)} - h_1 A}{h_2 - h_1}$$

Substituting,

$$S = \frac{L}{2} + h_1 \bigg/ \left(\frac{A\sqrt{(h_1 h_2)} - h_1 A}{h_2 - h_1} \right) + h_2 \bigg/ \left(A - \frac{A\sqrt{(h_1 h_2)} - h_1 A}{h_2 - h_1} \right)$$

Therefore

$$L = 2S - 2(\sqrt{h_1} + \sqrt{h_2})^2 / A$$

If $h_2 = 0$, then

$$L = 2S - 2h_1 / A$$

If $h_1 = h_2$, then

$$L = 2S - 8h_1 / A$$

The decision as to which condition should be used at a particular site can be made by solving either of the equations:

$$e = \frac{(q-p)L}{8} \quad \text{or} \quad e = \frac{(q-p)S}{8}$$

depending upon whether L or S is the known value. In either case, if e is found to be greater than h_1, then the equation for the first condition, i.e. when L is greater than S, should be used. If it is found that e is less than h_1, then the equation where L is less than S should be used.

Example of vertical curve length determination

A vertical curve is to be constructed between an ascending 3.5 per cent grade and a descending 4 per cent grade. The required safe stopping sight distance is 300 m, the dangerous object is 0.26 m above the carriageway, and the motorist's eye height is 1.05 m. Determine the minimum length of vertical curve that will satisfy this sight distance requirement.

Solution

$q = -4/100$, $p = +3.5/100$, and $S = 300$ m. Therefore

$$e = \frac{(q-p)S}{8} = \frac{[(-4/100) - (3.5/100)] \times 300}{8} = -2.82$$

However, h_1, the eye height, is equal to 1.05 m. Therefore the equation for $L > S$ will be used.

When $L > S$, the length of the vertical curve is given by

$$L = \frac{AS^2}{[\sqrt{(2h_1)} + \sqrt{(2h_2)}]^2}$$

However, $h_1 = 1.05$ m, $h_2 = 0.26$ m, $A = 7.5/100$, and $S = 300$ m. Therefore

$$L = \frac{(7.5/100) \times 300 \times 300}{[\sqrt{(2 \times 1.05)} + \sqrt{(2 \times 0.26)}]^2} = 1433 \text{ m}$$

Department of Transport recommendations   These are indirectly summarized in Table 6.30 with respect to vertical curve lengths. For both urban and rural situations, the curve length is determined from the formula $L = KA$ metres, where A is the algebraic difference in gradients (expressed as a percentage) and K has a value selected from the table for the design speed of the road. Except when joining two steep, long grades, the curvature length provided should be greater than the absolute minimum safe stopping sight distance; where economically feasible, desirable minimum lengths are provided on motorways.

Table 6.30 Recommended minimum K-values used in vertical curve length determinations[3]

Vertical curvature condition	K-values for design speeds (km/h) of					
	120	100	85	70	60	50
Crest						
passing	*	400	285	200	142	100
desirable minimum*	182	100	55	30	17	10
absolute minimum	100	55	30	17	10	6.5
Sag						
absolute minimum	37	26	20	20	13	9

*Not used with single carriageways

Sag curves

Whereas with summit curves the most important factor is the length of curve necessary for safety, there are at least four widely accepted criteria for determining the minimum lengths of sag curves. These are the vehicle headlight sight distance, minimum motorist comfort, drainage control, and general aesthetic considerations. In addition, since sag curves are often associated with highway underpassing structures such as bridges, in certain instances the curve length may be chosen to ensure the necessary vertical clearance and to maintain a safe sight distance.

There is still a considerable difference of opinion as to what value of radial acceleration should be used on vertical curves for *comfort design* purposes. The most commonly quoted values are between 0.30 and

0.46 m/s²; British practice is to use 0.30 m/s² for design speeds above 70 km/h.

If the vertical radial acceleration is assumed to be equal to $a\,\text{m/s}^2$, then

$$a = v^2/R = 3.6^2 V^2/R$$

and $R = V^2/13a$

where R = radius of the circle equivalent to the parabolic curve, v = vehicle speed (m/s), and V = vehicle speed (km/h).

Since the central angle $\varDelta$ of the equivalent circular curve is very small, and the circle practically coincides with the parabola,

$$L = RA = V^2 \varDelta/13a$$

Since $\varDelta = A$,

$$L = V^2 A/13a$$

where L = length of the sag curve (m) and A = algebraic difference in slopes, expressed as a decimal.

When a highway passes underneath a structure such as a bridge, the motorist's line of sight may be obstructed by the edge of the bridge. In such cases, and on unlit urban roads with design speeds $\leqslant 70$ km/h, the controlling factor should be the absolute minimum sight distance *requirement for safety* (British practice). Again, when calculating the required length of curve, two considerations have to be taken into account, as follows.

Firstly, when the required sight distance is less than the length of the sag curve, then

$$L = \frac{S^2 A}{8[C - (h_1 + h_2)/2]}$$

where L = length of the sag curve (m), S = sight distance (m), A = algebraic difference in tangent slopes, expressed in decimal form, C = vertical clearance to the critical edge of the structure (m), h_1 = vertical height of eye (m), and h_2 = vertical height of the hazardous object on the carriageway (m).

Secondly, when the required sight distance is greater than the length of the sag curve, then

$$L = 2S - \frac{8[C - (h_1 + h_2)/2]}{A}$$

In both of these equations, the critical edge of the structure is assumed to be directly over the point of intersection of the tangents. In practice, both equations can be considered valid provided that the critical edge is not more than about 60 m from the point of intersection.

Example 1 of sag curve length determination: comfort
Determine the minimum length of curve required to connect a descending

4 per cent grade to an ascending 3 per cent grade. The design speed of the road is 100 km/h and the acceptable radial acceleration is 0.3 m/s².

Solution
$a = 0.3$ m/s², $V = 100$ km/h, and $A = 7/100$. Therefore

$$L = V^2 A/13a = (100)^2 (0.07)/13(0.3) = 180 \text{ m}$$

Example 2 of sag curve length determination: clearance and safety
Determine the minimum length of valley curve required to connect a descending 4 per cent grade to an ascending 3 per cent grade. The vertical clearance is to be 5.1 m (British practice) and the required sight distance is 300 m. The height of eye for a commercial vehicle is 2 m and the hazardous object has a vertical height of 0.26 m.

Solution
Assuming that the sight distance is greater than the required length of curve, then

$$L = 2S - \frac{8[C - (h_1 + h_2)/2]}{A} = 2 \times 300 - \frac{8[5.1 - (2 + 0.26)/2]}{7/100} = 146 \text{ m}$$

General considerations

Proper design of the vertical alignment requires that considerations other than safety and comfort should also be taken into account. For instance, a smooth grade line with gradual changes should always be used in preference to one with numerous breaks and short lengths of grade. Roller-coaster types of profile should be avoided as they are dangerous as well as aesthetically unpleasing. Broken-back grade lines, i.e. a section composed of two vertical curves in the same direction separated by a short tangent length, should also be avoided; such a profile is particularly noticeable in valley topography where the full view of both vertical curves is not at all pleasing.

Where single level intersections occur on highway sections with moderate to steep gradients, the slope at the intersection itself should always be reduced. This will considerably help vehicles performing turning movements and may well serve to reduce potential accidents.

On long gradients, it may be desirable in certain instances to have a steeper slope near the bottom of the hill and lighten the slope near the top, instead of using a uniform sustained grade that may be only just below the maximum allowable. This procedure is particularly applicable to gradients on low-speed roads, where the approaching vehicles can accelerate into the rising sections, and on high-speed roads when a rising gradient occurs just after a falling one.

Vertical curvature superimposed on horizontal curvature, or vice versa, generally has, aesthetically, a very pleasing result. It should, however, always be carefully analysed for possible detrimental effects upon traffic. In particular, sharp horizontal curvature should never be introduced at or near the top of a pronounced vertical curve. At night-time, this can be especially dangerous since drivers may not notice the horizontal change in direction and severe accidents may occur. For similar reasons, sharp horizontal curves should never be introduced near the bottom of steep gradients because of the high vehicle speeds that can be expected. Not only is there the danger of vehicles overshooting such curves at night-time, but accidents involving skidding can also be expected during inclement weather as drivers attempt to negotiate the sharp bends.

When traffic conditions justify the provision of a dual carriageway, consideration should always be given to the feasibility of varying the width of the central reservation and possibly using two completely separate horizontal and vertical alignments. In this way, a superior design making the maximum possible use of the one-way feature of dual carriageways may be obtained at little additional cost.

Cross-section elements

For discussion purposes, the cross-section elements will be divided into two groups, viz. basic elements and noise barriers.

Basic elements

By basic cross-section elements are meant those features of the highway which form its effective width and which affect vehicle movement. The constituent parts of primary interest are the number and width of traffic lanes, the central reservation, shoulders, laybys, camber of the carriageway and, where necessary, the side-slopes of cuttings or embankments.

Figure 6.21 shows the various elements of some simplified highway cross-sections in urban and rural areas.

Traffic lanes

The number of traffic lanes to be used in specific situations is dependent upon the volume and type of traffic to be handled. Normally, however, the minimum number is two, new single lane roads being rarely constructed in developed countries today. Even though traffic volumes may be light, safety considerations and ease of traffic operation require two lanes.

While two-lane roads constitute the predominant part of the British highway system, there is also a substantial number of three-lane roads. Often these were constructed when the design volume exceeded the capacity of a two-lane road, but was not sufficient to justify a four-lane facility. Current British practice, however, in respect of the construction of new three-lane single carriageway rural highways, is to use them only where the provision of sufficient space to accommodate a dual carriageway is too costly or difficult environmentally, or where the best short-term solution is to widen an existing two-lane highway. Safety

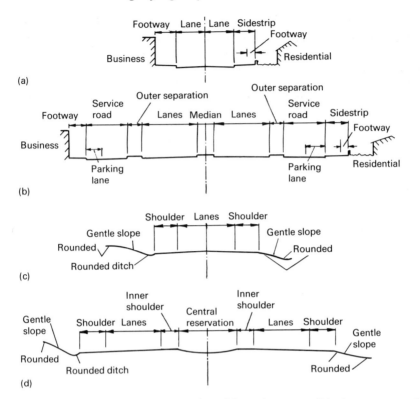

Fig. 6.21 Typical highway cross-sections: (a) two-lane street, (b) urban motorway, (c) two- or three-lane rural highway, and (d) rural motorway

considerations dictate that three-lane roads should only be constructed in rural areas where it is practical to provide nearly continuous overtaking sight distances. In urban or suburban areas, uncontrolled, three-lane, two-way facilities should never be constructed because of their high accident potential in heavy traffic conditions.

If it is expected that long-term traffic needs will eventually justify a four-lane rural road, but it is not required at this time, an economical procedure is to construct a three-lane road composed of two outer lanes of high-quality construction and a middle lane of lower quality and cost. Additional lanes of high-quality construction can then be added on the outside at a later time and the middle lane converted into a central reservation.

Four or more lanes are needed to enable vehicles to overtake on lanes not used by opposing traffic. Normally, these highways have dual carriageways separated by a central reservation. (At least one urban motorway in the USA, the Southside Expressway in Chicago, has a total of fourteen traffic lanes on four carriageways. These are composed of four through lanes on one carriageway plus three service lanes on another carriageway in each direction of travel.)

British governmental practice is not to assign a certain width to a lane, but rather to specify the carriageway width in relation to the traffic needs at a given location and then to assign a given number of lanes to that carriageway (see Table 6.31).

Table 6.31 General practice with regard to carriageway widths

Road type	Description of carriageway(s)	Carriageway width (m)
Urban		
Primary	Dual 4-lane	14.60
distributor	Overall width for 4-lane divided carriageway with central refuges	14.60
	Single 4-lane, no refuges	13.50
	Dual 3-lane	11.00
	Single 3-lane (for tidal flows only)	9.00
	Dual 2-lane (normal)	7.30
District	Single 2-lane (normal)	7.30
distributor	Dual 2-lane (normal)	7.30
	Dual 2-lane (if the proportion of heavy commercial traffic is fairly low)	6.75
Local	Single 2-lane	
distributor	in industrial districts	7.30
	in principal business districts	6.75
	in residential districts used by heavy vehicles (minimum)	6.00
Access	Single 2-lane, in residential districts (minimum)	5.50
roads*	2-lane back or service roads, used occasionally by heavy vehicles	5.00
	2-lane back roads in residential districts, used only by cars (minimum)	4.00
Rural	Single lane (Scotland and Wales mostly)	3.50
	Minimum width in rural junctions	4.50
	Single 2-lane (minimum)	5.50
	Motorway slip road	6.00
	Single or dual 2-lane (normal)	7.30
	Single 3-lane	10.00
	Dual 3-lane	11.00
	Dual 4-lane	14.60

*In industrial and principal business districts, use the values given for local distributor roads

High-quality roads are normally designed for standard carriageways with lanes typically 3.65 m wide, exclusive of space required for hardstrip, traffic islands, or central reservation. Particular circumstances may, however, give rise to an unavoidable need for a non-standard carriageway. For example, at a given location in an urban area, it may only be possible to obtain space for a carriageway width of about 14.6 m, in which case consideration of specific needs will enter the determination as to whether provision should be made for a 14.6 m undivided carriageway or a substandard dual carriageway of, say, 6.70 m width with a substandard 1.2 m wide central reservation.

Central reservations

Most dual carriageways have at least two traffic lanes in each direction divided by a central reservation at least 1.75 m wide, even in urban areas where space is restricted.

There is considerable difference of opinion between major highway authorities as to the *normal* width of the central reservation. For instance, in Britain, 4.5 m is the normal minimum width of central reservations in rural areas; in the USA, the equivalent figure is 18.3 m.

While it has been proved that the wider the central reservation the greater is the reduction in head-on collisions, it has not been possible to establish any overall relationship between the total number of accidents and the reservation width. It has been found, however, that an almost straight-line relationship exists between the median width and the percentage of vehicles involved in median accidents which actually crossed the central space[43]. It would appear that a central reservation width of 12–15 m is needed to bring an encroaching high-speed vehicle under control to avoid the possibility of it becoming involved in a head-on collision on the other carriageway.

Central reservations need not be of constant width. In fact, in rural areas, consideration should always be given to varying the width in order to obtain a safe, pleasing and economical design that fits the topography. Where possible, shrubs (but not trees) should be grown within the reservation area—not only are these aesthetically pleasing, but they also reduce headlight glare and act as crash barriers as they help to dissipate the energy of out-of-control vehicles. Care must always be taken, however, to ensure that the shrubbery does not reduce the necessary sight distance at intersections.

The reservation surfacing should be in direct contrast to the carriageways being separated, and be distinctly visible during day and night, in both wet and dry weather. For widths greater than about 1.8 m, grass is usually the most pleasant and suitable material. Below 1.8 m, however, grass is usually difficult as well as dangerous to maintain, and so use is often made of raised medians with contrasting bituminous or concrete surfacing; the median kerb should also be studded with reflector buttons or painted with reflectorized white paint to emphasize the contrast.

Shoulders

A shoulder is that portion of a roadway adjacent to the travelled way that is primarily used as a refuge area by parked vehicles. The need for such refuges is indicated by data from a study at the Mersey Tunnel[44] which showed that cars and lorries had *emergency* stops every 28 000 and 15 000 km, respectively. The provision of shoulders also gives a sense of openness which helps considerably towards maintaining driving ease, particularly in conditions of high traffic flow, and freedom from concern re lateral clearances (especially important for drivers of heavy commercial vehicles). Without shoulders, obstructions adjacent to the edge of the carriageway and/or vehicles parked on the carriageway cause a reduction in the effective width available for use by moving vehicles, thereby reducing the highway's capacity.

Well-designed and properly maintained full shoulders are a necessity on all rural and, where possible, suburban arterial roads carrying appreciable amounts of high-speed traffic. However, shoulder widths do vary in practice. About 3.35 m is needed by a lorry to enable a tyre to be changed without danger to the operator. Passenger cars require less space, and since these form the greater part of the total number of stoppages, shoulder widths of 3 m are usually recommended for major highway design purposes. This normally allows a 1 m gap between the parked car and the edge of the carriageway, which is ample from the point of view of safety. For economic reasons, however, hardstrip (verge) widths as low as 0.6 m are used on less important roads instead of full shoulders.

A shoulder should be capable of supporting vehicles under all conditions of weather, without rutting or shoving of the surface. If a motorist becomes bogged-down while parked, it is doubtful whether he/she will make full use of these facilities again. Also, skidding and overturning may occur if a vehicle is driven onto a soft shoulder at high speed and the driver then attempts to decelerate.

The best but unfortunately also the most expensive way of providing a stable shoulder is to extend the roadbase beyond the edge of the carriageway—not only is the shoulder made stable in this manner, but also added structural strength is given to the carriageway pavement. British practice[45] is normally to construct hardstrips to the same depth and of the same materials as the running lanes of the carriageway. In the case of hardshoulders, the same type of construction (flexible or rigid) is used; however, where the hardshoulder is designed to carry the minimum loading of 1m standard axles, the flexible roadbase or concrete pavement is made thinner than the adjacent running lanes, with a consequential step in subbase thickness. Only if the local subbase materials are likely to impede drainage at the subbase–roadbase interface (with consequential deterioration of the pavement and subgrade) is it usual practice for the hardshoulder to be constructed to the same thickness as the running lanes.

The appearance of the shoulder surfacing should be distinctly different from that of the carriageway, otherwise motorists will regularly use it as a traffic lane. Surfacings of grass are clearly delineated and aesthetically pleasing, but drivers are often afraid to use these shoulders for fear of being bogged-down. Probably the most effective type is a bituminous surfacing with either an added pigment material or different coloured stone chippings. Normally, the bituminous surfacing of a flexible hardshoulder is composed of a different thickness and specification from the running lanes, to suit the different traffic use. However, as it will be used as a slow traffic running lane when carriageway maintenance makes it necessary, the bituminous surfacing should have a texture and skid-resistance appropriate to that use.

The above discussion relates only to outside shoulders on divided highways. In some countries, however, it is not uncommon to construct full inner shoulders, adjacent to the central reservation. This is an undesirable practice as it encourages stopping motorists to decelerate (and accelerate) on the faster inner lanes, as well as being of course a significant extra construction cost. British practice in this respect is to provide only 1 m hardstrips in lieu of full inner shoulders on all dual carriageways, except for dual three-lane motorways where no hardstanding extra width at all is provided next to the central reservation.

Laybys, bus bays and bus stops

When economic considerations do not allow the use of shoulders, then laybys should be built at favourable locations along the highway. Great care must be taken, however, *not* to construct laybys where the sight distances from their exit and entry points are inadequate for safety.

British laybys are normally not less than 30 m long and 2.5 m wide. Obviously, however, they should be longer and wider where possible. If located on high-speed roads, they should be provided with adequate acceleration and deceleration lengths on either end. Their spacing along the highway should be related to the volume of traffic. Thus, for example, it is recommended that a 3 m wide by 100 m long layby be provided at about 1 km intervals on each side of three-lane and dual carriageway rural highways; between the laybys, a hardstrip 1 m wide is provided adjacent to the running carriageway, with 5 per cent tapers between each layby and the hardstrip. In the case of well-travelled and lightly-travelled single carriageway roads, 2.5 and 3.0 m wide laybys are normally provided at 1.5 and 5.8 km intervals, respectively, on either side of the carriageway; these laybys are provided with acceleration and deceleration tapers of lesser length than those on three-lane or dual carriageway roads.

Laybys are rarely provided on district and local distributor roads in urban areas: adequate space is usually not available. It is recommended, however, that laybys 3 m wide by 30 m long (excluding 16 m end tapers) be provided at intervals of not more than 1.5 km on each side of all-purpose primary distributors without hardshoulders. Laybys and bus bays are combined where possible; any such combined arrangement will normally need to be at least 45 m long (excluding 19.5 m end tapers) and between 2.75 and 3.25 m wide.

Where space permits, bus bays (3.25 m wide by at least 11 m long, with end tapers of 19.5 m) normally should be provided at bus stops; the final bus bay length used will depend upon the number of buses to be accommodated at any given instance. To lessen the risk of passenger queues being splashed in wet weather, the crossfall of the bus bay should be away from the kerb and towards special drainage facilities laid across the mouth of the bay.

Where the width available is inadequate for a bus bay, so that buses have to stop on the carriageway adjacent to the kerb, passing space for at least one line of traffic in the same direction of travel should always be available in addition to the effective space occupied by the bus.

Bus stops/bays on opposite sides of two-way single carriageway roads should be staggered by about 45 m, preferably so that buses can stop tail-to-tail and move off away from each other. In addition, they should be located midway between junctions (as this minimizes interference with turning traffic) and/or adjacent to pedestrian subways and bridges. Midblock sites are, however, often inconvenient for pedestrians, who normally prefer locations adjacent to intersections. Any bus stop at an intersection should preferably be sited on the exit side; if sited on the approach side, it should be far enough back from the stop line to ensure that: (a) a waiting bus does not obstruct visibility from the main road to the side road on the left, or vice versa, (b) vehicles waiting to turn left are not obstructed by the bus, (c) a bus that is required to turn right after moving-off has ample space to carry out the necessary weaving manoeuvre, and (d) waiting buses do not interfere with the efficient operation of traffic signals or the movement of traffic at a roundabout.

Bus stops/bays are normally spaced at intervals of not less than 2–3 per kilometre. In busy central areas, of course, they may be spaced more closely.

Camber

The term camber is used in highway engineering to describe the convexity of the carriageway cross-section. The main object of cambering is to drain water and avoid ponding on the road surface.

Early road-builders used much greater cambers than are used today, e.g. Telford used a camber with a side-slope of 1 in 30. These early roads had rough open surfaces and therefore severe cross-slopes were needed in order to remove the water quickly, before it could seep in the road pavement and foundation. Nowadays, however, these types of surfacing have been replaced by relatively impermeable ones, so that it has been possible to reduce the amount of camber very considerably and thereby increase the ease of driving. Today, both single and dual carriageway roads in Britain have average cross-slopes of 2.5 per cent[3]. Improved construction methods could mean that future years would see slopes as low as 1 in 60, e.g. one study[46] showed that the major benefit resulting from changing the cross-slope from 1 in 60 to 1 in 30 was a reduction in the amount of water which ponded in surface deformations of the pavement (the depth of water flowing across the road was little affected).

Modern two-lane roads desirably have either parabolic or circular cross-sections. These cross-sections have the advantage that the swaying of commercial vehicles is kept to a minimum as they cross and recross the crown of the road during an overtaking manoeuvre. On carriageways with three or more lanes, greater care has to be taken in deciding the manner in which the camber should be applied. If, for instance, the application of the camber takes the shape of a parabola, then the outer lanes may have an undesirably steep cross-slope which could seriously interfere with safe traffic operation. The equation of the parabola is such that the centre of the road will be very flat, while the desired cross-slope will be exceeded towards the outside. Common practice in this situation is to use a curved crown section for the central lane or lanes and to have a tangent plane section on each of the outer lanes. The cross-slope on the tangent planes is made the same or slightly steeper than that at the end of the curved section so that the accumulated water is more easily removed.

On dual carriageways, it is desirable for each carriageway to be cambered—not only does this minimize the sheeting of water during rainstorms, but also the difference between the low and high points in the carriageway cross-section is kept to a minimum. This latter advantage is a result of the smaller width which is sloped in a given direction and the avoidance of the higher rate of cross-slope which is necessary to get rid of the accumulated water when sloping a wider carriageway in one direction only. Changes from normal to superelevated sections are also easily made. The disadvantage of using cambered sections on both carriageways is that more inlets and underground drainage lines are required, with pick-up facilities near both edges of each carriageway.

Where carriageways are sloped in one direction to drain from the median space to the outside (i.e. British practice), savings are effected in drainage structures, and the treatment of intersecting roadways is easier. However, in areas subject to heavy rains or where the central reserve is used to store cleared snow, this procedure is most undesirable.

Another possible arrangement often suggested for dual carriageways is that each carriageway should have a one-way cross-slope draining towards the central reservation. This has the advantage that the outer lanes, which are most used by commercial traffic, are more free of surface water. In addition, there is the economical advantage that all of the surface water can usually be collected in a single drainage conduit within the reservation. However, a very serious objection to this procedure is that all the drainage must pass over the inner high-speed lanes. This can result in annoying and dangerous splashing on the windscreens of vehicles. Hence cross-sections with drainage concentrated in the central reservation are not generally used except on long bridge structures.

Side-slopes

Soil mechanics analyses make it possible to determine accurately the maximum slopes at which earth embankments or cuts can safely stand. In practice, however, these maximum values are not always used, flatter sections being preferred for reasons of safety and ease of maintenance.

Although the tendency is also to construct side-slopes that are as steep as possible for reasons of economy, this can be a short-sighted policy if future maintenance problems are not taken into account. Modern highway grading equipment can perform well on most slopes when cutting or filling. However, when the slope surfaces are being stabilized by spreading topsoil, and when they are being generally finished and rounded, it is virtually impossible to operate the necessary equipment efficiently on slopes greater than about 2 horizontal to 1 vertical. The flatter the side-slopes the easier it is to grow grass on them and the less chance there is of erosion. If the slopes are to be properly maintained and kept in pleasing appearance, the grading equipment necessary to do this can work most efficiently on slopes of 3 to 1 or flatter.

From a traffic safety aspect, the horizontal sight distance problem at curves in cuttings is considerably eased by the use of flatter side-slopes. In the case of embankments also, the flatter the side-slope the better; in this case, however, the three regions of the roadside shown in Fig. 6.22 are of considerable interest, viz.: (a) the top of the slope, i.e. the hinge point, (b) the front slope, and (c) the toe of the slope, i.e. the intersection of the front slope with level ground or a ditch.

The *hinge point* contributes to a loss of steering control since a vehicle leaving the carriageway tends to become airborne when crossing this point, particularly if the encroachment conditions (angle and speed) and embankment drop-off are severe. If the wheels are turned while airborne, and dig into the side-slope upon landing—this has the same effect as a sudden increase in friction coefficient—the potential for vehicle rollover is increased.

The *front slope* region is important in the design of long slopes where a driver can attempt a recovery manoeuvre or to reduce speed before impacting the ditch area. An erring driver's natural instinct is to attempt to return to the carriageway, and there is a front slope steepness at which a vehicle will roll during this recovery manoeuvre.

In many instances, the *toe of the slope* is close enough to the carriageway that the probability of a vehicle reaching a ditch area is high. In such situations, safe transition regions between the front and back slopes should be provided.

Research has shown[47] that vehicle rollover need not normally be a problem at the hinge-point region of an embankment with a slope of 2 horizontal to 1 vertical (or flatter), at encroachment angles up to 25 degrees and speeds up to 130 km/h. Rounding of the hinge points reduces the tendency of a vehicle to become airborne, thereby providing its driver with a greater opportunity to maintain control, and lessening the chance of a rollover.

If sufficient recovery distance is available, return manoeuvres can be accomplished on embankments of 3 to 1 or flatter at 130 km/h and an encroachment angle of 15 degrees without vehicle rollover. For this recovery to occur, however, the side-slope surface must be relatively uniform and have a high coefficient of friction (typically 0.6); if composed of soft material, as is most likely, vehicle rollovers can be expected for return manoeuvres attempted above 97 km/h. From a design viewpoint,

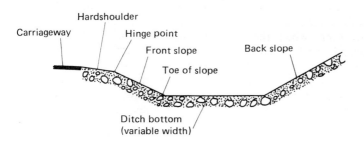

Fig. 6.22 The roadside side-slope regions

therefore, the provision of adequate surface drainage is necessary to alleviate sudden friction changes on the slope, whilst proper maintenance is required to eliminate any severe rutting that might be expected to occur.

Generally, the trapezoidal ditch configuration represents the safest cross-section for a vehicle to intrude upon at high speed, particularly if the ditch width is greater than about 2.4 m. With this type of cross-section, it is also safer to have the front slope as flat as possible, but no greater than 3 to 1. Less severe vehicle damage will be caused if the bottom corners are rounded to allow for a more gradual transition to and from the flat bottom.

Noise barriers

A stream of vehicles on a highway can be regarded acoustically as a distributed line source of noise. The noise level at any measurement point beside the stream is then dependent upon the strength of the source, the path length from the source to the measurement point, and any excess attenuation during propagation. The variables likely to affect source strength and propagation are listed below and discussed in Chapter 3.

Source strength variables are as follows:

(1)   number of vehicles on the highway,
(2)   acoustic emission-strengths of individual vehicles, e.g. vehicle speed, type and condition,
(3)   highway factors, e.g. highway grade and surfacing, whether surfacing is wet or dry,
(4)   driving conditions, e.g. free-flowing, acceleration.

Propagation variables are as follows:

(1)   distance to measurement point,
(2)   ground cover,
(3)   height of noise emission from vehicles,
(4)   height of propagation above ground surface,
(5)   screenings,
(6)   reflections,
(7)   meteorological conditions.

Predicting traffic noise

The *Noise Insulation Regulations 1975*[48] require a local authority to provide for noise insulation works to be carried out if the construction of a new or improved highway results in the following with respect to any adjacent dwelling or residential building.

(1)   The total expected maximum traffic noise level is not less than 68 dB(A) L_{10} (18-hour).
(2)   There is an increase of not less than 1.0 dB(A) in the total traffic noise from highways in the vicinity compared with the total traffic noise before the new/improved highway works were begun.

(3) In combining the maximum overall level of traffic noise from the new/improved highway and other highways in the vicinity, the new/altered highway effectively contributes not less than 1.0 dB(A) to the increase in traffic noise.

There are many procedures available to predict the traffic noise at locations adjacent to roadways (see, for example, reference 49). That which is used in Britain[50] assumes typical traffic and noise conditions during specified periods, whilst the source of the noise (the source line) is taken to be a line 0.5 m above the carriageway surfacing of the nearside traffic lane and 3.5 m in from the carriageway edge; the reception point at which noise is assessed at a building is taken as 1 m from the most exposed window or door in its facade.

The British prediction method is divided into two parts, as follows:

(1) the *prediction*, for an arbitrarily selected point (10 m from the edge of the carriageway) and within a given time period (06.00–24.00 h), of a basic L_{10} noise level for traffic on a normal working day on the length of road in question,
(2) the *assessment* of the resulting noise field, taking into account the principal factors affecting noise propagation at a distance from the highway up to 300 m.

The steps in the procedure are outlined in Fig. 6.23.

The *basic noise level* is obtained from data regarding the traffic flow rate, speed of traffic, composition of traffic, gradient of the highway and, where appropriate, carriageway surface.

The unadjusted basic noise level for both prevailing and future (i.e. the highest within fifteen years) traffic flows at a mean speed of 75 km/h, where the highway is level and the proportion of heavy vehicles is zero, can be determined from the formula:

$$L_{10} = 28.1 + 10 \log_{10} Q$$

where L_{10} = estimated basic noise level exceeded for just 10 per cent of the 06.00–24.00-hour day (dB(A)) and Q = traffic flow in both directions for normal roads on a normal working day (vehicles per 18-hour day).

In the case of 'abnormal' roads, i.e. where two carriageways are separated by more than 5 m or where the heights of the outer edges of the two carriageways differ by more than 1 m, the noise level produced by each of the two carriageways must be evaluated separately and then combined to obtain the unadjusted L_{10} for the roadway, using the formula:

$$L_{10} = \underline{L_{10}} + 10 \log_{10}(1 + 10^{-\Delta/10})$$

where $\underline{L_{10}}$ = the higher of the two unadjusted L_{10} values estimated for the two carriageways and Δ = difference between the two unadjusted noise levels, both of these being measured in dB(A).

In the case of the farside carriageway, the source line is assumed to be 3.5 m in from the farside kerb, and the effective distance from the kerb to

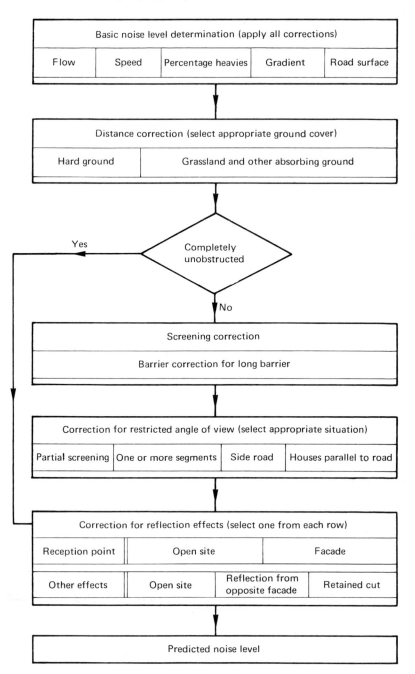

Fig. 6.23   Flowchart for predicting noise levels under straightforward single road situations[50]

be used in the distance correction is 3.5 m closer than this, i.e. 7 m in from the far edge of the farside carriageway.

The unadjusted basic L_{10} is next corrected for the traffic speed and the heavy vehicle content using the formula:

$$\text{correction (dB(A))} = 33 \log_{10}(V + 40 + 500/V) + 10 \log_{10}(1 + 5p/V) - 68.8$$

where V = mean 18-hour traffic speed (km/h) and p = percentage of heavy vehicles (> 1.52 t) over the 18-hour period.

When calculating the prevailing noise level (the 'before' level), the highway authority's actual measurement of speed at the site is normally used. The 'future' traffic speed, however, is normally the appropriate value taken from Table 6.32.

Table 6.32 'Future' traffic speeds, on given types of road, used in noise level calculations

Road type	Speed (km/h)
Roads not subject to a speed limit of < 96.5 km/h	
special roads (rural)*	108
special roads (urban)*	97
all-purpose dual carriageways*	97
single carriageways > 9 m wide	88
single carriageways ⩽ 9 m wide	81
Roads subject to a speed limit of 80 km/h	
dual carriageways	80
single carriageways	70
Roads subject to a speed limit of < 80 km/h but > 48 km/h	
dual carriageways	60
single carriageways	50
Roads subject to a speed limit of ⩽ 48 km/h	
all carriageways	50

*Excluding slip roads, which are estimated individually

The correction for the extra noise from the 'before' traffic on a gradient is derived from the following equation:

$$\text{correction (dB(A))} = 0.3G$$

where G = percentage gradient.

The 'after' correction for gradient, used when employing the highest mean speed within fifteen years, is given by

$$\text{correction (dB(A))} = 0.2G$$

In the case of one-way traffic or carriageways separated by more than 5 m, the correction applies only for the uphill flow.

The texture of the carriageway surface is normally not considered significant when predicting traffic noise using the British method, except in the case of deep random grooving (⩾ 5 mm) where the tyre-road interaction noise produces an increase which depends upon traffic composition. In such cases, the correction is determined from the following formula:

$$\text{correction (dB(A))} = 4 - 0.03p$$

where $p =$ percentage of heavy vehicles.

The final adjusted noise level determined after the application of the above corrections is then taken as the basic noise level from traffic on the highway—either that prevailing or as it will be within fifteen years, as appropriate.

The next step in the process involves determining a *distance correction* appropriate to the prevailing ground cover.

When there is unobstructed propagation over predominantly level ground, the surface of which is mainly (> 50 per cent) non-absorbent (e.g. paved, concrete or asphalt surfaces, or water), the distance correction is determined from the formula:

$$\text{correction (dB(A))} = -10\log_{10}(d'/13.5)$$

where $d' =$ minimum slant distance (see Fig. 6.24) from the effective source position to the reception point $= [(d + 3.5)^2 + (h - 0.5)^2]^{1/2}$, in which $d =$ horizontal distance from the edge of the carriageway to the reception point and $h =$ height of the reception point above the carriageway surface. d', d and h are measured in metres.

When predicting for reception points $\geqslant 4$ m above ground, the presence of low garden walls and fences, etc., can be ignored. When the reception point is below 4 m, hedges, short broken sections of low garden wall, paling fences, etc., can be ignored, but reasonably continuous walls and other permanent features must be taken into account in the calculations as short barriers which produce a partial screening effect.

If the surface between the noise source line and the reception point is mainly of an absorbent nature, e.g. grass, cultivated or planted, then extra attenuation (in addition to the distance attenuation) should be taken into account. The ground cover correction is progressive with distance, and particularly affects reception points close to the ground, as follows:

$$\text{correction (dB(A))} = -10\log_{10}(d'/13.5) + 5.2\log_{10}[3h/(d+3.5)]$$

$$\text{for } 1 \leqslant h \leqslant (d+3.5)/3$$

$$= -10\log_{10}(d'/13.5) \qquad \text{for } h > (d+3.5)/3$$

where d', d and h are as defined previously.

The next step in the process involves determining the *screening effect of intervening obstructions, barriers, limited angle of view of the road, noise reflection from nearby surfaces, etc.*

Where a purpose-built long barrier parallel to the highway, or an obstruction resulting from the site configuration, is interposed between the noise source and the reception point, an additional correction must be calculated and applied to the basic noise level corrected for distance

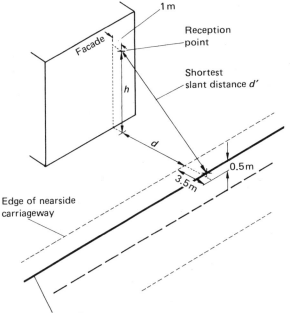

Fig. 6.24 The shortest slant distance from the source line to the reception point, as used in noise level determinations

according to the hard-ground correction described above. (If more than one single barrier is interposed, the barrier correction for each in turn is calculated and the lowest resultant noise level is used.) Any additional attenuation due to ground absorption is normally ignored in this instance since the near-ground noise rays are obstructed. The barrier correction, which is always negative, is as follows:

$$\text{correction (dB(A))} = A_0 + A_1 x + A_2 x^2 + \ldots + A_n x^n$$

where $x = \log_{10} \delta$, in which $\delta = (a + b - c)$, the path difference between the direct refracted rays in metres (see Fig. 3.30), and $A_0, A_1, A_2, \ldots, A_n =$ coefficients as given in Table 6.33.

The ranges of validity of this correction are as follows: for the shadow zone

$$-3.0 \leqslant x \leqslant +1.2$$

and for the illuminated zone

$$-4.0 \leqslant x \leqslant 0$$

Outside these ranges, the potential barrier correction is defined as follows: for the shadow zone

(1)   when $x < -3.0$, the correction is -5.0,
(2)   when $x > +1.2$, the correction is not defined,

Table 6.33 Coefficient values used in the correction for barriers, in noise level calculations

Coefficient	Shadow zone	Illuminated zone
A_0	−15.4	0
A_1	−8.26	+0.109
A_2	−2.787	−0.815
A_3	−0.831	+0.479
A_4	−0.198	+0.3284
A_5	+0.1539	+0.04385
A_6	+0.12248	
A_7	+0.02175	

and for the illuminated zone

(1)   when $x < -4.0$, the correction is -5.0,
(2)   when $x > 0$, the correction is zero.

Where only part of the road is shielded, the following procedure is adopted. Let θ_H be the total angle of view of the unscreened/unobstructed portion of the highway for which the ground between the road and the reception point is hard, let θ_S be the total angle of view of the unscreened part of the highway for which the intervening part of the ground is soft, and let θ_B be the total angle obstructed by barriers. Thus, for straight roads, $\theta_H + \theta_S + \theta_B = 180$ degrees ($= \pi$ radians).

(1)   The noise contribution at the reception point due to the part(s) of the highway covered by the angle θ_H is obtained by estimating the unobstructed noise level for hard-ground propagation using the shortest slant distance d' from the source line of the road and correcting this for the angle of view using the equation:

$$\text{correction (dB(A))} = 10 \log_{10}(\theta_H/\pi)$$

(2)   The noise contribution at the reception point due to the part(s) of the highway covered by θ_S is obtained by calculating the unobstructed noise level for soft-ground propagation and correcting this for the angle of view using the equation:

$$\text{correction (dB(A))} = 10 \log_{10}(\theta_S/\pi)$$

(3)   The combined contributions from θ_H and θ_S are then obtained using the equation:

$$L_U = L + 10 \log_{10}(1 + 10^{-\Delta/10})$$

where L_U = total L_{10} contribution from the unobstructed portions of the highway, L = the higher of the adjusted L_{10} values due to θ_H and θ_S, and Δ = difference between the adjusted L_{10} values due to θ_H and θ_S, all three of these being measured in dB(A).

To obtain the contribution from the screened parts of the road, L_B, the unobstructed noise level at the reception point for hard ground must be determined, and then a correction for long barrier attenuation applied as described before. A further correction for the angle obstructed is then applied as follows:

$$\text{correction (dB(A))} = 10 \log_{10}(\theta_B/\pi)$$

The predicted value for L_{10} (18-hour) is then obtained by combining L_U and L_B using the equation:

$$L_{10} = L + 10 \log_{10}(1 + 10^{-\Delta/10})$$

where L = the higher of the adjusted L_B and L_U values and Δ = difference between L_B and L_U, both of these being measured in dB(A).

In instances where it is necessary to take into account a large number of barriers (e.g. when calculating the noise level at a reception point behind a reasonably uniform row of houses which face onto a major highway), the procedure just detailed above is also applied. However, to save a lot of tedious calculations, the noise level may be assessed in terms of a hypothetical fractional opening Z, which is given by

$$Z = R/(R + b)$$

where R = average opening between buildings and b = average length of the buildings along the main road in the vicinity of the reception point, both of these being measured in metres.

In this instance, θ_H is assumed to be equal to zero, θ_S is taken as equal to $Z\pi$, and θ_B equals $(1 - Z)\pi$.

Where only a short length of highway is exposed (e.g. where a highway runs in the open between two tunnels), the distance and screening corrections are determined using the source line of the road, extended as necessary, and a correction applied (as previously described) for the angle of view, where θ is the unobstructed angle of view of the road at the reception point.

Where the length of highway is on a curve, it is usual practice to break down the curve into several straight-line segments, after which, using the shortest slant distance from each segment, the procedure described in the previous paragraph is applied. Separate contributions at the reception point are then combined using the equation:

$$L_{10} = 10 \log_{10} \sum_1^n 10^{L_n/10}$$

where L_{10} = combined noise level due to all n component noise levels L_1, $L_2, \ldots, L_n$, measured in dB(A).

Noise at a point down a side road that results from traffic on a main highway is taken as depending upon the angle of view of the source line, and this is generally governed by the size of the aperture formed by the buildings flanking the side road at the entrance to the main highway. The noise determination at the reception point in the side road is then made according to the procedure described for partial screening, using a hard-ground distance correction and ignoring any contribution from the screened part of the road. However, as is discussed later, the proximity of (reflection) walls along a side road will require the use of a facade correction factor at all points along that road.

Noise reflection from rigid surfaces adjacent to the reception point will increase the noise level above that estimated by the procedures described so far, i.e. these give a 'free-field' L_{10} (18-hour) noise level. Thus, to calculate the noise level at a point 1 m from a building facade—as is required by the 1975 Noise Regulations[48]—a correction of +2.5 dB(A) must be applied. (Other noise estimates along side roads lined with houses but away from their facades also require the addition of 2.5 dB(A) because of the proximity of the facades.)

Where there are facades such as houses or a noise barrier beyond the traffic stream along the opposite side of the highway, such that the fractional opening is less than 0.5, another correction of +1.0 dB(A) is required in addition to the facade correction of 2.5 dB(A) noted above. (The additional 1 dB(A) is only applied to side road calculations where there are substantial reflecting surfaces along the main highway opposite the aperture of the side road and within the angle of view of the reception point.)

Reflecting surfaces on both sides of the highway, as in a highway cut with retaining walls, cause the reflection of noise into the shadow zone, and the result is a reduced screening performance. The scale of the reduction in screening depends upon the depth of cut and the angle of the wall to the vertical. The adjustment required for this effect is as follows:

$$\text{correction (dB(A))} = F \times D$$

where F = correction factor for the angle of the wall = $\exp(0.019\phi^2)$, in which ϕ = angle of the wall to the vertical (degrees), and D = correction factor for the height of the wall = depth of cut.

This correction is in place of, and is not in addition to, the correction described above in relation to the reflection from the opposite facade.

Numerous examples are available in the literature (see, for example, references 50 and 51) illustrating noise predictions carried out for various conditions. The following example[50] shows the major procedures involved.

Example of noise level determination

Figure 6.25 shows the location of a standard elevated dual carriageway highway relative to the backs of existing houses. The following data are available in addition to those provided in Fig. 6.25. The mean 18-hour speed is 103 km/h. The proportion of heavy commercial vehicles is 50 per cent. The total flow is 50 000 vehicles per 18-hour day. The carriageway gradient is 2.2 per cent. The horizontal distance (across hard ground) of the reception point from the edge of the nearside carriageway is 67 m. The height of the reception point relative to the carriageway is minus 1.7 m. The height of the noise barrier above, and its distance from, the edge of the carriageway are 3 m and 3 m, respectively, and the path difference due to the barrier is 0.602 m. The height of the solid portion of the open box

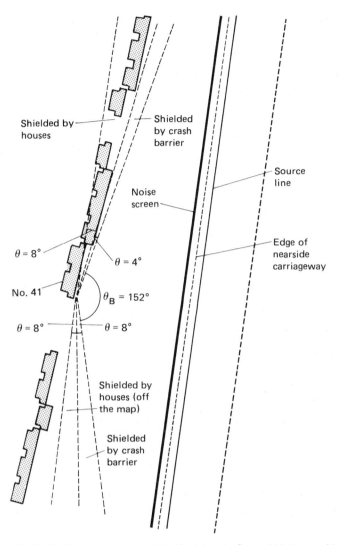

Fig. 6.25 Example noise problem involving an elevated highway with purpose-built barriers on both sides

safety barrier is 0.3 m, and the path difference is less than 0.001 m. The noise barriers and safety fences are located on both sides of the highway.

Determine the L_{10} (18-hour) noise level at a point 1 m from the rear facade of No. 41.

Solution
The unadjusted basic noise level is given by

$$L_{10} = 28.1 + 10 \log_{10} 50\,000 = 75.1\,\text{dB(A)}$$

Correct for traffic speed and heavy vehicle content:

correction
$$= 33 \log_{10}(103 + 40 + 500/103) + 10 \log_{10}[1 + (5 \times 50)/103] - 68.8$$
$$= 8.1 \, \text{dB(A)}$$

Correct for gradient:

$$\text{correction} = 0.3 \times 2.2 = 0.66 \, \text{dB(A)}$$

The adjusted basic noise level is given by

$$L_{10} = 75.1 + 8.1 + 0.66 = 83.9 \, \text{dB(A)}$$

Correct for distance:

correction for hard ground
$$= -10 \log_{10}[(67 + 3.5)^2 + (-1.7 - 0.5)^2]^{1/2}/13.5 = -7.2 \, \text{dB(A)}$$

Correct for angle of view of highway:

$$\text{correction for noise barrier} = 10 \log_{10}(152/\pi) = -0.7 \, \text{dB(A)}$$

$$\text{correction for safety fence} = 10 \log_{10}[(8 + 4)/\pi] = -11.8 \, \text{dB(A)}$$

Correct for screening by barriers:

correction for noise barrier
$$= -15.4 - 8.26x - 2.787x^2 - 0.831x^3 - 0.198x^4$$
$$- 0.1539x^5 - 0.12248x^6 - 0.02175x^7 - \ldots$$
$$= -13.7 \, \text{dB(A)}$$

where $x = \log_{10} 0.602$.

correction for safety fence
$$= -15.4 - 8.26y - 2.787y^2 - 0.831y^3 - 0.198y^4$$
$$- 0.1539y^5 - 0.12248y^6 - 0.02175y^7 - \ldots$$
$$= -5.0 \, \text{dB(A)}$$

where $y = \log_{10} 0.001$.

Adjust for distance, angle of view and screening:

(1)   for noise barrier section

$$L_{\text{B}}' = 83.9 - 7.2 - 0.7 - 13.7 = 62.3 \, \text{dB(A)}$$

(2)   for safety fence section

$$L_{\text{B}}'' = 83.9 - 7.2 - 11.8 - 5.0 = 59.9 \, \text{dB(A)}$$

(3)   combining for noise barrier and safety fence gives

$$L_{10} = 62.3 + 10 \log_{10}(1 + 10^{-2.4/10}) = 64.3 \, \text{dB(A)}$$

Correct for reflection effects:

$$\text{correction for reception point, 1 m from facade} = +2.5 \, \text{dB(A)}$$

$$\text{correction for reflection from barrier on farside} = +1.0 \, \text{dB(A)}$$

Final predicted noise level at the back of No. 41:

predicted noise level $= 64.3 + 2.5 + 1.0 = 67.8$ dB(A)

and rounding to the nearest whole number gives

predicted L_{10} (18-hour) noise level $= 68$ dB(A)

Noise protection and insulation
Basically, there are five ways by which the effects of traffic noise can be ameliorated, as follows.

(1)   At the source—this entails the control of the emission of noise from individual vehicles (see Chapter 3 for a brief discussion re this method of control).
(2)   Through planning control of adjoining land uses—this involves the siting of more tolerant land uses next to busy highways. Typical noise-tolerant uses include wide landscaped stormwater channels, playing fields and parks, car parks, light industry, warehousing, offices, trades/services areas, and retailing facilities. For obvious reasons, this method of noise control is most applicable to rural areas and new/redeveloped urban areas.
(3)   Through the use of distance—as land values and maintenance costs continually increase, noise control by distance alone becomes an expensive technique, especially in urban/suburban areas. In practice, setbacks beyond a distance of about 35 m are rarely justified as the additional noise reductions are not very great, i.e. the additional alleviation tends to approximate the rule-of-thumb relationship of 3 dB(A) reduction per doubling of distance.
(4)   Amelioration of the effects of traffic noise will result through proper attention being paid to this problem in the architectural design and site planning of the buildings being protected.
(5)   Through highway geometric design and/or the use of noise barriers[51, 52]—in order to offer as much protection as possible, any barrier should be sited either close to the reception point or close to the noise source. The barrier ideally should be sufficiently long to obscure completely the highway as seen from the area being protected, because of the effect that the angle subtended by the barrier has upon the noise level at the reception point. The normal range of barrier height is 1–3 m; barriers less than 1 m are limited in value, whilst those greater than about 3 m often represent a significant visual intrusion upon the landscape.

Acoustic barriers are most effective if made continuous and without gaps. The minimum mass required in any given situation can be determined from the equation:

$$M = 3 \, \text{antilog}[(A - 10)/14]$$

where $M =$ mass of barrier (kg/m^2) and $A =$ potential noise attenuation, measured in dB(A) (taken as positive).

Grassed or planted earth mounds are the most aesthetically pleasing form of outdoor barrier, especially in rural-type environments. For space reasons, however, it may be necessary to use mesh reinforcement or some form of revetment on the traffic side of the mound. Whatever the type of barrier, a freely-flowing alignment with the top edge following the general ground slope is preferred. Furthermore, the barrier should begin and end at existing features on the highway boundary, e.g. at bridges or where fence-lines, walls, or hedges intersect.

Side winds can be a hazard to traffic passing the ends of a noise barrier. Thus the ends are generally reduced to a height of about 1 m (in steps of not more than 0.5 m) to ensure a gradual reduction in side pressure. Wherever there is the likelihood that a noise barrier may be struck by an errant vehicle, it is usual practice for a safety barrier to be installed between it and the carriageway for highways with design speeds of 80 km/h or more.

Details regarding the principles underlying the structural design of noise barriers, including wind loading determinations, are available in the technical literature[53].

Intersection design

Highway intersections occur in a multiplicity of shapes. They can, however, be divided into the seven basic forms shown in Fig. 6.26. Handling the traffic movements indicated by these junction forms most usually requires the application of channelization practices, traffic management techniques, and/or traffic control devices; in certain instances, it may require the additional construction of roundabouts or junctions with grade-separation.

The main factors influencing the choice of a particular intersection layout are cost, capacity, delay to vehicles, aesthetics and, particularly important, safety.

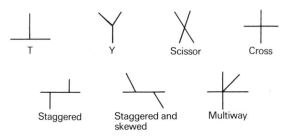

Fig. 6.26   Basic forms of intersections at-grade

Basic design principles

In practice, the development of intersection layouts is very much influenced by governmental recommendations and standards, as well as of course the previous experience and knowledge of the designer. Underlying most recommendations and standards relating to safety are the following fourteen basic design principles with which the highway engineer should be familiar.

(1) *Minimize the carriageway area of conflict.* This principle is illustrated in Fig. 6.27(a) which suggests how the large area in which possible conflicts can take place at a conventional Y-intersection can be modified. When elongated islands are placed appropriately in the intersection, the size of the uncontrolled area is significantly reduced so that there is less chance of confusion and less chance for potentially dangerous vehicle movements to occur.

Another common example of this problem in countries with inherited road systems is a four-way intersection in which two opposing arms are

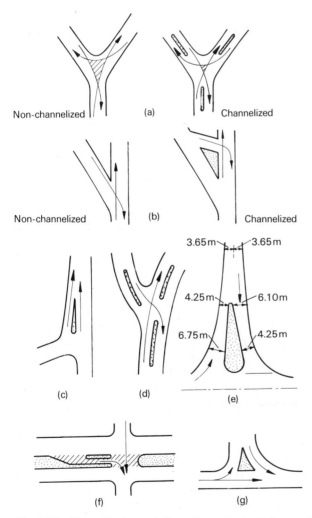

Non-channelized   (a)   Channelized

Non-channelized   (b)   Channelized

3.65 m   3.65 m

4.25 m   6.10 m

6.75 m   4.25 m

(c)   (d)   (e)

(f)   (g)

Fig. 6.27 Channelization techniques illustrating basic intersection design principles

slightly offset from each other. Normally, realignment of the intersection so that the arms are directly aligned (or, preferably, widely spaced) makes the intersection significantly safer.

(2)   *Control the angle of potential vehicle conflict.* Figure 6.27(b) illustrates the very dangerous situation where the traffic streams on two opposing roadways meet at about a 30-degree angle. There are obvious risks here, e.g. the entering motorist from the minor road has particular difficulty in seeing to the left, which increases the chance of conflict with traffic travelling in both directions on the major road. Furthermore, should an accident occur between opposing vehicles, the risk of a fatality is high since the relative speed is high, and most of the kinetic energy will be expended in damaging the vehicles and their occupants.

As the angle of potential collision swings away from head-on, accidents generally become less severe. Most authorities are of the opinion[54] that roads intersecting at about 90 degrees are the most desirable, and when the angle of intersection is reduced to less than about 60 degrees, the accident hazard is considerably increased. Besides reducing the potential accident severity, a right-angle arrangement also provides the entering motorist with favourable conditions for estimating the speed and position of traffic on the through road, whilst the time and distance required by the entering vehicle to cross the conflicting traffic stream are both reduced.

(3)   *Control relative speed.* This applies particularly to merging traffic streams. Two traffic streams going in the same direction at, say, 80 km/h, and merging at an angle of 30 degrees, have a relative speed of 42 km/h. If the angle of merge is narrowed to 10 degrees, the relative speed is reduced to 14 km/h, so that, in the event of an accident occurring, the relative damage caused should be significantly less. Furthermore, merging in this way causes less disruption to traffic.

Figure 6.27(c) shows how a flat intersection angle enables merging vehicles to accelerate easily and run parallel to a major traffic stream until a safe entry gap appears. At angles greater than about 15 degrees, this manoeuvre is more difficult, longer gaps are required, and some form of traffic control, e.g. a stop sign, is desirable.

(4)   *Control high speeds.* Traffic carrying out cutting manoeuvres should do so at low speeds. Figure 6.27(d) suggests how the speed of an entering minor road traffic stream can be reduced by causing it to bend substantially, whilst Fig. 6.27(e) illustrates how it can be reduced by funnelling the traffic into a gradually narrowing opening. As well as causing the motorist to feel hemmed in, so that his or her response is to reduce speed, funnelling also prevents overtaking from taking place in the potential conflict area.

(5)   *Provide protection for vehicles leaving or crossing the main traffic stream.* Vehicles leaving a main road should preferably decelerate and, if necessary, stop in road space separate from that used by through traffic; this minimizes the potential for rear-end collisions. Furthermore, vehicles crossing high-volume traffic streams should be able to complete the manoeuvre in two stages.

The shadowed area in Fig. 6.27(f) provides decelerating and/or storage space for vehicles waiting to turn to the right and cross the opposing traffic stream. This layout also has the safety feature that a motorist attempting to cross the main highway only has to be concerned with taking one decision and obtaining one safe gap in each traffic stream at a time.

(6) *Clearly define the travel paths to be followed.* Figure 6.27(a) also illustrates this important principle, the objective of which is to minimize the free choice of the driver in respect of the path to be followed. Figure 6.27(g) is another example; in this case, a prohibited movement is discouraged, i.e. a triangular island is deliberately shaped to encourage motorists entering a one-way traffic stream on the main road to turn left only and follow the correct direction of travel.

(7) *Reduce the number of points of conflict.* The number of potential points of conflict can be reduced by prohibiting certain traffic movements (see Chapter 7), by channelization (see Fig. 6.27), or by eliminating some arms from the intersection.

In rural areas especially, consideration should always be given to using two separated (staggered) T-junctions instead of straightover four-way intersections. The right–left stagger—whereby crossing traffic first turns right out of the minor road, proceeds along the major road, and then turns left without stopping—is preferred to the left–right stagger. In the latter case, opposing queues of right-turning vehicles from the major road could have to wait (side by side, if the inter-junction spacing is small) on the major road until suitable cutting gaps in the opposing streams become available. The spacing between two staggered right–left intersections will vary, mainly according to location (e.g. whether urban or rural) and the design speed of the road; thus on high-speed rural roads, a time spacing of 3 seconds might be desirable, whereas in lower-speed areas, a distance spacing as low as 40 m might be acceptable.

In the case of left–right staggered intersections, it will normally be necessary to widen the intervening carriageway to provide for decelerating, right-turning vehicles and/or storage space for these vehicles.

(8) *Favour predominant or high-speed traffic flows.* Priority of movement should normally be given to the major traffic movement. Thus, for example, in Fig. 6.27(b), it is the minor traffic movement that is deflected and required to stop before it enters the major traffic stream.

As well as generally improving the capacity of the junction, safety is also increased by this rule. For example, drivers who travel for long, uninterrupted distances at high speed will often be slow to react to a sudden change in alignment or to the entry of a high-speed vehicle from a minor road. On the minor road, adequate warning of a change in the driving environment should be provided, however.

(9) *Provide proper and safe locations for traffic control devices.* In good intersection design, the possible use of traffic control devices and other highway furniture should always be considered, e.g. the design of a junction to be controlled by signals may differ significantly from one requiring only channelization and signs.

Large volumes of traffic having complex turning movements can often be handled very efficiently by a combination of traffic signal control and island channelization. If vehicle-actuated detectors are placed in the vehicle channels and the signs are located on the islands, then vehicle conflicts and delays can be reduced substantially.

(10) *Protect pedestrians.* All of the islands shown in Fig. 6.27 also fulfil the purpose of providing refuges for pedestrians so that crossing of the road can be carried out in separate movements. Where possible, particularly on wider roads, central island refuges should be provided for pedestrians.

(11) *Provide reference points for drivers.* Motorists should be provided with reference points such as traffic islands and STOP/GIVE WAY lines at intersections which indicate where, say, the lead vehicle in a minor road traffic stream should stop until a suitable entry gap appears in the main road stream.

(12) *Control or restrict access in the vicinity of an intersection.* No minor roads or driveways should be permitted within the immediate area of influence of a newly designed intersection. If such access points exist, then preferably they should be closed. If closure is not possible, for unavoidable practical reasons, then central reservation/channelization techniques should be used to prevent the entering motorist from crossing the traffic flow, i.e. the entering vehicle should always be forced to merge with the nearside traffic stream.

(13) *Provide advance warning of change.* Motorists should never be suddenly faced with the unexpected. Thus advance signing of intersections ahead can be particularly important on minor roads, where visibility is restricted on all road types, and on high-speed highways, so that vehicles can be encouraged to slow down. Channelization islands should not be located on the top of crest curves nor on short horizontal curves, as motorists can experience difficulties when unexpectedly confronted with raised islands on what they would normally assume to be the 'natural' vehicle pathways.

(14) *Illuminate intersections for night-time use.* Obviously, this is not possible at all intersections, especially in rural areas. However, priority for lighting should always be given to intersections with heavy pedestrian flows, heavy vehicular flows, and where raised channelization islands are introduced into what might otherwise be considered the natural vehicle pathways.

At-grade intersections

As with the highway between intersections, a number of geometric design elements need to be considered at an intersection to ensure that its final layout is both functional and safe. These are covered in detail in the governmental design manuals, so that the following discussion only attempts to give an overview of some of the main factors involved in the design of at-grade intersections (excluding roundabouts which are discussed later).

Design vehicle

The turning capabilities of vehicles influence the shape of the kerblines and the width of the carriageway at junctions. The off-tracking of the following rear wheels of buses and heavy commercial vehicles requires larger corner radii and extra lane widths to enable these vehicles to negotiate intersections safely, without having to stop and manoeuvre in order to avoid intruding beyond the designated carriageway space.

British practice is to make allowance for the swept turning paths of long vehicles where they can reasonably be expected to use a junction; consideration is also given to the manoeuvring characteristics of these vehicles in the design of staggered junctions. The swept turning paths are normally generated by a 15.5 m long articulated vehicle, whilst manoeuvring at staggered junctions is assessed using an 18.0 m long draw-bar trailer combination (see reference 17 for example junction designs). These design vehicles comply with the current and possible future *Vehicle Construction and Use Regulations*.

As an aid to the highway designer, templates have been developed which allow the turning paths of vehicles to be checked against intersection layouts as they are developed.

An important point to consider in respect of corner design in urban areas is whether parking is to be permitted on the intersecting streets. If parking is permitted, then a smaller corner radius may be used as the wheel tracks on the approach arm will normally be further from the kerb; however, if parking is not permitted and moving vehicles are placed in the parking lane, a much wider swing will be made by heavy vehicles into the intersecting street unless a large radius is provided.

Capacity

The number of through traffic lanes and speed change/storage lanes provided at a junction should be sufficient to provide adequate capacity throughout its anticipated life. In the case of at-grade intersections which are eventually to be signalized or grade-separated, this may involve the design of separate construction stages before the ultimate design of the intersection is achieved.

If one or more lanes on an arm of an intersection are to be dropped, they should be carried through the junction and dropped on the farside. This provides greater capacity through the intersection where it is most needed.

Horizontal alignment

As noted previously, the desirable intersection angle of intersecting roads at a junction is about 90 degrees. When roads intersect at angles less than, say, 60–70 degrees, consideration should be given to realigning the minor road.

Sight distance

The intersection, and traffic on its approach arms, should be seen and appreciated by approaching drivers. Thus, for example, at major/minor priority junctions, the practice is for the major road through traffic to be provided with the desirable minimum stopping sight distance on the approaches to and through each junction.

Drivers on the minor road should have unobstructed visibility to the left and right along the main road for a distance dependent upon the major road traffic speed. The visibility should be available on a line between two points 1.05 m above the carriageway as follows:

(1) a point 9 m along the centreline of the minor road from the continuation of the line of the nearer edge of the carriageway of the major road (for lightly-travelled minor roads, but not housing estate roads, the dimension may be reduced to 4.5 m in difficult circumstances),
(2) a point at a length given in Table 6.34, measured along the nearer edge of the major road carriageway from its intersection with the centreline of the minor road.

Table 6.34 Main road distance measurements used in the determination of visibility splays[17]

Main road design speed (km/h)	Distance (m)
120	295
100	215
85	160
70	120
60	90
50	70

Parking should not be permitted within visibility splays as this will obstruct visibility. Where it may be necessary to erect essential traffic signs within such visibility areas, great care should be taken to minimize their obstructive effect.

Channelization islands
The term 'channelization' has come to be used in highway intersection design to refer to the situation where directional islands are used to divert vehicles into definite travel paths so that the safe movement of traffic is facilitated, vehicle conflict points are reduced, and traffic friction points are minimized. To enable the traffic islands to be seen clearly, they should preferably be bordered by raised kerbs, have an area of at least 4.5 m² and, where possible, be provided with illuminated signs or bollards at suitable places, e.g. apexes to islands. Islands smaller than about 4.5 m² are often defined by carriageway markings alone, i.e. ghost islands. Figure 6.27 shows some channelization techniques used at intersections.

There are no internationally accepted criteria which indicate exactly whether or not, or how, particular intersections should be channelized. Hence, every intersection must be considered on its own merits when determining whether the use of these techniques is advantageous. Comprehensive examples of very many forms of channelized intersection, and detailed comments on their usage, are readily available in the literature[17, 55].

Auxiliary lanes

Speed change lanes are provided at intersections in order to allow through vehicles to proceed relatively unhindered by turning vehicles. Also, the extra carriageway widths serve to reduce accident severity by enabling turning vehicles to merge with, and diverge from, the main traffic streams at low relative speeds.

In urban areas, the carriageway is often flared at an intersection to provide storage space for vehicles waiting to turn, thereby increasing capacity. Whilst not speed change lanes in the proper sense, these storage lanes can perform a similar function on all-purpose urban roads, if they are properly designed and made sufficiently long.

Diagrammatic examples of different shapes of acceleration and deceleration lanes are shown in Figs 6.28 and 6.29. They are essential at all intersections on high-speed, high-volume roads, whether they be at-grade or associated with grade-separated interchanges. They are desirable at intersections on all other roads, but economics may prevent their application except in special circumstances.

Deceleration lanes normally have a priority of construction over acceleration lanes, since without them vehicles leaving the through carriageway would have to slow down within a high-speed traffic lane; this movement is well recognized as a cause of rear-end collisions. Of the three forms of deceleration lanes shown in Fig. 6.28, the types used most effectively by drivers are (b) and (c). The continuous-turn manoeuvre illustrated in Fig. 6.28(c) is particularly well adapted to interconnections between high-speed roads with large volumes of turning movements. The length of a deceleration lane is dependent upon the speed at which vehicles can manoeuvre onto it from the main carriageway, the rate of safe deceleration, and the turning speeds of vehicles after traversing the lane (which can vary from near zero to as high as 120 km/h).

Acceleration lanes permit entering vehicles to increase speed in order to enter upon the main carriageway at the speed of its traffic. If the main road traffic is very heavy, a long acceleration lane also provides the

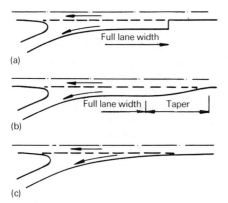

(a)

(b)

(c)

Fig. 6.28 Diagrammatic examples of deceleration lanes

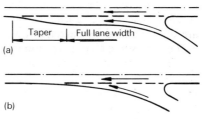

(a)

(b)

Fig. 6.29 Diagrammatic examples of acceleration lanes

entering traffic with space to manoeuvre whilst it awaits merging gaps in the main traffic stream. As much as possible of the acceleration lane should be adjacent to and flush with the main carriageway. Also, there should be no vertical kerb between the acceleration lane and the shoulder at the end of the lane, so that a merging vehicle that cannot find a gap in the main traffic stream is provided with a safety outlet, and can overrun onto the shoulder if necessary. Figure 6.29(b) illustrates the continuous-turn type of acceleration lane most favoured by motorists. In either case, the length of the lane parallel to the carriageway is dependent upon the speed of the traffic on the through carriageway, the rate of acceleration of the merging vehicles, the design speed of the merging minor road, and the volumes of through and entering traffic (see references 16, 17, 18 and 31 for design values).

British practice is to provide diverging/deceleration lanes at all grade-separated intersections, at interchanges, and at at-grade junctions on all-purpose highways with design speeds of 85 km/h and above as follows:

(1) on the nearside of the carriageway for left-turning vehicles when the volume of turning traffic is greater than 600 vehicles *AADT*, or the heavy goods vehicle content exceeds 20 per cent and left-turning traffic exceeds 450 vehicles *AADT*, or (at any design speed) the junction is on an up- or down-gradient >4 per cent and the left-turning vehicles exceed 450 *AADT*,
(2) for right-turning traffic on the offside, at gaps in the central reservation of dual carriageway roads—where they also serve the function of a storage lane,
(3) on heavily-travelled single carriageway roads (when ghost island markings may be used as delineators instead of traffic islands).

Merging/acceleration lanes are normally provided at all grade-separated junctions and interchanges, and at at-grade intersections on dual carriageway roads when the left-turning minor road flow is greater than 600 vehicles *AADT*; however, they may be provided at flows as low as 450 vehicles *AADT* where the heavy goods vehicle content exceeds 20 per cent, or where the merging lane is on an up-gradient >4 per cent.

Intersection spacing
The whole question of what is an appropriate intersection spacing is complex and confusing[54] and there is little international agreement on appropriate values. A number of researchers and governmental authorities have produced prescriptive spacing values for different types of highway, but these should be treated with caution because of the widely varied conditions under which they are applied.

In principle, short inter-junction road sections are desirable in residential areas to provide ready access to homes and to ensure that speed and traffic volumes remain low, i.e. the junctions act as 'chokes' on the free movement of traffic. The converse should be the case for, say, motorways and all-purpose dual carriageway primary distributors, whose function it is to provide for even and relatively rapid movement over longer distances to, from, and within the urban area.

A major factor controlling inter-junction spacing on any high-speed, high-volume highway is the minimum distance required to allow true weaving to occur, i.e. the distance required to allow two traffic streams moving in the same general direction to cross each other by successive merging and diverging manoeuvres. This problem arises on carriageways between adjacent junctions, between successive merging and diverging lanes at intersections, and on links within free-flow interchanges. Lengths of carriageway between service areas and intersections on major highways also constitute weaving areas.

British design practice in respect of the spacing of grade-separated junctions and interchanges on high-speed, high-volume highways in rural areas is that they should be at least 2 km apart. In extreme cases, where the traffic forecasts are at the lower end of the range for the carriageway width in question (see Table 6.7), a lesser distance down to 1 km may be acceptable.

Figure 6.30 is intended to cover the design situation with respect to weaving on highways with grade-separation. It relates the total weaving flow to the design traffic concentration, i.e. to D/V, where D is the design flow for the main carriageway (the allowable peak hourly flow per lane on the carriageway or link carrying the major traffic flow) upstream of the weaving area, in vehicles per hour per lane, and V is the design speed, in kilometres per hour under the same conditions. The large graph in this figure is first used to determine a minimum weaving length on the basis of the combined weaving volumes; the length thus determined is compared with the absolute minimum weaving length allowable for the chosen design speed (obtained from the smaller graph), and the greater of the two lengths is then taken as the minimum acceptable length of weaving section subject to the minimum signing requirements.

As well as the minimum weaving length, the weaving width is also a major consideration in the design of an inter-junction weaving area. British practice in this respect is contained in the following formula:

$$N = \frac{Q_{nw} + Q_{w1}}{D} + \frac{Q_{w2}}{D}\left(\frac{2L_{min}}{L_{act}} + 1\right)$$

In the above, the symbols are defined as follows. N is the number of lanes. Q_{nw}, Q_{w1} and Q_{w2} are the total non-weaving flow, the major weaving flow and the minor weaving flow, respectively, measured in veh/h. D is the design flow for the main carriageway upstream of the weaving area, measured in veh/h per lane. L_{min} and L_{act} are the minimum and actual lengths of the weaving area available, respectively, both measured in metres (see also Fig. 6.30 for L_{min}).

When calculating the number of traffic lanes, a fractional part will inevitably require a decision as to whether to round up or down. In some instances, it may be possible to vary the position of a junction and thus increase or decrease the weaving length so that the fractional part will converge towards a whole number of lanes. In other cases, a high fractional part with a high weaving volume suggests rounding up, whereas rounding down might be more appropriate if the fraction is small and

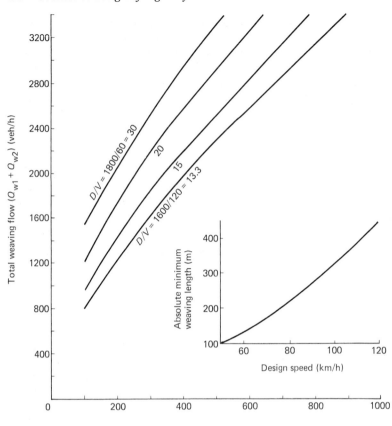

Minimum length of weaving section (m)

Fig. 6.30 Minimum weaving length relationships used in the design of motorways and high-quality, all-purpose roads[31]

the weaving flow is low. The composition of the traffic is also important, e.g. urban (commuter) drivers are more efficient than motorists on recreational routes. The reliability of the predicted flow is another major factor affecting the decision to round up or down.

The manner in which the weaving length and width are determined is best illustrated by the following example.

Example of weaving area design

The predicted 30th highest hour traffic volumes through a weaving section between interchanges on an urban all-purpose 7.3 m dual carriageway in the 15th year after opening are as follows.

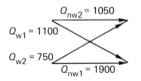

The predicted heavy vehicle content is 10 per cent throughout. The new road is to act as a primary distributor and partly also as a link between residential and industrial development. The weaving section is straight and a length of 400 m is available. The average gradient through the section plus 0.5 km upstream of the merge is less than 1 per cent. The design speed is 80 km/h and the peak hourly design flow is 1600 veh/h per lane. Determine the number of lanes required between the intersections.

Solution
Step 1 Determine the minimum acceptable weaving length.
  For a design flow of 1600 veh/h per lane and a design speed of 80 km/h, $D/V = 1600/80 = 20$. From the large graph in Fig. 6.30, the minimum weaving length is 215 m for a total weaving flow of 1850 veh/h. From the small graph in Fig. 6.30, the absolute minimum weaving length is 220 m for a design speed of 80 km/h. Use the greater value of 220 m.

Step 2 Determine the number of lanes required within the weaving section.
  Using the previously discussed equation,

$$N = \frac{(1900 + 1050) + 1100}{1600} + \frac{750}{1600}\left(\frac{2 \times 220}{400} + 1\right) = 3.51 \text{ lanes}$$

The decision as to whether to round up or down is not at all obvious. In this instance, however, a large proportion of drivers will be local who regularly travel between the residential and industrial areas, and thus are likely to weave efficiently. As the junction is located within an urban environment, there is likely to be a severe restraint on land availability for an extra lane. The horizontal alignment is straight. All other factors being equal, therefore, an appropriate decision might be to round down to three lanes.

Roundabout intersections

In one sense, roundabouts can be considered as a form of channelized intersection in which vehicles are guided onto a one-way roadway and required to move in a clockwise direction (in Britain) about a central island. At one time, the roundabout intersection was considered to be the answer to many of the problems associated with intersections on highways. This is evidenced by the multitude of roundabouts to be found on British highways at this time. In fact, roundabout intersections have particular advantages and disadvantages, and the decision as to whether a roundabout should be used at any individual location requires an understanding of these.
  Where roundabouts are properly used and designed, the efficient flow of traffic is promoted by the orderly movement of vehicles about the central island. There is only minor delay to traffic due to speed reductions and stopping. Furthermore, the possibility of having vehicle conflicts can be considerably reduced (see, for example, Fig. 6.31). Since all traffic

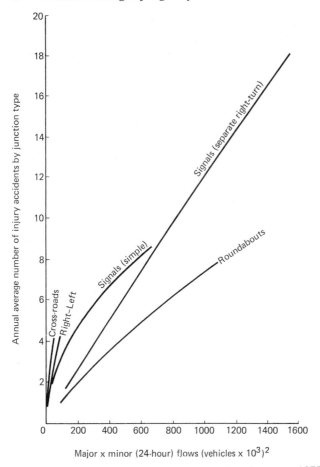

Major x minor (24-hour) flows (vehicles x 10^3)^2

Fig. 6.31   Injury accidents at junctions on dual carriageways, 1975–80[20]

streams merge and diverge at small angles, accidents which do occur rarely have fatal consequences, damage being usually confined to vehicles only.

Data from existing mini and double roundabouts indicate that the conversion to these layouts of intersections that were formerly either major/minor priority or traffic signal controlled resulted in savings in the number of personal-injury accidents, with additional reductions in the proportion of fatal and serious ones[56].

Roundabout intersections are not as readily adaptable as are traffic-signal-controlled intersections to the long-term stage development of a highway. If constructed to meet the long-term needs, they usually result in over-design when compared with the immediate traffic requirements. For traffic control reasons (e.g. they interfere with platoon movements), roundabouts should not be provided in areas with an existing or proposed area-wide urban traffic control system.

For safe and efficient operation, there is need for warning and directional signs at roundabouts. At night-time, the central island and the entrances and exits should be well lit. In addition, aesthetically pleasing landscaping of the central island is required.

Design rules for mini roundabouts

The main objective of efficient intersection design is to achieve high capacity with safety within the junction area. This in effect means that the layout encourages motorists to make full use of the space available. The following guideline rules may be enunciated in respect of the design of single island roundabouts operating under offside-priority control.

(1) *The design should provide adequate entry and exit widths.* Bearing in mind that vehicle speeds are at their lowest upon entry into the roundabout, i.e. as entering vehicles are about to accept gaps in the circulating traffic, this rule means that entrances should be flared if adequate 'capacity' is to be provided at the GIVE WAY line. British practice is for the lane widths at the GIVE WAY line to be at least 2.5 m, and tapered back. The best entry angle is about 30 degrees. (The flare length should be at least 5 m in urban areas, whilst 25 m is adequate in rural areas.) Each exit should allow for an extra traffic lane over that of the limb downstream.

(2) *The design should provide adequate circulation width, compatible with entry widths and turning movements.* If the circulating carriageway is too restricted in relation to the traffic that needs to use it, congestion will result. Its width should normally lie between 1.0 and 1.2 times the maximum entry width.

(3) *The design should provide for the deflection of traffic passing through the roundabout.* Entry path curvature, which is a measure of the amount of entry deflection to the left imposed on vehicles at entry, is one of the most important determinants of safety at a roundabout. Experience at a number of (formerly) conventional weaving roundabouts that were converted to mini roundabouts to increase capacity has indicated that poor conversion design can result in an increase in the number of accidents. These accident increases have been attributed to higher vehicle speeds within the intersection as a result of over-reducing the size of the central island without providing for adequate traffic deflection. On this basis, it is essential that traffic be forced to slow down on the approaches to roundabouts, particularly if they are located on high-speed highways, so that, under all flow conditions, vehicles traverse the intersection at a safe speed[23]. Speed reduction can be effectively achieved by ensuring that the crossing movements at intersections are adequately deflected by introducing a stagger between the entrance and exit arms (on the same carriageway), or by creating large/subsidiary traffic deflection islands in the entry[19, 20].

In addition, all proposed layouts should be checked against excessive speeds by ensuring that no through vehicle path (assumed 2 m wide) is on a radius greater than 100 m, which is the curvature corresponding approximately to 50 km/h with a sideways force of 0.2g.

(4)   *The layout provided should be simple, clear and conspicuous, so that the important features of the design (and other vehicles) are easy to see and easy to use.* The important features of the design should be easily comprehensible to the motorist. Complex intersections cause confusion and increase the potential for accidents.

Figure 6.32 illustrates a ring junction design which was constructed at Colchester. This is a compact design whereby the usual one-way circulation about a single large central island is replaced by two-way circulation about a central 'core' (in this case, a small square island only 5 m across), with a three-arm mini roundabout at the mouth of each arm. This type of design has a number of advantages: it results in high capacity (5 per cent more than any other equivalent roundabout design at Colchester), all opposing right-turns are non-hooking, the path lengths of the right-turn movements are shortened, and the crossing conflicts between the various traffic streams are separated and therefore made easier. Its major disadvantage is that the layout is complex, e.g. its many GIVE WAY lines cause difficulty for a minority of drivers. As a result of this complexity, ring roundabouts are not at present in wide usage in Britain.

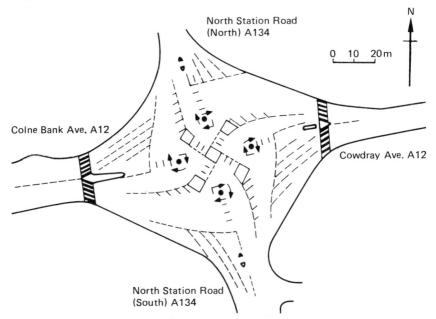

Fig. 6.32   Colchester's experimental ring roundabout

Usage of offside-priority designs in rural areas

It has become popular for highway engineers to recommend the use of mini roundabouts as low-cost solutions to the problems of congested major/minor junctions in rural and semi-rural locations. The following procedure[57], developed to help decide when to use this solution, is based on a comparison of the existing delays to vehicles passing through the original junction, with those future delays which might be expected to follow the installation of the non-conventional roundabout.

Delays at an ordinary major/minor T-intersection are normally inflicted only on minor road vehicles, which have to yield priority to the major road traffic streams. With an offside-priority roundabout system, however, the previously unimpeded straight-ahead major road traffic has both to reduce speed and to be prepared to give way to minor road vehicles on its right.

The delay to major road traffic due to the physical layout and markings associated with a roundabout's design is known as the fixed or geometric delay and, unlike congestion delay, is present throughout the day. Research has shown that this delay can be calculated from the following formula:

$$D_{24g} = Q_{24}(1 - 2R)D_g/6000$$

where D_{24g} = total anticipated 24-hour fixed delay (h), Q_{24} = anticipated total daily junction flow (veh/24 h), R = ratio of minor to total flow (daily average), and D_g = average geometric delay per vehicle (hundredths of a minute) obtained from the empirical equation $D_g = 0.432V - 15.5$, in which V = average of the mean of the approach and exit speeds (km/h).

Unlike fixed delay, most congestion delay occurs during peak traffic periods. The following empirical relationship has been derived to provide an estimate for the total 24-hour congestion delay at an offside-priority mini roundabout:

$$D_{24c} = 2 \times 10^{-4} Q_{24}^2 P_3/Q_p$$

where D_{24c} = total anticipated 24-hour congestion delay (h), Q_{24} = anticipated total daily junction flow (veh/24 h), P_3 = ratio of flow in the three highest hours to the 24-hour flow, and $Q_p = 0.8K(\Sigma W + \sqrt{A})$ (veh/h). Here ΣW = sum of the basic road widths used by traffic in both directions on all approaches (m), A = area within the junction outline (including islands) which lies outside the area of the basic cross-roads (m²), and K = a site factor dependent upon the roundabout type and number of arms (e.g. typically 60 and 45 for single island 3- and 4-arm roundabouts, respectively).

The total 24-hour delay at the existing T-intersection can be estimated from the following empirical formula:

$$D_{24x} = D_3/8P_3^2$$

where D_{24x} = total 24-hour junction delay (h), D_3 = total junction delay during the peak three hours (h), and P_3 = ratio of flow in the peak three hours to the 24-hour flow.

If D_{24x} is significantly greater than $D_{24g} + D_{24c}$, then it can be concluded that the changeover of the intersection should be investigated further.

Alternatively, it may be assumed that, at the 'breakeven' flow, the flow–capacity ratio of the roundabout junction is about 0.5 (as is, in fact, often the case). Then, by algebra, it can be shown that the 'guideline' daily flow required to generate delays which would equal those observed at the major/minor T-junction, at the existing flow level, approximates

$$Q_{24G} \simeq 7500D_3/P_3^2[10D_g(1-2R)+18]$$

where Q_{24G} = guideline flow (veh/24 h) and D_3, P_3, D_g and R are as defined before.

Thus if the introduction of the mini roundabout means that a higher guideline flow than that which is actually occurring is required to balance the 'before' and the 'after' delays, it can be concluded that a net benefit will result from the conversion.

High-speed road designs   Roundabout intersections on high-speed rural highways pose a particular problem in that they require extremely long weaving lengths if a low relative speed difference between vehicles entering and within the junction is to be ensured. However, if the excessive area requirements are to be avoided, and large volumes of traffic still accommodated, then a substantial reduction in speed through the intersection must be expected and accepted. Unfortunately, statistics show that all roundabout junctions on high-speed rural roads are particularly prone to the kinds of accident in which speed is a contributory factor. Analysis of such accidents showed that motorists on dual carriageway rural roads, in particular, tend to become so 'speed-adapted' that they have difficulty in reacting to, and slowing down sufficiently on the approach to, a roundabout.

To overcome the high-speed driver problem, research was carried out which showed that the placement of a series of 0.6 m wide thermoplastic 'bars' (ninety in total) across the carriageway for about 400 m on the approach to a roundabout—with each spaced at exponentially reduced intervals, e.g. 7.69 m between the first and second, and 2.73 m between the eighty-ninth and ninetieth—gave motorists the illusion that their speeds were increasing as they came closer to the roundabout, which in turn encouraged them to slow to a more realistic speed more quickly. Tests of this road-marking pattern—which is now commonly installed on high-speed dual carriageways in Britain which have no major junctions or severe bends in the 3 km immediately prior to the conventional/non-conventional roundabout—showed that it resulted in overall reductions of 56–62 per cent per annum in speed-associated injury accidents[58].

Intersections with grade-separation

When at-grade highways cross, the number of vehicles that can pass through the intersection is controlled by the characteristics of the junction rather than by those of the highways themselves. Not only do they provide many opportunities for vehicle conflicts, with the resultant expected accidents, but at-grade intersections also reduce vehicle speeds and increase operating costs. It is when these difficulties become unduly great that intersections with grade-separation become most advantageous.

Grade-separated structures generally have very large initial costs when compared with single level intersections. The main situations which justify the very considerable extra expenditure can be summarized as follows.

(1) *At intersections on motorways*—the construction of a highway with complete control of access automatically justifies the use of grade-separated structures in order to ensure the free movement of high-speed traffic.

(2) *To eliminate existing traffic bottlenecks*—the inability of an important at-grade intersection to provide the necessary capacity is in itself a justification for a grade-separation on a major highway.

(3) *Safety considerations*—some at-grade intersections are accident prone, regardless of the traffic volumes they carry. For instance, many lightly-travelled rural roads having high vehicle operating speeds have relatively large numbers of accidents at certain intersections. In these locations, land is relatively cheap and so it may be possible to construct fairly low-cost grade-separations and so eliminate these accidents.

(4) *Economic considerations*—at at-grade major road junctions, very considerable economic losses can be incurred due to intersectional frictions and the resultant delays to traffic. These are usually in the form of increased costs for fuel, tyres, oil, repairs and accidents, as well as the increased time costs of the road users. If these intersections are converted to ones with grade-separation, the very considerable long-term economic gain to the community may by far outweigh the burden of the initial capital costs.

(5) *Topographic difficulties*—at certain sites, the nature of the topography or the cost of land may be such that the construction of an at-grade intersection is more expensive.

Grade-separation without slip roads

In essence, this type of structure is simply a bridge or series of bridges which enable the traffic streams on the intersecting highways to cross over each other without any vehicle conflicts taking place. This type of intersection is most often constructed in rural areas where a minor road crosses a major road and the turning movements are not sufficient to justify expenditure on interconnecting ramps or link roads.

In urban and suburban areas, grade-separations of this type are also used to cut down on the total number of intersections on major streets. In this way, overall traffic safety and efficiency of movement is increased by concentrating the turning traffic at a limited number of locations where adequate ramp facilities can be built. They may also be constructed at locations in urban or rural areas where the site conditions are so difficult that it is not economically feasible to connect the roadways.

Grade-separation with slip or link roads

This is a system of interconnecting slip or link roads in conjunction with a grade-separation or grade-separations which provide for the interchange of traffic between two or more roadways on different levels. British practice[16] is to divide these intersections into two groups, viz. grade-separated junctions and interchanges.

A *grade-separated junction* is a form of grade-separation that involves the use of an at-grade junction (i.e. either a major/minor junction or a roundabout) at the commencement or termination of slip roads. A motorist entering a slip road from, say, a motorway has already made the decision to leave the mainline traffic and, accordingly, both money and land are saved by designing the slip road to a lower design speed, e.g. typically 70 and 60 km/h in rural and urban areas, respectively.

An *interchange* does not involve the use of an at-grade junction; it provides for uninterrupted traffic movement between mainline carriageways by the use of link roads. As a consequence of their function, link roads are normally designed to a higher design speed than slip roads, viz. 85 km/h in rural areas and 70 km/h in urban areas. (To avoid confusion, it might be noted that a slip road or a link road that turns through 270 degrees is also called a loop road.)

As with the at-grade intersections, there are very many types of junction with grade-separation which are used in various situations. Since, however, their basic purpose is to provide an easy and safe means by which vehicles may transfer from one roadway to another, it is possible to classify the many types according to the manner in which they perform this function. The simplest way of so doing is to classify them according to the number of approach roads or intersection legs they serve. Thus they may be considered as three-, four-, and multi-way. Figure 6.33 illustrates several forms of junctions with grade-separation which fall within these general classifications.

Three-way junctions If one of the intersecting legs of a three-way intersection is an approximate prolongation of the direction of approach of another, and if the third leg intersects this prolongation at an angle of between 75 and 105 degrees, the intersection is called a T-intersection. The equivalent interchange is called either a *T-interchange* or a *trumpet interchange*. If one leg of the intersection is a prolongation of the approach of another, and the third leg intersects this prolongation at an angle less than 75 or greater than 105 degrees, it is called a Y-intersection. The equivalent interchange is also called a *Y-interchange*.

Examples of traffic movement at three-way, two-level T- and Y-interchanges are shown in Fig. 6.33. Both utilize a single bridge structure and illustrate the situation where the greater volume of interchange traffic is given preferential turning treatment at the expense of the lower turning volume which has to use the semi-directional loop. These designs are most suitable for connecting two major roads, or a major road to a motorway, provided that the loop movement is relatively small. If the loop movement is heavy, then extra bridge structures may have to be constructed so that both turning movements are favoured equally. Where possible, three-way interchanges should be designed to enable future conversion to four-way ones without alteration.

Four-way junctions The simplest type of four-way, two-level junction

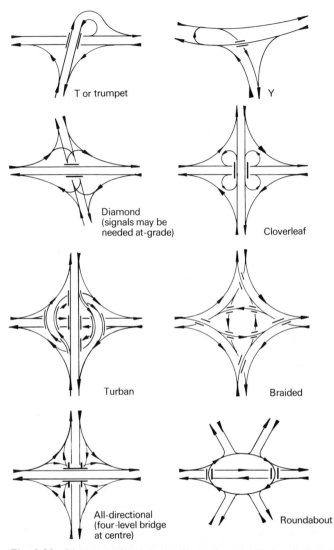

Fig. 6.33 Diagrammatic examples of some basic traffic movements at junctions with grade-separation

with grade-separation is the *diamond*. Consisting of a single bridge and four one-way slip roads, it has the particular advantage that it can be located within a relatively narrow land area since it needs little extra width beyond that required for the major road itself. As Fig. 6.33 illustrates, this junction has direct high-speed entrance and exit slip roads on the main road and at-grade terminals on the minor road. Thus its use is confined to intersections of major and minor highways. Its narrow width enables it to be easily used in urban areas where the high cost of land might make it impractical to use interchange forms.

A major advantage of the diamond configuration is that it is very easily understood by motorists. Furthermore, a greater slip road capacity can be relatively easily obtained under normal circumstances, by widening either at the slip road exit onto the minor road (for storage purposes) or throughout its length. The main disadvantage of this type of junction relates to the conflicts that can occur at the intersections where the slip roads meet the crossing minor road. These junctions may need a pair of closely-spaced traffic signals with relatively long cycle times if the minor road carries fairly heavy volumes of traffic.

In practice, of course, the combinations of layout possible under the diamond classification are literally legion (see, for example, references 34, 59 and 60).

The *cloverleaf* is a four-way, two-level single structure interchange having no terminal right-turns at-grade, i.e. right-turning traffic is handled instead by loop slip roads, thereby obviating the need for traffic signals on the crossing highway. Internationally, it is often regarded by motorists as the ultimate answer to intersection problems. It has the great advantage of being very uncomplicated to use, while the eight turning movements are accomplished with no direct vehicle conflicts. Nevertheless, there are a number of features about this type of interchange which limit its usefulness.

The first and most important is that if a cloverleaf is used at the junction of two high-speed, heavy-volume highways, an excessively large area of land may be required to enable the loops to handle the traffic at relatively low speed differentials. British practice is to use minimum radii of 75 m for loops that are both onto and off motorway mainlines; minimum radii of 50 and 30 m are used for loops that are off and onto mainline all-purpose roads, respectively. Short radii can be dangerous[61]—whilst there is no clear evidence that interchanges with properly designed loop roads necessarily have a greater accident potential than those having straighter link roads, it is essential that good visibility and adequate warning signs be provided on the approaches to and across minimum radius loops so that drivers can brake before entering the curves (thereby minimizing the need to brake within the loops and the possibility of spinning out of control).

A second undesirable feature of the cloverleaf is that vehicles waiting to make right-hand turning movements must negotiate a 270-degree semi-direct turn. Not only can this represent a relatively difficult design problem but, in addition, as vehicles leave a particular loop upon entering the main highway, it is necessary for them to weave their way through other vehicles attempting to enter the adjacent exit loop. Therefore, when traffic volumes are heavy, the area required for the cloverleaf may have to be considerably enlarged to provide adequate weaving lengths between entrance and exit points on the main carriageways.

Another point that might be noted in respect of cloverleaf interchanges is that there is a view held by many highway engineers to the effect that loop roads do not operate very satisfactorily when widened to more than one lane. Furthermore, from a safety aspect, pedestrian movements along cross-streets in urban areas are more difficult to handle with this type of interchange.

The *turban*, the *braided* and the *all-directional* multilevel interchanges are particular types with all right-turning, as well as left-turning, movements made directly. Thus they fall into the category known as 'directional', free-flow interchanges, whereas the cloverleaf might be categorized as 'semi-directional'. Directional interchanges are the highest type of interchange and are most suitable at the intersections of motorways carrying high volumes of traffic. When compared with the semi-direct-loop types of interchange, they reduce vehicle travel distance, increase the speed of traffic operation, have greater capacity and, more often than not, eliminate any weaving problems.

Directional interchanges are very expensive to construct because of the type and number of bridge structures required. However, they can need relatively small land areas and thus certain types can be used in urban locations, provided that the aesthetics of a multilevel bridge structure are environmentally acceptable. In rural areas, traffic volumes are rarely sufficiently heavy to justify direct connections in all quadrants of an interchange; direct connections are often made in one or two quadrants, and the remaining turning movements are handled by loops.

Multiway junctions The most common type of grade-separated junction in Britain is the two-bridge roundabout. It is particularly advantageous when the intersection has more than four approach arms (Fig. 6.33). With this type of grade-separation, it is usual to have the main highway underpassing or overpassing an at-grade roundabout intersection. Vehicles enter and leave the main road on diagonal slip roads in a manner similar to that illustrated for diamond junctions with grade-separation.

The grade-separated roundabout has the great advantage that it can be adapted from an existing at-grade roundabout that is overloaded. The grade-separation frees the movement of the major through traffic stream, while the minor and interchanging traffic is confined to the roundabout. There is a great number of existing at-grade roundabout intersections on British highways, and so it is quite likely that many of these grade-separated junctions will be constructed in the future. Roundabout junctions can require relatively small overall areas (e.g. a fully-directional, free-flow interchange can require over five times as much land as a compact, three-level roundabout type), and require less carriageway area than many types of interchange (e.g. cloverleaf). However, it should be noted that the safety of the junction as a whole is limited by the manner in which the intersections of the slip roads with the minor road are handled, e.g where drivers are likely to leave a fully-grade-separated highway after a long stretch of driving, a 'dumbell' roundabout arrangement on the minor road is to be preferred to conventional priority junctions.

General design guidelines for grade-separated junctions and interchanges
The design of a junction with grade-separation is a complex matter, and

each has a unique combination of problems arising from the need to satisfy traffic demands, existing highway networks, topography, land use, and the physical, social and political environments. The problems are at their most complex in urban areas, albeit they are not easily resolved in rural areas either. Stock solutions to major junction design needs are rarely possible, and the depth of investigation required is perhaps best reflected by noting that an expenditure of up to five man-years on preliminary design is not unusual[62].

At an early stage in the design of a rural junction with grade-separation, the following points should be considered[63]:

(1)   the requirements of traffic flows through the junction,
(2)   the general siting, depending upon engineering, topographical, environmental, landtake, and operating and capital costs,
(3)   the minimum area of land required for the grade-separation structure, allowing for possible savings by allocating enclosed land for agricultural or maintenance compound use but including the cost of access.

Of these, the determination of the siting and configuration is probably the most complicated.

Preliminary steps involved in the process generally include the following:

(1)   by inspection, deciding upon a range of site options, noting those which are obviously inappropriate on environmental and engineering grounds,
(2)   the carrying out of operating cost assessments for the main traffic flows between common points on each approach route,
(3)   the determination of operating cost assessments for significant turning movements in a number of different layouts,
(4)   assessments of the orders of cost, landtake, etc., for each option,
(5)   assessments of the effect of removal/insertion of various traffic movements, and of variations in the geometric standards (e.g. design speed and link length) provided for them,
(6)   the refinement of the analysis in an iterative way for appropriate site options and layouts.

Particular steps/points involved in the landtake examination phase of the preliminary design process include the following.

Non-free-flow solutions should be considered before alternatives which incorporate less restraint. Non-free-flow means that vehicles in the design year peak hour at the junction with grade-separation can expect to 'yield' and 'hesitate in groups' without extensive queueing. Non-free-flow solutions generally have lower capital costs and landtakes.

The decision as to whether to use a free-flow or a non-free-flow configuration depends mainly upon the site and upon the balance of the turning flows. For example, large and well-balanced turning movements—especially if they contain upwards of 20 per cent heavy commercial vehicles—are best catered for by a multilevel, free-flow layout. However, if the major flows are straight ahead, whilst the turning movements are relatively low in volume (e.g. less than 15-20 per cent of the total approach flow) and not predominantly right-turners or heavy vehicles, then a three-level roundabout junction may be appropriate. If there are heavy single turning movements, an extension of the three-level roundabout layout to include an independent direct link may be appropriate. Then, again, in cases where the noise caused by accelerations at a three-level intersection is unacceptable, a free-flow junction with tight loops may be an appropriate solution.

Junctions with grade-separation should be designed to cater primarily for the major traffic flows. The through vehicles in such intersections should be able to maintain the design speed selected for the main motorway route. Major traffic flows should be uninterrupted by weaving if possible; unfortunately, weaving often occurs on some three-way interchanges where the major flow at a merge is often the joining flow. If weaving has to occur, it should be spread over as long a distance as possible to reduce the possibility of multiple, sudden, potentially dangerous manoeuvres at high speed.

Connectors for turning movements should normally be designed safely, economically and to a lower speed than mainline routes. British practice[64] in respect of the design speeds of connectors is summarized in Tables 6.1 and 6.2. Note that a typical link road design speed is 85 km/h (as related to a 120 km/h mainline design speed), although speeds down to 60 km/h are appropriate for slip roads carrying relatively low volumes of traffic in difficult terrain. In urban areas, even lower connector design speeds may be used, for cost and/or environmental reasons. If special 'short cut' turning facilities to link carriageways within an interchange are provided to reduce both travel distance and time for emergency and maintenance vehicles, their design speed may be as low as 25 km/h, as the usage of these links will be restricted to official vehicles.

Slip roads that are part of grade-separated junctions are normally one-way. Two-way slip roads, such as may be used at a half-cloverleaf junction have higher accident rates than one-way roads, and are therefore normally separated either by a physical central reserve with safety fences or by a solid double white line road marking. The accident risk for one-way slip roads at junctions is similar whether the mainline traffic is over or under. However, to match mainline vehicle speeds on merging and reduce them on diverging, the preferred treatment is to design off-slips up hill and on-slips down hill, with the minor road over the mainline. Diverge and merge slip roads need not be of the same length.

Links within interchanges may require advisory speed limits to warn motorists regarding the safe negotiating speed. Only one level of speed limit should be used within an interchange as steps down in speed limits will only confuse the driver. Single lane links obviously are much cheaper than two-lane links for interchanges with structures of substantial length. Consequently, two-lane, one-way links are normally only provided where the traffic flows are near the top of the range (see Table 6.7). A disadvantage of single lane links is that they require closure during maintenance activities.

In the case of 270-degree loops, generous curvatures can be wasteful in land and capital costs and do not necessarily lead to low operating costs; however, the extra distances travelled at high speed very often outweigh slower speeds on shorter, tighter loops.

All turning movements of a minor nature should be accommodated, if it is reasonable to do so, to retain flexibility. Some designs, selected for other reasons, may easily, and with little expense on landtake, enable minor traffic movements to be inserted within the overall layout. However, their inclusion should not be permitted to distort the choice of layout unduly.

Difficult sites should be avoided. Where possible, sites should be avoided which require long, steep gradients which will slow heavy commercial vehicles and cause them to bunch into platoons. If such sites cannot be avoided, then grade-separation of the heavy vehicles should be considered so as to avoid conflict between the heavy and light vehicles, particularly at the merge and diverge areas. Sites on hilltops should also be avoided because of their adverse effect upon the visual amenities of the area. Furthermore, motorists may have difficulty in reading direction and other signs at the junction, silhouetted as they will be against the skyline, i.e. they should preferably be located against a dark background.

An additional significant cost at difficult sites can arise from the need for lighting. In practice, motorway lighting is normally provided at rural intersections which contain unusual and/or complicated design features, or sections on long bridge structures containing substandard features such as narrow hardshoulders or central reservation. Other factors which might influence the provision of lighting are steep gradients up to the junction and the susceptibility of the site to foggy conditions.

Learn from existing designs. An excellent paper is readily available in the literature[65], which describes many of the major interchanges actually constructed in Britain. It is strongly recommended to the reader that as many of these sites as possible be visited to see at first hand the different types of interchanges and the factors which influenced their selection.

Computer-aided highway design

The discussion so far in this chapter has concentrated on the fundamentals of geometric design, and relatively little has been said about the manner in which the detailed design process might be carried out. Historically, these used to be carried out with the aid of first a slide rule, and then a calculating machine, with the routine calculations involved in a given design occupying a major proportion of the work-time of many highway engineers. However, beginning in the late 1950s, as computers began to become more widely available, highway designers grasped the opportunity to use the (then) new technology to perform the more tedious aspects of their work more quickly and more accurately, as well as automatically. The result in the first instance was a wasteful development and proliferation of similar computer programs which often duplicated each other and, being machine-dependent, were not portable from design office to design office.

Unlike the situation which still exists today in many countries, the advantages of coordination in this technical area were quickly recognized in Britain, and in 1966 a Joint Computer Panel was set up by the Department of Transport and the County Surveyors Society with the aim of producing an integrated set of computer programs for highway design. The first such integrated set of programs, which was an amalgamation of what were considered to be the best of the existing programs, was completed in May 1967; this was entitled the *B*ritish *I*ntegrated *P*rogram *S*ystem for Highway Design. *BIPS1*, as it became known, has since been amended on a number of occasions. Two other program suites, the *GENESYS* subsystem known as *HIGHWAYS/1* and the *MOSS* system, became available in 1974 and 1975, respectively. These three suites are readily available in Britain today and the young highway engineer is urged to seek out the most up-to-date versions.

The *BIPS* programs are issued by the Department of Transport, and are available for use on all makes of computer. The *GENESYS* (which is an abbreviation of *GEN*eral *E*ngineering *SYS*tems) subsystem *HIGHWAYS/1* was issued by GENESYS Ltd in versions for a number of machine types. The *MOSS* (an abbreviation of *MO*delling *Sy*Stems) system was issued by the Moss Consortium of three local authorities, viz.: Durham, Northamptonshire and West Sussex County Councils, and versions are available for a number of machine types.

In addition to the above three seminal program suites, which cover the complete design process, there are a number of other good proprietary design programs that are available via particular bureaux. Many of these proprietary programs have the ability to pass data to, and receive output from, the above three main suites.

The main program suites now available offer broadly similar facilities which allow the highway design to be developed in the following stages, viz.: ground models, ground cross-sections, horizontal alignment, preliminary design and noise assessment, superelevation design, road cross-sections, full cross-sections, junction design, quantities, and plotting. Excellent summaries of the basic processes involved are readily available in the literature[66, 67].

Selected bibliography

(1) McLean JR, Review of the design speed concept, *Australian Road Research*, 1978, **8**, No. 1, pp. 3–16.

(2) McAllister IW, Relevance of standards, *Proceedings of Seminar X on Road Design*, PTRC Report P131, pp. 204–215, London, Planning and Transport Research and Computation (International) Company Ltd, 1975.

(3) *Road Layout and Geometry: Highway Link Design*, Departmental Standard TD9/81, London, The Department of Transport, August 1981, and as amended January 1985.

(4) Almond J, Speed measurements at rural census points, *Traffic Engineering & Control*, 1963, **5**, No. 5, pp. 290–294.
(5) Greenshields BD, A study of traffic capacity, *Proceedings of the Highway Research Board*, 1934, **14**, pp. 448–474.
(6) Lighthill MJ and Whitham GB, On kinematic waves, II: a theory of traffic flow on long crowded roads, *Proceedings of the Royal Society*, 1955, **A339**, No. 1178, pp. 317–345.
(7) Duncan NC, A note on speed/flow/concentration relations, *Traffic Engineering & Control*, 1976, **17**, No. 1, pp. 34–35.
(8) Duncan NC, A further look at speed/flow/concentration relations, *Traffic Engineering & Control*, 1979, **20**, No. 10, pp. 482–483.
(9) Gerlough DL and Huber MJ, *Traffic Flow Theory*, Special Report 165, Washington DC, The Transportation Research Board, 1975.
(10) Duncan NC, *Rural Speed/Flow Relations*, TRRL Report LR651, Crowthorne, Berks., The Transport and Road Research Laboratory, 1974.
(11) *Speed/Flow Relationships on Suburban Main Roads*, London, Freeman, Fox and Associates, January 1972. (A report of a study carried out for the Road Research Laboratory.)
(12) *Speed/Flow Formulae for Rural Roads*, Leaflet LF170, Crowthorne, Berks., The Road Research Laboratory, 1971.
(13) Forsgate JA and Hammond HN, *Speed/Flow Relations on Recreational Roads*, TRRL Report LR638, Crowthorne, Berks., The Transport and Road Research Laboratory, 1974.
(14) Bampfylde AP, Porter GJD and Priest SD, *Speed/Flow Relationships in Road Tunnels*, TRRL Report SR455, Crowthorne, Berks., The Transport and Road Research Laboratory, 1979.
(15) Farthing DW, Some factors affecting rural speed/flow relations, *Traffic Engineering & Control*, 1977, **18**, No. 1, pp. 12–18.
(16) *Traffic Flows and Carriageway Width Assessment*, Departmental Standard TD20/85, London, The Department of Transport, November 1985.
(17) *Junctions and Accesses: The Layout of Major/Minor Junctions*, Advice Note TA20/84, London, The Department of Transport, November 1984.
(18) *Junctions and Accesses: Determination of Size of Roundabouts and Major/ Minor Junctions*, Advice Note TA23/81, London, The Department of Transport, December 1981.
(19) *The Geometric Design of Roundabouts*, Advice Note TA42/84, London, The Department of Transport, August 1984.
(20) *The Geometric Design of Roundabouts*, Departmental Standard TD16/84, London, The Department of Transport, August 1984.
(21) *Traffic Flows and Carriageway Width Assessment for Rural Roads*, Advice Note TA46/85, London, The Department of Transport, November 1985.
(22) *Choice Between Options for Trunk Road Schemes*, Advice Note TA30/82, London, The Department of Transport, July 1982.
(23) Kimber RM and Coombe RD, *The Traffic Capacity of Major/Minor Priority Junctions*, TRRL Report SR582, Crowthorne, Berks., The Transport and Road Research Laboratory, 1980.
(24) Addendum No. 1 to *Roads in Urban Areas—Metric Corrigendum*, Technical Memorandum H12/73, London, The Department of the Environment, August 1973.
(25) Webster FV and Newby RF, Research into relative merits of roundabouts and traffic signal controlled intersections, *Proceedings of the Institution of Civil Engineers*, 1964, **27**, pp. 47–75.

(26) Ministry of Transport, *Urban Traffic Engineering Techniques*, London, HMSO, 1965.
(27) Blackmore FC, Priority at roundabouts, *Traffic Engineering & Control*, 1963, **5,** No. 2, pp. 104-106.
(28) Blackmore FC, *Capacity of single level intersections*, RRL Report LR356, Crowthorne, Berks., The Road Research Laboratory, 1970.
(29) Millard RS, Roundabouts and signals, *Traffic Engineering & Control*, 1971, **13,** No. 1, pp. 13-15.
(30) Mellors AR, Which form of control to use at a road junction?, *Proceedings of Seminars Q and R on Road Design and Highway Maintenance*, pp. 29-34, PTRC Report P144, London, Planning and Transport Research and Computation (International) Company Ltd, 1977.
(31) *Layout of Grade Separated Junctions*, Departmental Standard TD22/86, London, The Department of Transport, March 1986.
(32) Louis LJ, Sight distance requirements of rural roads—a review, *Australian Road Research*, 1977, **7,** No. 2, pp. 32-44.
(33) Haslegrave CM, *Measurement of the Eye Heights of British Car Drivers Above the Road Surface*, TRRL Report SR494, Crowthorne, Berks., The Transport and Road Research Laboratory, 1979.
(34) *A Policy on the Geometric Design of Rural Highways*, Washington DC, The American Association of State Highway Officials, 1965.
(35) Neilson ID, Kemp RN and Wilkins HA, *Accidents Involving Heavy Goods Vehicles in Great Britain: Frequencies and Design Aspects*, TRRL Report SR470, Crowthorne, Berks., The Transport and Road Research Laboratory, 1979.
(36) Crawford A, The overtaking driver, *Ergonomics*, 1963, **6,** No. 2, pp. 153-170.
(37) Troutbeck RJ, Overtaking design sight distances for rural road design, *Proceedings of the Australian Road Research Board*, 1980, **10,** Part 4, pp. 84-98.
(38) Brock G, *Road Width Requirements of Commercial Vehicles When Cornering*, TRRL Report LR608, Crowthorne, Berks., The Transport and Road Research Laboratory, 1973.
(39) Hill GJ, Prediction of vehicle swept paths, *The Highway Engineer*, 1978, **25,** No. 12, pp. 14-19.
(40) County Surveyor's Society, *Highway Transition Tables*, London, Carriers, 1969.
(41) Ackroyd LW and Bettison M, Effect of maximum motorway gradients on the speeds of goods vehicles, *Traffic Engineering & Control*, 1971, **12,** No. 10, pp. 530-531.
(42) Hills P and Prince P, Optimum gradient standards for major rural highways, *Proceedings of Seminar M on Road Design*, pp. 313-340, PTRC Report P169, London, Planning and Transport Research and Computation (International) Company Ltd, 1978.
(43) Hurd FW, Accident experiences with transversable medians of different widths, *Highway Research Board Bulletin*, 1956, No. 137, pp. 18-26.
(44) Bartlett RS and Chhotu SR, *An Analysis of Vehicle Breakdowns in the Mersey Tunnel*, TRRL Report LR484, Crowthorne, Berks., The Transport and Road Research Laboratory, 1972.
(45) *Hardshoulder and Hardstrip Design and Construction*, Technical Memorandum H4/76, London, The Department of the Environment, 1976.
(46) Ross NF and Russam K, *The Depth of Rainwater on Road Surfaces*, RRL Report LR236, Crowthorne, Berks., The Road Research Laboratory, 1968.

(47) Marquis EL and Weaver GD, Roadside slope design for safety, Paper No. 11 922 of the *Transportation Engineering Journal of the ASCE*, 1976, **102**, No. TE1, pp. 61–73.

(48) *Noise Insulation Regulations 1975*, Statutory Instrument 1975, No. 1763, London, HMSO, 1975.

(49) Brown AL, Prediction of noise levels from freely-flowing road traffic: an evaluation of current models, *Australian Road Research*, 1978, **8**, No. 4, pp. 3–14.

(50) Department of the Environment, *Calculation of Road Traffic Noise*, London, HMSO, 1975.

(51) Lassiere A, *The Environmental Evaluation of Transport Plans*, London, The Department of the Environment, 1976.

(52) *Guide on Evaluation and Attenuation of Traffic Noise*, Washington DC, The American Association of State Highway and Transportation Officials, 1974.

(53) *Noise Barriers, Standards and Materials*, Technical Memorandum H14/76, London, The Department of the Environment, 1976.

(54) Cameron JMW, *The Influence of the Layout of the Road Network on Road Safety: A Literature Review*, Technical Report RF/3/77, Pretoria, SA, The National Institute for Transport and Road Research, April 1977.

(55) Committee on Channelization, *Channelization: The Design of Highway Intersections At-grade*, Highway Research Board Special Report 74, 1962.

(56) Lalani N, The impact of accidents on the introduction of mini, small and large roundabouts at major/minor priority junctions, *Traffic Engineering & Control*, 1975, **16**, No. 12, pp. 560–561.

(57) Marlow M, *Conversion of Rural and Semi-rural Major/Minor T-junctions to Offside Priority*, TRRL Report LR883, Crowthorne, Berks., The Transport and Road Research Laboratory, 1979.

(58) *Transverse Yellow Bar Markings at Roundabouts*, Circular Roads No. 17/78, London, The Department of Transport, 1978.

(59) Ministry of Transport, *Roads in Urban Areas*, London, HMSO, 1966, as corrected in Technical Memorandum H12/73, October 1973.

(60) *A Policy on Design of Urban Highways and Arterial Streets*, Washington DC, The American Association of State Highway Officials, 1973.

(61) Hewitt RH, Road transition curves for an accelerating vehicle, *Journal of the Institution of Highway Engineers*, 1971, **18**, No. 3, pp. 7–16.

(62) Simpson D, Major motorway interchanges: a review of existing sites, the new DOE recommendations and possible future trends, in the seminar as reference 2, pp. 173–191.

(63) *Design of Rural Motorway to Motorway Interchanges: General Guidelines*, Technical Memorandum H6/75, London, The Department of the Environment, April 1975.

(64) *Layout of Grade Separated Junctions*, Advice Note TA48/86, London, The Department of Transport, March 1986.

(65) Williams OT, Some considerations in the design and operation of multi-level interchanges, *The Highway Engineer*, 1974, **21**, No. 5, pp. 12–20.

(66) Baker AB, The use of the computer in highway design: the state of the art, in the seminar as reference 42, pp. 279–288.

(67) Robinson A, Comparison of the latest British highway design suites, BIPS3—HIGHWAYS 12—MOSS, in the seminar as reference 30, pp. 59–73.

7
Road safety and traffic management

This chapter is concerned primarily with measures to improve the safe and efficient movement of pedestrian and vehicular traffic, subject to the constraints of environmental preservation and public acceptability. Road safety and traffic management works are normally fairly low in capital cost, and give high rates of return in relation to the benefits gained from their implementation.

The types of activity that may be considered as falling within the scope of this chapter are literally vast. For example, they may range from publicity campaigns aimed at influencing motorists and pedestrians to behave more responsibly, to the installation of complex area-wide traffic control schemes designed to reduce central area congestion and improve traffic flow outside the central area, to the painting of white lines on the carriageway at a junction to ensure a safer turning movement. The intention in this chapter is to present some of the more important of these techniques, and to discuss in detail the basic factors pertinent to their usage.

Traffic management policies and strategies are not considered in this chapter—they are discussed in Chapter 3. Similarly, traffic management measures dealing with the control of standing vehicles are dealt with in Chapter 4.

Accident statistics and trends

The desire to travel in safety is strongly rooted in mankind. Nonetheless, the history of the road is also a history of conflicts and accidents involving man, beasts and, until fairly recently, animal-drawn vehicles. In more recent times, the 'natural hazards' of the past have been replaced by the statistical probability of conflict which arises from the sheer intensity of modern traffic, and from the disparate nature of the foot and wheeled traffic using today's carriageways.

Nationally and internationally, it can be said that deaths and injuries from accidents on the highway have now reached epidemic proportions. In the whole world, it is estimated that 0.25m deaths and over 10m injuries occur as a result of road accidents every year[1]. Road accidents are now the greatest cause of death for young people in the age group 15–25 years.

In Britain, the risk of a person being involved in an injury accident is once in 57 years, in a fatal accident once in 2500 years, and an accident not involving injury about once in nine years. The relative infrequency of these accidents explains why people in general tend to accept the annual road toll as being part of their way of life, each thinking 'it will not happen to me'—but it can, and does, to many, e.g. in 1983, there were 5618 people killed and 310 679 injured on Britain's roads[2].

Accidents are a major drain on the national purse. In 1983, the cost to the economy of road accidents in Britain was estimated to be about £2380m. The average cost per fatal accident in 1983 was £167 160, whilst serious injury, slight injury and damage-only accident costs averaged £9455, £1240 and £515, respectively. The average costs per casualty in 1983 for fatal, seriously injured and slightly injured casualties were £150 045, £6950 and £170, respectively.

International comparisons

When the numbers of road fatalities, motor vehicles and populations of different countries are compared, the following common features emerge.

(1) General trends are not really that much different in different countries, so that the relationship between road fatalities, licensed motor vehicles and population can be generally expressed by the following empirical formula:

$$\text{fatalities} = 0.0003[(\text{vehicles})(\text{population})]^{1/3}$$

This formula, which is based on results[3] from 68 different countries, suggests that each country has an 'inbuilt' toleration level with respect to what is acceptable regarding road accidents, particularly fatalities. In other words, road users become more careful and authorities more stringent as the dangers on the highway increase.

(2) As the number of vehicles increases, the number of motorist fatalities and casualties increases much faster than the number of pedestrian casualties.

(3) The ratio of single vehicle accidents to total accidents decreases as the motor vehicle population becomes greater.

(4) The ratio of slight to total casualties increases when the number of registered vehicles increases.

Table 7.1, which compares death rates for Britain with those for the USA and three Common Market countries of roughly similar populations and degrees of motorization, shows that conditions in Britain compare relatively well with those in some other countries.

Notwithstanding the above, absolute comparisons of fatality and casualty rates in different countries must be treated with considerable care, as they can contain results arising from such diverse factors as differing traffic compositions, variations regarding the proportion of travel occurring in built-up areas, and different qualities of street lighting, highway standards, vehicle legislation, etc. Furthermore, the accident reporting procedures can be very different, e.g. in Britain, a road fatality is described as being due to a road accident if death occurs within thirty

Table 7.1   International comparison of road accident fatality rates in 1982[2]

Country	Vehicles per 100 population	Road deaths per 10⁵ population	Car user deaths per 10⁸ car-km	Pedestrian deaths per 10⁵ population
Britain	36	11	1.1	3.4
France*	48	25	3.0	4.4
Germany	46	19	1.9	4.2
Italy*	43	15	2.0	3.0
USA	71	19	1.5	3.1

*Data adjusted to allow for differing definitions of the term 'fatality'

days; in Italy, seven days; in France, six days; in Austria, three days; in Japan, one day; whilst, in Portugal, a road fatality is only considered to occur at the scene of the accident or immediately afterwards.

Trends in Britain

Figure 7.1 summarizes the scale and extent of the accident and casualty problems in Britain since 1926, in relation to the growth of road usage by traffic. As is clearly demonstrated by these data, the years 1965 and 1966 were watershed years in respect of the containment of the road accident and casualty situation. Much of the credit for this can be attributed to various education, enforcement, and engineering measures, although trends have also been influenced by factors beyond the control of road safety workers. Particular legislative decisions which have probably contributed to the change are given in Table 7.2.

Table 7.2   Legislative measures which have helped to contain the growth in accident and casualty numbers

Year	Measure
1965	112 km/h speed limit imposed on all motorways and other previously unrestricted roads
1967	Fitting of seat belts made compulsory for all new cars
	It became an offence to drink and attempt to drive with over 80 mg of alcohol per 100 ml of blood
1968	Vehicle tyres must have 1 mm of tread
1969	Three-year-old vehicles required to obtain a test certificate before being relicensed
1971	Sixteen-year-olds banned from driving motor cycles, scooters or three-wheelers
1973	Safety helmets made compulsory for users of two-wheeled motor vehicles
1975	Statutory duty to promote road safety placed on local authorities in England and Wales
1978	High-intensity rear fog lamps made a mandatory fitment to most vehicles manufactured from 1 October 1979
1983	Wearing of seat belts made compulsory for front-seat occupants of cars and light vans

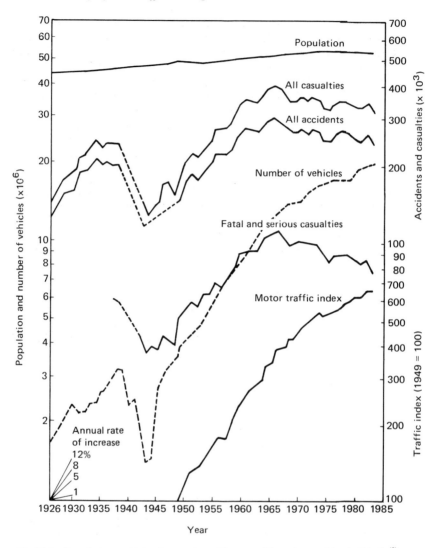

Fig. 7.1   Population, vehicles licensed, accidents, traffic and casualties in Britain[2]

The total number of deaths on the British highway system has remained relatively stable over the past half-century. With the exception of the abnormal World-War-II-affected period, the number of people killed in road accidents each year since 1931 has generally been within the range 5000–7000. No matter what measure one uses, this clearly indicates that the roads have become safer in respect of fatalities. This should not be viewed with complacency, however; whilst much of the credit for the containment can be attributed to better roads and improved safety measures, a significant part is also most likely due to the higher standards of medical care which have been developed and applied to accident cases over the years, e.g.

the proportion of deaths ascribed to fracture of the skull has decreased significantly since the mid-1960s[4]. Another factor which has probably contributed to this containment is the relative lessening in the numbers of pedestrians and cyclists using the carriageways.

In 1977, nearly 266 000 road accidents involving 34 800 casualties were reported to the police; Table 7.3 shows where these occurred and who were involved. Table 7.4 summarizes the types of collisions involved in the 239 400 accidents which occurred in 1983, whilst Table 7.5 expresses the conditions under which these accidents occurred. Analysis of the data behind these statistics reveals the following results (which tend to be consistently repeated from year to year).

(1)   *Road users injured*—over one-half of those killed in urban areas are pedestrians and nearly three-fifths of those killed in rural areas are car occupants; in total, one-fifth of all casualties are pedestrians, two-fifths are cyclists, and two-fifths are car occupants.

(2)   *Environment*—three-quarters of all accidents are in urban areas, two-thirds of urban accidents are at junctions, one-third of all accidents occur in darkness, one-third of all accidents happen on wet roads, and relatively few accidents occur on motorways.

(3)   *Vehicles and collisions*—three-quarters of all accidents involve cars, nearly one-third of urban accidents involve a pedestrian and only one vehicle, one-third of rural accidents involve only one vehicle, and about 9 per cent involve three or more vehicles.

Table 7.3   Accidents in 1977[5]: (a) where they occurred (as a percentage of the total number of accidents) and (b) who were involved

Location	Urban areas (48, 64 km/h)				Rural areas (80–112 km/h)			
	M roads	A+A(M) roads	B roads	Other roads	M roads	A+A(M) roads	B roads	Other roads
At junctions (within 18.25 m)	<0.1	25.6	5.9	19.4	0.1	4.8	1.0	1.5
Elsewhere	<0.1	10.9	2.9	12.9	1.2	7.6	2.0	4.1
Total	<0.1	36.5	8.9	32.3	1.3	12.4	3.0	5.6

(a)

Casualty	Urban areas (48, 64 km/h)		Rural areas (80–112 km/h)	
	Killed (%)	All casualties (%)	Killed (%)	All casualties (%)
Pedestrian	52.6	26.7	12.8	3.7
Cyclist (pedal or motor)	24.1	30.4	20.3	19.0
Car occupant	19.6	34.9	58.7	66.6
Goods vehicle occupant	2.1	3.2	6.3	8.1
Public service vehicle occupant	1.0	4.3	1.0	1.7
Others	0.6	0.4	1.0	0.9

(b)

Table 7.4 Percentage of accidents in 1983 involving particular types of collision[2]

Collision type	Urban areas (48, 64 km/h)	Rural areas (80–112 km/h)
1 vehicle + pedestrian	28.5	4.5
1 vehicle only	13.2	33.1
2 vehicles	51.9	51.7
3 or more vehicles	6.4	10.7

Table 7.5 Percentage of accidents in 1983 taking place under particular road surface conditions[2]

Visibility condition	Dry	Wet	Snow/ice
Daylight	51.7	17.8	0.9
Darkness	17.0	11.8	0.8

Pedestrians

After remaining constant throughout the 1950s, the number of pedestrian fatalities took a relatively sharp upward turn in the 1960s, e.g. 3153 in 1966 versus 2225 in 1957. After remaining at a plateau of about 3000 per annum through the late 1960s and early 1970s, the number of pedestrian deaths declined again in the 1970s, to 1914 in 1983. In each year of the period since 1968, about 20 per cent of all road casualties have been pedestrians; the proportion of fatalities that were pedestrians dropped from 39 per cent to about 35 per cent over this period.

Analysis of the pedestrian casualty statistics shows that, typically, about two-fifths are children ($\leqslant 15$ years) and one-quarter are adults of retirement age ($\geqslant 65$ years). Table 7.6 shows a similar trend, and pinpoints the particularly inexperienced 5–9-year-olds (when they begin to use the roadway alone) as being the most vulnerable of the young pedestrians, and the 70+ ages (when agility, judgement, and recuperative powers are reduced) as being the most vulnerable for older people.

Table 7.6 Fatal and seriously injured casualties per 10^5 population, by age and class of road user, for 1983[2]

Age group	Pedestrians	Pedal cyclists	Two-wheeled motor vehicle users	Car users	Other vehicle users
0–4	30	1	—	9	—
5–9	81	18	—	12	1
10–14	73	41	3	13	3
15–19	45	30	216	103	8
20–29	24	11	82	102	10
30–39	15	6	21	54	7
40–49	15	6	16	44	6
50–59	20	6	11	38	6
60–69	28	5	5	34	3
70+	55	3	2	28	6

Pedal cyclists

The number of pedal cyclist fatalities fell steadily between the mid-1930s and the mid-1970s. Furthermore, in the decade immediately prior to 1974 (the year of the 'energy crisis'), the total number of cyclist casualties also fell steadily. Much of the reason for these falls was undoubtedly the decline in the use of the bicycle.

Coincident with the decline in the use of the bicycle as a mode of transport, and the consequent fall in the number of casualties, there was a rise in the risk to which the pedal cyclist was exposed as a result of increasing motor traffic. Thus, whilst the statistics for 1959 and 1974 show a drop of about thirty thousand in the *number* of cyclists injured on the roads, the injury *rate* rose from 383 to 492 per 10^8 km (an increase of 28.5 per cent) over the same period. In other words, it became more risky to ride a bicycle on the highway.

Following the 1973–74 energy crisis, pedal cycle usage began to grow again for a while. Cyclist casualties grew also, by about 11 per cent in both 1975 and 1976. Between 1976 and 1983, casualties then grew by nearly a further one-third, to 31 022. In December 1983, 9–10m bicycles were estimated to be in use in Britain[2].

In 1978, children represented 22 per cent of the population but accounted for 40 per cent of the cyclist casualties. Table 7.6 shows that the 10–14 age group (when inexperienced cyclists start to use the roads) is the most vulnerable. Further evidence suggests that the casualty rate for children is substantially higher per unit distance travelled than the casualty rate for adults; the same data also suggest that boys have a higher casualty rate per unit distance travelled than girls.

Motor cycles

Prior to the 1973–74 period, usage of two-wheeled motor vehicles had been declining, but with the energy crisis it began to rise again relatively dramatically, i.e. from 3.75×10^9 veh-km in 1972 to 7.97×10^9 veh-km in 1982[6]. Between 1972 and 1982, the number of motor cycle user casualties also grew by 65 per cent.

Some 21 per cent of the road fatalities in Britain in 1982 were motor cyclists or their passengers; most were young men. As Table 7.6 shows, it is the inexperienced 15–19-year-old motor cyclists who are most vulnerable.

Most motor cyclists who are killed receive head injuries. The risk of death or serious injury due to head injury alone is 20–30 per cent less for those wearing a safety helmet than for those without. Full-face helmets reduce the chance of head injury as compared with open-face helmets.

Consistently, about one-half of the motorized two-wheeler casualties occur in accidents involving cars, whilst the casualty proportion involving no other vehicle is approximately one-quarter[7]. In the overwhelming majority of accidents involving a car and a motor cycle, the car is at fault[8]; it appears that road users as a whole have a perceptual set towards larger vehicles and in a very real sense simply do not 'see' a motor cycle on the carriageway. About one-quarter of the single vehicle accidents involve collisions with parked, unattended cars, and another one-third happen while negotiating bends or corners at junctions; the rest are an assortment of incidents such as hitting kerbs, losing control on bumpy/slippery roads (especially roadworks), and animals running into the road[9].

There is a disproportionately high rate of involvement in fatal and serious accidents by motor-cycle-type vehicles. This is reflected in the casualty statistics in Table 7.7.

Table 7.7   Casualty rates per 10^8 veh-km, by class of road user and severity, for 1983[2]

Casualty	Killed	All severities
Pedal cyclists	6.3	595
Two-wheeled motor vehicle riders	13.6	932
Car drivers	0.5	33
Bus/coach drivers	0.1	15
Goods vehicle drivers		
light	0.3	20
heavy	0.2	12

Goods vehicles

In 1969, the number of goods vehicle drivers and their passengers who were accident casualties were 14 939 (256 fatalities) and 9109 (149 fatalities), respectively; by 1983, the corresponding numbers were reduced to 6751 (107) and 3127 (54), respectively. Over the same period, the casualty rates for goods vehicle drivers were reduced from 52 to 20 per 10^8 km for light goods vehicle drivers (< 1.52 t), and from 29 to 12 per 10^8 km for heavy goods vehicle drivers (> 1.52 t), whilst the fatality rates declined from 0.7 to 0.3 per 10^8 km for light goods vehicle drivers and from 0.7 to 0.2 per 10^8 for heavy goods vehicle drivers.

Many factors have contributed to these risk reductions, but for heavy goods vehicles they must include restrictions in the hours of driving, the licensing and testing of drivers to a higher standard, and the thorough annual testing of vehicles, as well as improved vehicle safety features.

Heavy goods vehicles make a relatively disproportionate contribution to most categories of road user casualty. Indeed, for pedestrians and all occupants of smaller vehicles, the probability of an injury being fatal or serious in an accident involving a heavy goods vehicle is very much greater, e.g. although only about 5 per cent of cyclists who are seriously injured are hit by heavy goods vehicles, about one-fifth of their fatalities occur in this type of accident.

Public service vehicles

Both the accident rate and the numbers of PSV user casualties and fatalities for buses and coaches have remained relatively constant with a slight decline throughout the 1970s. What is somewhat disturbing, however, is that public service vehicles appear to play a disproportionately large part in involvement in fatal and serious pedestrian accidents. The reason for this may be related to the fact that bus routes tend to concentrate on high-activity pedestrian areas, e.g. central areas and shopping centres.

The above should not be interpreted as meaning that public service vehicles should not be allowed into pedestrianized areas, however. For example, an investigation[10] into two 'closed' streets which had buses operating amongst pedestrians on kerbless surfaces found that the interaction between pedestrians and slow-moving buses was such that it resulted in a pedestrian-free space ahead of each moving vehicle, which extended ahead much further than the vehicle's stopping point. Furthermore, the statistics suggested that the introduction of kerbless surfacings into accident-free streets previously used only by buses and pedestrians did not make the accident situation any worse.

Cars and taxis
In 1983, occupants of cars and taxis accounted for about 37 per cent of the fatalities and 42 per cent of the casualties incurred on the roadway. As with pedal cyclists and motor cyclists, it is the inexperienced users of motor cars who suffer the peak casualty rates (see Table 7.6). Most car-occupant fatalities and casualties occur on rural roads.

Factors contributing to accidents

The above statistics and trends describe the scale of the accident problem, but in general do little to convey its nature. In contrast, the main contributory factors to accidents as determined in a major on-the-spot analysis of 2042 accidents are summarized in Fig. 7.2. In total, there were 2.3 contributory factors per accident. These factors were distributed such that the human element contributed to 95.0 per cent of all accidents, road factors to 28.0 per cent, and vehicle factors to 8.5 per cent. The road user was the sole contributor in 65.0 per cent of accidents; in contrast, road and vehicle factors were usually linked with a road user factor.

The main road user factors are as follows:

(1)　perceptual errors, e.g. driver or pedestrian looks but fails to see, distraction or lack of attention, misjudgement of speed or distance,
(2)　lack of skill, e.g. inexperience, lack of judgement, wrong action or decision,
(3)　manner of execution, e.g. deficiency in actions (i.e. too fast, improper overtaking, failure to look, following too closely, wrong path), deficiency in behaviour (i.e. irresponsible or reckless, frustrated, aggressive),
(4)　impairment, e.g. alcohol, fatigue, drugs, illness, emotional stress.

As road environment deficiencies are usually associated with driver

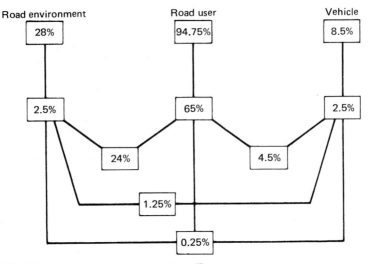

Fig. 7.2   Contributions to road accidents[1]

error, features can be grouped which show similarity in the difficulties they present to the driver, as follows:

(1)   adverse road design, e.g. unsuitable layout and junction design, poor visibility due to layout,
(2)   adverse environment, e.g. slippery road, flooded surface, lack of maintenance, weather conditions,
(3)   inadequate road furniture or markings, e.g. insufficient and/or unclear road signs and road markings, poor street lighting,
(4)   unexpected obstructions, e.g. roadworks, parked vehicles, other objects.

The vehicle defects which make a major contribution to accidents are mainly of the kind which can develop in a relatively short space of time due to the lack of regular maintenance by the user of the vehicle. Defective tyres and brakes are prominent in this list of contributory defects.

Another detailed analysis[11] of pedestrian accident data in terms of the relative involvement of drivers and pedestrians has also given the information in Table 7.8 regarding the relative blameworthiness of these two classes of road user.

Table 7.8   Who are at fault in pedestrian accidents[11]

Degree of involvement	Drivers (%)	Pedestrians (%)
Primarily at fault	41	65
Partially at fault	19	14
No blame allotted	40	21

Road user characteristics

The most complex and least understood element of any road problem is the human one. In the light of the very great contribution to road accidents which results from human failings, it is useful to look at some of the characteristics that are often considered to be important in such accidents.

Vision

Good vision is a prerequisite for safe driving, as it accelerates the process of perception–reaction to traffic situations. The present vision test, which is a prerequisite for a British private vehicle driving licence, is simply one of static visual acuity, whereby a motorist must demonstrate an ability to read correctly in good daylight a clean vehicle registration plate with 79 mm high letters at a distance of 20.4 m (or at 22.9 m for characters of 89 mm in height). However, this vision test can be considered inadequate, since it is a test for visual acuity only at the time of the initial test, and reveals nothing about other important visual factors such as depth perception, field of vision, night blindness and colour blindness.

Visual acuity is the ability to focus quickly and to see clearly without a blur. The Department of Transport's number plate test is equivalent to an acuity standard, using the Snellen notation, of between 6/12 and 6/9. (With the Snellen notation, the greater the denominator, the larger is the physical size of the test object and the poorer is the acuity.) Given that an acuity of 6/6 is usually referred to as a 'normal' acuity, a 6/12 acuity means that people are only able to read at 6 m letters on a standard test chart that should normally be read at 12 m. Field experiments have shown that a standard traffic sign with 127 mm high letters can be read at about 85.5 m by a driver with 6/6 vision, whilst drivers with 6/12 vision must approach to within 34.5 m of the sign in order to read it.

Only a minuscule percentage (perhaps 0.1 per cent) of drivers fail the Department of Transport's visual acuity test at the time of the driving test[12]. A study[13] of 1368 drivers carried out at 25 sites found, however, that between 1 and 3 per cent of all drivers should fail the vision test at any given moment. This latter study also found that acuity declined with increasing age, particularly when the age of 40–45 years was exceeded, e.g. 5 and 10 per cent, respectively, of drivers aged 65 years or over failed to meet the criteria of 6/12 and 6/9. Since evidence exists of a weak association between acuity and accidents for drivers over 55 years, it would appear that a case could be made for recommending that middle-aged and elderly drivers should be more strongly urged to maintain a good level of acuity.

As well as having good visual acuity, a driver should also have good *depth perception*. This visual skill requires good teamwork by both eyes to enable the driver to judge relative distances and to locate objects correctly in space. Many overtaking accidents might possibly be avoided if drivers realized and accepted that their depth of perception capabilities were weak, and that they needed to take extra care when overtaking other vehicles.

A third important visual faculty is the *field of vision*. Although a person's most acute vision is subtended by a 3-degree cone, it is still fairly satisfactory up to about 20 degrees. This is why it is considered good practice to locate traffic signs within a 10-degree cone, which is the level beyond which the visual acuity for legibility rapidly decreases; this is roughly the area covered by the width of the hand when held at arm's length. Although good acute vision is limited to about 20 degrees, most drivers have sufficient peripheral vision 'to enable them to perceive objects contained in a cone between 120 and 160 degrees, which is why the attentive driver whose eyes are focused on the road ahead is able to notice and avoid side hazards.

Some drivers have what might be termed 'tunnel vision', i.e. vision that is most acute in the direction straight ahead, but little or nothing is noticed to the right and left or up and down. A driver who has tunnel vision, and neither knows of it nor realizes its importance, is a potential danger to other road users. Such drivers are particularly dangerous on high-speed highways where overtaking manoeuvres are continually being carried out, and on busy city streets where vehicles are parking and unparking, and pedestrians are likely to step unexpectedly onto the carriageway. Early detection of tunnel vision need not necessarily lead to the disqualification of a driver; rather the driver should be trained to move his or her head continually when driving in order to overcome this liability.

A driver should also have good *night vision*. This demands three important visual skills, as follows.

(1) *The ability to see efficiently under low illumination levels* The recognition of detail is easiest when the illumination level is that which is available in ordinary daylight; this unfortunately normally cannot be provided at night-time. In addition, many drivers also suffer from 'night blindness' in that dusk blots out for them as much as does late darkness for others.

(2) *The ability to see against headlight glare* There is considerable variation in the degree of 'glare-out' experienced by individuals subjected to the same amount of glaring light, and in the time required for the recovery of sensitivity. Normally, the eye adapts itself quickly when going from darkness to light, but takes longer (up to 6 seconds) to adjust from light to darkness, e.g. when a vehicle enters an unlit tunnel or after a driver is dazzled by opposing headlights. Older persons are most susceptible to glare-out, as are people with poor visual acuity.

From a preventative aspect, it should be noted that the angle between the line of vision and the glare source is of the utmost importance in respect of glare-out. When the glaring light is only 1 degree to the side of the line of vision, the effect is three times that when the angle is 5 degrees.

(3) *The ability to distinguish between various colours, particularly between green and red* Although it has not been possible to establish a

relationship between colour blindness and accidents, the heavy reliance placed upon colour in traffic signs and signals emphasizes the importance of this visual skill. One important step taken to alleviate this problem has been to standardize the positions of the different coloured lamps in traffic signals.

Hearing

There appears to be no relationship between poor hearing and driving accidents. In fact, American data suggest that most deaf mutes are very careful drivers. Hearing sensitivity is apparently more closely related to vehicle operating practices, e.g. judging when to change gears, than to road safety.

Sex

It is often suggested that men are better or worse drivers than women, but until relatively recently there was little firm evidence in the British accident literature to support or disprove this hypothesis. Between March 1970 and February 1974, however, the Transport and Road Research Laboratory carried out an in-depth on-the-spot accident study[14] which examined the driving behaviour of 2654 car drivers involved in 2036 accidents, and the resulting human errors which led to those accidents.

Whilst the study revealed some interesting variations in the sorts of errors made and the types of accidents which occurred, there was no statistically significant difference in the level of responsibility for accidents in which they were involved between men and women. In the case of male drivers, 60 per cent of them were judged as being primarily or partially to blame for all of the accidents in which they were involved, whereas the corresponding figure for female car drivers was 56 per cent. Similar results were obtained when these accidents were broken down into two-car and one-car accidents.

Overall, the analysis of the data, in relation to the making of errors leading to accidents, suggests that males drive faster than females, and are more ready to drink and drive. The female is less experienced in the art of driving, is more easily distracted, sometimes fails to look before taking action, e.g. at junctions, and even when she does look, often does not see the hazards.

Effect of biorhythms

According to the biorhythm theory there are three predictable cycles which apply to all human beings and govern their physical, emotional and intellectual well-being. These are a 23-day physical cycle, a 28-day emotional or sensitivity cycle, and a 33-day intellectual cycle; they begin at the moment of birth and never vary in length. It is often said that there are 'critical' days at the beginning and middle of each cycle for physical, emotional or intellectual abilities, and drivers are purported to have a higher risk of road accidents on those particular days. The risk is said to increase further when critical days in two cycles coincide, and to be greatest when three such days coincide.

A comprehensive statistical examination of insurance company accident data[15] involving 112 560 drivers and many possible groups of critical days has shown, however, that there are no grounds for believing that there is an increase in highway accident risk on supposed 'critical' days in any of the biorhythm cycles.

Effect of alcoholic drinks

The effect of alcohol is that of a general anaesthetic. It is a drug which depresses the central nervous system, affecting first the brain and then the spinal cord. As a result, perceptions are blunted, coordination is impaired and, in particular, the driver's power to evaluate his or her own performance is blunted. The driver so affected does not compensate sufficiently for his or her slower mental processes and tends to take more risks, e.g. by carrying out ill-judged overtaking manoeuvres.

Even moderate amounts of alcohol prolong the reaction times to the sensations of sight, touch and hearing. This is of considerable importance to the driver, since increasing the reaction time significantly increases the distance required to stop a vehicle. Alcohol also impairs night vision, as it increases the time required for the eyes to adjust from light to darkness.

Upon intake, alcohol is first absorbed into the bloodstream and then distributed through all the fluids and tissues of the body in proportion to their water contents. The level of intoxication attained by any individual then depends upon the concentration of alcohol in the brain and central nervous system—not just upon the total amount of alcohol taken into the stomach—and this is closely related to the alcohol concentration in the blood. This latter concentration is dependent upon the rate at which alcohol is absorbed into the blood and the rate at which it is burnt up by the body or excreted in the urine, breath or perspiration.

Many factors affect the rate of absorption and elimination of alcohol from the body, and these cause variations not only between individuals but also in the same person at different times. The rate of absorption into the bloodstream depends upon the individual's constitution, the time since the last meal, the food consumed, and how quickly the alcoholic beverage is taken—beer and light wines are absorbed less quickly than spirits, and all drink is absorbed more slowly if taken on a full stomach or with food. Alcohol is eliminated from the blood at a more or less standard hourly rate of about 15 mg of alcohol per 100 ml of blood; again, however, there are variations between individuals, and the elimination rate tends to be faster if the initial concentration is high. Tests have shown that 0.5 litre of typical British beer will take 2–3 hours to disappear from the blood, and 4 litres will take about 9 hours. If the latter content of alcohol is taken in spirit form, it may take 15 hours to disappear.

It is not possible to state how much any particular person needs to drink before becoming 'incapable' of driving. It is possible, however, to give the minimum amounts that the average person must drink in order to attain a certain alcohol concentration in the blood. These data, shown in Table 7.9, are based on tables published by the British Medical Association[16].

Table 7.9 Minimum intake of alcohol in the form of beer or whisky needed to raise the blood–alcohol level in a 70 kg man to a given concentration

Blood–alcohol concentration (mg/100 ml)	Beer or stout (pints)	Single whiskys
20	$\frac{1}{2}$	1
40	1	2
55	$1\frac{1}{2}$	3
75	2	4
100	3	6
150	4	8

The accident risks associated with certain concentrations of alcohol are suggested in Table 7.10 which lists some of the results of three comprehensive investigations into the problem associated with drinking and driving. While the results obtained in these studies are comparatively different, they all show quite clearly that it is considerably more dangerous to drive after drinking heavily.

The most comprehensive of these studies (at Grand Rapids) not only found that drivers were significantly more likely to be involved in accidents by the time they reached 80 mg/100 ml, but also that the accident involvement curve rose so steeply after the 80 mg/100 ml level that the risk was 10 and 20 times greater than normal at 150 and 200 mg/100 ml, respectively. The Grand Rapids study also concluded that the increase in accident risk associated with high alcohol levels is greater for young and elderly drivers than for middle-aged ones, whilst at low alcohol levels it is greater for those who drink rarely than for regular drinkers.

A later British study[18] found that of the motor vehicle drivers killed in 1976 who had blood-alcohol readings in excess of 80 mg/100 ml, 30 per cent were aged between 16 and 19 years, 46 per cent were aged between 20 and 39 years, and the remaining 24 per cent were aged 40 years or more. Also, in the peak danger period about midnight[19], 70 per cent of pedestrian fatalities had readings in excess of 80 mg/100 ml, whilst 50 per cent were above 150 mg/100 ml; in addition, the risk of an accident increased with age, being highest for those aged 70 years or more.

Table 7.10 Relative risks to drivers associated with drinking alcohol[17]

Blood–alcohol concentration (mg/100 ml)	Location of study		
	Toronto, Canada	Grand Rapids, USA	Bratislava, Czechoslovakia
30	1.0	1.0	1.0
30–99	1.5	1.4	7.0
100–149	3.0	6.0	31.0
>150	10.0	15.0	128.0

Drugs

In recent years, concern has begun to be expressed regarding the effects that drugs other than alcohol may have upon driver performance. However, the variety of drugs and the diversity of their effects upon different individuals are so great that their identification and quantification in the human body is inordinately complex. Consequently, there is very little information available about the drug–road-accident relationship.

The British study to which reference is made above[18] also concluded, firstly, that the numbers of men and women who take therapeutic drugs before driving are sufficiently substantial to suggest that the possibility of a significant relationship should not be dismissed without further investigation and, secondly, that the drugs most commonly used (in order of usage) are sedatives/hypnotics, tranquillizers and analgesics. In the same study, in an accident sample of 1216 driver casualties, 27 per cent of the men and 17 per cent of the women who admitted taking drugs before their accidents, also admitted taking alcohol within the 6-hour period prior to the accident.

Accident proneness

It is commonly held that some groups of drivers are more prone to accidents than others, and many papers have been published giving the results of studies into this assumption. For example, it is now well established that drivers under 30 years are very much more likely to have accidents than drivers over 30 years, that women are less likely than men to have certain types of accident, that middle-class drivers are less likely than working-class drivers to suffer accidents, and that poorly-trained drivers are more likely to do so than highly-trained ones.

Evidence would also now appear to be gaining strength to the effect that there is another subset of drivers characterized by certain personality features which are supposed to make them likely to drive more dangerously, and consequently more likely to be involved in accidents. One of the seminal studies[20] in this area has suggested, for example, that accident repeaters react against authority, have aggressive tendencies, and may be irresponsible and socially maladjusted. More recent work[21] has indicated that male drink–driving offenders tend to be under 50 years; single, divorced or separated; in semi-skilled or unskilled manual socio-economic groups; unemployed; consume in excess of 180 g of alcohol (about 12 pints of beer) per week, and have previous motoring convictions.

An excellent research report, including a review of the literature, on this complex study area is readily available in the literature[22].

Fatigue

Obviously, a tired driver has longer perception–reaction times and is more likely to commit an error of judgement on the roadway. For example, long-distance truck drivers who drive less than 55 hours per week are less likely to be involved in traffic accidents than are those who drive longer hours, especially in the early hours of the morning[23]. What is not so clear, however, is the effect of fatigue caused by such features as poor living conditions and extreme anxieties. For example, two large insurance companies in the USA reported in 1949 that about 60 per cent of all long-distance lorry accidents happened in the first 3.5 hours of driving. The US National Safety Council reported similarly that most driver-asleep accidents happened after the driver was at the wheel for only a few hours.

For least tiring driving, the motorist should take a rest after, at the very most, one hour's driving. Drinking coffee helps to ward off drowsiness and fatigue; this is substantiated by laboratory tests which indicate that caffeine slightly speeds up reaction times.

Road environment: safety considerations

Whilst the highway environment may be the sole contributor to only about 2.5 per cent of accidents, when put together with road user errors, it accounts for another 24 per cent. It is useful, therefore, to consider some of the road-related factors which may be considered to influence the accident situation.

Highway type

Relative risks associated with the road environment are reflected in Table 5.15 and Table 7.11 below. From these data, it can be seen that a motorist is much more likely to be involved in an accident when travelling on town streets than on rural roads. This can be attributed mainly to the greater numbers of diversions and decisions to which the urban driver is subjected, e.g. at junctions and/or due to the presence of pedestrians and cyclists.

The fatality rates for urban and rural roads appear not to be greatly dissimilar at first sight. However, if the pedestrian and cyclist statistics are removed from the urban data, then it can be deduced that the motor vehicle fatality rate for rural roads is much higher than that for urban roads. This is due primarily to the higher speeds which occur on rural highways, so that when accidents occur they have a greater severity. Whilst the accident rate on rural roads has been significantly improved over the past twenty years in particular, there are still numerous locations that are inadequate

Table 7.11 Casualty rates on different types of highway[1]

Highway type	Casualties per 10^8 veh-km	
	Fatal	All injuries
Motorways	0.84	26
Rural roads (80–112 km/h)		
A-class	2.90	79
others	1.90	89
Urban roads (48, 64 km/h)		
A-class	3.10	191
others	2.50	192

from a safety viewpoint. Indeed, it would seem that drivers are prepared to accept short-term increases in risk levels at sites with adverse design conditions, e.g. bends with reduced visibility and hump-backed bridges, rather than reduce their speeds to safer levels[24].

Motorways have both the lowest total injury and fatality rates. Again, this can be attributed to the high standards of design of these facilities, and to the relative lack of interference from other vehicles crossing the highway, pedestrians, cyclists, etc.

Intersections

Two-thirds of all urban injury accidents and one-third of all rural injury accidents happen within about 18 m of an intersection (see Table 7.3). Consequently, one of the most fruitful applications of highway traffic engineering lies in the improvement of road junctions. Very often, minor improvements can be carried out which will reduce accidents and improve road safety beyond all proportion to their capital cost.

Given that there are some 5.77 and 1.19 intersections per kilometre of urban and rural road, respectively, in Britain[25] and that there are about 330 000 and 130 000 intersections in urban and rural areas, respectively[26], the scale of the correction problem appears at first to be very daunting. However, by carrying out proper accident investigation procedures, troublesome sites can be fairly readily identified and appropriate correction measures applied. The decision as to what remedy to use is aided by an understanding of certain accident relationships at junctions (see also Chapter 6).

Analyses[27, 28] of accidents at *three-way intersections* have indicated the following basic points.

(1)   Right-turn movements are much more dangerous than are left-turn movements. Figure 7.3 shows that 25 per cent of the accidents studied were the result of collisions between vehicles turning right from the main road and vehicles travelling in the same direction on the main road. Seventeen per cent involved cutting vehicles which entered the main highway in order to turn right, and came into collision with main road vehicles travelling in the opposite direction. This emphasizes the desirability of using channelization techniques to segregate vehicles entering and leaving the main through highway. Thus, for example, the painting of simple hatched 'ghost-island' markings on wide single carriageway roads in rural areas, to allow vehicles to wait whilst making a right-turn, has been shown to result in accident reductions of about 40 per cent[29].

(2)   Slight right-hand splayed junctions (see Fig. 7.4) are less likely to have accidents than are either left-hand splays or a square layout. A slight right-hand splay layout enables a right-turn to be made easily *from* the main road without the vehicle necessarily having to slow down, and it also forces drivers about to turn right *onto* the main road to slow down on the minor road.

(3)   The frequency of accidents is proportional to the square root of the

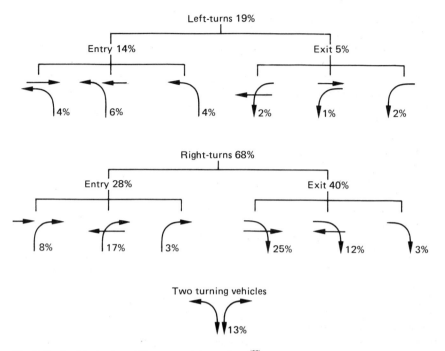

Fig. 7.3 Accidents at rural three-way intersections[27]

flows at the junction and not, as might be expected, to either the sum or the product of the flows. This bears out what traffic engineers have intuitively surmised for some time—that motorists are more careful at busy intersections. This relationship can be used in the proper circumstances to justify removing access or restricting turning movements onto particular roads.

The formulae relating the number of accidents to traffic flow at a simple three-way junction are

$$A_r = \frac{4.5(q_r Q)^{1/2}}{10\,000}$$

and $$A_1 = \frac{7.5(q_1 Q)^{1/2}}{10\,000}$$

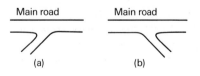

Fig. 7.4 Splay junctions: (a) left-hand and (b) right-hand

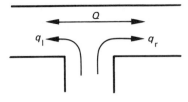

Fig. 7.5   Vehicle streams to which Q, q_r and q_l refer when predicting accidents at three-way junctions

where Q, q_r and q_l are the flows per day in each direction along the major road and about the right and left shoulders of the junction, respectively (see Fig. 7.5). To illustrate their use, consider two identical three-way intersections that are fairly close to each other on a major highway. If both minor roads carry the same amount of traffic, and this is allowed unrestricted access to the major road, then the total number of accidents that can be expected is

$$A_1 = 2 \left(\frac{4.5(q_r Q)^{1/2}}{10\,000} + \frac{7.5(q_l Q)^{1/2}}{10\,000} \right)$$

If, however, one of the minor roads is refused access, and its traffic is diverted to the other minor road, then the number of expected accidents is

$$A_2 = \frac{4.5(2q_r Q)^{1/2}}{10\,000} + \frac{7.5(2q_l Q)^{1/2}}{10\,000}$$

This latter figure A_2 is 30 per cent less than the previously expected A_1.

Another analysis of accidents at urban three-way intersections[30] showed the same general trends as are given above. In this study, however, the most dangerous manoeuvres involved vehicles turning right from the minor road and either merging with the farside stream of traffic or crossing the nearside traffic stream, in that order.

Numerous data are available to show that *four-way junctions* are more dangerous than three-way intersections, e.g. it is reported[31] that cross-road intersections on main radial roads in London are at least four times more dangerous on average than T-junctions. Data such as these explain why it is now common practice to consider changing a straightover four-way intersection to two three-way staggered junctions whenever the accident statistics appear to justify it; average accident reductions of 60 per cent have been achieved by this improvement[29].

As noted previously (see Chapter 6), a right–left stagger—whereby crossing traffic first turns right from the minor road, proceeds along the major road, and then turns left without stopping—is safer than a left–right stagger.

Using GIVE WAY and STOP signs, to ensure that one roadway is given priority over another, can be a very cheap and effective way of reducing accidents at four-way (and three-way) junctions[32]. These signs, particularly when used with channelization islands, also inform the minor road drivers that there is a junction ahead, thereby encouraging them to slow up well in advance. The installation of offset islands on the minor road approaches of rural four-way intersections—whereby the islands are sited on the centrelines of the minor road approach lanes, so that they are not opposite each other, and angled so as to guide the vehicle easily into the intersection—has been shown to reduce accidents on average by 50 per cent[29].

Where properly carried out, the conversion of junctions in urban areas from major/minor priority control to traffic signal or offside-priority (small or mini) roundabout control has resulted in accident reductions averaging 30 per cent or more.

The results of a study of nearly 500 accidents at *conventional weaving roundabouts* in London are given in Table 7.12. The most notable feature of these data is that nearly one-half of the accidents involved only one vehicle. Two-thirds of the vehicle accidents which occurred while entering the roundabout happened at night. Over one-half of the single vehicle accidents involved motor cycles—most of which skidded—and one-seventh involved pedestrians.

These statistics emphasize the need to illuminate all types of roundabout intersections so that they are more visible to the road user at night; the driver, in turn, is then encouraged to slow down prior to the junction. The accident statistics for weaving roundabouts have also been compared with those for priority-controlled roundabouts, and there is ample evidence to show that the introduction of the priority rule reduced accidents (by up to 40 per cent).

Table 7.12 Accidents at conventional weaving roundabouts[33]

Type and location	Accident (%)
Single vehicle accidents	
entering roundabout	22
on roundabout	20
leaving roundabout	7
Two-vehicle accidents	
both entering roundabout	8
one entering and one on roundabout	17
one entering and one leaving roundabout	3
both on roundabout	10
one on and one leaving roundabout	11
both leaving roundabout	2

Accesses

As with intersections, other access points on highways also can be significant contributors to accidents—and for the same reasons.

Whereas a junction or intersection can be defined as a meeting point between two or more highways, an access is most simply defined as any means of entry to (or exit from) a highway from any premises. As such an access can vary in scale of use from the heavily-travelled entrance to a hypermarket, which may have to carry as much traffic as a major intersection, to a rarely used entrance to a field-gate.

For safety reasons, it is highly desirable to reduce to the minimum possible the number of access points onto existing or new heavily-travelled and/or high-speed roads, as well as having only the minimum number of intersections. Why this applies to both urban and rural roads is explained by Fig. 7.6. This figure, which is taken from an American study with a very large database, shows clearly that the total accident rate increases linearly with the number of commercial accesses permitted onto the highway. For example, increasing the number of accesses per kilometre from 5 to 15 on dual carriageway roads resulted in an increase of 36 per cent in the accident rate for rural locales, and 47 per cent for urban locales; the corresponding increases on two-lane highways are 36 and 38 per cent, respectively.

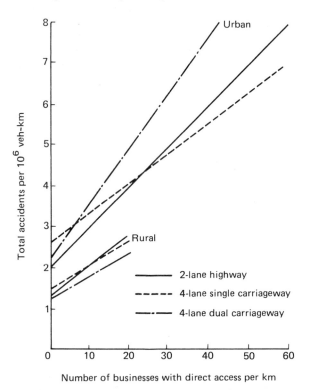

Fig. 7.6 Total accident rate on non-interstate highways, by number of businesses per kilometre (reported in reference 34)

Urban locale

In those cases where a distinction has been made between 'old' and 'new' urban residential areas, it has been found[35] that accident rates (particularly child accident rates) can be higher on road networks in old-type areas, e.g. in Britain, 1.3–1.8 times greater, in Sweden, 2.0–4.0 times higher. This indicates why old residential areas are often priority targets for remedial safety action. However, to improve road safety in these locales normally requires making the whole network safer, and not just some points on the network.

Old urban areas tend to be situated closer to city centres and their streets used for parking by non-residents. The street network is often complex, and does not allow for the efficient control of traffic. In some instances, a major through route may cross the residential area; in others, heavy volumes of through traffic seek to access the central area by filtering through the narrow street system. Old areas often include a mix of land use activities, and frequently tend to be inhabited by older people. Where children are plentiful, the streets often also become playgrounds as green spaces and proper play facilities are normally limited.

Apart from road networks that are a function of their age of development, it has also been suggested that residential area accident patterns vary according to the density of housing and national conditions. Thus, it is reported[35] that one French study found that vehicle accidents at junctions were more numerous in areas with apartment buildings, whilst pedestrian accidents tended to be less frequent. Low-density areas with large plots and curvilinear streets with no on-street parking have been found in the USA to encourage high-speed car-to-car and car-to-obstacle accidents, whilst high-density areas with narrow, grid-pattern streets lined with parking induce pedestrian and bicycle hazards and accidents at junctions.

Seasonal, daily and weather effects

A statistical comparison of casualty totals and of daily figures for motor vehicle travel in 1959 showed that the casualty rates per 10^8 vehicle-kilometres tended to be higher in Winter and on Saturdays. In the same year, the daily number of fatalities varied from 2 to 73, whilst the total daily number of casualties varied from 408 on 18 January to 1618 on 4 July and 2553 on 24 December.

At the individual junction level, there is considerable variation[30] in the hourly counts of conflicts throughout the day with, as might be expected, the number of conflicts being closely related to vehicle flow levels for the same hours.

Although fog is a common event, *thick fog* (visibility < 200 m) occurs in Britain relatively infrequently. Furthermore, an individual outbreak is usually confined to a small area of the country, and at any given location will rarely prevail for more than ten days per year. It occurs least in coastal areas and most in central England where, on the whole, it tends to be fairly infrequent and patchy rather than widespread. Measurements taken during the only four days of thick fog that was nationally widespread in the period 1963–66 found[36] that: (a) traffic flow was reduced by about 20 per cent, (b) accidents in darkness and those involving pedestrians were reduced slightly, and (c) overall, there was no change in the fatal and serious accident rates per unit of traffic, whereas the slight injury and total accident rates increased by about 70 and 50 per cent, respectively.

Fog impairs visibility by means of light absorption and scattering. As a consequence of light scattering, the fog itself becomes illuminated; this produces a veiling effect which is a major cause of variable visibility under different lighting conditions.

American work[37, 38] suggests that fog has mixed effects in that it is a hazard with which drivers attempt to cope, but not all are successful in doing so. Thus, for example, speeds on highways are reduced (but only by 8–13 km/h) and more uniform, i.e. there are decreases in the number of very short and very long headways; these attest to the fact that drivers recognize fog as a hazard which requires increased caution when driving. Accidents which occur in fog have an increased probability of involving either a single vehicle or more than three vehicles. The fatality rate in fog tends to be higher than that in non-fog conditions.

The most hazardous driving situations arising in fog include lane blockage from a stopped or slowly-moving vehicle, lateral tracking errors that result in a vehicle leaving the roadway or compromising safety in adjacent lanes, and driver inability to comprehend traffic signs or signals. Of these, the blocked-lane hazard is generally considered to be the most dangerous.

A measure of the influence of *wet weather* on highway safety can be gained from the fact that the percentage of accidents involving skidding is much higher on wet roads than on dry ones. For example, one study[39] showed that 1 in 3 accidents occurring on wet roads involved skidding, which is more than twice the rate for accidents in the dry.

An analysis of the age and sex of drivers involved in skidding revealed some difference between the sexes, but for both the indications were that, with increasing age and experience, drivers learn to avoid skidding to some extent. Motor cycles are about ten times more likely to be involved in personal-injury skidding accidents than are cars, due to the difficulties that arise when braking in an emergency and to the greater vulnerability of motor cyclists to injury. When the carriageway is wet, cars with greater engine capacities tend to have higher skidding rates. Similarly, there is a slight trend towards higher skidding rates for motor cycles with larger engine capacities—with the notable exception of scooters. Scooters have the highest skidding rate notwithstanding that they have engine capacities of less than 150 cc.

The skid-resistance properties of tyres and road surfaces are ultimately derived from the manner in which tractive forces are generated at the tyre–road interface. It seems well confirmed[40] that frictional forces arise from four independent sources, as follows.

(1)   Adhesive bonds exist between two dry and otherwise uncontaminated surfaces in contact—sliding consists in continually fracturing these bonds.

Under inclement weather conditions, adhesional friction between tyres and roads rarely exists. Water and other contaminant films are attached to rubber and road surfaces virtually everywhere, and are not completely forced out of the contact area in the finite time of contact.

(2)   The tearing of rubber which occurs on very rough surfaces, even when those surfaces are lubricated, expends kinetic energy, producing a friction force of a relatively small magnitude.

(3) The hysteresis or damping loss property of rubber produces a resistance to sliding over rough surfaces. The reason for this is that the rubber is subjected to repeated distortion and deformation as it passes over the rough carriageway surface and, since the road surface does not strain, a sheer stress is developed in the tyre contact-patch. If the sheer stress is high and/or the resistance is low, some slip will occur. (It is this slip mode that is minimized in a stiff-belted tyre with the result of long tread life.)

Many investigators contend that wet rubber friction is mainly due to the hysteresis or damping loss component of friction.

(4) Viscous drag occurs when a liquid fills the gap between two sliding surfaces.

From the above, it can be seen why road surface roughness in general has always been recognized by highway engineers as a critical factor in any consideration regarding safety on the highway. There is ample empirical evidence to the effect that sharp, angular, coarse aggregates provide road surfaces with good skid-resistance, especially at high speeds.

Most skidding accidents occur where drivers must brake or corner. Simply by improving the resistance to skidding at locations where these accidents tend to cluster, considerable reductions in accidents can be achieved at a very low cost, e.g. the resurfacing of a number of roads which were slippery in wet weather reduced wet road accidents by 80 per cent and all road accidents by 45 per cent.

Accidents at night

An international survey[41] of road accidents gave the following common characteristics with regard to those occurring during hours of darkness.

(1) Fatal accidents at night represent between 25.0 and 59.0 per cent of all accidents, with a median value of 48.5 per cent.
(2) The most serious problem is that of the private car and its driver and occupants; young drivers (<25 years) are statistically over-represented in night-time accidents.
(3) These accidents often involve only one vehicle.
(4) A high proportion of accidents involving pedestrians also occur at night; however, children are not involved to any great extent.
(5) Driving speeds are often excessive.
(6) Drivers of commercial vehicles are not present to any great extent in night-time accidents.

A major analysis[42] of over 78 700 night-time injury accidents in Britain concluded as follows.

(1) The dark accident rate (1.64 per 10 veh-km) was 1.33 times the daylight rate.
(2) The accident rate at dawn was 1.5 times the daylight rate (due largely to the coincidence of dawn with the morning hours of peak travel in winter), whereas that at dusk was of the same order as in daylight.

(3)   In darkness, about 20 per cent more accidents occurred on wet roads than would be expected if the roads were dry.
(4)   Three-quarters of the dark accidents occurred on roads that were illuminated, whilst the relative accident rate in illuminated urban areas was about 2.5 times that in unilluminated urban areas.

The fact that some three-quarters of the dark accidents occurred on roads already illuminated should not be interpreted as meaning that road lighting is not an effective safety measure. Numerous studies have shown that there are significant reductions in all types of accident wherever non-motorway roads are properly illuminated, e.g. a before-and-after analysis[43] of nineteen trunk road sites subject to a 112 km/h speed limit found that fatal and serious accidents decreased by some 60 per cent, and all injury accidents by about 50 per cent, as a result of installing road lighting. This result suggests that most lighting effort in future will need to be applied towards improving conditions in urban areas, whether through better quality public lighting or by other means such as better identification of road users, vehicles or the environment. In this respect, it has been estimated[44] that for average road surface luminance $(\bar{L})$ values within the range 0.5–2.0 candelas/m², an increase of 1 cd/m² is associated with a 35 per cent lower accident rate by the relationship

$$\frac{\text{dark accidents}}{\text{day accidents}} = 0.66 \exp(-0.42\bar{L})$$

Measures proposed to improve the night-time accident situation generally include the following.

(1)   Improve road guidance and retroflect/light traffic signs for night-time use.
(2)   Provide special paths for bicycles and pedestrians.
(3)   Encourage pedestrians and cyclists using the carriageway to wear white clothing/retroflective devices.
(4)   Encourage pedestrians to travel on the side of the road facing oncoming traffic.
(5)   Ensure regular correction of vehicle headlamp aiming and malfunctions.
(6)   Restrict vehicle speeds at night.
(7)   Enforce legal measures to reduce drink–driving.
(8)   Improve road lighting at pedestrian crossings, intersections and other high-risk locations in both urban and rural areas.
(9)   Use bright and rough carriageway surfaces.

Potential accident savings

From the previous data, it can be seen that the greatest potential for accident savings lies in influencing human behaviour. Paradoxically, however, human behaviour can be very difficult to change. In many instances, there is no known remedy for the causes of certain human errors.

On the basis of *proven remedies*, i.e. measures for which there is strong evidence of potential benefits regardless of blame, an approach has been developed whereby the scale of realistic accident or casualty reductions can be estimated when applied to certain target groups. The result of this work was applied to the road accident situation pertaining in Britain in 1977, and the potential savings identified are summarized in Table 7.13. These are the possible ultimate savings at some future date when all known options in each area are implemented. (No interactions between benefits have been assumed in these estimates, so the total potential is less than the sum of the individual components.)

Table 7.13 Potential for accident and injury reduction[1, 11]

Options	Potential savings (%)
Road environment	
Geometric design, especially junction design and control	10.5
Road surfaces in relation to inclement weather and poor visibility	5.5
Road lighting	3.0
Changes in land use, road design, and traffic management in urban areas	5.0–10.0
Overall	20% of accidents
Vehicle safety measures	
Vehicle maintenance, especially tyres and brakes	2.0
Anti-lock brakes and safety tyres	7.0
Conspicuity of motor cycles	3.5
Seat-belt wearing	7.0
Other vehicle-occupant measures	5.0–10.0
Overall	25% of casualties
Road user and road usage	
Restrictions on drinking and driving	10.0
More appropriate use of speed limits	5.0
Propaganda and information	Up to 5.0
Enforcement and police presence	Up to 5.0
Education and training	Up to 5.0
Other legislation, e.g. restrictions on parking	Up to 5.0
Overall	33% of accidents

For very many road safety problems there are alternative remedies, and the measure finally decided upon in any given instance will usually depend upon its ease of application and economic considerations. The following few examples illustrate some of the difficulties involved in determining priorities for alternative actions[11].

(1) Lower speed limits in rural areas may reduce the need for better surfaces to alleviate wet weather problems, but observance of speed limits may be more difficult and costly to achieve than changes in road surface texture.

(2)   Stricter drink–driving law enforcement may reduce the incidence of excess speed and loss of control of vehicles, especially in the hours of darkness.

(3)   Within the area of highway engineering alone, there are alternatives of geometric design or control, surfacings, signs or markings, which individually or in combination may provide a satisfactory solution to intersection conflicts.

(4)   The dark accident problem can be alleviated by any or all of the following: road surface texture, road lighting, clearer definition of road alignment and obstacles (by reflectorization, etc.), and vehicle lighting.

(5)   Increased seat-belt wearing will alter the ratio of injury to non-injury accidents, and may in consequence change the priorities for dealing with different situations, e.g. roads with different speed limits, or accidents by day or night, for which the severity of injury differs.

(6)   Urban/rural/motorway situations often demand different priorities, e.g. pedestrian and bicycle casualties predominate in urban areas, whilst vehicle–vehicle conflicts predominate on rural roads and lead to car occupants being the most vulnerable road users.

Road safety education

It is often said that highway traffic engineers should be expert in all aspects of the four Es of their subject area, viz.: engineering, environment, education, and enforcement. In practice, most engineers seek expertise in only the first two of these, and often neglect the very real values and benefits to be gained from the knowledge and proper utilization of the other two, especially education.

There is no standard definition of the term 'education' as used in the context of road safety. To avoid confusion, the following discussion will subdivide this overall term into two components whose titles more properly reflect the objectives of the programmes to which they refer, viz.: attitude-changing programmes and education and training programmes.

Attitude-changing programmes

Attitude change is usually the objective of a road safety campaign that employs mass communication techniques. It is also sometimes the objective of local driver improvement programmes developed by traffic authorities for those who come to their notice through being involved in road accidents or traffic offences. Classical examples of nationwide attitude-changing programmes are the drink–driving and seat-belt campaigns.

Drinking and driving

The Licensing Act of 1872, which made it an offence in Britain to be 'drunk while in charge on any highway or other public place of any carriage, horse, cattle or steam engine' (extended in 1925 to include 'any mechanically propelled vehicle'), and all of its successors prior to 1967,

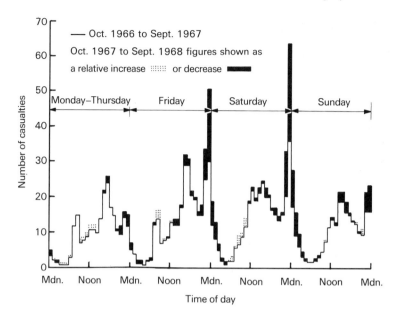

Fig. 7.7 Average number of fatal and serious casualties (all road users) by hour of day in Britain, 1966–68[45]

were relatively ineffective as traffic safety measures. The main reason for this was the lack of a legal definition for the word 'drunk'.

A major step forward in respect of changing attitudes towards the drinking and driving problem occurred with the passing of the Road Safety Act of 1967, which made it an offence for anybody to be in charge of a vehicle whose blood-alcohol or urine-alcohol content was above 80 mg/100 ml or 107 mg/100 ml, respectively. This Act received very wide publicity and its immediate effect was to bring about the biggest reduction in road accidents known to have occurred anywhere in the world following drinking and driving legislation. Thus, for example, in the following year alone, the total road casualty toll fell by 11 per cent and deaths by 15 per cent. These mainly resulted from a 34 per cent reduction in casualties between 10 p.m. and 4 a.m. (40 per cent on Saturday night/ Sunday morning), the main drinking and driving hours (see Fig. 7.7). All this occurred with no diminution either in the amount of traffic or in the total alcohol consumption; a fall in business at country inns was very temporary.

It has since been estimated that the Road Safety Act of 1967 resulted in at least 5000 lives and 200 000 casualties being saved over the period 1967–74.

However, not too long after the implementation of the legislation, its dynamic beneficial effects began to wear off (see Table 7.14). The governmental inquiry that looked into this situation nearly a decade later[46] made the following conclusions.

Table 7.14 Percentage of persons involved in road fatalities with blood–alcohol levels in excess of the legal limit[18]

Fatality	Year					
	1967*	1968	1970	1972	1974	1976
Motor vehicle drivers	32	20	23	30	36	38
Motor cycle riders	13	8	18	19	28	24
All	27	17	21	26	33	33

*Until September 1967

(1)   The minimum penalty—one year's disqualification—was an adequate deterrent for the first offender in that for most such drivers it involved heavy incidental expense and a cramping of life-style, and for some it meant loss of livelihood and the need to find new employment.

(2)   High-risk offenders, e.g. second offenders, should be required to apply to court for restoration of their licences after the end of the ordinary disqualification period and demonstrate that they were taking steps to control their drinking.

(3)   Police must have full discretion to stop and test any motorist if a purposeful and effective enforcement strategy is to be maintained, i.e. it was not that the law was inadequate or the penalties too weak but that they could not be expected to deter when drivers no longer expected to get caught.

(4)   A permanent and intensive programme of public education regarding the dangers (including punishments) associated with drinking and driving should be immediately mounted.

The above data have been presented in order to emphasize a fundamental point regarding attitude-changing legislation. It is that legislation that places restrictions on people's behaviour, resulting in accident savings, can be effective *provided that: (a) powers are given to the police and the Courts to enforce the legislation, and (b) continuous and ample publicity is given to the implications of breaking the law.*

Seat belts

Numerous studies have shown that the most severe and life-threatening injuries to car occupants involved in road accidents are those to the head and body. It is also well established that the majority (about 60 per cent) of such injuries are the result of frontal impacts by cars with roadside obstacles such as fixed objects, e.g. trees, lamp standards and walls, or with the fronts, sides or rear ends of other vehicles. The greatest need for protection in car accidents, therefore, is to prevent/reduce injury to heads and bodies caused by occupants continuing to move forward at very nearly the full speed of the car before the impact, and striking (usually) the windscreen or instrument panel. This is the primary function of a restraining harness such as a seat belt; a secondary function is to hold occupants in their seats, particularly during overturning accidents.

Seat belts in cars were first seriously considered as an important safety device in Britain in the late 1950s. The early belts were either full harness (with two shoulder straps and a waist belt), lap belts, or single diagonal belts; they were relatively little worn, however, as they were either too cumbersome or too ineffective.

Following an evolutionary process, a three-point belt, i.e. a lap and a diagonal shoulder belt, was developed which is now a standard attachment on most new cars. With this belt, the shoulder strap is not permanently anchored to the upper part of the central door pillar, but continues through a metal loop attachment down to a spring-loaded reel that is also fixed low on the pillar. This inertia-reel belt system allows the motorist to lean forward to reach controls or glove compartments, but when the car strikes another, or rolls over, a sensing mechanism locks the reel and the belt restrains the wearer.

There is no doubt whatsoever but that the wearing of seat belts reduces injuries and saves lives. This is very clearly illustrated in Fig. 7.8 which is based on data from a comprehensive study carried out in the Australian State of Victoria following the introduction of compulsory seat-belt legislation.

In relation to this figure, it should be noted that the great majority of Victoria's population live in the Melbourne conurbation, almost all cars were fitted with seat belts at the time of the study, legislation making compulsory the wearing of these belts was passed in December 1970, and the law was vigorously enforced from then onwards by the police. In this instance also, there is no doubt but that the effectiveness of the

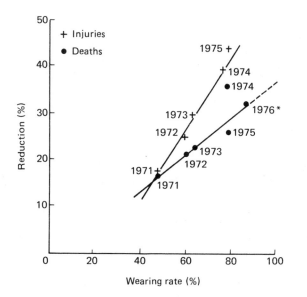

*Reduction of injuries for 1976 not given in reference 47

Fig. 7.8 Death and injury reductions in Victoria, Australia, versus the wearing rate of seat belts in Melbourne[47]

attitude-changing legislation was considerably influenced by a continuous and widespread publicity campaign espousing the value of a seat belt in reducing the severity of an accident, and by the vigorous support to the campaign given by the law-enforcement agencies.

In relation to the value of seat belts, it might also be noted that one British analysis of injury accidents involving heavy goods vehicles[48] only determined that, in 80 per cent of the collisions, the direction of impact upon the vehicle carrying the injured occupant(s) was frontal; approximately one-half of the serious and fatal injury accidents were incurred to the body (particularly the abdomen) as a result of trapping by intrusion of the cab structure, and a further 29 per cent resulted from the occupants being ejected. None of the heavy goods vehicles examined in this study was fitted with seat belts. Had seat belts been worn, it was estimated that all or most of the injuries would have been prevented in about 25 per cent of the cases, and the number or severity of the injuries might have been reduced in a further 15 per cent.

Education and training programmes

The repertoire of perceptual, motor and informational skills which must be acquired in order for a person to operate, say, a vehicle safely on the highway can be defined as driver *training*, whereas a background knowledge of road rules, physical laws, vehicle mechanics, and factual information about road safety may be included in a programme of driver *education*[9]. A similar definition can be developed in respect of road safety education and training for other users of the carriageway, e.g. the young, the elderly and cyclists. It is left to the reader to judge which of the following examples of road safety programmes are in fact 'education' and which are 'training' programmes. In practice, little attempt is generally made to differentiate between the two.

Child safety

Only about 35 per cent of the primary schools and 17 per cent of the secondary schools in Britain have formal programmes of teaching relating to road safety. Furthermore, only a small proportion of the head teachers in these schools consider that the school should play the main part in the road safety training and education of children; they place this responsibility squarely on the shoulders of parents. While parental responsibility is regularly stressed in road safety campaigns, the fact remains that, however seriously parents take this, many simply do not know how to teach road safety, e.g. they may not understand that a child's ability to absorb such training and put it to effective use varies widely with age, temperament and mental attitude.

In 1971, the Government launched a campaign to teach children the main principles of how to cross the road safely, using words that were easily understood by children. The campaign was aimed chiefly at parents through press and television and at children through television, posters, cinema and a brochure. This campaign, which launched the *Green Cross*

Code, cost £570 000 over a three-month period; the brochure component of the campaign was particularly successful, with over seven million of these being distributed. The effectiveness of the campaign was assessed by comparing actual child pedestrian casualties during the campaign with those which might have been expected in the absence of any special external factors. This analysis[49] determined that total child pedestrian casualties were significantly below those which might otherwise have been expected, with a probable reduction of about 1250 casualties, i.e. 11 per cent.

The organization of cycle training for children was firmly established by the Royal Society for the Prevention of Accidents (RoSPA) in 1947; such training is currently carried out in 20 per cent of primary schools, usually by police officers but sometimes by teachers or road safety officers. In 1959, a *National Cycling Proficiency Scheme*, sponsored by the Government, was introduced; this programme is now utilized by over 10 000 schools, and is taken by some 250 000 children each year. In some schools, training takes place during school hours; more usually, it happens after school or at weekends. Whilst cyclist training is generally more effective if carried out on the roads rather than on the playground[50], road training has its inherent dangers and is therefore not utilized as much as perhaps it might.

Unfortunately, little definitive information is available on the road exposure of child cyclists, e.g. the amount of time spent cycling or how this varies with the age of the child, so it is difficult to estimate the effect of this scheme upon child cyclist casualties. One evaluation of the NCPS training course[51] did conclude, however, that it had brought about an improvement in cycling behaviour. Perhaps the most telling effect is reflected in some statistics provided by RoSPA, viz.: whilst child cyclists increased from less than one-half of the child population in 1959 to more than two-thirds in 1977, and the number of motor vehicles on the road increased from 8.2m to 18.0m, the number of child cyclist casualties decreased from 14 613 to 9709 during that same period.

Driver education/training

A gap often appears to exist in road safety education and training between when cycle training is usually completed and the young adult learns to ride a moped/motor cycle or drive a car. It is often argued, therefore, that driver education/training programmes should be introduced into schools to cover this situation.

It is generally agreed that the main aims of such programmes should be to influence attitudes, inculcate a sense of responsibility and consideration for other road users, and give pupils the background knowledge they will later need as drivers. These objectives, for example, are implicit in the *Schools Traffic Education Programme* introduced by the Institute of Motor Cycling to help meet the needs of pupils staying longer at schools when the school leaving age was raised to sixteen years. Training in the riding of motor cycles is also available through a scheme sponsored by the Royal Automobile Club and the Auto Cycle Union.

Courses in driver education and training at schools have been a feature of academic activity in the USA for many decades. A major reason for the introduction of such courses was that they were part of a 'package deal' by which State Governments obtained federal government finance for highway development and safety programmes.

The effects of driver education/training in schools are difficult to assess, particularly in terms of accident reduction benefits. One major evaluation of the international literature on the subject[8] concluded as follows.

(1)   Such courses in high schools normally include both education and training, with a bias towards the former, i.e. the amount of training generally given is negligible in relation to the complexity of the skill to be acquired.

(2)   There was no evidence that the courses then carried out had any extra effect upon subsequent accident records, when compared with other formal and informal training methods.

(3)   The number of offences against traffic regulations incurred by graduates of such courses is significantly lower (typically by 20–50 per cent) during the first few years of completing the programmes.

A 1971 study[52] determined that, while only 11 per cent of Britain's secondary schools had driver or pre-driver education/training programmes, about one-half of the head teachers considered that driving should be taught at schools. A later study concluded that, whilst driver training in schools may have little significant effect upon the accident situation, the Department of Transport's driving test standard was reached by pupils taking these courses with a lower number of hours behind the wheel, a greater proportion of these pupils (voluntarily) wore seat belts than the average, and they generally had a greater knowledge of driving matters.

In contrast with the USA, most drivers in Britain receive their initial instruction at *commercial driving schools*. A comparatively low level of car ownership, together with the early development of a centralized testing authority with specially trained examiners and a nationally accepted Highway Code, created the conditions whereby it became economic to develop and pursue formal courses of instruction aimed specifically at the Department of Transport's test. Studies of the effectiveness of these commercial schools have concluded[8] as follows.

(1)   The techniques of safe driving are acquired painstakingly over a period of several years, and no real method of teaching them to a beginner has yet been devised.

(2)   There is no evidence that the formal commercial courses available to learner drivers have any value in road safety terms over informal training methods.

(3)   A commercial school tends to have a greater success in preparing a person to pass a driving test than does a private individual.

As a broad generalization, it can be said that there are two distinct types of *driver improvement* programmes, viz.: defensive driving courses which are concerned with 'how to stay out of trouble', and advanced driving courses which are concerned with 'what to do when you get there'. Both types of courses share the characteristic that they are usually taken voluntarily by people who are sufficiently concerned about safe driving to devote time and money to an activity which they expect will improve their performance.

Defensive driving courses appear to be effective in reducing accident involvement, with reductions of up to 33 per cent being achieved by drivers with previous records at least as good as the population average[8]. Typically, the traffic regulation violation records of graduates of these courses are also improved. When carried out by the police, an additional benefit of these courses is the opportunity to form a more harmonious relationship with members of the driving public[53].

In contrast, there is no evidence to suggest that advanced driving courses (which concentrate on the skilful handling of vehicles) have any effect upon accident involvement. Indeed, it has been suggested that drivers who receive training in such advanced driving skills as controlled skidding are inclined to incorporate these techniques into their normal driving pattern, instead of reserving them for emergency use. Nevertheless, those courses which do include a strong segment dealing with defensive driving should logically be of value.

Accident investigations

More than 60 per cent of those who suffer road accidents do so within 16 km of their own home[54]. Thus, while accident prevention is partly a national problem which can be tackled via national road safety education programmes which seek to change peoples' behaviour, it is also very much a local problem where the emphasis is usually on minor changes to the physical environment.

A zero level of accidents is ideally desirable—but practically impossible to attain. Accident reduction targets must therefore be based on what is socially acceptable, and practically and economically achievable. A practical five-year target accepted by most local authorities is a reduction in accidents in their areas of 10–15 per cent. This is based on work which has demonstrated that a minimum accident reduction of 33 per cent can be achieved by the low-cost treatment of selected sites, and between one-third and one-half of all accident sites are susceptible to this type of treatment. In addition to being practical, any phased accident reduction programme must also be cost-effective; thus the Department of Transport considers that the first year economic rate of return for any individual small road improvement scheme should be at least 50 per cent. This is a conservative target, i.e. more usual rates of return lie between 50 and 500 per cent.

Accident studies

Any phased accident reduction programme involves the determination of priorities for site improvement. Sometimes this will require the resisting of pressure for particular schemes which arouse considerable public interest. The best insurance against this kind of pressure is the carrying out of professional accident studies which can be used to demonstrate to the public, as necessary, the relative merits of different schemes. In this way, the public can be encouraged to participate in the decision-making process and be reassured that the local authority is carrying out the best course of accident reduction.

Six basic steps have been recommended[55] as the basis for a detailed accident study at selected locations in a community. They are as follows.

(1)   Obtain adequate vehicle accident records.
(2)   Select high-accident-frequency locations in order of severity.
(3)   Prepare collision diagrams, and sometimes physical-condition diagrams, for each selected location.
(4)   Summarize the facts.
(5)   Supplement accident data with field observations during the hours when most accidents have been reported.
(6)   Analyse the summarized facts and field data, and prescribe remedial treatment.

Collecting accident records

The first step in any phased accident study programme is the establishment of a databank of accident records within the study area. Such records are the indispensable tools of traffic safety engineering. The factors which cause and influence the severity of accidents are very numerous, and hence it can be very difficult in many instances to determine the true causes. Usually, it is only by patiently collecting and objectively analysing the records that the traffic engineer can determine whether and what corrective measures are feasible.

In Britain, this information is readily available from the police authorities, who are required to record accident details on standard STATS 19 forms (which may be added to in order to meet local needs). Methods of storing and processing the data vary considerably from local authority to local authority. The most usual forms of databank are[54]: (a) copies of the STATS 19 forms filed in date, serial number or road order, (b) brief summary cards prepared manually from the STATS 19 forms and similarly filed, (c) punched cards suitable for manual or mechanical sorting prepared from the STATS 19 forms and similarly filed, and (d) computer stores maintained on magnetic tapes or disks.

In addition, very many urban local authorities maintain pictorial representations of accident clusters, be they at locations such as intersections, bends, along stretches of road, or within certain localities. Such 'black spot' maps may be developed using pins with differently coloured heads, to signify different severities of accident, on wall maps. (A historical record of these maps can be maintained by photographic means.) Pin maps are particularly useful as a public relations and education tool when housed in local road safety centres well patronized by the general public.

Of more practical value to the traffic engineer is manual plotting on transparencies laid over Ordnance Survey area sheets for urban areas and strip maps for rural areas. If only the main road network and those roads recorded as having accidents are placed on the transparency, then the accident clusters stand out sharply. Furthermore, up to 3–5 years of accidents can be recorded on one sheet due to the relatively large scales used.

In large local authorities, it is generally accepted that the most effective method of initial appraisal is to process automatically the selected accident detail on the computer. In this way, ranked lists of urgent problems can be developed economically at regular intervals, e.g. every month.

Selecting dangerous locations

On the basis of the collected evidence, it is possible to isolate locations, be they intersections, stretches of roadway, or local areas, having clusters of accidents. There is no uniform procedure by which these may be ranked. Possibly the most commonly used criteria are the average annual accident total taken over three years (usually applied to high-risk spots such as intersections or small areas) and the accident rate per million vehicle-kilometres (applied to road lengths).

A relatively simple method of ordering dangerous locations in the first instance, that is widely used in many countries, is to assign a rating number to each accident and the weighted result is then used instead of the total number of accidents in isolating the trouble spots. This method accepts accident severity as the most important consideration by assigning, say, twelve points to each fatal accident, three points to a personal-injury accident, and one point to a damage-only accident.

Whatever the actual method finally used, it is common practice also to prepare priority site lists subdivided according to: (a) cluster problems and (b) specific factor problems, for which established remedial actions are possible. Locations in which the former may arise are, for example, as follows: high-risk spots or nodes such as bends, junctions or crests, high-risk road sections, high-risk through routes (i.e. a successive number of links or nodes), and high-risk localities or grid squares. Examples of the latter are as follows: darkness, loss of tyre adhesion, overrunning STOP/GIVE WAY lines, restart from STOP/GIVE WAY lines, conflict at traffic-signal-controlled junctions, nose-to-tail collisions involving vehicles waiting to turn right from the main road, excessive speed at roundabouts, single vehicle accidents at bends and crests, head-on collisions at bends and crests, and collisions at uncontrolled intersections.

Preparing collision diagrams

Having been assured that an accident problem of some significance exists at a particular location, the traffic engineer will need to get an initial 'feel' for the accident situation there. The collision diagram is a most valuable tool in this exercise, since it indicates graphically the nature of the accident record at any given section.

Figure 7.9 is one example of a collision diagram drawn for an intersection. Normally produced on a 1:500 scale plan of the problem

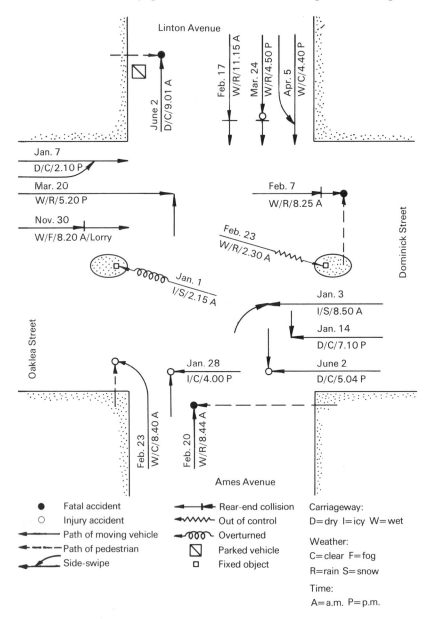

Fig. 7.9   Example of a collision diagram at an intersection

area, the diagram shows by arrow indications the movements which led to each accident. The date and hour of each collision are shown alongside one of the arrows. If weather and visibility accidents are important at the location in question, these are also indicated. Colour coding can be used if further information is required, e.g. whether the vehicle skidded or whether the injuries were severe or minor.

In many instances, the collision diagram will be supplemented with a scaled drawing or photograph illustrating the location of road signs and markings, pedestrian crossings, traffic signals, bus stops, parking locations, sight obstructions, and fronting land uses. This is particularly helpful if it is considered that physical features are influencing the accident experience.

Summarize the facts
It is very probable that at this stage certain dominant trends will become apparent which will suggest physical features which may need correction. For instance, if a large number of accidents involve skidding, then it may be that the carriageway requires resurfacing.

In some instances, however, the point of origin of the sequence of events leading up to an accident may be a more important indicator of the correcting action required than either the point of actual collision or the terminating positions of the vehicles involved. For example, excessive speeds at an inadequate bend may be the cause of motorists losing control of their vehicles, and whether they eventually recover, leave the carriageway, or collide with another vehicle is then a matter of chance. On the collision diagram, however, the accidents may be noted as occurring perhaps 100 m from the bend.

In-depth field investigation
Visiting the site in question, preferably during the hours when the greatest number of accidents occurs, will usually provide further evidence regarding the remedial action required. In order to appreciate the road user's point of view more fully, it is usual to start some distance from the site and drive along the conflict paths.

Items which may need to be observed and noted at the site range from the compatibility of the layout, signs and carriageway markings with the current recommendations of the Department of Transport, to topographical features such as the sky, colour of buildings, vegetation foliage, and road alignment, to the traffic distribution. Indeed, so many features need to be considered that it is inadvisable to rely upon memory and normal powers of observation. Instead, a check-list is best pre-prepared for use at the site.

Appendices 3.1-3.3 of reference 54 contain excellent examples of comprehensive check-lists that can be used both at the site and in the office to help determine the remedial actions that may need to be carried out.

Final analysis
On the basis of the information obtained from the office and field studies, it should be possible for the traffic engineer to propose positive recommendations leading to a significant reduction in accidents. In most cases, it will be found that these recommendations will involve the application of standard road safety and traffic management measures, e.g. improved signing and road markings, better sight distances, limitation of parking, carriageway resurfacing, prohibition of certain turning movements, better lighting, the installation of traffic signals or pedestrian crossings, or possibly the need for channelization of an intersection.

Whatever the proposed improvement, it is axiomatic that it should be cost-effective. One way of ensuring this is to carry out a simple economic analysis of the proposed remedial action, with a view to determining the discounted economic rate of return. Assuming the annual benefits are constant over the life of the scheme, the rate of return can be determined from the formula:

$$r = 100 \left(\frac{BV - C}{C} \right) = \frac{100[(A \pm M \pm O)V - C]}{C}$$

where r = discount rate of return (%), B = benefits accruing in the first year of the scheme, V = benefit multiplier to allow for the life of the scheme (obtained from Table 7.15), A = accident savings in the first year, M = change in annual maintenance cost, O = savings in journey costs in the first year, and C = capital cost of the improvement.

When the rates of return have been determined for most of the priority list (or 'league' table) previously determined, a new priority list may be determined, this time on an economic basis.

Table 7.15   Extract from net discount table[54]

Current net discount rate (%)	Length of scheme life (years)					
	1	2	3	5	10	15
7	0.935	1.808	2.624	4.100	7.024	9.108
8	0.926	1.783	2.577	3.993	6.710	8.559
9	0.917	1.759	2.531	3.890	6.418	8.061
10	0.909	1.736	2.487	3.791	6.145	7.606

Example of the economic analysis process used in priority determinations
A scheme has an estimated life of fifteen years. The capital cost is £30 000, and the estimated annual savings in accidents are £7500. There are also annual savings in journey costs as a result of the improvement of £2000, whilst the extra annual maintenance cost is £1000. The current net discount rate is 7 per cent. Determine the rate of return.

Solution
From Table 7.15 the benefit multiplier is 9.108. Hence

$$r = \frac{100[(7500 - 1000 + 2000)9.108 - 30\,000]}{30\,000} = +158\%$$

The discounted rate of return carries a positive sign and is well in excess of the target minimum of 50 per cent.

Controlling vehicle speeds

Excessive speed is the single most widely *blamed* cause of accidents. Whilst this assumption needs to be clarified, there is no doubt whatsoever but that high speeds are major contributors to the fatal accident statistics. As a result, it is now usual practice in most countries to place upper speed limits on streets and highways.

Properly, the term 'excessive speed' should be used to refer to the situation where a vehicle travels at a speed greater than that which the traffic and road conditions allow. With this definition, a speed in excess of 18 km/h may be regarded as unsafe in a residential area[56], whereas 120 km/h on a non-crowded rural motorway can be a safe cruising speed. If drivers could be relied upon to adjust their speeds according to the prevailing circumstances, there would be no need for legislation to control speeds. Unfortunately, however, a substantial proportion of the driving population cannot be relied upon and so speed restrictions are necessary.

Consequences of high speeds

An early arrival at a destination and the psychological thrill of driving at high speed are two considerations which encourage the driving of vehicles at great speed. What are not always so obvious, however, are the positive reasons why higher speeds are not so desirable.

Stopping and overtaking distances
A major consequence of increasing speed is that the total distance required to bring a vehicle to rest is considerably increased (see Chapter 6). In urban areas in particular, therefore, vehicle speeds should be reasonably low so that the motorist will be able to bring the vehicle to a stop when necessary to avoid colliding with the many-to-be-expected interferences on the carriageway.

As also discussed previously, the distance required to overtake a vehicle in safety is substantially increased the higher the overtaken vehicle's speed. This is a most important consideration on rural roads because so many of these roads were never really designed and, in consequence, have inadequate hills and curves. Many needless accidents happen at these locations because motorists attempt to overtake blindly.

Even where the highways were scientifically designed, the original design criteria may not meet modern requirements. A basic design criterion which illustrates how standards change is the one for eye height. In Britain, the distances required for safe overtaking are currently measured between two consecutive points 1.05 m above the carriageway; prior to the mid-1960s, it used to be 1.14 m—which was the eye height of an average driver in a small sports car prior to World War II. Design and styling trends in the motor industry over the past twenty-five years mean, therefore, that on certain otherwise well-designed roads many cars do not now have available to them the sight distances required for safe overtaking at high speeds.

Vehicle separation

The minimum safe separation between successive vehicles—which depends upon the type of the following vehicle, its braking efficiency, and its driver's perception–reaction time—also increases with vehicle speed. The separations which should be maintained by vehicles, in the state in which they are normally met on the road, in order to give an overall 95 per cent chance of avoiding a rear-end collision when the vehicle ahead slams on the brakes, are given in Table 7.16. As can be seen, not only do the required clearances greatly increase with speed but, in addition, commercial vehicle requirements are considerably greater than those for cars.

Observations on the road have shown that vehicles travel much closer together than the values suggested in Table 7.16, and separations become increasingly insufficient at higher speeds. This is the main reason why so many accidents involve the following vehicle colliding with the rear of the preceding one.

Table 7.16 Desirable clearance distances between moving vehicles in order to avoid rear-end collisions

Speed (km/h)	Desired clearance distance (m)		
	Car behind car	Commercial vehicle behind commercial vehicle	Commercial vehicle behind car
48	24.4	39.6	48.8
64	39.6	64.0	79.2
80	54.9	94.5	118.9
90	73.2	131.5	167.7

Resistance to skidding

The skidding resistance offered by the carriageway surface when wet decreases as the vehicle speed increases. Results of tests made on a variety of surfaces throughout a range of speeds up to 161 km/h have shown that the skidding resistance at the highest speeds sometimes fell to less than half its value at 32 km/h, depending upon the surface being tested. Indeed, some roads have surfaces which are perfectly satisfactory at 32 km/h but become slippery when a vehicle is travelling at 125 km/h or more.

Sign legibility
As will be discussed later, a given size of road-sign lettering can become inadequate at higher speeds. This means that at high speeds the message of a sign may be missed or not fully understood by the driver within the distance available for acting upon it.

Pedestrian risk
The pedestrian's estimate of the speed of a vehicle becomes less reliable when the speed is high. The data in Table 7.17 represent the results of one study which attempted to determine the reliability of the onlooker's estimate of the speeds of vehicles. In this study, pedestrians stationed at the side of the test road were asked to push buttons at the last instant at which they deemed it safe to cross the carriageway, in the face of an opposing vehicle. As can be seen from this table, the accuracy of the pedestrians' judgement deteriorated as the speed was increased; the sharpest deterioration took place as the vehicles travelled at speeds of 80–96 km/h.

Table 7.17 Onlookers' misjudgements of the speeds of oncoming vehicles, based on 100 observations at each speed

Effect of misjudgement	Speed of oncoming vehicle (km/h)				
	32	48	64	80	96
Would have been hit	2	7	9	11	22
Would have been a near miss	8	7	7	12	13

Actual observations on the road have confirmed the results shown in Table 7.17. These further showed that the speeds of smaller vehicles and those travelling on the farside of the road were underestimated more often.

Fuel usage
As noted previously (see Chapter 1), higher speeds result in greater fuel utilization, e.g. the fuel consumption of a typical 1350 cc car is about 14 l/100 km when travelling at 130 km/h, but declines to about 7 l/100 km at 65 km/h. One exercise[57] which examined this problem concluded that there would be substantial fuel savings if only drivers were willing to reduce their speeds on rural motorways and dual carriageways. (With a speed limit of 96 km/h, for example, the fuel savings to the nation would be about 127 000 t per annum; with a speed limit of 80 km/h, the annual savings would increase to about 237 000 t.)

Miscellaneous effects
The risk of mechanical failure in a vehicle due to metal fatigue, overheating, burst tyre, etc., is more likely at sustained high speeds than at low ones. Loss of control, when something unexpected does occur, is much more likely at the higher vehicle speeds.

Mandatory speed limits

From the previous discussion, it is clear that there can be a substantial deterioration in the principal factors controlling safety when vehicle speeds are increased. It is logical therefore to conclude that accident numbers and severity should be decreased if speeds are restricted on streets and highways where safety requirements make it desirable. The validity of this conclusion is suggested by the historical data in Table 7.18, and by more recent published work in the literature[59].

Table 7.18   Effects of urban speed limits upon accidents in various locations[58]

Location	Date imposed	Limit (km/h)	Percentage change	
			Fatal and serious	All injury
Britain	1935	48	−15	−3
Britain (various sites)	1945–53	48	—	−10
Sweden	1955	50	−11	—
Northern Ireland	1956	48	−23	−24
Netherlands	1957	50	−10	−6
Germany	1957	50	−30*	−18
London (various sites)	1958	64	−28	−19
Jersey	1959	64	−47	−8
Switzerland	1959	60	−21*	−6

*Fatalities only

Types of speed limit

Two types of speed limit are in general use throughout the world. For want of better terminology, they will be termed the 'reasonable' and 'absolute' speed limits.

Reasonable limit   With this type of restriction, no numerical speed limit is specified but instead dependence is placed upon the driver to adjust speed to the roadway conditions. Under this restriction, a motorist driving on a motorway in heavy fog at, say, 48 km/h might be summoned for exceeding the reasonable speed limit; on a clear day with little traffic on the motorway, this same motorist would be well within the law when travelling at 112 km/h.

In theory, the reasonable speed limit is ideal. It is flexible and allows the enforcing officials to adjust the limit according to the conditions. In addition, it is a concept which the motorist likes, i.e. there is nothing more aggravating than to be on an open road but forced to maintain an artificially low speed.

In practice, however, this type of restriction is inefficient. To be successful it requires the cooperation of the motoring public and this cannot always be achieved. In addition, it relies heavily upon the judgement of the enforcing officials as to what constitute reasonable speeds for the conditions—thus in essence making them both police officers and judges. At worst, this can lead to abuse; at best, it leads to understandable and degrading arguments in court.

Absolute limit   With this type of restriction, a numerical speed limit is specified for a road or group of roads. If a vehicle exceeds this limit, it breaks the law and there is little that the motorist can do to challenge it in court. While this is a decided advantage from the enforcement aspect, it lacks flexibility and often results in ruffled motorist feelings due to unreasonable speed restrictions being placed on obviously higher-speed roads.

Throughout the world the trend is towards utilizing absolute speed limits on roads. In general, the following trends are discernible.

(1)   Speed limits are applied to both urban and rural roads all through the year.

(2)   Different upper speed limits are used for different types of highway in different environmental locations. In addition, it is now often suggested that lower (as well as upper) speed limits should be imposed on high-speed roads of motorway calibre. A typical lower speed for a rural motorway would be 64 km/h. Its purpose would be to reduce accidents resulting from vehicles moving at such slow speeds that other vehicles are impeded and have to take risks in order to overtake and maintain reasonable speeds.

(3)   Different speed limits are used for different types of vehicle. Thus speed limits for commercial vehicles on rural highways are generally 8 to 16 km/h less than those for passenger cars. Cars pulling caravans, oversize vehicles, etc., have even lower speed limits imposed on them.

(4)   Different speed limits may be in force during day and night.

Observance of absolute limits

A measure of the extent to which drivers respect speed regulations can be obtained by comparing the number of vehicles which travel at higher speeds before and after the initiation of a speed limit. Observations made on the highways of a number of countries indicate that the proportion of vehicles travelling faster than the limit is always considerably reduced after the speed limit comes into operation. Complementarily, another study showed that, when the speed limit was raised from 48 to 64 km/h on some major roads in the London area, the proportion of vehicles exceeding 64 km/h rose from 9 to 14 per cent.

The single greatest hindrance to motorist observance of speed restrictions is the establishment of a general speed limit on specific roads or road sections where obviously it is inappropriate. This can be avoided, however, if a speed distribution study is carried out on the highway before imposition of the speed limit. Assuming that the great majority of drivers travel at reasonable speeds, then, say, the 85th percentile speed might be used as a guide to the most desirable speed limit. In no instance, however, should the posted speed limit ever exceed the design speed of the road.

Certain highway sections may require special speed regulations that are different from the general limits imposed. An obvious example of this is the section having an excessively high accident record, where investigations into the causes of these accidents reveal that the existing speed restriction should be either raised or lowered. Transition sections with 'medium' speed limits should always be interposed when a high-speed rural highway is about to enter a built-up area. Construction zones, as well as being carefully signed, should also be posted for suitably low speeds.

Recommendations regarding the choice of general speed limits for particular locations are available in the literature[60].

Persuading drivers to reduce speed
On the whole, speed limits are considered useful by the public, and drivers do not support the view that motorists should be left to decide appropriate speeds for themselves[61]. Nevertheless, there are always some members of the motoring public who require some urging to ensure that they travel at sensible speeds. Generally, there are two approaches used to persuade drivers to reduce speed.

The first approach might be described as the *deterrence method* in that it relies upon the presence of the police for the enforcement of the general speed regulations. Typically, the police utilize 'speed traps' to check vehicle speeds at, for example, locations where it is known that speed limits are regularly exceeded or where accidents associated with speeding have been identified. Most commonly, the actual measurements are made with a radar speedmeter, and speed limit violators are then prosecuted at a later time. Alternatively, police patrol cars may be used to pace the traffic flow. Although expensive in personnel utilization, this latter method can be very effective when distances have to be policed, on congested roads, and in foggy weather conditions, since the presence of police cars tends to act as a positive deterrent to high speeds.

In recent years, greater attention has been given to the development of a traffic management approach to controlling speeds at particular locations using *self-policing methods*, i.e. the methods used do not depend upon the deterrent powers of the police for their enforcement. Examples of these are rumble bars, bar markings, and road humps.

Continuous high-speed driving has a marked effect upon drivers' estimates of their own speed, which leaves them particularly vulnerable to accidents at the interface of high-speed and lesser-quality highways. Thus accidents often happen at the ends of motorways, at roundabouts, severe bends, etc., even though the motorist may slow down to leave the fast stretches of road. What happens in fact is that, although their senses may tell them otherwise, the drivers have not reduced speed sufficiently, so that they suddenly find themselves travelling at unsafe speeds for the new road conditions.

One way of helping to reduce speed further is to place a series of *rumble bars* (8–13 m long) across the carriageway to alert motorists to the changed driving conditions over a distance of, say, 200 m prior to the danger site. A surface which produces a high noise level has been developed at the Transport and Road Research Laboratory that is quite effective in this respect. This surface, which consists of 13–19 mm roadstones held in an epoxy resin binder, produces an increase in noise level of 10 dB(A) at most speeds in cars and vans; this is considered 'most noticeable' by motorists. Table 7.19 shows some average speed reductions produced by rumble bars at particular sites.

Table 7.19 Effect of rumble areas upon average vehicle speeds[62]

Finish of rumble area	Speed reduction (km/h)	Level of significance (percentage)
30 m from dual roundabout	5.7	0.1
95 m from apex of 48 km/h dual carriageway bend		
left-hand lane	2.9	0.1
right-hand lane	2.2	0.1
200 m from dual carriageway traffic lights		
left-hand lane	1.0	n.a.
right-hand lane	2.2	1.0

Whilst rumble bars can be effective, the noise produced necessarily limits their usage to locations which are not sensitive to a sound detriment of this nature. Another persuasive method now in wide usage on high-speed dual carriageways with roundabout intersections is the application of *bar markings* to the carriageway for a distance of about 400 m before the roundabout (see also Chapter 6). The effect of this road-marking pattern is to give motorists the illusion that their speeds are increasing as they come closer to the roundabout; thus they are encouraged to slow to a more realistic speed more quickly. Tests have shown that, properly used, this persuasive approach can result in overall reductions of 56–62 per cent per annum in speed-associated injury accidents[63].

Both rumble bars and bar markings are most effective on high-speed rural highways; their usage in urban areas is very limited. In urban areas, however, the situation regularly arises where a posted speed limit is not sufficiently effective in discouraging vehicles from being driven too fast along roads that are particularly susceptible to accidents, e.g. near schools, childrens' playgrounds, and old peoples' homes, or from using as 'short cuts' certain streets that are not suitable for use by through traffic, e.g. streets in residential areas. One very effective way of ensuring that desired speed limits are not exceeded (and through traffic is discouraged) at these locations is to construct *road humps* across the carriageway[64, 65].

There are two basic types of road hump: (a) those that are narrow enough in the direction of travel to be straddled by the wheels of all normal vehicles, and (b) wide humps which cannot be straddled except by a minority of large vehicles. Narrow humps administer a sharp jolt to the vehicle suspension, except at low speeds when the crossing time is long enough for the vehicle body to deflect upward as each axle passes over the hump. Whilst narrow humps can be safely crossed at quite high speeds without undue discomfort, the driver is usually afraid to do so for fear of damaging the vehicle or losing control. Wide humps result in a less severe ramp effect and have a longer crossing time; a greater height may also be used without fear of grounding low-slung vehicles. The overall effect of a wide hump is to cause a vehicle body deflection rather than a rapid deflection of tyres and suspension.

A speed control hump has been designed[66] which is considered suitable for use in residential streets with a 48 km/h speed restriction. This hump has a maximum height of 102 mm, is 3.7 m wide, and has the shape of a segment of a circle in cross-section. To make it more visible to drivers at night, the hump's surface is painted with white road paint and a layer of ballotini, i.e. small glass beads as used in road markings, sprinkled over the paint; alternatively, a more permanent solution is to use a white mastic asphalt as a surface dressing for the humps.

Extensive testing of road humps in residential areas gave the relationships shown in Fig. 7.10. Analysis of the test data also revealed the following points.

(1) On some roads, the number of vehicles exceeding the speed limit was reduced from 50 per cent to less than 5 per cent.

(2) Traffic flow reductions averaged 37 per cent, but varied according to the locality and the availability of alternative routes.

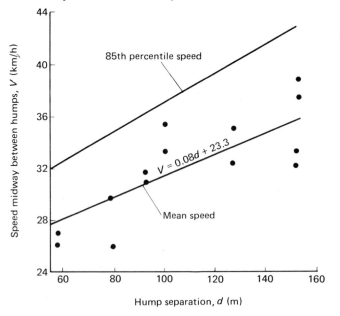

Fig. 7.10 Effect of road humps upon vehicle speeds on 48 km/h residential streets

(3) The changes in speed led to reductions in kerb delays and increases in the inter-vehicular gaps required by pedestrians to cross the road in safety.

(4) Beneficial environmental impacts were noted by residents, particularly in respect of reductions in traffic noise.

(5) A reduction of 61 per cent in accident casualties was averaged on humped roads, whilst an increase in the number of accidents on the surrounding alternative roads was not great enough to be statistically significant.

(6) Delays were caused to fire vehicles and ambulances (up to one minute on any one journey).

Overall, road humps are a most effective traffic management and safety measure for use on residential streets. Their usage is most applicable to streets that are less than about 0.8 km long, have slow entry speeds due to bends or intersections, no easy alternative on similar adjacent roads, no bus services, and where motorists can be warned of the presence of the humps well in advance.

Advisory speed signs

Many road accidents occur because of excessive speeds at inadequate bends (on rural roads particularly), at unexpected obstructions on the carriageway, e.g. roadworks, or in wet/foggy conditions on high-speed roads. Advisory speed signs can be used to encourage drivers to slow down to speeds that are suitable for the road conditions prevailing.

Difficulty arises in defining exactly what is a safe speed for use at bends. One suggestion[67] is that the selected speed should be consistent with the maximum radial acceleration for the bend in question. Another, more simple, proposal[68] is that the advisory speed should be the 85th percentile 'before' speed as observed for free-moving private cars; for level rural roads, this can be estimated from the following equation:

$$S_{85} = 99.34 - 2.48X$$

where $S_{85} = $ 85th percentile 'before' speed (km/h) and $X = $ curvature of bend (degrees).

For temporary hazards such as roadworks or foggy conditions, it is not practicable to set down hard and fast rules regarding the sign legend and speed, or the exact location of signs, barriers and lamps. Obviously, the exact details depend upon the situation at hand; in the case of roadworks, they should be located sufficiently in advance of the new construction to allow the oncoming vehicle to slow safely and easily to the desired speed. The real problem yet to be solved, however, is not how to design and locate the signs but rather how to persuade the motorist to accept the advice contained on them.

Over half of the motorway system in Britain is equipped with illuminated signs of the matrix type. On rural motorways, these signs are normally mounted in the central reservation at intervals of 2–3 km. The legend on the signs is controlled through a computer by police at motorway control offices, and can be adjusted to suit the traffic conditions. Measurements of the effect of an 80 km/h speed limit displayed on one of these signs prior to roadworks on the M4 motorway, taken under both wet and dry weather conditions, have shown that the sign had relatively little effect upon driver behaviour. For example, when the 80 km/h speed limit was displayed, the mean car speeds at the test site only reduced from 116.0 to 108.0 km/h in dry weather, and from 113.5 to 109.0 km/h in wet weather[69].

Protecting the pedestrian

In the period from April 1972 to March 1973, the (then) Department of the Environment collected information on virtually all journeys made outside buildings during a particular week by a sample of households containing a total of 17 000 individuals. This study showed that people spent about 20 minutes per day travelling by foot, and that the average distance travelled was 1.32 km. On this basis, it was estimated[70] that this implied a pedestrian accident rate of about 310 accidents per 10^8 km walked; this rate was greater than that for car drivers but less than that for motor cyclists. It was also found that during daylight hours the average number of pedestrian accidents was approximately proportional to the product of pedestrian and vehicle flows. In addition, Fridays and Saturdays were relatively more dangerous than other days for pedestrians, and Winter was more dangerous than Summer. Finally, the most vulnerable pedestrian populations were the post-60 and pre-15 age groups.

The most common difficulties experienced by elderly pedestrians include an inability to assess the speeds and distances of vehicles, failing eyesight, and a reluctance/inability to learn to cope with modern traffic[71]. The high accident statistics for the elderly pedestrian on Fridays and Saturdays, especially in Winter, are not unrelated to the greater effect of alcohol upon their senses. Older people are also more vulnerable in terms of fatalities because their recuperative powers are less.

Amongst the young, the high accident rate can be perhaps expected, but not accepted, because of the impetuousness of youth and the use made of streets as playgrounds by children, e.g. a London study in 1969 found that 36 per cent of children were playing in the street at the time of their involvement in an accident, as were 12 of the 47 child pedestrian fatalities. Ice-cream vans, which attract children to hurry across the road without taking proper care, can be a serious danger in residential areas particularly; the limited evidence available in the early 1970s suggested then that up to 5 per cent of child casualties could be associated with buying ice cream from mobile vans[72].

Traffic management is primarily concerned with two aspects of pedestrian control. The first and more important of these is pedestrian safety. Given the vulnerability of the pedestrian when involved in a conflict with a motor vehicle, and the varied causes underlying these accidents, the traffic engineer tackling this problem inevitably has to become involved in activities ranging from the promotion of safety education, to the provision of pedestrian footways and crossings, to the development of environmental areas from which vehicles are wholly or partially excluded. The second aspect of traffic management, which must not be forgotten, is the need to alleviate the effect of pedestrian movement upon the ease of traffic flow. British law allows the pedestrian to have unrestricted access to all parts of the highway system with the exception of motorways, so that at congested locations, or when the footpath is cracked and uneven, or when the pedestrian simply wishes to take the shortest distance across the road, he or she is at liberty to step onto the carriageway at any time or at any place. In central areas of cities, this right is often exercised to the extent that vehicular traffic is brought to a standstill as continuous streams of pedestrians cross the road at particular locations.

Footpaths and protective barriers

Ideally, *footpaths* should be planned as a secondary network of 'streets' for pedestrians only, so that their users are completely separated from vehicular traffic. This applies not only to shopping and business areas but also to residential areas where pathways should be located so as to give convenient access from homes to shops, schools, playgrounds, old people's hostels, post offices and popular meeting places such as clubs and public houses. They should also be linked to bus stops, coach stops and railway stations. In practice, the segregated footpath approach is most readily adopted in new residential areas, whereas existing urban areas usually have to make maximum use of footpaths adjacent to the carriageway.

General British practice with respect to footpaths is summarized in Table 7.20. Footways adjacent to shopping frontages in town centres usually have to accommodate more pedestrians and perambulators than any others, and they should therefore be sufficiently wide for both free movement and 'window-shopping' without people being jostled. One guide which can be used to check pedestrian footpaths adequately is that the capacity of an open footway may be taken as 33–49 persons/minute after deducting approximately 1.0 m 'dead width' in shopping streets and 0.5 m elsewhere. There is no published guide as to when the surface quality of a footpath is unsatisfactory but logic suggests that, in order to induce people to remain on the footpath, its surface quality must be at least equal in merit to that of the adjacent carriageway.

At dangerous and/or congested locations such as at bus and railway stations, thronged intersections, schools or along footpaths on busy shopping streets, *guardrails* can be used both to prevent pedestrians from spilling onto the carriageway and to channel the stream of pedestrian traffic towards formal pedestrian crossings. The guardrails should be inset about 0.5 m from the kerb in order to give adequate clearance for passing vehicles, and provide a place of refuge for those persons who stray onto the carriageway and find themselves in peril on the wrong side of the railings,

Table 7.20   Recommended minimum footway widths adjacent to roadways[73]

Road type	Recommended minimum footpath widths
Primary distributor	
urban motorway	No footways
all-purpose road	3 m*
District distributor	3 m in principal business and industrial districts*
	2.5 m in residential districts*
Local distributor	3 m in principal business and industrial districts*
	2 m in residential districts*
Access road	Principal means of access
	3 m in principal business districts*
	2 m in industrial districts*
	2 m normally in residential districts*
	3.5–4.5 m adjoining shopping frontages
	Secondary means of access
	1.0 m verge instead of footway on roads in principal
	business and industrial districts
	0.6 m verge instead of footway on roads in residential areas

*If no footway is required, provide verge at least 1.0 m wide

with no quick access to the footpath. To be effective at busy locations such as intersections and inter-junction pedestrian crossings, the guardrails should extend for at least 10 m on either side of the crossing.

Pedestrian crossings

If the flow of vehicles is slight, pedestrians can be generally left to decide in their own time when to cross the road with the aid of suitable gaps in the traffic stream. As traffic gets heavier, however, it becomes increasingly necessary to provide formal pedestrian crossings. These can be either surface crossings or segregated crossings.

Surface crossings

There are two main types of surface crossing, viz.: (a) uncontrolled crossings such as zebra crossings, where the right-of-way is automatically given to the pedestrian on the crossing, and (b) controlled crossings, where the right-of-way is given to the pedestrian by the police (e.g. at police-controlled crossings) or by traffic lights (e.g. at traffic signals and pelican crossings).

Experiments with the use of uncontrolled pedestrian crossings were first initiated in London in 1927; their success was such that in 1934 the (then) Minister for Transport (Mr L Hore-Belisha) made formal regulations regarding their usage, and they then began to be used elsewhere in the country. As a result, there are now well over 10 000 zebras in use in Britain today; in addition, there are many thousands in use elsewhere in the world. A *zebra crossing* is simply an uncontrolled portion of the carriageway where the pedestrian has legal priority over the motor vehicle. The crossing strip is outlined by parallel lines of studs and marked with alternate black and white thermoplastic stripes parallel to the centreline of the road; the beginning and end of each crossing are marked by flashing yellow beacons. Longitudinal 'zig-zag' centrelines and edgelines are placed on the carriageway for nearly 19 m on either side of the crossing strip; a GIVE WAY line is located about 1 m from the crossing at which vehicles should stop to allow pedestrians to cross.

Whilst zebra crossings have generally proved their worth over the years, the carriageway areas beside them have been found to be particularly hazardous, e.g. in 1974, it was considered that the risk to a pedestrian using the crossing was one-third that experienced within about 50 m of the crossing and one-half that measured elsewhere on the carriageway[74]. The danger is magnified if the zebra crossing is located within 18.25 m of an intersection[75], where the complexity of decision-making is greatest for both the pedestrian and the motorist. The purpose of the zig-zag lines (introduced in 1971) is therefore: (a) to discourage pedestrians from crossing the road adjacent to the crossing, (b) to improve drivers' visibility of the crossing, and (c) to indicate to motorists where they are not permitted to park or overtake other vehicles that may have slowed down or stopped for pedestrians (i.e. in 1970, some 25 per cent of pedestrian accidents at crossings resulted from overtakings). This simple traffic management measure has significantly reduced the numbers of vehicles overtaking, cars parking, and pedestrians crossing within the zebra's zone of influence.

Police-controlled crossings are the simplest form of controlled surface crossings. Marked simply by two parallel lines of studs, they function only as formal crossings when a police officer or traffic warden is present to control traffic.

Although the capital costs of such crossings are negligible, they are normally only installed at sites where positive control is required to balance the needs of pedestrians and vehicles at certain times of the day (usually peak periods). For practical personnel reasons, however, it is unlikely that use of this type of crossing will be developed much further in future years.

Traffic signals are used in a number of ways to control pedestrian movement across the carriageway. By far the most widely used procedure is simply to allow the pedestrians to cross with the green lights when opposing vehicular traffic is brought to a standstill at a signalized junction. However, if the pedestrian crossing movements are heavy, additional facilities may be incorporated in the signal phasing to help them.

Figure 7.11(a) shows the basic phasing arrangements normally used at a four-way intersection. Figure 7.11(b) illustrates how Phase B can be split into two parts so that pedestrians can cross the North–South road in complete safety during peak pedestrian periods without interfering with the main traffic movement on the East–West road; commonly, 10 seconds of crossing time is given to pedestrians with this arrangement, and a push-button activator is not necessary.

When pedestrian traffic seeks to cross the arms of the intersection for

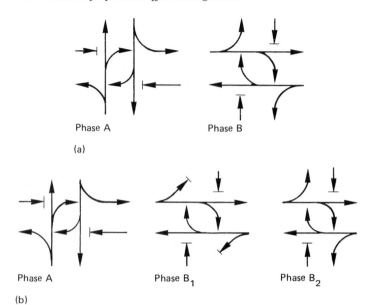

Phase A   Phase B

(a)

Phase A   Phase B₁   Phase B₂

(b)

Fig. 7.11 Standard phase arrangements at signalized four-way intersections: (a) basic arrangement and (b) modified arrangement with split pedestrian phase

most of the day, the signal phasing can be designed so that a full pedestrian phase may be brought into operation by push buttons located on signal posts. With this facility, all vehicles are halted by red lights and 'green man' lights indicate to pedestrians that it is safe to cross the road.

Whilst separate pedestrian phases can be very satisfactory for those who are on foot, they can be very costly in terms of vehicle time at important traffic intersections. On the one hand, the pedestrian phase has to be sufficiently long to ensure that it is completely safe for the pedestrian to cross; on the other hand, the consequent reduction of time available for other traffic movements may necessitate a substantial lengthening of the signal cycle so that the vehicular traffic can be handled properly. The net result may be that the signal cycle becomes too long so that pedestrians will not wait for the period allotted to them, and try to cross during traffic phases; alternatively, if the cycle length is reduced to satisfy the pedestrian requirements, the free movement of vehicles will be impaired.

First introduced by regulation in 1969, the *pelican* (*pedestrian light-controlled*) *crossing* is a push-button-actuated crossing that is used when signal control is required at sites away from intersections. For drivers, the only difference between the pelican signals and the normal traffic signals is a flashing amber period which replaces the familiar red–amber period before the green; the flashing amber means 'give way to pedestrians— proceed only if the crossing is clear'. For pedestrians, the only difference is a flashing 'green man' period before the start of the red period; this means 'do not start to cross—the lights are about to change'.

Pelican signals remain at green to drivers until the pedestrian presses the button to secure a crossing phase. However, since they operate on a fixed-time basis, vehicle detectors are normally located in the carriageway prior to the signals so that green extension times can be granted to vehicles, depending upon the driving conditions[76].

Whilst it is much more expensive than a zebra crossing (by two to four times), a pelican crossing is particularly useful at locations where it is necessary to interrupt heavy and/or fast traffic flows to allow pedestrians to cross, or where the pedestrian flow is so heavy that breaks are needed to allow vehicles to proceed. They can also be useful in solving special problems at isolated sites, e.g. to replace a zebra crossing in a contraflow bus lane to give better protection for pedestrians. An added advantage of the pelican-signal-controlled crossing is that its operation can be more easily incorporated into area-wide traffic control schemes.

Simple numerical criteria[74] can be used to consider whether the installation of either a zebra or a pelican crossing is appropriate, viz.:

$$PV^2 \geqslant 10^8 \quad \text{(for a crossing without a pedestrian refuge)}$$

or $$PV^2 \geqslant 2 \times 10^8 \quad \text{(for a crossing with a refuge)}$$

where P = average number of pedestrians per hour crossing (in both directions) a 100 m length of road centred on the site of the proposed crossing and V = average number of vehicles per hour passing (in both directions) the proposed site.

Both sets of data used in the above formulae are based on averages of the same 4-hour counts which give the maximum value of PV^2, even though the peak flows of pedestrians and vehicles may not coincide.

Pelicans are normally preferred to zebras when the vehicular traffic flow exceeds 500 veh/h on roads without a refuge (750 veh/h on roads with) or when the pedestrian flow exceeds about 1100 pedestrians per hour. The extra cost of the crossing is usually justified by accident savings (at sites with heavy vehicle flows) and by reduced delays in traffic (at sites with heavy pedestrian flows).

In practice, pelican or zebra crossings are usually considered unnecessary if the total vehicle flow is less than 300 veh/h (400 veh/h with a refuge) because pedestrians experience little delay or difficulty in finding safe gaps in which to cross in light traffic; indeed, at low traffic volumes, it is unlikely that the pedestrians will use the crossing anyway. Similarly, a minimum flow of about 50 pedestrians per hour is considered necessary before contemplating a formal crossing, i.e. an unused crossing will be given scant attention by regular travellers using the route and it could become very dangerous if used by the unwary.

Segregated crossings

Ideally, all road crossings should be of this type since, as the title implies, there is no possibility of conflict as all pedestrians traverse the carriageway by means of a subway or bridge. Unfortunately, this ideal type of crossing also happens to be the most expensive; typically, subways and footbridges are of the order of 35 and 25 times, respectively, the capital cost of a zebra crossing, and 3.5 and 2.5 times, respectively, the cost of installing a full set of traffic signals with a pedestrian phase at a junction. Segregated crossings are particularly effective on fast highways, on roads carrying high volumes of traffic, and at roundabouts where the concentration of vehicles is so high that there are insufficient gaps for pedestrians to cross.

General conditions to be taken into account when deciding whether to install a segregated crossing include vehicular and pedestrian flows, alternative types of facilities and their relative adequacies, physical problems in siting and access, capital costs, accident records, type of road and its environment, and the economic benefits to be derived from its installation. Whatever the outcome of analyses involving the above, they are of little value if the crossing is not used when constructed. Factors which most typically influence utilization are, in order of importance to the pedestrian, directness of the route, ease of negotiation, interest of specific features, general environmental appeal, and safety. From this, it can be seen that a segregated crossing that is justified solely on the basis of being, for example, 'safe and environmentally appealing' may be relatively little used, as these two particular criteria are fairly low in the pedestrian's perceived list of priorities. This is reflected in Fig. 7.12 which shows the average proportionate use made of a selected group of segregated crossings with good design features as compared with the mean observed use of a random sample of facilities.

Assuming that the facility is clean, well-designed and well-illuminated, pedestrians will still only use a segregated crossing provided that the route via the crossing is quicker than the surface route[78]. In the case of a footbridge, it is not possible to realize almost complete usage until the crossing time is about three-quarters of that by the surface route.

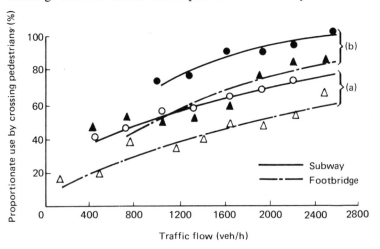

Fig. 7.12 Footbridge and subway usage[77]: (a) all sites and (b) sites with good locational/structural qualities

Table 7.21 Walking speeds of pedestrians on ramps, stairs and on the level[80]

Type of pedestrian movement	Average speed of adults (m/s)
Level walk	1.52
Slopes up and down, up to 1 in 10	1.22–1.37
Stairs	0.15 (vertically)

Subways require only a very small time saving in order to ensure 100 per cent usage.

Before constructing a segregated crossing, proper studies should be carried out to ensure that it meets the basic directional movements of the potential users. Ideally, it should be sited at a location that automatically guarantees a swifter passage; alternatively, the need for guardrails to lengthen the pedestrian path via the surface route should be examined. Desirably, there should be no traffic signals, stop sign, or existing pedestrian bridge within about 200 m of the proposed location[79]. Preliminary estimates of the likelihood of the success of a particular type of crossing under varying traffic flows can be made on the basis of walking speed and delay data such as are shown in Table 7.21 and Fig. 7.13.

Details of good design features of subway crossings are readily available in the literature[82, 83].

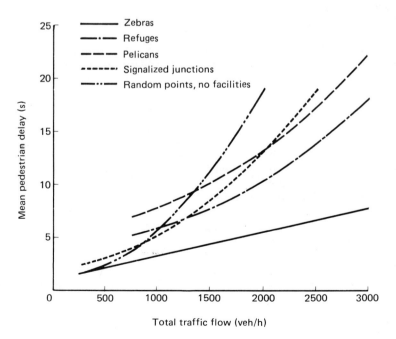

Fig. 7.13 Mean pedestrian delays associated with different surface crossing facilities[81]

Greater priority for pedestrians: precincts

There are many ways of improving streets within which pedestrian activities concentrate. Basically, they all have in common the following objectives[84]:

(1)   reduction of pedestrian–vehicle conflict,
(2)   improved control of traffic movement,
(3)   enhancement of the environment,
(4)   revitalization of the area.

The practical implementation of these objectives may involve the partial or total closure of streets.

Generally, pedestrian precinct schemes can be divided into three main types, viz.: central area/suburban schemes, regional shopping centres (hypermarkets), and residential schemes. The following discussion is concerned only with central area/suburban schemes and residential schemes.

Central area/suburban schemes

Most existing shopping and commercial centres have developed along main traffic routes and at their junctions. These centres have thrived, mainly because of the accessibility provided by the roads on which they were located; now, however, many of these roads are unable to continue providing good accessibility in comparison with other new areas, and so the attractiveness of the older centres is diminishing. The development of priority for pedestrians has been proposed as a practical method by which this problem can be alleviated, i.e. by providing attractive traffic-restricted areas and maximizing the activities and resources already available within the centres, yet retaining the accessibility already available to them.

Precinct schemes inevitably involve controversy, and good public consultation is essential to the proper development of any proposal and its likely acceptability to occupiers/owners of buildings and businesses flanking the proposed precinct area, to civic leaders who must eventually make the final decision as to whether or not to proceed, and ultimately to the general public who will decide whether or not to patronize it. Suppressing realities, and overemphasizing the advantages versus the disadvantages, may progress a marginal scheme but it will almost certainly prejudice the credibility of future proposals.

Over the years, a number of different types of precinct[85] have been developed. They can be generally distinguished according to the different levels of restriction upon vehicles (see Fig. 7.14 and Table 7.22).

Pedestrian and interrupted precincts, which involve the full closure of some streets, are the most widely developed form of precinct. Designed for pedestrians, the designated streets are normally left uncovered, although sheltered areas may be extended. Kerbs are removed and the whole area paved to the same level from wall to wall. Significant landscaping is usually carried out, and the amenities for children considerably increased.

Normally, these precincts are most suitably located in central areas. The

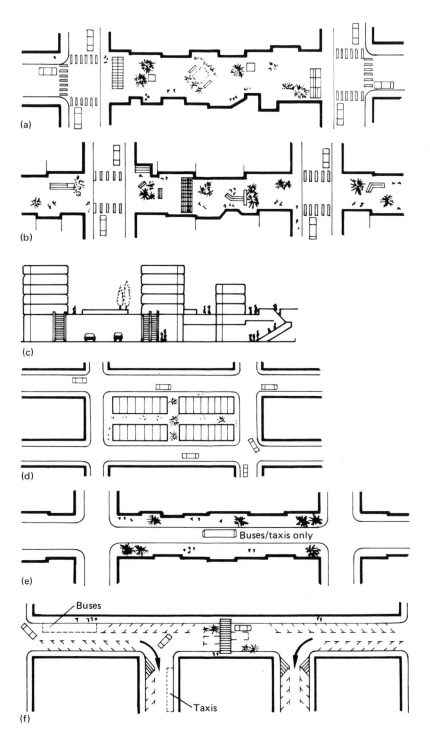

Fig. 7.14 Types of precinct with greater priority for pedestrians: (a) pedestrian, (b) interrupted, (c) multilevel, (d) displaced, (e) public transport and (f) parking precincts

Table 7.22 Effects of creating central area or suburban precincts with greater priority for pedestrians

Aspect	Type of precinct					
	Pedestrian	Interrupted	Multilevel	Displaced	Public transport	Parking
1 Street environment	Significantly improved	Significantly improved	Unaffected	Can be worsened	Improved	Improved
2 Pedestrian–vehicle conflict	Eliminated	Significantly reduced	Eliminated	Eliminated	Significantly reduced	Reduced
3 Need for comprehensive design for displaced traffic	Yes	Yes	No	No	Yes	Yes
4 Inter-street pedestrian movement	Eased	Eased	Unaffected	Eased	Eased	Eased
5 Development potential						
drive-in activities	Eliminated	Eliminated	Eliminated	Eliminated	Eliminated	Continued
non-drive-in activities	Maximized	Maximized	Maximized	Maximized	Improved	Improved
6 Kerb parking	Eliminated	Eliminated	Unaffected	Unaffected	Eliminated	Increased
7 Access by public transport	Eliminated	Significantly reduced	Unaffected	Unaffected	Improved	Continued
8 Normal street access for business servicing	Eliminated	Eliminated	Unaffected	Unaffected	Eliminated	Continued
9 Emergency vehicle access	Possible	Possible	Difficult	Possible	Easy	Easy
10 Need for extensive construction/ remodelling	Minor	Minor	Major	Major	Minor/ negligible	Minor/ negligible

need for extra car parking space is very much dependent upon the size of the central area, whether the anticipated patrons are likely to be mainly office workers or shoppers from the suburbs, and the extent to which car parking space is already available. Most businesses benefit from the pedestrianization—with the exception of those dependent upon vehicles for their development, e.g. service stations, 'fast-food' bars and wholesalers.

The use of a street or group of streets for pedestrians only means that displaced traffic must be handled by other links in the network which previously carried fewer vehicles. In selecting the alternative routes, the following will normally need to be considered[85]:

(1)   the capacity of the proposed alternative routes, both at mid-block and at intersections, to carry the additional traffic,
(2)   whether routing traffic onto a particular route will cause a change in the character of that road, or its adjacent development, through increased traffic volumes or increased noise,
(3)   the ability of the pavements on each alternative route to carry the increased pavement loading,
(4)   the suitability of the horizontal and vertical geometry of each proposed route to through traffic, especially heavy commercial vehicles and buses,
(5)   the presence of any height or weight restrictions,
(6)   the desirability of separating the circulation movement around the precinct from through movements,
(7)   the necessary linking of the circulation system to the through routes,
(8)   the maintenance of satisfactory public transport routes and services.

Where additional traffic capacity is required in the network, it may be possible to achieve this by prohibiting parking, and/or introducing one-way operation, and/or improving junction layouts in combination with better traffic control.

The displacement of public transport can be minimized by ensuring that there is provision for convenient services at the perimeter of the pedestrianized areas. In the case of large precincts, it may be appropriate to give priority to public transport and increase its attractiveness by allowing it to penetrate the pedestrianized areas at low speeds; in such instances, however, a special form of public service vehicle should be employed.

Multilevel and displaced precincts are off-street by definition, and so their traffic effects are mostly of the generation type. Normally, they also require significant new construction, and very often are enclosed. Except in the very large town centres, additional parking is usually considered by the precinct developers to be essential to ensure suitable economic returns for the finance invested, and hence this type of precinct may be considered more appropriate for non-central areas where extra parking can be more easily provided.

Public transport precincts permit public service vehicles (and possibly taxis) within their boundaries. Usually, the carriageway width is reduced, whilst the pedestrian areas are enlarged.

Parking precincts represent a return to the old High Street concept whereby people are encouraged to bring their cars and extra (controlled) street parking is provided. Slow travel speeds are mandatory, and vehicles without destinations there are discouraged from entering these precincts. The carriageway width can usually be reduced, and the pedestrian areas enlarged. An added advantage of this type of precinct is the ease with which it can be accessed by the disabled and the handicapped.

Residential precincts

In recent years, particular attention has been given to the possibility of making traffic much more compatible with housing and residential areas. The idea has been gaining ground that it would be much more sensible to structure lightly-travelled residential areas, and their streets, in a manner which accords equal movement rights to all inhabitants and visitors, whether they be motorists, motor cyclists, pedal cyclists, adult pedestrians or children. This approach appears to have been most readily accepted in the Netherlands where at least 800 residential precincts (or 'Woonerven') have been developed since enabling legislation was passed in 1976.

The main feature of the Dutch 'Woonerf' is that the residential/pedestrian function of the area clearly predominates over any provisions for traffic. Characteristic street features can be described as follows[86]:

(1)   the creation of an attractive welcoming streetscape which appeals to pedestrians, e.g. variety in paving materials, landscaping, street furniture and parking racks for bicycles,
(2)   the restriction of car parking to a limited number of sites marked on the roadway,
(3)   the creation of constrictions in the roadway wherever children often play, to make them safer,
(4)   the removal of features which suggest that motor traffic has priority, e.g. no separate carriageways or long, straight kerbs,
(5)   the use of road humps and narrow sections of roadway which change the character of the street, and ensure that traffic speeds are low,
(6)   the redesign of an area as an entity so that pedestrians have full access but it is less accessible to vehicular traffic.

British practice in respect of shared vehicle and pedestrian surfaces is as yet much more cautious, and generally their usage is only recommended for cul-de-sacs or short access loops[87].

Cycle routes, tracks and lanes

At no time since World War II has so much interest been shown by the general public in bicycles and cycling as at present. Some of the many *advantages* to be gained by the use of bicycles are as follows[88].

(1)   Cycling requires only moderate effort, except on long, steep slopes.
(2)   Bicycles need no fuel, give off no fumes, and make almost no noise.
(3)   Bicycles offer door-to-door mobility and are not constrained by public transport timetables.

(4)   About sixteen bicycles can be parked in the space required for one car.

(5)   In congested urban conditions, the bicycle is as quick for short journeys as are other forms of transport.

(6)   A bicycle can be obtained at economical cost and is economic to operate.

(7)   In isolation from other traffic, the low operating speed of a bicycle results in a high level of safety.

(8)   The bicycle offers a health bonus.

The bicycle also has some significant *disadvantages*, as follows.

(1)   Safety is a problem when bicycles have to mix with other traffic.

(2)   Security is a problem because a bicycle or its accessories can be easily stolen or damaged.

(3)   There is no protection from the weather for the rider.

(4)   There is a certain amount of physical discomfort in warmer areas, especially when there is a lack of adequate facilities to shower or change at places of work.

(5)   The distance travelled is restricted by the physical condition of the rider and the time available to make the journey.

Whilst the numbers of bicycles 'on the road' grew very significantly in the 1970s, particularly after the 'energy crisis' of 1973-74, e.g. in 1975, approximately 12m bicycles were sold in Britain as compared with about 5.25m in 1967[89], the actual usage has not grown correspondingly because of the high-risk factor associated with cycling on the public highway. It is now generally accepted that, before cycling is again fully accepted as a useful transport mode, the cyclist accident rate must be reduced dramatically. This has generated renewed interest by highway and traffic engineers in ways by which cyclist safety can be improved, as well as considering how cycling can be made a more viable transport mode.

Where cycling accidents occur

As noted previously, more than 80 per cent of cycling accidents involve a motor vehicle, and more than 40 per cent of cycling casualties are aged under fifteen years. Furthermore, some 80 per cent of cycling accidents take place during daylight hours, and about two-thirds take place at or close to an intersection.

An important analysis of cycling accidents by a RoSPA Working Party is reported[90] to have identified seven major types of accident involving a motor vehicle (see Table 7.23).

These accident findings suggest that the highest priority should be given to cyclist training on the road for children in the 10-15 year age group. Furthermore, special attention should be given in these courses, and in publicity campaigns, to problems associated with turning right into a side road, emerging from a footpath/driveway onto a carriageway, and coping with parked vehicles. Particular attention should be paid to defensive cycling training whereby cyclists are encouraged to watch out for vehicles

Table 7.23   Major types of accident involving a motor vehicle and a bicycle

Type of accident	Percentage
Cyclist turning right into a side road, motorist going straight ahead	21
Cyclist emerging from a footpath/driveway, motorist going straight ahead	18
Cyclist emerging from a side road, motorist going straight ahead	17
Cyclist and motorist going straight ahead on same road	17
Motorist emerging into path of cyclist going straight ahead	11
Cyclist colliding with a parked vehicle	8
Motorist turning into side road, cyclist going straight ahead	7

emerging from side streets and for vehicles turning right into side roads across their paths.

Above all, these data suggest the need to separate the bicycle from the motor vehicle wherever possible; this normally involves the laying out of preferred cycle routes which may incorporate low-volume (motor vehicle) 'safe' roads that are conducive to cycling, designated cycle lanes on existing carriageways, and/or separate cycle tracks. At junctions, they imply that cycle crossings should preferably be grade-separated (subways), or at least channelized (markings), so that irrational movements are minimized (see, for example, reference 91).

Planning cycle routes

Generally, cycle routes can be divided according to their function. Thus *strategy routes* are the longer-distance routes which link large areas (e.g. urban regions) together; as such they form the cycling equivalent of a primary distributor road system, and naturally tend to abut or complement primary distributor roads. *Area routes* serve as links between local areas and surrounding shared open spaces such as river parks and sports fields. *Neighbourhood routes* serve as short-distance links, often between well-defined origins and destinations within a particular local area; typically, these routes radiate from schools, the local shopping centre, or community facilities. *Recreational routes* tend to have no specific origins and destinations, i.e. they simply provide for pleasure cycling within a recreational area; they also link parks and other recreational zones.

Ideally, the selection of a cycle route should be a logical outcome of a planning process whereby information is gathered and analysed relative to probable use including the types of trips to be served, e.g. whether commuter, recreational, or neighbourhood-type travel[91]. In practice, however, the process is not so straightforward. Whilst, for example, some transport studies may aid in the establishment of general demand corridors, and thus help to locate strategy routes, they are generally limited by the coarseness of their data and the fact that most bicycle travel involves relatively short trips.

Three criteria have been described[92] as essential when planning non-recreational routes for cyclists in an urban area; these are as follows.

(1) All routes should be as direct as possible, and certainly as direct as alternative footpaths or roads.
(2) The network should link all the major points of attraction, e.g. schools, shopping and employment.
(3) Cycle routes should be separate from major roads and, where possible, the local road system.

As a result of applying these criteria, a basic network of neighbourhood routes would typically consist of a series of spokes of a wheel radiating from the centre, the lengths of the spokes being determined by typical journey distances for purposes associated with the centre; then, when spokes from adjoining centres coincide and meet, a continuous area or strategy route is formed.

Essential survey information requirements used in the identification and determination of suitable cycle routes are as listed below[93]:

(1) inventory of infrastructure—existing cycle tracks, lanes, routes and parking stands, foot/bridle paths, towpaths and parks, disused rights-of-way, bridges and tunnels not on the road network, back-street network, major road junctions, etc.,
(2) levels of use of the items under (1),
(3) accident records of the items under (1).

Also of interest in this process are data on cycle-ownership levels, cycle trip rates and lengths, and the attitudes of both cyclists and other road users.

Definitions: cycle tracks and lanes

A cycle track is legally defined in the Highways Act of 1959 as a way constituting or comprised in a highway, over which there is a right-of-way on pedal cycles with or without a right-of-way on foot. Thus a cycle track either can be a highway in its own right, i.e. separate from the road system, or it can form part of a regular highway for motor vehicles.

For the purpose of this discussion, however, it will be assumed that the term *cycle track* refers only to those portions of cycle routes that are physically separated from carriageways used by motor vehicles. The term *cycle lane* will be used to describe that portion of a highway carriageway that may be normally reserved for cyclists.

Elements of design

A well-designed cycling facility is also a well-located one. It is important to keep this in mind since the design of a cycle track or lane is most usually intertwined with its location.

Firstly, and perhaps most obviously, the cycle route and its component parts must increase the *safety* of its users. Preferably, this means separating the cyclist from the motor vehicle, and designing the tracks/lanes so that they have adequate clearances, widths and sight distances. Good junction design is a prerequisite. Proper signings and markings are also very important (see reference 94 for a summary schedule of British signing and marking practices).

Notwithstanding the obvious importance attached to the pursuit of safety, there is ample evidence to the effect that since cyclists have to develop their own muscle power, they are not willing to deviate very far from what they perceive to be the shortest or fastest route to their destination. Thus *convenience* is another critical locational/design feature. An important part of a convenient design is the vertical alignment, i.e. too many steep hills will cause cyclists to divert to alternative highway routes.

Attractiveness to cyclists is another significant criterion, i.e. the cycle route must appear 'right' to potential users. Thus, for example, particular attention needs to be paid to the design of unusual features such as subways[83] so that cyclists do not have any feelings of insecurity or claustrophobia when using them. Locations within surroundings which leave the cyclist with a satisfying visual experience should be sought wherever possible. Furthermore, the bicycle facility should be designed and maintained so as not to be unsightly in itself; cyclists are particularly sensitive to uneven and unswept surfaces, so that ease of access by maintenance vehicles is another important design consideration. Where at all possible, the cycle route should be illuminated so as to minimize its attractiveness to vandals and muggers.

The difficulties inherent in obtaining new rights-of-way in existing urban areas are such that the ability to make maximum *economic use of the existing infrastructure* is a most important consideration in route location and design in such areas. This involves measures such as selecting lightly-travelled (by motor vehicle) back-streets for use by cyclists where appropriate, designating lanes on more heavily-trafficked wide carriageways for the use of bicycles only, and determining when to accept reduced design standards (e.g. forcing the cyclist to slow down and/or dismount briefly) in order to ensure both safety and economy. There are no absolute guides available as to when special cycle facilities are warranted; as a general guide, however, a separate (lane or track) facility may be provided either where bicycle volumes are $\geqslant 200$ per day and motor vehicle volumes are $\geqslant 2000ADT$, or where the same bicycle volumes occur on roads with motor vehicle speeds $\geqslant 64\,\text{km/h}$[91].

Geometric design

Current British practice with respect to cycle route dimensions is available in the literature[95]. The following discussion therefore concentrates on features underlying the geometric design of a cycle facility.

Design speed and capacity   The speed at which a cyclist travels is dependent upon many factors, not least being the age and physical condition of the rider, the type of bicycle and its gearing arrangement, weather conditions, and the geometric features of the cycle track/path.

Whilst many bicycles have the capability of being ridden at speeds in excess of 50 km/h, most cyclists normally travel within the range 11–24 km/h, with the higher speeds being achieved on downhill sections. Typically, therefore, a suitable design speed would be 24 km/h, with about 32 km/h being appropriate for long downgrades; only in exceptional circumstances could a design figure less than 16 km/h be considered.

As yet, there appear to be no uniformly agreed capacity data for bicycle facilities. Table 7.24 summarizes some values quoted in the literature from various European countries; these (upper limit) data refer generally to travel at grades less than 2 per cent at speeds in the range 13–19 km/h.

Table 7.24 Reported capacities of cycle tracks/lanes[96]

| Traffic direction | Estimated capacity (cycles per hour) | | | | Country |
	1 lane	2 lanes	3 lanes	4 lanes	
One-way	2530				Britain
	1700–2000	3400–4000			Netherlands
		2000	3500		Germany
Two-way	850–1000		1700–2000		Netherlands
		1500	2500	4000	Germany
			10000		Britain

Widths and clearances European standards recognize the following dimensions[96] for a bicycle and cyclist: handle bar width = 0.59 m; bicycle length = 1.75 m; pedal clearance = 0.15 m; vertical space occupied by a bicycle and rider = 2.25 m. In addition, any minimum width specification must allow for lateral and vertical clearances to obstructions and lateral movements between cyclists.

Thus the clear space below any vertical obstruction (e.g. tree branch or roof of underpass) should be not less than 2.4–2.5 m above the cycleway surface, and should preferably be higher.

If a manoeuvring space of about 220 mm is allowed on either side of a single cyclist, then this would suggest a minimum cycle lane width of at least 1 m. In the case of two cyclists travelling abreast or passing each other, the minimum surface width required should never be less than about 2 m. Recommended lateral clearances to fixed obstructions vary considerably from country to country, e.g. the Netherlands = 0.48 m and Germany = 0.24 m. If a figure of 0.25 m is assumed as the very minimum clearance acceptable, then clear space widths of at least 1.5 and 2.5 m are required for one- and two-lane facilities, respectively. Where no physical barrier is provided between cycle lanes and motor vehicle lanes, additional clearances will need to be added, e.g. heavy vehicles travelling at 80 and 112 km/h need to be physically separated by at least 1.2 and 2.3 m, respectively, if the aerodynamic forces developed in passing are not to knock the cyclist to the ground.

Horizontal curvature   The minimum radius of curvature must always be consistent with the design speed of the cycle track/lane. In practice, of course, cycle lanes on shared carriageways follow the regular road alignment, and since these are designed to accommodate motor vehicles, they are normally more than adequate for cyclists. An empirical formula that is widely used for curve design on separate cycle tracks is[97] as follows:

$$R = 0.24V + 0.42$$

where R = unbraked radius of curvature (m) and V = design speed (km/h).

This relationship gives a design radius of 6.2 m for a design speed of 24 km/h. Normally, curves designed in this way are not superelevated. However, as there is a tendency for cyclists to lean into bends, there is a case for either widening on curves or increasing the clearance to fixed objects adjacent to the track. Typically, a maximum widening in the range 0.5–1.0 m may be allowed for this purpose[92].

Maximum grade   Level cycling is preferred by most riders; in practice, however, many bicycle facilities will not be level because of the nature of the existing terrain. The grade that can be permitted is very much dependent upon the length of the hill section, the characteristics (and physical condition) of the cyclist, the characteristics of the bicycle (e.g. type of bicycle, gear ratios, weight and tyres), wind speed and direction, and quality of the road surface. As all of these determinants are variable, there are no absolute values that can be specified with surety.

Figure 7.15 is a useful design guide in respect of the grade–distance relationship.

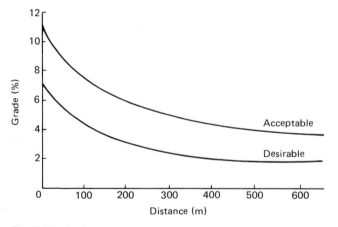

Fig. 7.15   Cycle track grade–distance criteria (based on 9.6 km/h 'steady-state' speed)[98]

Sight distance   As with conventional roadways, all types of cycle tracks and cycle lanes should be designed so that adequate sight distance is always available for safe stopping. Design values for bicycle facilities can be calculated in the same way as for highways (see Chapter 6) using the formula:

$$S = 0.278tV + \frac{V^2}{254\mu \pm G}$$

where S = safe stopping sight distance (m), t = perception–reaction time (s), V = design speed (km/h), μ = coefficient of friction, and G = grade.

Recommended values (in the USA) for t and μ are 2.5 s and 0.25, respectively[91]. On this basis, the required sight distance on a level cycle track with a design speed of 24 km/h is 25.75 m. An eye height of 1.4 m and an obstruction height of 0.1 m have been suggested[92] as appropriate for evaluating vertical curvature designs for cycle tracks in Britain.

Aids for the motorist

Traffic management aids for the motorist can be most easily divided into two groups: those that help the driver to perform the driving task safely, and those that are there to help and protect in the event that he/she gets into trouble. Aids that fall into the former category include traffic signs, carriageway markings and delineators, and anti-glare fences; safety fences and incident detection fall into the latter group.

Roadside traffic signs

In this age of the motor car, more and more people travel to further and further places. One of the problems associated with these movements is the difficulty which motorists from different countries, speaking with different tongues, have in understanding traffic regulations on highways far away from home. As a result, the tendency has developed whereby nations have begun to move towards the international standardization of traffic signs.

As of yet, however, there is no worldwide uniformity with respect to road signing. In geographic areas where the language problem is of little importance, e.g. in the USA[99], the tendency has been to make use of signs which place a greater reliance upon word legends, although significant use is also made of code-symbols. In order to overcome the language problem, other countries (particularly in Europe) have, wherever possible, adopted signs which communicate their message by means of ideographic representations rather than inscriptions that may be incomprehensible to non-local motorists.

General powers to control traffic signs in Britain were first taken[100] in the 1930 Road Traffic Act. During the 1940s and 1950s, Britain kept basically to the system first introduced in 1933, whilst Europe as a whole shifted towards a system of signing based on a United Nations Protocol agreed at Geneva in 1949. Following increasing criticisms regarding the inadequacies of the 1933 system for modern road conditions, Britain acceded to the Geneva Protocol and this was reflected in the 1964 Traffic Signs Regulations. The British commitment to the principles of international signing was reaffirmed in the regulations which came into effect in November 1975; these regulations reflect an international agreement on signs developed at Geneva in 1971 (and a protocol on road markings in 1973) under the auspices of the UN Economic Commission for Europe (ECE).

Current British practice re road signs is described in a series of publications[101] prepared by the Department of Transport.

Definition
Legally, traffic signs are 'any object or device (whether fixed or portable) for conveying to traffic on roads, or any specified description of traffic, warnings, information, requirements, restrictions or prohibitions of any description specified by regulations ... and any line or mark on a road for so conveying such warnings, information, requirements, restrictions or prohibitions'. Thus, by definition, traffic signs are composed of roadside signs, traffic signals, carriageway markings, retro-reflecting road studs, and other such indications on or adjacent to the roadway.

Some of the more important of these items are discussed separately in this section. The present discussion is concerned primarily with roadside traffic signs.

Principles of signing
Clear and efficient signing is a key element of traffic engineering. Road users rely upon signs for information and guidance, whilst highway authorities depend upon them for the efficient working/enforcement of traffic regulations, for traffic management and control, and for the promotion of road safety.

In order to obtain the greatest efficiency of usage from highway signs in general, and roadside signs in particular, the following principles of signing have been enunciated[102].

(1) Signs must be designed for the foreseeable traffic conditions and speeds on the roads on which they are to be used.
(2) They should be conspicuous, so that they will attract the attention of drivers at a sufficient distance and be easily recognizable as traffic signs.
(3) In order that the driver's attention is not distracted from the task of driving, signs should contain only essential information and their significance should be clear at a glance.
(4) Sign-lettering should be legible from sufficiently far away to be read without diverting the gaze through too great an angle.
(5) Signs should be placed so that they are obscured as little as possible by vehicles and other objects.
(6) They should be designed and sited so that, after reading the sign, the driver is left with sufficient time to take any necessary action with safety.
(7) Signs should be effective by day and night.

Two main categories of roadside traffic sign have been developed in Britain on the basis of the above principles. These are based on the concept that the shape and colour of signs should provide the main division of function; particularly that a red triangle should signify a warning, a red circle should signify a prohibition, a blue circular disc should signify a prescribed action which must be taken, and rectangular signs should give information. The division of function is shown in Table 7.25.

Table 7.25 Division of traffic sign functions[103]

Main category	Subcategory	Identification
Informatory	Network	Advance direction signs
		Direction signs
	Warning	Red triangle
		White markings (broken line)
	Other	Rectangular plates
Regulatory	Mandatory	White on blue disc
	Prohibitory	Red ring
		White markings (continuous)
		Rectangular waiting and loading plates
		Yellow markings

Informatory signs
Information signs guide the road user along established routes, inform regarding intersecting highways, direct to towns, villages and other important destinations, identify rivers, parks and historical sites, etc.; generally, they help the road user along the way. Most information signs are rectangular in shape, but the simpler direction signs usually have one end pointed. Unlike most other types, information signs do not lose effectiveness by over-use and should be erected wherever there is any doubt.

The following discussion relates to the factors which must be taken into account when designing and locating information signs. Whilst this is primarily concerned with direction signs, which account for only about 20 per cent of all traffic signs, many of the principles discussed here also apply to the design of warning, mandatory and prohibitory signs.

Lettering A driver trying to read a direction sign starts scanning it as soon as the words are legible. When the sought-for word is found, his or her gaze is returned to the carriageway ahead. If there are N words on the sign, then this operation takes about $(N/3+2)$ seconds. Of course, on average, the actual time may be about half of this because the required word is not always last; however, in the worst case, this relationship is applicable.

During the period of scanning and reading, the vehicle is drawing closer to the sign and the driver's eyes must diverge further and further away from the straight-ahead position, reaching a maximum with the finding of the required word. In Britain, the maximum *divergence from the line of sight* is used as a measure of the amount of distraction produced by the sign. Visual acuity falls sharply at points more than 5 degrees from the line of sight, and so it is generally accepted that the maximum divergence should be limited to 10 degrees, so that the edge of the carriageway up to about 15.25 m ahead of the vehicle is still within 5 degrees of the line of sight.

The visual acuity of drivers is a factor of considerable importance. Research has shown that the great majority of drivers can read lower-case letters with an x-height of 25.4 mm at a distance of about 25.5 m, so it is possible to use this relationship as the basis of letter-height design on signs. (The x-height of lower-case script is the height of letters such as u, m or x). Width to height ratios ranging from 1:1.33 to 1:1 are optimum for upper-case alphanumeric characters, with numerals being slightly narrower than letters; widths less than 66 per cent of character height should be avoided because of the marked decrease in legibility when viewing angles become acute[104].

Figure 7.16 illustrates a typical situation where a sign is positioned so that it makes a lateral displacement of S metres from the driver's path. In order to meet the maximum divergence criterion of 10 degrees, the driver must have completed reading the message by the time point B is reached; this is at a distance of $S \cot 10°$ from the sign. If the vehicle speed is V km/h and it takes $(N/3+2)$ seconds to read the sign, then the driver must begin reading at point A, which is $0.278V(N/3+2)$ metres from point B. Thus the driver must begin reading at a total distance from the sign of

$$d_{AB} + d_{BC} = S \cot 10° + 0.278V(N/3+2)$$

If the driver can read a 25.4 mm x-height letter at a distance of 15.25 m, then the required *letter height* is

$$x = \frac{S \cot 10° + 0.278V(N/3+2)}{15.25} \times 25.4$$

where x = height of lower-case letter (mm), S = offset distance to the centre of the sign, in metres (in the case of a multilane highway, the vehicle should be assumed to be in the fast lane), V = speed of vehicle (km/h), and N = number of words on the sign.

In these equations, the speed value V should normally be the design speed of the road and not, as is suggested by some, the average speed. It is logical to assume that if the road has a design speed of, say, 80 km/h then the signs should also be tied to that speed. On 'old' roads, the existing 85th percentile speed may be considered a suitable value to use.

There has been a good deal of argument about the relative merits of *upper- versus lower-case letters*, and conflicting claims are often made

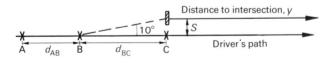

Fig. 7.16   Calculation of letter size for advance direction signs

about the merits of the two types. One investigation which compared the distances at which signs of equal area but with various types of lettering could be read[105] showed that there was no significant difference between the results obtained with good lower-case, upper-case without serifs, and upper-case with serifs lettering. As a result, the choice can be practically considered as a matter of preference or based on aesthetics.

A practical point to consider is that a sign with lower-case lettering is narrower and taller than an equal-area, equally-legible sign with upper-case lettering. Thus, when there is a restriction on the width of a sign, e.g. on footpaths in towns or on narrow country roads, there is an advantage in using lower-case letters.

Layout Bigger signs cost more money, so it is necessary to be aware of the most economical layout for directional signs. Two examples of sign layout are illustrated in Fig. 7.17. Note that both signs have exactly the same total area but the letters in the stack layout are larger.

By far the best results are obtained with the stack type of layout; for a given sign area, stack signs are legible at greater distances than diagrammatic signs[106]. Stack signs are most properly used at straightforward intersections, whilst the introduction of diagrammatic signs at complex junctions leads to a reduction in erratic manoeuvres and 'errors' at these intersections.

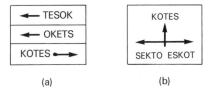

(a) (b)

Fig. 7.17 Examples of sign layout: (a) stack layout and (b) map layout

Colour The choice of sign colour is to a large degree limited by technical and aesthetic requirements, as well as by cost. To be easily read, letters should be of light colour on a dark background, or of dark colour on a light background. If the sign is a small one, e.g. of area $< 1.85\,\mathrm{m^2}$, the target value of the sign is increased by using dark letters on a light background (see Table 7.26). In contrast, light letters on a dark background are suitable for large signs, e.g. on motorways.

From an economic aspect, light lettering on a dark background is to be preferred since it is cheaper to reflectorize letters rather than the whole

Table 7.26 Areas at which colours are equally conspicuous at a distance of 230 m[107]

Background colour	Sign area (m²)
White	1.49
Red	1.67
Blue	1.86
Green	2.04
Black	3.34

background area of a sign. British practice is for directional signs on motorways to have white lettering on a blue background. The primary route network of trunk and principal routes utilizes white lettering on a dark green background and route numbers in yellow. The remaining A-class (non-primary) principal roads and B-class roads have black lettering on a white background.

Siting From the point of view of reflectorization, the closer the sign is to the edge of the carriageway, the greater the relative intensity of reflectorization. However, desirable though it might appear to be, signs should not be located too close to the highway edge; a sign that is too close will not only become spattered with mud, thereby severely reducing its reflectivity, but it may also constitute a hazard to traffic. Generally, the edge of a directional sign should never be closer than, say, 1.2 m to the carriageway edge in the case of a high-speed road without shoulders, or 0.6 m to the edge of a hardshoulder in the case of a high-speed road with shoulders. With lower-speed roads, British practice is for the sign to be as close as 0.45 m to the edge of the carriageway.

The lower-edge of the sign should be as close as possible to the eye height of the driver; typical practice is for it to be between 0.9 and 1.5 m above the highest point of the carriageway alongside, with the higher mounting being used where excessive spray is likely to soil the sign. Of course, if the sign is erected transverse to a footpath in a built-up area, there should be a clearance of at least 2.1 m to allow for pedestrians to walk beneath.

Most traffic signs are set transverse to the line of travel of highway vehicles—with the obvious exceptions of plates detailing parking restrictions (which are usually parallel to the kerb), some direction signs (which must point approximately in the direction to be taken), and some signs which must be on both sides of the carriageway (e.g. speed limit signs). Experience has shown that in rural areas specular reflection from traffic signs can be troublesome to oncoming vehicles; to minimize this effect, signs are therefore normally set at an angle of 95 degrees from the left-side (the approach side) edge of the carriageway, so that they face slightly away from the beam direction of the headlights of vehicles when they are within 200 m of the signs.

In order to allow a driver sufficient time to comply in safety with the message given, the sign itself must be sited at the correct distance before the location to which it refers. This distance will normally vary with the design speed/speed limit of the highway. Thus, for example, advance direction signs should be sited sufficiently far in advance of the intersection for the motorist to make the appropriate manoeuvre—which may necessitate stopping—without endangering him- or her-self or others. If the sign is located as in Fig. 7.16, the driver will finish reading the sign at a distance of at least $S \cot 10°$ or $5.7S$ metres from the sign. If the sign is assumed to be Y metres from a junction, then the distance $(Y + 5.7S)$ metres must be equal to or greater than the minimum safe stopping sight distance.

On heavily-travelled multilane roads, and at complicated intersections, overhead signs can be used advantageously to direct traffic; these signs are normally mounted at least 5.5 m above the lanes to which the information applies. Advantages of overhead signs are a clear view of the message— most important in urban areas—greater mounting height giving greater visibility at curves and hills, an ability to provide larger lettering, and the elimination of the need for the driver to move his or her eyes horizontally away from the line of travel. Overhead signs have the disadvantage of being very costly, as compared with roadside signs, because of the specially designed larger structures needed to support them. To avoid being lost against the normal sky, overhead signs usually have white letters imposed on a dark background.

Usage   Informatory signs can be grouped according to type, as follows:

(1)   advance direction signs—approaching junctions,
(2)   direction signs— at junctions,
(3)   route confirmatory signs—leaving junctions,
(4)   temporary direction signs—at roadworks, etc.,
(5)   miscellaneous—e.g. alternative routes for heavy goods vehicles,

or according to a place-name classification system (see Table 7.27).

Advance direction signs have only one place-name for each direction, as a general rule. On primary routes, this name should be the next place of major traffic importance along the immediate route, i.e. the primary destination.

Once a place-name has appeared on an advance direction sign, it should appear on all subsequent signs of this type until the destination is reached. Hence the number of destinations to be signed on any one route should be as small as possible in order to keep place-names on any one sign to the absolute minimum.

On motorways, the directional signing at each junction generally only shows the most important destinations. More information is often included on primary and non-primary directional signing, with primary destinations being given greater prominence than other destinations on individual signs.

Table 7.27   Place-name classifications[108]

Classification	Number in class	Examples
Regional	11	London, The North, South Wales, The Lakes
Super-primary	21	Birmingham, Dover, Brighton
Primary	361	Slough, Staines, Crawley, Heathrow
Special	—	Ring Road, City Centre (named), Town Centre
Local	—	All other destinations

Confirmatory signs reassure the driver after passing through the junction regarding the correctness of the route. Primary destinations up to about 250 km ahead can be shown on these signs, well before they appear on advance direction signs. Again, once a place-name has appeared on a route confirmatory sign, it should reappear on all subsequent confirmatory signs until the destination is reached.

Environment   It can be difficult reconciling the signing needs of the road user with the aesthetic requirements of the surrounding area. From an aesthetic point of view, a sign should normally be as unobtrusive as possible; from the driver's point of view, a sign should be as conspicuous as possible. A graphic artist designing a sign may wish it to be larger than might be necessary for legibility, so as to make it more attractive, e.g. by providing generous margins and bigger spaces between words; the highway authority is concerned with the cost of the sign and therefore prefers it to be as small as possible.

It is fortunate that the interests of amenity and the traffic engineer often coincide. Normally, the engineer states the minimum requirements that will meet the road user's needs, and the designer then produces the most economical design within these limitations. Thus, for example, current British practice re signs for highways designed for 80 km/h or more[109] is that those which require driver decisions should be limited to a choice between two alternative routes at any time, and the number of place-names associated with each should be as small as possible, and not exceed a total of four on any one sign; this degree of control obviously helps to keep the sign smaller than it might otherwise be. In urban areas, smaller signs can be used because of the lower speeds; they should be used (with light backgrounds) wherever large signs would be out of scale with the surroundings. In addition, they should be maintained in good condition and removed when no longer required.

Support posts for information signs need not be distinctive; in fact, they should blend into the background. Thus it is usual practice in Britain for posts, bracing and fixing clips, and the backs of signs to be a nondescript grey colour.

Warning and regulatory signs

The most numerous and probably the most important of the signs are the warning and regulative ones. The major signing systems in use throughout the world utilize combinations of shapes and colours to distinguish various classes of these signs. As noted previously, the British system conforms generally to the European system as defined at Geneva in 1971.

Warning signs are used to alert the road user regarding potential dangers ahead. They indicate a need for extra caution by road users, e.g. by requiring a reduction in speed, or some other such manoeuvre. To be most effective, warning signs should be used sparingly; over-frequent usage to warn re conditions that are otherwise readily apparent will tend to bring these signs into disrepute and detract from their effectiveness.

In the British system, most warning signs are distinguished by an equilateral triangle with a red border encompassing a black symbol—this is usually a pictogram of the potential hazard—superimposed on a white background. Typically, they are used at such locations as approaches to intersections not previously indicated by advance direction signs, dangerous bends or hills, concealed or unguarded level crossings, and near schools, pedestrian crossings, and converging lanes, and at other locations where the driver requires warning of hazardous conditions on or adjacent to the carriageway. Table 7.28 summarizes British practice in respect of the usage of warning signs.

Perception of a sign does not necessarily imply legibility, and for *mandatory signs* the form and colour of the sign are more important than its text[110]. Most mandatory signs are circular with white or light-coloured symbols on a blue background. Important exceptions are the STOP and GIVE WAY signs which have distinct shapes and colours, as well as capital letters, in order to produce a more forcible impact on the road user.

Mandatory signs give definite instructions when it is necessary for the motorist to take some positive action, e.g. 'turn left'. The octagonal STOP sign is used only at junctions where the visibility is so bad that it is imperative for the driver to stop on every occasion. This sign should not be used indiscriminately; if used where stopping is rarely necessary, its impact on drivers is depreciated and it will tend to be ignored with potentially serious consequences for road safety. The inverted triangular GIVE WAY sign is used at intersections where control is not exercised by traffic signals, police or STOP signs, but where there is the need for drivers on minor roads to proceed so that they do not cause inconvenience or danger to traffic on major roads, e.g. on minor roads at junctions in heavily-travelled rural areas.

As the name implies, *prohibitory signs* generally give definite negative orders which prohibit the motorist from carrying out particular manoeuvres, e.g. 'no right-turn' and 'no entry'. With the exception of the waiting restriction and no entry sign, all prohibitory signs are circular

Table 7.28   British practice in respect of warning signs

85th percentile approach speed of cars, x (km/h)	Typical roads	Distance of sign from hazard (m)	Recommended clear visibility of sign (m)
$x \leqslant 32.0$	Narrow rural or very narrow urban roads carrying <1500 vehicles per day and <350 commercial vehicles per day	45	60
$32.0 < x \leqslant 48.0$	Urban and other rural roads of a local character	45	60
$48.0 < x \leqslant 64.0$	Urban and rural two-lane single carriageway roads	45–100	60
$64.0 < x \leqslant 80.5$	Urban motorways and high-standard two- or three-lane rural roads with few junctions	110–180	75
$80.5 < x \leqslant 96.5$	Dual and single carriageway roads of ⩾ three lanes	180–245	75
$x > 96.5$	Motorways and high-standard all-purpose dual carriageway roads	245–305	105

with a red or white centre; the symbols or inscriptions are black or dark blue. In addition, the three turning prohibitory signs have diagonal bars across their faces.

Carriageway delineation

There are conflicting versions as to when and where the first road-marking delineator was used. It would appear, however, that a road existed in the 1960s in Mexico[111] in which the paving material showed a centreline of contrasting colour; this roadway was constructed about the year 1600. The centreline was of a white material and appears to have been designed to separate the flow of traffic in opposite directions.

Whatever their exact origin, modern carriageway delineators (including markings) have definite and important functions to perform in a proper scheme of traffic management. In many instances, they are used to supplement the regulations (or warnings) of traffic signs or signals. In other situations, they are used to obtain results, entirely on their own merits, that cannot be obtained with other devices.

Carriageway delineators have, however, several definite limitations to their effectiveness. They may be obliterated by snow or dirt, obscured when the volume of traffic is heavy, and some are not readily visible when the carriageway is wet. Markings are not very durable when subject to heavy traffic wear and must be replaced at frequent intervals; in addition, they cannot be utilized at all on unsurfaced carriageways.

The following discussion will concentrate on the main longitudinal delineators which feed information continuously to the driver. Basically, there are three types of delineator used for this purpose: line markings, road studs, and post-mounted delineators. Before discussing these, however, it is useful to consider how they aid the motorist in carrying out the driving task.

Reasons for longitudinal delineation

Generally, it can be said that the various forms of longitudinal delineation have three main functions: (a) to characterize the road, (b) to provide route guidance, and (c) to act as a tracking reference.

By *characterization* is meant the situation where particular forms of delineation are used to provide the driver with information about the nature of the road which leads him or her to expectations regarding the ease of the driving task. For example, if the centreline on a carriageway were to be marked in yellow, then it could indicate to the driver that this is a two-way carriageway, whereas a white centreline (or lane-line) might be used to indicate a one-way carriageway.

The term *route guidance* is used to define the situation where carriageway delineation might be used to direct the motorist into one of, say, two alternative routes. Consider, for example, a simple Y-junction at which all three legs appear to be of the same route hierarchy so that the motorist entering the junction along its stem has difficulty in discerning which of the two remaining legs is the major route. If, however, a particular form (or colour) of delineation were to be used on the stem and continued through the appropriate (left or right) leg, it would be more clear to the driver which route to follow.

Common practical examples of delineation being used for route guidance purposes are the standard treatments at motorway exits (see Fig. 7.18) and entrances, and lane drops which involve merging (e.g. a dual carriageway with four traffic lanes merging into a two-lane single carriageway) or diverging (e.g. a three-lane carriageway on a motorway where one lane peels off at an exit, and two lanes continue on the main route).

In relation to the *tracking reference* function, it should be remembered that a motorist has two basic types of motion continually under control to ensure that the vehicle's path remains with the road at night or day: these are speed and direction. While at first one might think that speed is judged by looking at the vehicle's speedometer, the fact is that, in practice, the driver prefers to use subjective judgement by noting the rate at which objects by the side of the road are passed—and at night, particularly on rural roads, these reference objects are not visible. Of greater real importance is directional control since[112] 'in driving, the task is not exactly that of following a line, but rather of remaining at a constant lateral displacement from it, the line being either the kerb or a white guideline'. It has been shown by experiment that the centre and edge of the carriageway are two of the principal locations used by drivers in directional guidance, and hence strong emphasis at these locations, especially at curves at night, can greatly simplify the driving task. Furthermore, the higher the vehicle speed, the more useful is this longitudinal contrast delineation.

Also at night, the uncertainty caused by the glare of oncoming vehicles can be heightened by a lack of delineation. For example, a driver meeting an oncoming vehicle is inclined to look at the carriageway edge so as to avoid looking directly at the headlights. If, however, the road edge is indistinct, the driver must glance back periodically at the centreline to check the vehicle's lateral position—but, in so doing, his or her eyes become more 'light-adapted' so that, on looking back again, the carriageway edge appears even more indistinct so the driver must look again at the centreline to regain his or her lateral bearings. The situation can, in fact, develop to the stage (in heavy traffic flow) where the driver cannot discern the centreline because of the glare, and has to look directly at the oncoming vehicle's headlights in order to become oriented—which explains why the deliberate delineation of the edge of the carriageway must help this situation.

Line marking
The two most widely used line-marking materials are paints and plastic compounds (cold applied and thermoplastics).

There are very many *paints* available for road-marking purposes; these can be classified by the type of base, i.e. alkyd, epoxy, rubber, vinyl, water base and high polymer. Drying times vary according to the type of

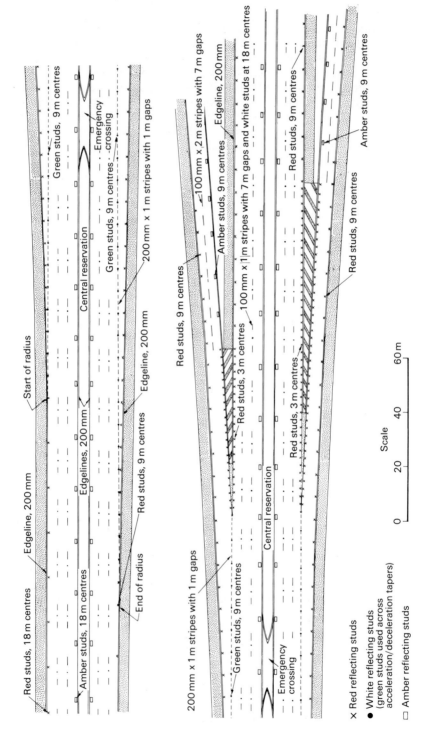

Fig. 7.18 Delineation practices on British motorways

Red studs, 18 m centres

Edgeline, 200 mm

Amber studs, 18 m centres

Edgelines, 200 mm

Green studs, 9 m centres

Central reservation

Green studs, 9 m centres

Emergency crossing

Red studs, 9 m centres

End of radius

Start of radius

Edgeline, 200 mm

200 mm × 1 m stripes with 1 m gaps

Red studs, 9 m centres

200 mm × 1 m stripes with 1 m gaps

Green studs, 9 m centres

Emergency crossing

Central reservation

Red studs, 3 m centres

Red studs, 3 m centres

Red studs, 9 m centres

100 mm × 1 m stripes with 7 m gaps

100 mm × 2 m stripes with 7 m gaps

Edgeline, 200 mm

Amber studs, 9 m centres

Amber studs, 9 m centres

100 mm × 1 m stripes with 7 m gaps and white studs at 18 m centres

Red studs, 9 m centres

Red studs, 9 m centres

Amber studs, 9 m centres

Scale

0   20   40   60 m

× Red reflecting studs

● White reflecting studs
  (green studs used across
  acceleration/deceleration tapers)

□ Amber reflecting studs

paint, laid thickness, and atmospheric and roadway conditions; typically, however, they range from 10–60 minutes.

Between 80 and 90 per cent of the road lines laid in Britain are *thermoplastic*, whereas on the Continent and in the USA the reverse is generally true. Thermoplastics are of two general types: extruded hot plastic and spray-on. Both are intended for long-life installations and require special application equipment. In addition, thermoplastics are more durable on bituminous surfacings than on concrete roads.

Thermoplastic has the following advantages.

(1)   It has a longer life as compared with paint.

(2)   It fills the interstices of rough-textured roads, whereas paint soon wears from the surface-dressing peaks and the interstices fill with dirt.

(3)   Its high temperature of application enables it to fuse with a bituminous road surface, sometimes even when the road is cold or slightly damp.

(4)   It is proud of the road surface, and this assists visibility on a wet night by facilitating drainage of the water film.

(5)   It contains 60 per cent sand and a binder which ensures good skid-resistance as it erodes.

(6)   The material has a rough surface when laid which aids immediate diffusion.

(7)   It sets almost immediately after being laid; typically, traffic can travel on it as little as 5 s after application[113].

The disadvantages of thermoplastic are as follows.

(1)   It has a greater initial cost as compared with paint.

(2)   Rapid application on a large scale is more difficult because of the large bulk of material which has to be melted down.

(3)   Care is needed to avoid an undue build-up of thickness by successive applications, as this can be hazardous to motor cyclists.

(4)   Adhesion is usually poor on concrete road surfaces.

(5)   Under hot weather conditions, thermoplastic has a tendency to creep.

(6)   On dirty roads carrying light traffic, thermoplastic discolours more readily than does paint.

The *reflectorization* of a paint or thermoplastic road marking is achieved by the addition of tiny glass spheres (ballotini) which are premixed and/or dispensed ('dusted') onto the surface of a line material as it is being laid. Premixing, very often followed by dusting, is the usual practice with thermoplastic; dusting is the more common practice with paint, although reflectivity only exists as long as the beads remain in place on the surface. An advantage of premixing is that as the binding material becomes worn by traffic, further beads are exposed to reflect the light from the vehicle's headlights back to the driver.

While the excellent reflectivity properties of lines containing ballotini are well recognized in dry weather at night, they are limited in their effectiveness during rain, fog and (obviously) snow. In the case of rain and fog, both experiments and experience have shown that the reflectivity of the lines decreases as rain falls, i.e. the water films covering the beads reflect the light thrown from vehicle headlamps away from the driver rather than back. When there is sufficient moisture to submerge the beads—as most easily happens with paint—the lines can become practically invisible.

In Britain, white materials are used for road markings intended for moving traffic, i.e. centrelines, lane-lines and edgelines.

Centrelines Centrelines indicate the division of travelled way carrying traffic in both directions. They are usually denoted by broken single lines; these dashed lines act as traffic separators which may be crossed at the discretion of drivers. The underlying design principle is for the dashed lines to be 'weak', i.e. short with long spacings in between, on straight and safe sections of highway, and for them to be 'strengthened', i.e. made longer within the same overall module, as road sections are approached where increased caution is required but the line may still be crossed. At locations where line-crossing is most dangerous, the dashed warning line becomes a solid continuous line.

Centrelines are normally provided on two-way surfaced single carriageway highways with an even number of lanes. At sites such as a carriageway-width transition or where an extra uphill traffic lane is provided, the 'centreline' may not be actually located at the geometrical centre of the road.

Two contiguous parallel lines are normally used as a centreline at horizontal and vertical curves on two- and three-lane highways where overtaking is prohibited because of restricted sight distances or other dangerous conditions. The parallel lines may be both continuous or one may be broken and the other continuous. If the nearside line is broken, it indicates that drivers on that side may cross the lines at their discretion; if the nearside line is continuous, they may not cross.

Figure 7.19 shows how a no-overtaking zone can be very simply established on a vertical curve. Two workers each holding a 1.05 m high vertical target walk along the carriageway in direction A at a constant distance (the minimum overtaking sight distance) apart. When B's target can no longer be seen through a 'peep-hole' aperture on A's target, A_s and B_f are marked on the carriageway; these are the start of the continuous line in direction A and the finish of the continuous line in direction B, respectively. When B's target again comes into A's sight, A_f and B_s are marked; these are the finish of the solid line in direction A and the start of the solid line in direction B, respectively. Both lines are then spotted out and marked in accordance with standard practice.

Lane-lines Lane-lines are particularly useful in the organization of traffic into its proper channels, and in increasing the efficiency of carriageway usage at congested locations. They are normally used on rural highways with three or more lanes, on one-way carriageways, at the approaches to important intersections and pedestrian crossings, and on congested urban

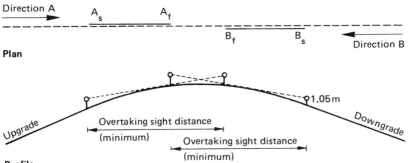

Fig. 7.19 Establishing no-overtaking zones on a vertical curve

streets where the carriageway will accommodate more lanes of traffic than would be the case without the use of lane-lines. Lane-lines are usually indicated by broken lines, the marks being considerably shorter than the gaps. At locations requiring extra caution, stronger lane-line markings may be used to help to alert the driver to the conditions prevailing.

Edgelines These are continuous or broken longitudinal line markings which indicate to drivers the location of the edge of the carriageway on unilluminated rural roads, laybys and intersections. Simple logic and the studies which have been carried out would appear to suggest, with regard to delineating the edge of the carriageway, the following points[114].

(1) Drivers like edgelines. More important, they feel that they make the driving task easier and more secure, and hence they can enjoy the drive much more.
(2) Edgelines do not cause increases in the number or severity of accidents. It is likely that they reduce accidents, particularly at junctions, *provided that* the junctions are at least partially unmarked.
(3) Edgelines do little for the motorist during the day, when the contrast between the carriageway and the side of the road is such as to make it readily distinguishable.
(4) When the shoulder or verge is not in good contrast with the carriageway surface, a properly located edgeline can be very effective in ensuring that the motorist stays on the designated travelway both by day and night.
(5) On a normal width, single carriageway road at night, an edgeline may result in vehicles moving small distances in a lateral direction towards the centreline. The amount of movement will probably be influenced by the location of the edgeline relative to the edge of the carriageway. There is no evidence to suggest that this small amount of lateral movement is a cause for concern in relation to accidents.
(6) Under normal dry weather circumstances by day or night, edgelining has no practical effect upon speed. No data are available regarding either speed or placement under wet weather conditions.

(7) Whether the edgeline is dashed or continuous has little or no effect upon either vehicle placement or speed. This assumes that the dashed line gives the general appearance of a near-continuous line when the motorist is travelling 'with the traffic' on the roadway.

(8) Edgelining cannot but result in less roadside maintenance being required.

(9) Edgelining should help to reduce pedestrian accidents in rural areas, by more clearly designating the part of the roadway where the person on foot may not walk.

British practice is for two types of edgeline to be used on all-purpose roads; both are white in colour and 100 mm wide. The first is a continuous solid line, whilst the second is a dashed line on a 4.5 m module using a 1 m mark. The lines are laid with the inside edge of the marking 220 mm from the nearside edge of the carriageway, and may not be continued across junctions or laybys. The more emphatic solid line is used: (a) where the demarcation between the carriageway and verge is particularly bad, (b) on roads prone to fog and mist, (c) where headlamp dazzle is severe on heavily-trafficked two- and three-lane roads, (d) at sudden changes of carriageway width, (e) on approaches to narrow bridges, and (f) on approaches to bends indicated by warning signs. Since the Department of Transport requires that a traffic lane should be at least 2.75 m wide, only roads of 5.8 m or more in width may carry both a centreline and edgelines.

Road studs

First used about 1936, road studs can be divided into three main groups, viz.: cat's-eye, corner-cube, and ceramic. Road studs are generally retro-reflective for night visibility; in fact, many engineers are convinced that the 'ideal' answer to the wet reflection problem in snow-free areas is to use reflectorized road lines in combination with reflectorized road studs. Non-reflective road studs are designed for daytime delineation (as well as marking pedestrian crossings and parking bays). Road studs come in a host of shapes and sizes, e.g. round, rectangular, triangular, dome-shaped and wedge-shaped, and are made from a variety of materials, e.g. steel, cast-iron, glass, rubber, ceramics and plastics (polyester, acrylic).

Cat's-eye reflecting road studs (reflex lens type) are used extensively in Britain. Each cat's-eye consists of two parts: a metal base embedded in the road and a separate rubber pad insert, into each side of which (for two-way roads), or in one side (for one-way carriageways), two longitudinal biconvex reflectors are fixed. As vehicle tyres pass over the rubber pad, its centre part is depressed so that the faces of the reflectors are automatically wiped by the front part of the pad, thus giving the stud its well-known self-cleaning property. The length of time before the insert rubber pad must be replaced depends very much upon the speed and density of the traffic, as well as upon the lateral location of the road stud on the carriageway; however, measurements have shown that the reflectivity of a cat's-eye on a centreline of a high-speed road can fall to 50 per cent of its original value after twelve months.

Corner-cube road studs are bonded directly to the road surface by means of an epoxy resin. They are composed of a plastic shell containing a reflector face (or two faces, if used for dual guidance purposes on a two-way road) made up of numerous reflective 'corner-cubes'. The reason for the name corner-cube is that the individual reflectors each consist of three sides of a cube, and a headlight ray is reflected through all three sides before returning to the eye of the motorist. If the reflector is to work effectively, the angle between each side of the cube (and the adjacent one) must be within a few minutes of 90 degrees; if it is not, the light will be reflected back in too wide a cone with consequent loss of intensity.

When both are new, corner-cube reflectors (reflective area $= 2000\,\text{mm}^2$) are optically more efficient than cat's-eye reflectors (reflective area $= 130\,\text{mm}^2$). Objective measurements have shown that a corner-cube road stud can return as much as twenty times more light than a cat's-eye at a distance of 100 m or more; however, when viewed more obliquely, e.g. at distances of 30 m or less, its superiority is not as great. Even though abrasion from tyres soon causes the face of a corner-cube reflector to become etched with a network of fine scratches which cause diffusion of much of the light (typically, the reflecting power can be reduced to about 10 per cent of its original value in a year), yet at long distances it can still be considerably brighter than a cat's-eye after the same period of time.

Reflecting road studs are reported[115] as having the following advantages over conventional line markings.

(1) They provide increased reflectivity under wet weather conditions.
(2) Their durability and life are much greater than those of painted lines.
(3) The vibration and audible tone produced by vehicles crossing the road studs creates a secondary warning.
(4) Replacement is generally less frequent than for painted lines, and hazardous repainting operations under heavy traffic conditions can often be avoided.
(5) The use of different coloured reflex lenses allows the imposition of directional control upon the motorist, e.g. in conveying a 'wrong way' message.
(6) They can be used as transverse rumble bars.

Disadvantages of road studs are as follows.

(1) They have a relatively high initial cost.
(2) They are susceptible to damage during snow-clearing operations.
(3) They are not generally practicable unless the roadway has a relatively high-quality surface which does not require an immediate overlay or surface dressing.

Experience would suggest[113] that under the most severe conditions the service life of a corner-cube reflective marker is as low as six months to one year. For most motorway locations, life expectancy varies from three to eight years, and to over ten years for rural roads with low traffic volumes. In Britain, road studs must conform to a standard specification[116].

In the USA and Australia, *ceramic road studs* are used for daytime visibility and to supplement the corner-cube studs at night when wet. They are often placed on a 3 mm butyl pad in order to provide a yielding underbase. Ceramic road studs, which are expected to last in excess of ten years, provide little night delineation in dry weather, i.e. they become covered with grime, pitted and badly tyre-stained from traffic during extended periods of hot dry weather.

Figure 7.18 shows the line-marking and road-stud delineation practice on British motorways, including at junctions. Note the different types of line and stud used at different locations, each intended to convey a message to the motorist regarding that position.

Post delineators

The purpose of this type of delineator is to assist the night driver by outlining the horizontal and vertical road alignment. Historically, post delineators tend to have developed as simple concrete posts which were first painted white, and then reflectorized devices were fastened to them so as to face oncoming vehicles. In recent years, however, most countries have moved away from the use of rigid posts for safety reasons, and replaced them with guide posts of such materials as (flexible) neoprene, polyvinyl or rubber, or breakaway wooden posts.

Whilst they are not widely used in Britain, post delineators of varying forms and shapes are extensively used on the Continent and in the USA for night and poor weather delineation where road markings are ineffective. They are used to supplement (not replace) line markings, and are not used where the highway is illuminated. These delineators define the highway alignment most effectively when placed close (both vertically and horizontally) to the edge of the carriageway. If too close, however, their effectiveness is reduced very quickly by dust and mud splash (see Table 7.29) and they must be hand-cleaned very regularly. Typically, therefore, the reflectors are mounted on posts about 1 m above the ground and set back 1 m from the carriageway edge of roads without shoulders.

Table 7.29   Effect of dirt upon the reflective power of post-mounted reflectors on the central reservation of a motorway[117]

Weeks exposed	Reflective power (percentage of clean value)
0	100.0
1	17.6
3	5.7
5	1.4
9	0.4
11	0.7
13	1.0
15	0.5
17	0.5

Note: The height of the reflectors above the road was 850 mm and the offset distance from the road to the reflector posts was 600 mm.

Different coloured reflectors are normally used on posts on opposite sides of the road. The reason for differentiating between left and right is primarily to enable the motorist looking a long distance to determine the manner in which the highway is curving; if the same type of reflector is used on either side, the two lines can be indistinguishable from a distance. If white reflectors are used on both sides, there is the particular danger of their being confused with opposing headlights.

A disadvantage of post delineators is that they interfere with roadside maintenance. This can be alleviated by painting the tops of the posts with reflectorized paint so as to make them more easily seen by maintenance crews; because of the low intensity, the painted tops will not normally detract from the positive delineation pattern for the average driver.

Anti-dazzle screens

As noted previously, an important consideration affecting a driver's vision at night is the dazzle from oncoming vehicle headlights on unilluminated highways. In very heavy traffic, the result can be that a driver is subjected to dangerous 'glare-out'. On heavily-travelled dual carriageways, anti-dazzle screens are sometimes erected on the central reservation to alleviate this problem.

The main design criterion for an anti-dazzle screen is that it should cut off the light from oncoming vehicles directed towards the driver at oblique angles (up to 15-20 degrees). From an environmental aspect, it is also desirable that the screen should not be very noticeable, and that as much open vision as possible is maintained in a sideways (perpendicular) direction. For it to be effective in screening the light from all types of oncoming vehicles (including heavy commercial vehicles), the screen must reach to at least 1.73 m above the carriageway.

A vane-type of anti-dazzle screen which meets these criteria has been tested by the Transport and Road Research Laboratory.[118]. This screen, which uses vertical plastic vanes of a dull green colour mounted above a central reservation safety fence at 0.8 m centres, was not noticed by between one-quarter and one-third of the drivers on a motorway where it was evaluated. More important, however, the number of night accidents observed on the screened section of highway was 44 per cent lower than that expected from the evidence on the unscreened control length; this represented a saving of 0.9 accidents per kilometre per year. Other subsidiary effects were a reduction in the number of vehicles using main-beam headlamps, and an average lateral movement of 150-250 mm away from the screen.

It might be noted that hedges of *Rosa Multiflora Japonica* have been planted in wide central reservations in some countries to serve as anti-dazzle screens, as well as improving the environment.

Fixed roadside hazards

Many road accidents involve collisions between out-of-control vehicles and individual fixed objects adjacent to the carriageway. Many of these accidents involve fatalities. Typical fixed objects that are struck by vehicles include light and electricity poles, sign supports, trees, bridge abutments, piers and parapets.

Research has shown that on average 85 per cent of all out-of-control vehicles striking an obstacle do so within 10 m of the roadway. A simple solution to the roadside hazard problem therefore is to ensure that no such obstacles are allowed within this 10 m vulnerability zone. In practice, however, this is very often not possible, even in the case of new highways.

In the case of *poles and posts*, the number of accidents may be reduced by increased spacing, joint use by different utility authorities, and/or selective resiting. Frangible light poles and sign posts have also been developed and are in use in the USA[119]; these break near the bottom under impact, thereby reducing the damage caused to vehicles and injury to their occupants. A breakaway electricity pole has also been developed which is claimed[120] to be able to reduce accident severity without causing the downfall of electric wires.

The removal or relocation of such fixed obstacles as bridge abutments, piers or parapets, retaining wall ends, heavy overhead sign supports, etc., is impracticable, and so considerable research has been initiated since the late 1960s (particularly in the USA) towards developing energy-absorbing vehicle-impact systems, commonly known as *crash cushions*[121, 122], which seek to reduce the severity of such collisions. Crash cushions are protective systems which protect errant vehicles from impacting massive fixed hazards by smoothly decelerating the vehicles to a stop when they hit head-on.

Impact-attenuation systems can be divided into two broad categories, depending upon the basic principles by which they function, viz.: energy-absorption systems and momentum-transfer systems. With the former, the kinetic energy of the impacting vehicle is absorbed by progressive crushing (plastic deformation) of the crash cushion material or by means of hydraulic action, against a rigid support, i.e. a cushioning effect is provided. With the latter, the momentum of the impacting vehicle is transferred to an 'inertia barrier' or expendable mass of material located in the vehicle's path; no rigid support is required as the kinetic energy is not absorbed but is transferred to the cushion.

Some of the cushions that have been tested and found to be effective in the USA include:

(1)   an array of empty steel drums that absorb impact energy by progressive crushing,
(2)   nylon cells containing water which dissipates the impact energy by being forced out at controlled rates through orifices,
(3)   lightweight vermiculite concrete 'helicells' which absorb the impact energy by controlled crushing,

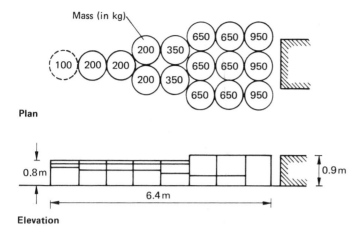

Plan

Elevation

Fig. 7.20 Crash cushions of sand-filled barrels

(4) frangible plastic barrels which are designed to shatter on impact—the barrels contain varying amounts of dry sand which is scattered as the momentum of the vehicle is transferred to the sand (see Fig. 7.20),

(5) tyre–sand inertia barriers composed of sand-filled scrap tyres—the sand and tyres are scattered as the momentum of the vehicle is transferred to the cushion.

These crash attenuators are designed to limit the average permissible vehicle deceleration rate to that which is numerically equal to twelve times the acceleration of free fall during impacts at speeds of up to 96 km/h by vehicles with a mass range of 0.90–2.04 t.

American studies have shown crash cushions to be effective in reducing the severity of collisions with fixed objects to the extent that of those collisions in which fatal or serious injuries might have been expected, 75 per cent resulted in minor injury or property damage only, 18 per cent resulted in injuries requiring hospitalization, and only 7 per cent resulted in fatalities.

Longitudinal safety fences

Roadside safety fences are a longitudinal protective system designed to reduce the severity of off-carriageway single vehicle accidents. Occasionally, they may also be used to protect pedestrians and 'bystanders' from vehicular traffic. Where properly installed, safety fences serve to deflect erring vehicles away from dangerous locations; they also have a desirable psychological effect in providing the nervous motorist with a feeling of security when traversing apparently dangerous highway sections.

Safety fences can be divided into two main groups, viz.: guardrails (or edge barriers) and crash barriers. Guardrails tend to be used along highways that are located on high embankments with steep side-slopes, or along verges where fixed obstacles are continuously close to the edge of the carriageway; they may also be used to deflect erring vehicles away from individual fixed hazards such as bridge abutments. Crash barriers are located within central reservations and are primarily intended to prevent intruding vehicles from crossing over into the opposing traffic stream.

Principles of safety fence design
Ideally, a safety fence should present a continuous smooth face to an impacting vehicle, so that the vehicle is redirected, without overturning, to a course that is nearly parallel to the fence face and with a lateral deceleration which is tolerable to the motorist. To achieve these aims, the vehicle must be redirected without rotation about either its horizontal or vertical axis (i.e. without 'spinning out' or overturning), and the rate of lateral deceleration must be such as to cause the minimum risk of injury to the occupants.

In practice, the happenings at a safety fence are so complicated that it has not yet been possible to devise a theoretical treatment which represents what actually does occur. As a result, safety fence research is usually carried out in full-scale road tests. The following discussion must therefore be regarded as a theoretical description based on a greatly simplified model of what occurs during an actual collision.

Lateral rotation As indicated in Fig. 7.21(a), assume that a vehicle of mass m is in collision with a safety fence at an angle θ. The horizontal forces acting on the vehicle are the normal reaction of the fence R which tends to turn the vehicle anticlockwise, and $(F + \mu R)$ which tends to turn it clockwise. In this diagram, the sliding friction between the vehicle and the fence is represented by μR, and F is the average value of the other forces produced by, for instance, striking posts or other obstructions.

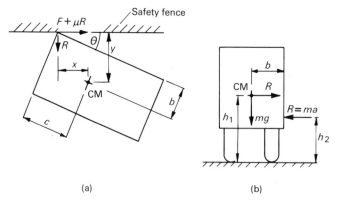

(a)   (b)

Fig. 7.21 Principal forces acting when a vehicle strikes a safety fence: (a) horizontal forces and (b) vertical forces, where m = mass of vehicle, h_1 = height to the centre of mass (CM), h_2 = height of safety fence, a = lateral deceleration, and g = acceleration of free fall

If the fence is to fulfil its function of redirecting the vehicle parallel to the fence face, then the net effect of the moments of these forces must be clockwise. This condition is given by

$$\frac{F+\mu R}{R} < \frac{x}{y} = \frac{c\cos\theta - b\sin\theta}{c\sin\theta + b\cos\theta}$$

Thus it can be seen that if F or μR is la. ge, the condition may not be satisfied and the vehicle may spin out of control with a clockwise rotation. When $\tan\theta = c/b$, the vehicle will spin out from the fence regardless of the value of $(F+\mu R)$. For a typical car, the value of this critical angle is about 70 degrees. If, however, $F=0$ (i.e. the vehicle does not strike any posts), then for $\mu = 0.01$ (steel on steel) and $\mu = 0.3$ (steel on concrete), $\theta = 64$ degrees and 54 degrees, respectively. In other words, as long as the impact angle is kept below 54 degrees for a concrete fence, and 64 degrees for a steel fence, then spin-out will only occur if appreciable retardation is imparted to the vehicle by posts or other fence components—which is why, to ensure that retardation is small however severe the impact, support posts should be made weak in the direction of the line of the fence.

Lateral deceleration Whilst an impact is taking place, the safety fence may deflect and the vehicle will crumble to a certain extent. If it does not bounce off or climb but instead swings and scrapes along the fence, then the centre of mass of the vehicle will move through a total lateral distance of $(y-b+d)$, where d is the sum of the deflection of the fence and the lateral crumpling of the vehicle. The most important quantity from the point of view of the motorist is the average lateral deceleration of the vehicle; this is given by

$$a = \frac{(v\sin\theta)^2}{2[c\sin\theta + b(\cos\theta - 1) + d]}$$

where v is the approach velocity.

This equation is actually a reflection of the risk of injury to the motorist. Death can result if the human body is subjected to excessive deceleration—and the amount which it can withstand is in turn influenced by whether or not the person is wearing a safety harness.

If the fence consists of a rail or cable with supporting posts, the tension in the fence is given approximately by

$$T = \frac{ma}{2\sin\alpha}$$

where α is the angle of the rail or cable relative to its original position.

Overturning If the impacting vehicle is assumed to be a rigid body, the principal vertical forces acting on it are as shown in Fig. 7.21(b). From this, it can be seen that the vehicle will tend to overturn if

$$ma(h_1 - h_2) > mgb$$

i.e.
$$a > \frac{gb}{h_1 - h_2}$$

Hence if the effective height of the safety fence is equal to or greater than the centre of mass of the vehicle, then the vehicle cannot overturn.

Types of safety fence
The multitude of crash barriers available commercially can be divided into three main types: steel beam, flexible wire, and rigid concrete fences. Details of British practice with respect to these are readily available in the technical literature[123].

Steel beam fences   The two main types of tensioned beam safety fence used in Britain are the tensioned corrugated beam and the rectangular hollow section beam. Of these, the corrugated beam is probably the more widely used, not only in Britain but also elsewhere in the world.

Corrugated mild steel beams are normally not less than about 300 mm deep, and formed so that the traffic face has a central trough at least 75 mm in depth. Typically, a beam is mounted at an overall height of 760 mm above and parallel to the edge of the adjacent carriageway. The beam is attached by shear bolts to mild steel Z-section posts, and tensioned between anchorages sunk in the ground to which the beam ends are sloped down. Upon impact by a vehicle, the bolts fracture, so that the posts can be knocked over without appreciably affecting the original height of the beam. A clearance of about 1.2 m is desirable behind the single beam fence most used in Britain, to allow for safe deflection upon impact; however, a double beam fence, i.e. one with a beam on each side of the standard post, reduces the safe deflection space to about 0.6 m, whilst further stiffening can be obtained by halving the post spacing.

Major advantages of a steel beam fence are that it presents a broad face to traffic, is effective with a wide range of vehicles, and is suitable for installation at most sites. The impact-duration time varies from type to type, but generally lies between the time for concrete and cable fences.

Special attention must be paid to the end treatment of these barriers, as an end-impact can be of a particularly serious nature. Ideally, this problem should be overcome by continuing the barrier beyond the sphere of influence of the traffic.

Wire rope fences   Cable fences in various forms have been in use since the late 1920s as a means of easily and efficiently stopping an out-of-control vehicle. The main advantage of this barrier is that, because of its great flexibility, a cable can 'slowly' decelerate a crashing vehicle and redirect it most easily along a path parallel to the safety fence. In addition, it is comparatively simple to fix the height of the different cables so as to cater for the greatest number of different vehicles in use today; however, care has to be taken to ensure that the cable height is not so great that it rides up over the bonnet of a small car, e.g. a sports car. Its continued satisfactory operation is dependent upon the correct height of impact being maintained; this requires the ground on either side of the cable to be hardened and level.

The greatest disadvantage of a wire rope fence is that it cannot be used at locations with narrow clearances because of its significant potential for deflection under impact, e.g. the fence used in Britain requires a deflection clearance of at least 1.8 m. Thus cable fences cannot normally be used on curves of radius less than about 610 m because of the tension in the rope and the nature of the posts; because of their greater flexibility, usage of these fences is normally also unsuitable where there are lighting columns in the central reservation.

A wire rope fence is probably most useful at sites where minimal air resistance is needed, e.g. to prevent snow accumulations on mountain roads.

Concrete fences With metal fences, which are designed to deflect upon impact, the friction generated between the impacting vehicle body and the barrier metal is expected to furnish the major redirection and deceleration force; maximum friction is assured when the metal beam is placed at a height which prevents 'vaulting' over the fence or 'pocketing' of the front wheel.

With a concrete barrier that has properly contoured sides, however, the concept is significantly different, i.e. at impact angles in the normal range, redirection is accomplished by the vehicle wheels and not by the body, whilst energy absorption results from compression of the suspension system and not from deformation of the vehicle body[124]. The manner in which concrete crash barriers function during impact is reflected in Fig. 7.22. This figure shows the basic cross-section of the solid contoured wall, constructed of mass concrete, that the Department of Transport considers suitable for use on heavily-trafficked urban roads where both road and central reservation widths are restricted. This design is a variation of the 'safety shape' crash barrier that has been in use on New Jersey freeways since about 1955.

Note that the design is such that the initial contact of an erring vehicle's tyre is with the 75 mm high vertical kerb; this tends to slow the vehicle and may offer some redirection. Once this initial resistance is overcome, however, the front wheels will climb the slope of the barrier and one or two wheels and the side of the vehicle lifted perhaps 325 mm above the adjacent hard surface; this lifting action absorbs the energy component perpendicular to the barrier and overcomes the overturning moment by compressing the vehicle suspension system. At flat-angle impacts, this is supposed to happen with no contact between the crash barrier and the vehicle body. If the speed and angle are sufficiently high, the wheel will continue onto the near-vertical upper barrier wall, which will complete the redirection and deceleration of the vehicle and return it to the ground parallel to the barrier. Ideally, this redirection is accomplished with the minimum damage to both the vehicle and the

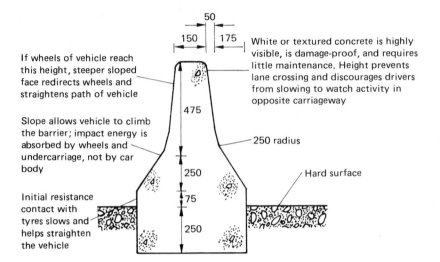

If wheels of vehicle reach this height, steeper sloped face redirects wheels and straightens path of vehicle

White or textured concrete is highly visible, is damage-proof, and requires little maintenance. Height prevents lane crossing and discourages drivers from slowing to watch activity in opposite carriageway

Slope allows vehicle to climb the barrier; impact energy is absorbed by wheels and undercarriage, not by car body

250 radius

Initial resistance contact with tyres slows and helps straighten the vehicle

Hard surface

Fig. 7.22 Performance characteristics and typical cross-section dimensions of the New-Jersey-type safety fence used in Britain (dimensions are given in mm)

safety fence, and without intolerable deceleration forces acting on the occupants of the vehicle.

Additional advantages claimed for this type of fence include the following.

(1) It forms a practical base for light posts and sign supports.
(2) Electrical power cables can be enclosed within the concrete in raceways placed during construction.
(3) It provides some protection against headlamp glare.
(4) Adjacent areas are at least partially shielded from highway noise, particularly the noise component from the tyre–carriageway interaction.
(5) There are various opportunities available for the development of designs which are of significant aesthetic interest.

Practical applications
As noted previously, safety fences can be described as either edge barriers or median barriers. Edge barriers are primarily used to prevent vehicles from hurtling off the roadway and down the sides of steep embankments. They should not be used where it is economically possible to provide embankment slopes of 4 horizontal to 1 vertical or flatter, since a driver forced onto such slopes has a much better chance of regaining control over the vehicle than he or she has of walking away uninjured from a collision with a safety fence. Except at laybys, these fences should always be placed at a fixed distance from the edge of the carriageway so as to avoid confusion as to their location during bad weather. Edge barriers should be highly visible if they are to be fully effective.

The separation of carriageways by a central reservation serves to reduce, but normally does not eliminate, collisions between opposing vehicles. If head-on collisions are to be nearly eliminated, central reservations of about 15 m are normally required to enable erring drivers to regain control of their vehicles. In Britain, it is the rare highway that has such wide medians; in fact, most motorway construction is provided with central reservations of 4.5 m. The inadequacy of this narrow width has given rise to a call for median barrier fences as an economical way of eliminating accidents which occur because vehicles traverse the median and collide with vehicles on the other carriageway.

Notwithstanding that it is now government policy to install median barriers on all new British motorways, it is doubtful whether they are justified on all roads at all locations. This can be explained by reference to Fig. 7.23. As can be seen, a substantial number of these accidents involved vehicles intruding onto the central reservation; of the 220 involved, 93 traversed it entirely. Of these 93 movements, *only 12 resulted in collisions with vehicles on the opposing carriageway.* Now the important factor to be considered is what would have happened if a safety fence had been installed within the median. Firstly, it is certain that all 93 crossover movements would have resulted in collisions with the safety fence, while some of the remaining 127 would almost certainly have hit the barrier also. Secondly, it can reasonably be expected that there were many unreported cases where vehicles entered the central reservation but their drivers regained control and were able to steer back onto the carriageway before an accident happened. If a barrier had been located on the median, then again it is very probable that some of these vehicles would have struck it.

Thus the technical problem can be reduced to determining whether the savings in accidents involving collisions on the opposing carriageway are in excess of any likely increase in the overall accident rate at the location in question.

It has been shown[126] that the erection of central safety barriers on all-purpose dual carriageways as a general policy would be likely to reduce the number of fatalities by an estimated 15 per cent, the numbers of serious and slight accidents would change very little, whilst the number of non-injury accidents would probably show an increase of about 14 per

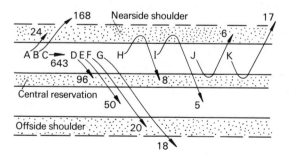

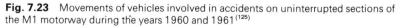

Fig. 7.23 Movements of vehicles involved in accidents on uninterrupted sections of the M1 motorway during the years 1960 and 1961[125]

cent. However, the study also pointed out that particular lengths of highway can exhibit very different accident patterns, so that the overall results might not give a reliable indication of possible benefits in individual cases.

Incident detection on high-speed roads

Once accidents or vehicle-breakdowns occur, it is important that they be detected as quickly as possible. If detection is slow, lives may be lost and/or severe traffic congestion may occur unless rescue and alleviation procedures are brought quickly into action. This problem has led to the development of traffic surveillance and control systems.

The requirements of a traffic surveillance and control system as far as incidents on high-speed roads are concerned have been defined[127] as including the following abilities:

(1)   to detect incidents promptly,
(2)   to convey accurate information about the position and nature of the incident, e.g. the number of lanes blocked and the occurrence of injuries,
(3)   to dispatch promptly suitable aid to the site of the incident,
(4)   to warn other motorists of the hazard ahead and influence their movement to cause, for example, speed reduction, lane changing, or diversion onto the adjacent street network,
(5)   to control entry of vehicles onto the highway,
(6)   to direct motorists on the surrounding road system, including those who have previously been directed off the motorway and those who had intended to join it.

Surveillance concepts that have been used to determine the time, of occurrence and nature of traffic incidents include[128] electronic surveillance, close-circuit television, aerial surveillance, emergency telephone boxes, cooperative motorist-aid programmes, citizen-band radio and, most commonly, police and service patrols. Incident removal and control normally involves the provision of one or more of the following:

(1)   emergency services such as police, fire and ambulance,
(2)   repair and tow services,
(3)   alternative control strategies,
(4)   driver information.

Details of surveillance and control systems used in various countries are available in, for example, references 127 and 128.

Improving traffic flow

Very many of the strategies that are implemented in order to improve road safety, reduce fuel consumption, better the environment, etc. (see Chapter 3) also act to improve traffic flow. Similarly, many of the established measures taken to improve traffic flow also have useful byproducts in terms of fuel conservation, less pollution, fewer accidents, etc. Important traffic management measures in this latter respect include restricting turning movements at intersections, the implementation of one-way-street systems, tidal-flow operation, the closing of side streets, and the use of traffic signal (including urban traffic control) systems.

Restriction of turning movements

Whilst many junctions operate most efficiently during off-peak periods, serious congestion which is caused by *right-turning traffic* can often occur during peak periods. This can be particularly serious when opposing right-turning vehicles 'lock' and introduce temporary stoppages of all movements through the intersection. Even though locking may not occur, a few right-turning vehicles can cause a disproportionate loss of capacity.

If there is a heavy right-turning movement, the vehicles may be accommodated at intersections with signal control by inserting an extra phase in the cycle; this, however, should be avoided wherever possible since it usually results in a fairly long signal cycle with consequent delays. Early cut-off and late-start signal arrangements, which allow extra time for the right-turning traffic either before or after the opposing straight-ahead movements, are usually preferred since they result in less overall delay than would a separate phase.

In many instances, it is better to ban right-turning traffic entirely during all or part of the day, rather than attempt to provide directly for it. There are three commonly used alternative routing procedures which allow right-turning vehicles to complete this manoeuvre without actually making a right-turn at the critical intersection.

(1)   Figure 7.24(a) illustrates the diversion of the right-turning movement to an intersection further along the road where there is more capacity. This routing is most useful for dealing with a difficult right-turn from a minor road onto a major road; the right-turning movement then takes place at a minor–minor intersection.
(2)   Figure 7.24(b) illustrates the diversion to the left before the congested intersection in what is known as a G-turn. This is most suitable for a right-turn off a major road, since it results in a left-turn off the major road and a straightover movement at the critical intersection. The diversion involves two right-turns at minor intersections and care has to

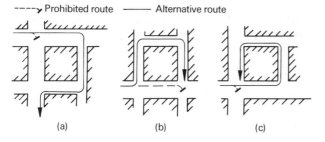

Fig. 7.24   Re-routing right-turning traffic

be taken to see that these do not create extra difficulties. A particularly important consideration is careful signing, so that non-local motorists do not overshoot the initial left-turning.
(3)   Figure 7.24(c) illustrates what is known as a Q-turn. This is a diversion to the left beyond the intersection which requires three left-turns. Although this is regarded as the least obstructive diversion, it does require the motorist to travel twice through the original intersection, thereby increasing the total volume of traffic handled there.

The main difficulty associated with the introduction of right-turn restrictions is that of finding alternative routes that are suitable not only with regard to width and structure, but also with regard to amenity. Problems can also arise with buses which, because of their need to serve particular objectives, are usually preferred to return to the original route as quickly as possible; hence they are extra sensitive to turning restrictions. In any particular instance where this is critical, it may be possible for the buses to be exempted from the right-turn ban. Although this may cause some minor confusion, a few vehicles turning right without opposition from other right-turners can usually be accommodated without loss of efficiency.

Of particular importance is the attention that must be paid to the signing of the diversions. Normally, the turning restrictions are indicated to the motorist by conventional signs mounted on posts, but, if the prohibitions are in force only during the peak periods of the day, then they can be advantageously indicated by overhead neon signs which light up only during the critical congestion periods.

Left-turns are not normally considered to be obstructive to traffic flow at intersections, and hence they are rarely banned. Left-turn bans may, however, be utilized occasionally to reduce the possibility of pedestrian–vehicle conflicts when the number of persons crossing the minor street is unusually heavy. In these cases, of course, the ease with which vehicles can move through the junction is also improved.

One-way streets

With one-way-street operation, motor vehicle movement on any given carriageway within the system is limited to one direction. One-way streets are generally considered to be one of the simplest and most economical tools available for the relief of traffic congestion without expensive reconstruction or excessive policing. It is also generally considered that their most effective usage is in and about the central areas of cities and towns, where the feasibility of applying more extensive and expensive road construction is usually limited.

Advantages of one-way streets
The primary reason for making a street one-way is to improve traffic movement. Although one-way operation is normally also accompanied by a reduction in accidents, safety is rarely the main reason for its introduction.

Overall, the introduction of a one-way system is generally associated with the following.

Increased capacity The conversion from two- to one-way operation can increase the capacity of a street from zero to 100 per cent, depending upon the conditions prevailing locally, e.g. distribution of traffic, turning movements, and street widths. Some design flow figures for one-way streets are given in Table 6.12; if these data are compared with the two-way data given in Table 6.11, the direct effects of making streets one-way can be deduced.

Motorists find it more convenient and less confusing to drive where all vehicles are moving in the same direction, thereby enabling more efficient usage to be made of the carriageway. In addition, odd lanes which could not be utilized under two-way working can now be fully used under one-way operation; this is of considerable importance in Britain because so many towns have inherited streets which are not wide enough for four lanes of traffic but have widths well in excess of that required for two lanes.

Slow-moving or stationary vehicles are also more easily overtaken when one-way operation is in effect. More important, traffic congestion at busy intersections is very considerably reduced because of the elimination of some right-turning movements and because any extra road width can now be used more efficiently to speed vehicle movement. The development of a one-way system also allows for more efficient traffic signal timing which, in turn, results in smoother traffic flow, a reduction in stops, and a consequent reduction in air pollution[129].

Increased speed The more even flow of traffic allowed by the removal of opposing vehicles also permits higher operating speeds. The higher speeds can be taken advantage of by linked signal systems which can be designed to benefit from them. For example, on the Baker Street/Gloucester Place scheme in London, the average speeds were more than doubled from 12.9 to 27.4 km/h as a result of changing from two-way to one-way operation, while the volume of traffic was only slightly increased[130].

Not only are vehicle speeds increased, but both the journey times and the variability of journey times are also reduced. This latter factor is of considerable importance to public transport facilities since regularity of service helps to attract and keep passengers. For instance, following the introduction of the Baker Street/Gloucester Place one-way scheme mentioned above, the coefficient of variation of bus journey times—this is the ratio of the difference between the maximum and minimum journey times to the average journey time—dropped from 0.37 to 0.17, indicating greater consistency in the journey times due to the initiation of one-way operation.

Increased safety The introduction of a one-way-street scheme generally results in a reduction in the number of accidents, particularly between-intersection accidents. Accidents of the head-on variety are eliminated

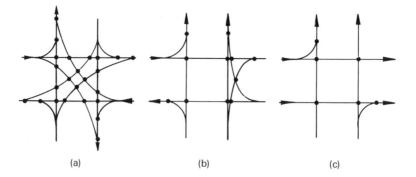

Fig. 7.25 Potential points of conflict at an intersection: (a) two two-way streets, 24 conflicts, (b) one one-way and one two-way street, 11 conflicts, and (c) two one-way streets, 6 conflicts

because of the removal of an opposing traffic stream. Accidents due to bad road lighting should also be reduced since there is now no headlight glare problem. Certain types of accident at intersections are greatly reduced because of the reduction in the number of possible points of conflict (see Fig. 7.25).

Pedestrian accidents at intersections can also be reduced, since pedestrians can be given a fully-protected crossing while traffic emerges from a side road (see Fig. 7.26) without loss of time in the signal cycle. In addition, signal-controlled pedestrian crossings can often be provided at other points without interfering with vehicle progression. On one-way streets controlled by linked signals, pedestrians are usually able to make safe crossings at intermediate intersections during gaps created by the signal timing.

Economic savings   One-way-street operation is also generally held[131] to give economic savings which are normally well in excess of the initial costs of the schemes; these savings arise from a reduction of motorist journey times and a saving in delays. A computer-based traffic assignment and queueing model called *CONTRAM*, i.e. *CON*tinuous *TR*affic

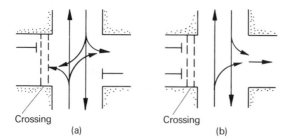

Fig. 7.26 Pedestrian crossings: (a) on two-way streets, the crossing may require a separate signal phase to enable safe crossing of the road, and (b) on one-way streets, the crossing is fully protected while side traffic is moving

*A*ssignment *M*odel[132], is available which can be used to predict the economic and environmental effects of introducing traffic management schemes such as one-way-street systems (as well as closing a street to traffic, or the adoption of different traffic signal control plans).

One-way systems should also result in economic savings due to a lessening of the need for police-officer control at congested intersections and streets. Definitive studies regarding savings in police man-hours due to the introduction of one-way systems have yet to be published in Britain. However, it is reported that, about the year 1950, some 725 of the 3500 km of streets in Philadelphia were one-way and that these one-way streets required 75 per cent less police-officer control than was required by the two-way streets. A similar report came from the City of New York which at that time had 2400 km of one-way streets; these required 50 per cent less enforcement than did the two-way streets.

Improved parking Normally, the parking problem is also the beneficiary of significant relief when one-way operation is introduced into the central area of a city. At the very least, parking and unparking manoeuvres tend to be less dangerous and obstructive when all vehicles face in the same direction. Driving across opposing traffic in a two-way street in order to park on the 'wrong' side of the street is eliminated with one-way operation. In many instances, it may be possible to allow parking on one side of streets hitherto considered too narrow for parking, just as on wider streets parking on both sides may become allowable once they become one-way.

Disadvantages of one-way streets
Whilst the advantages of one-way operation normally by far outweigh the disadvantages, it is possible for the disadvantages to be extremely significant. Some of the more obvious disadvantages are as follows.

Increased travel distance With the installation of a one-way scheme, it is usually necessary for motorists to travel further to reach their destinations. This is because a pair of parallel one-way streets becomes a dual carriageway with buildings on the central reservation, and each block of buildings becomes a roundabout; thus the motorist cannot head directly to a destination but must take a longer and more indirect route. (Obviously, the effect of this increase in distance should be offset by the decrease in journey time which results from increasing the journey speed.)

Another important inconvenience is that public transport stops for opposing directions of travel will have to be relocated on different streets. Cases will arise where these stops may be so far apart that potential bus passengers will baulk at walking distances they consider to be excessive.

Loss of amenity One-way-traffic operation sometimes involves vehicles using residential streets with consequent loss of amenity to the surroundings. This can be very detrimental when there are large numbers of heavy commercial vehicles present in the traffic stream. Hospitals and schools can also be affected adversely by the changeover.

Loss of business   The introduction of a one-way system is often opposed by local trading interests who fear that business will be detrimentally affected. While this must always be taken into account, it is rarely that there is evidence to support this contention. This factor can be expected to be more deeply felt when buses are taken away from parades of shops or when the increased speed of traffic effectively bisects a shopping street which hitherto had permitted the movement of pedestrians from one side to the other.

Increased severity of accidents   One-way operation tends to increase the numbers of vehicles weaving between lanes and to result in higher speeds. While there is usually a reduction in the total number of accidents, the higher speeds at the moment of impact will normally cause an increase in the severity of non-head-on accidents.

Possible confusion   The introduction of a one-way scheme often results in confusion to motorists, particularly just after the scheme has been initiated. With the use of intensive publicity and proper law enforcement, it will be found, however, that local motorists soon change their driving habits; strangers to the system may still be confused, however, and may violate the one-way rule so that head-on collisions can result.

Difficulties of introduction
The most obvious prerequisite to a one-way scheme is that the existing road system should be capable of being modified. For instance, if part of the traffic is displaced from one street, there must be a complementary street available to take the removed vehicles. With the gridiron pattern of streets, this can usually be considered as automatically assured; with the more irregular patterns normally prevalent in the urban areas of Britain, the availability of a complementary street cannot be taken for granted, e.g. in a linear town that is strung-out along one main roadway. In other instances, the complementary street may be much narrower than the main street, and the result is an unbalanced, inefficient flow of traffic.

Probably the greatest problems associated with establishing one-way operation occur at the ends of the one-way roads. Ideally, each complementary pair should converge to form a Y-intersection, but of course this is not always possible. If both of the parallel roads end on a cross-street, then this will be required to carry a considerable extra burden of traffic, and the two intersections will have to handle much more turning traffic; indeed, if the cross-street is an important street, it may be required to carry an unbearable burden, with the result that congestion ensues. In many instances, the only way to avoid this is to extend the one-way system beyond the area originally intended.

Interconnection between two complementary roads can also raise problems. The two streets should not be too far apart—usually not more than about 125 m—or be at such different levels that heavily-travelled connecting roads have steep gradients. Care must also be taken to ensure that the connecting roads do not become overloaded or unusable.

Signing It is absolutely essential that the one-way scheme should be thoroughly signed at all points where the motorist may have a decision to make. NO ENTRY signs are required at all terminals of one-way streets, and ONE-WAY and/or TWO-WAY traffic signs should be placed at the entrances and exits of all the intersections within the scheme. Where necessary, supplementary NO LEFT-TURN or NO RIGHT-TURN signs should also be displayed.

These signs should be located where they can easily be seen by the motorist. In most cases, this will necessitate signs on both sides of the one-way carriageway.

Tidal-flow operation

With the ever-increasing use of the private car for travelling to and from work, all towns now experience the familiar morning and evening rush periods. This situation can result in the regular under-utilization of one-half of the street width, with over-utilization and congestion on the other half; thus, in the morning, the inbound lanes may be filled to capacity, with vehicles in the outbound lanes able to move freely, whilst, in the evening, the reverse may be true. The term 'tidal flow' has been given to traffic exhibiting this unbalanced characteristic at peak periods.

Tidal-flow lane operation refers to the traffic management process whereby the total carriageway width is shared between the two directions of travel in near proportion to the flow in each direction. As a result, the lanes assigned to a given direction of travel vary with the time of the day, i.e. use is made of under-utilized lanes to provide extra capacity in the peak direction of flow during peak periods.

Introduction requirements

For maximum efficiency of operation, the number of traffic lanes allocated to each direction of travel should correspond as closely as possible to the ratio of the flows at the peak periods. In addition, adequate provision must be made for the lighter flow on the two-way road.

This latter factor means that streets with three or four lanes are best adapted to complete reversal so that they are completely inbound in the morning and outbound in the afternoon. A three-lane road with two lanes reserved for the heavier flow and one for the lighter will usually give about the right ratio of lanes for many traffic situations, but, if peak period clearway restrictions are not applied and enforced, the single lane may be frequently blocked and the light traffic halted. The four-lane road is also more suitable to complete reversal because of the difficulty of keeping a single lane continually open. In addition, it is rarely that the ratio of three lanes to one represents the ratio of the traffic flow.

On a five-lane road, two-way tidal-flow operation can be readily carried out, since three lanes can be applied to the heavier flow and two lanes to the minor flow. During off-peak hours, the median lane should be converted into a temporary central reservation and no traffic allowed on it.

On a six-lane road, four lanes are usually given to the heavier flow during the peak hours. During the off-peak hours, traffic can be allowed on three lanes in each direction. If flows are sufficiently low during off-peak hours, it may be possible to allow parking on the outside lanes.

Whenever there is a need for tidal-flow operation on a particular route, there is also a need for the imposition of clearway restrictions. At the very least, parking should be prohibited on the side of the major flow during the critical hours. Right-turns from the predominant flow should be forbidden at important intersections because of the restrictions which they place on the free movement of traffic. Depending upon the volume of the minor flow, right-turning may, however, be allowed at non-critical intersections.

A consequence of tidal-flow operation is that central pedestrian refuges have to be removed.

A positive feature of tidal-flow operation is that it lends itself to the linking of the traffic signals in the predominant direction of travel. The linked system will, however, need to be adjusted according to whether it is peak period or off-peak traffic flow that is being carried.

Controlling the reversible lanes
Since they were first introduced in Los Angeles in 1928, considerable ingenuity has been exercised in developing ways of draining and then controlling the reversible lanes so that tidal-flow operation can be safely and efficiently carried out. The three most commonly used procedures require the use of signs, traffic-light signals, and movable barriers.

Signing This widely used method requires permanent signs to be placed at intervals along the route so that drivers and pedestrians alike are informed regarding the number of lanes operating in each direction at stated times of the day. Immediately prior to the change in direction, police patrol cars are used to clear the way of opposing traffic.

This procedure requires the smallest initial outlay of capital and so is viewed with favour in many cost-conscious communities. It is most successfully utilized on reversible streets with few visitors in the traffic stream, as regular users can be relied upon to know the rules and keep the lanes operating correctly once they have been established. The main drawback of this method is that police vehicles are continuously tied down during the peak periods.

Frequently, traffic cones, prefaced with KEEP LEFT signs, are placed along lane-lines to differentiate between the directions of travel; again, however, while this can be very successful on a temporary basis, it is most uneconomical over a long period of time because of the manual labour required to lay and remove the cones.

In addition to the normal fixed signs, various types of mechanically-changed signs have evolved. These range from an electrically-driven system of shafts and gears which swings a signboard through 180 degrees (from its hidden position behind another sign) into position over the tidal-flow lanes, to sliding signs[133], to rotating sign segments in prism shape which are used to display different messages. One interesting proposal is for the use of a holographic sign, whereby the message area of the sign is projected optically as a real or a virtual image in space above or beside the roadway so that no structural support is required.

However they are displayed, the signs should be illuminated internally during the periods of tidal flow, the message being varied according to whether it is the morning or evening peak. These signs attract attention and serve better the needs of the irregular users of the tidal-flow route.

Overhead traffic signals Traffic-light signals suspended over the centre of each reversible lane at the beginning and end of each reversible section are the most flexible means of controlling tidal-flow operation, as well as being the most common method of control. Intermediate signal lights can be suspended as necessary at intersections along the route. The signals can be of the normal pattern with a face for each direction of travel or, as is more desirable, of a pattern comprising, say, a green arrow or a red cross spelled out in energized retro-reflectors, e.g. a driver facing a downward-pointing green arrow is allowed to drive in the lane beneath the arrow, but cannot do so when facing a red cross. If the signal shows red in both directions—e.g. during off-peak hours—it means that the reversible lane is being used as a temporary central reservation.

Signals can very easily drain opposing traffic from a reversible lane. The easiest way is to bring on the red signal consecutively from the far end while keeping the running direction amber until all the opposing signals are red. The transitions should be carried out at times that are well clear of the peak periods. When compared with signing control, traffic signals are a relatively expensive method of traffic management. While the initial cost of installing the lights may be high, however, the long-term costs are low.

Movable barriers The simplest type of movable barrier used to separate lanes of opposing traffic is the *traffic cone*. Whilst traffic cones do not prevent the passage of vehicles, and are labour-intensive when laying and/or removing, they are very effective in controlling the entrances and exits of tidal-flow lanes, as well as delineating along the lane-lines.

Next in simplicity, and similar in application to traffic cones, are various types of vertical *posts or stanchions* that are inset in holes along the lane-lines. Typically, the posts are of a polyvinyl chloride plastic-type of construction, and when struck by a vehicle they flatten without damaging the vehicle and generally without being dislodged. Whilst these dividers avoid some of the problems associated with traffic cones, they still have the problem of requiring a relatively large labour force for placement. Mechanically-operated post dividers have been developed, but so far many have been proved not to be very dependable and to have high maintenance costs.

Probably the most famous *rigid divider* is the system installed in Chicago's Lake Shore Drive about 1935. This uses hydraulically-operated 'fins', i.e. structural steel boxes (508 mm wide by 406 mm deep by 7.62 m long) placed in a trench that is flush with the carriageway surface when in the lowered position; when the dividers are required, double-acting hydraulic jacks raise the fins 203 mm above the carriageway to act as a barrier. The fins are placed along every other lane-dividing-line of the eight-lane urban motorway for a distance of 3.54 km, so that it can be operated as six lanes inbound and two outbound, or four lanes in each direction, or six lanes outbound and two inbound. The dividers are raised very slowly so that vehicles can easily steer clear as they rise.

Movable rigid barriers are normally very expensive to install and maintain in good working order. Generally, it is considered that their costs are only justified when the capital and maintenance costs are favourably balanced against the savings in property acquisition and road construction that result from not having a wider carriageway. For example, in the Chicago illustration noted above, the eight existing lanes with tidal-flow operation are able to do the work of twelve lanes on a conventional dual carriageway.

Closing side streets

In many urban areas where there are lightly-travelled side streets along an important route, traffic flow can often be eased by closing a number of the side streets to vehicular traffic. The conditions under which this traffic management procedure might be utilized are most simply illustrated by comparing the advantages and disadvantages associated with closing side streets.

The main *advantages* of closing side streets are as follows.

(1) *Improvements in journey time and running speed* Vehicle running speeds and journey times are closely associated with the number of side street connections along a route. Normally, if some of the streets are closed, there is a reduction in the factors restricting vehicle movement and vehicle speeds will be increased and journey times decreased.

(2) *A reduction in the number of accidents* Many accidents happen at intersections, and a disproportionate number of those occurring in urban areas happen at lightly-travelled junctions. Many of these would be avoided if side streets were closed. Similarly, a substantial reduction in accidents can be expected when cross-road intersections are converted to staggered junctions.

(3) *Usage of linked signals* When utilizing linked signals, it is desirable that the signalized intersections be, say, at least 275 m apart. On heavily-travelled roads, it may be worthwhile considering the closure of a number of intermediate side streets in order to make use of linked traffic signals.

(4) *Usage for parking* In certain instances, a closed side street can also be used as a car park, especially for long-term parking. Care has to be

taken, however, that as a result large numbers of vehicles do not attempt to exit onto the main street at one time and without control. If this happens, the resulting restriction to flow may outweigh the other advantages.

(5) *Usage for pedestrian precincts* The advantages associated with turning busy shopping streets into pedestrian precincts are numerous and obvious. In housing areas, where children use the streets as playgrounds, the closing of certain side streets to vehicular traffic will obviously have a tremendous beneficial effect in terms of accident reduction, both pedestrian-vehicle in the streets in question, and vehicle-vehicle at the intersections. More important, however, is the improvement to the living environment, particularly in the older residential areas where pleasant old homes so often are allowed to become dilapidated and rundown as the neighbourhood becomes identified as a traffic thoroughfare.

The main *disadvantages* of closing side streets are as follows.

(1) *Congestion at intersections* When particular side streets are closed, then the displaced vehicles will move to other streets. The net result may be that the increase in volume at the remaining side road connections may necessitate signal control measures that otherwise might not be necessary.

(2) *Increased parking on main roads* When side roads which provide rear accesses to buildings fronting on a major road are completely closed to vehicular traffic, there is an automatic tendency for these vehicles to park on the main road. This can result in decreases in traffic speed and flow on the major road and increases in accidents associated with parking and unparking of vehicles.

(3) *Interference with other management measures* In certain instances, it may be more advantageous to use a side street as part of a right-turn diversion route rather than close it entirely.

(4) *Non-availability as an alternative route* When major road intersections are congested or when roadworks are being carried out, it is often found that certain side roads compose the only alternative route for traffic.

(5) *Non-availability of quick access for emergency vehicles* In times of emergency, ambulances, police vehicles, fire vehicles, etc., can be delayed in gaining access to closed streets.

Traffic signals

Road-traffic signalling devices have been used to help to improve traffic flow on roadways for well over one hundred years. The first traffic signals, which were manually operated, were installed at the intersection of Bridge Street and New Palace Yard outside the Houses of Parliament in London in 1868. This installation, which incorporated semaphore arms and red and green lights illuminated by gas, was short-lived, however, due to an explosion which injured the policeman on duty and brought the experiment to a hasty conclusion. After this unfortunate episode, interest in traffic signal development languished until about 1914 when an electrically-controlled traffic signal device was used in Cleveland, Ohio; it is this device that can perhaps be truly regarded as the basis of modern traffic signals. In 1926, the first modern type of signal was introduced into Britain with the installation of an automatic signal system, with a fixed-time operating sequence, in Princes Square, Wolverhampton; a vehicle-actuated installation was developed at the junction of Cornhill and Gracechurch Street in the City of London in 1932.

Since these early days, the traffic signal has become one of the principal tools of the highway engineer engaged in traffic management. Well-designed traffic signals, when properly located and operated, usually have one or more of the following *advantages*.

(1)   They provide for the orderly movement of traffic. Where proper physical layouts and management measures are used, they can increase the traffic-handling capacity of a congested intersection.
(2)   They reduce the frequency of right-angle and pedestrian accidents.
(3)   Under conditions of favourable spacing, they can be coordinated to provide for continuous or nearly continuous vehicle progression in linked or area-wide traffic control schemes.
(4)   They can be used to interrupt heavy traffic at given intervals in order to permit other vehicles or pedestrians to cross speedily and in safety. As they are confident of eventually gaining the right-of-way, drivers are willing to tolerate longer delays at a red light than at a STOP sign.
(5)   When compared with the cost of police control, they represent a considerable economy at intersections where there is a need for some definite means of assigning right-of-way.
(6)   Signals have standard indications which drivers can very easily follow. At night or in foggy weather, they are more easily understood than the hand signals of a police officer.
(7)   Unlike manual controllers, who frequently delay main road traffic while straggling vehicles negotiate an intersection, automatic signal control is impartial in assigning right-of-way.

Although automatic traffic signals are usually preferable to police control, it should be noted that their *disadvantages*, which are as follows, are magnified when the signal installation is unwarranted, badly designed or improperly maintained.

(1)   They can increase *total* vehicle delay at intersections. This is especially noticeable during off-peak hours on major streets which previously had priority over side streets. The traffic throughput from the major street during peak periods may also be reduced, as the movement time through the junction must now be shared with minor traffic.
(2)   When improperly located and/or timed, they may cause unnecessary delay, thereby increasing motorist irritation and causing decreasing respect for this and other traffic management devices.
(3)   They usually cause an increase in the frequency of rear-end collisions.

(4) They are not normally capable of granting right-of-way to emergency vehicles such as ambulances and fire engines.

(5) Failure of the installation, although infrequent, may lead to serious and widespread traffic difficulties, especially during peak traffic periods.

Basic features

Before discussing traffic signals as such, it is convenient to define and discuss some of the terms and basic features associated with this form of intersection control.

Traffic signal Generally, a traffic signal[134] may be described as a complete installation at a junction, which includes signal heads (containing different coloured lanterns), poles, wiring, control mechanisms, etc. Specifically, the term is also used to refer to an optical device (the signal head) which is operated electronically by a controller and displays a prescribed message which causes pedestrian and vehicular traffic to be alternatively directed to stop and go.

Signal head A typical main signal head is composed of three lanterns arranged vertically above each other, with a red lens on top, amber in the middle, and a green lens at the bottom. The lenses are normally 203 mm in diameter, and each is illuminated from behind by an independent light source.

The standardization of lantern location is important as it ensures that colour-blind drivers are always aware of which lamps are 'burning' at a given time.

The normal sequence of lantern operation is red, red–amber shown together, green and amber. The length of the red–amber period is now standardized at 2 seconds, and the amber at 3 seconds. The function of the red–amber period is to indicate to stopped motorists (and to pedestrians) that the lights are about to change to green; they should therefore be prepared to enter the intersection as soon as the lights change, and so wasted cycle time can be kept to a minimum. The amber period similarly warns approaching vehicles of a coming change in the signal indication so that they can slow down safely; at the same time, it acts as a clearance interval for vehicles or pedestrians within the junction, as well as for those moving vehicles that are so close to the stop line that to halt suddenly would be dangerous.

Filter signals Filter signals, normally mounted alongside main signal heads, permit the movement of vehicles in the direction shown by the green arrow even though the main signal is showing red. These signals should be used with care, however, as: (a) they can create problems for pedestrians crossing the road from which the filtering vehicles emerge, and (b) there is a risk of collision with vehicles in any traffic stream with which the filtering vehicles are merging.

Traffic signal controller This is the complete timing mechanism which controls the operation of the traffic signals. Nowadays, this mechanism may be electromechanical, electronic or solid-state in construction. Some of the different types of controllers are as follows.

(1) An *automatic controller* is a self-operating mechanism which operates the traffic signals automatically. Automatic controllers are usually fitted with a facility switch which enables the signals to be changed manually by depressing and releasing a push button.

(2) A *pretimed controller* is an automatic controller for supervising the operation of traffic signals in accordance with a predetermined fixed-time cycle.

(3) A *traffic-actuated controller* is an automatic mechanism for supervising the operation of traffic signals in accordance with the varying demands of traffic. The controller receives data from detectors located on one or more approaches to the junction and, on the basis of these data, allocates green time by predetermined methods.

(4) A *master controller* is an automatic controller for supervising a system of secondary controllers, maintaining definite interrelationships, or accomplishing other supervisory functions. The master controller's computer operates on the basis of data provided by traffic detectors located strategically throughout the street system, and selects an appropriate system program from its repertory for implementation.

(5) A *local controller* is a mechanism for operating traffic signals at an intersection, or two or three adjacent intersections, which may be isolated or included in an area-wide signal system. The very considerable advances that have been made in computer technology over recent years have now resulted in the development of minicomputers for use as local controllers.

Traffic detector A traffic detector is any device by which vehicles or pedestrians can inform a traffic-actuated controller of their presence.

Time cycle The time cycle is the period of time required for one complete sequence of signal phases. The cycle has a variable and non-uniform length when operated by a traffic-actuated controller; in the case of a pretimed controller, it has a uniform predetermined length.

Generally, short cycle lengths are to be preferred as they reduce traffic delay; unless operated as part of a deliberate traffic restraint policy (see Chapter 3), cycle lengths should rarely be allowed to exceed 120 seconds.

Traffic phase The traffic phase is that part of the time cycle allocated to any traffic movement or any combination of traffic movements receiving the right-of-way. The simplest form of phasing is two-phase, whereby phase A provides for one street whilst phase B provides for the other cross-street. Phases may be split, they can overlap, or be timed concurrently. Figure 7.11 shows two simplified phase arrangements as might be used at four-way junctions.

Generally, it is good practice to try to hold the number of phases to a minimum as, logically, each phase reduces the amount of green time per hour that is available for other phases. Delays to traffic may also be increased by extra phases due to additional starting delays and the need for extra amber periods and longer cycles.

Vehicle-extension period The minimum green period is the shortest period of right-of-way which is given to any phase; it is normally at least long enough to allow vehicles waiting between the detector and the stop line to get into motion and clear the stop line. With traffic-actuated signals, this minimum period can be extended by vehicles which cross the detector during the green period. The length of the vehicle extension period depends upon the speed of each vehicle, as measured at the detector, and is automatically varied to enable the vehicle to reach 3–6 m beyond the stop line. These green time extensions are not directly additive; instead, the time in the controller is only reset to a new value if the new extension is for a period that is longer than the unexpired time of the previous extension. When the interval between vehicles crossing the detector becomes greater than the vehicle-extension period, the right-of-way is transferred to vehicles waiting for another phase.

To limit the delay on waiting phases when there is a continuous flow of traffic, the green period of the running phase is assigned a maximum length; when this has expired, the signal controller can automatically change the right-of-way regardless of the demand for further extensions. This maximum period is timed from either the beginning of the green period of the running phase (if vehicles are waiting on other arms of the junction) or the moment that the first vehicle crosses the detector on one of the other arms (if vehicles are not waiting).

Intergreen period This is the time between the end of the green period of the phase losing the right-of-way and the beginning of the green period of the phase gaining the right-of-way. Thus it includes amber time plus any all-red time and the overlapping red/red–amber time between the two green periods.

For a simple two-phase signal control, the intergreen period is therefore 4 s, as the red and red–amber periods of the two phases normally overlap for 1 s. If, however, due to the 'busyness' of the junction it is decided to insert a 7 s all-red period to allow for safer pedestrian movement, then the intergreen period would be 12 s, i.e. 3 s amber plus 7 s all-red plus 2 s red/red–amber overlap.

Lost time This is the total time during the cycle which is not effectively used for vehicle movement. It is made up of the time when all signal heads show red or red–amber, plus a 'waste' allowance of 2 s per change of phase to allow for the tailing-off of vehicle movement during the amber period and the starting delays at the beginning of the green period. At each change of phase, therefore, the lost time amounts to one second less than the intergreen period.

Traffic signals at isolated intersections

Traffic signals used at isolated intersections can be classified into the following groups: (a) pretimed and (b) traffic-actuated. The latter can be subdivided into fully- and semi-traffic-actuated signals. In addition, the signals can be either operated independently—by far the more common approach—or coordinated as parts of linear (linked) or area-wide (urban traffic control) systems.

Pretimed traffic signals With basic pretimed traffic signal operation, the controller assigns the right-of-way at a junction according to a predetermined schedule or a series of such schedules. The time interval for each component of the signal cycle is fixed in length and normally based on historic traffic flow patterns. The usage of these signals is most effective at intersections where traffic patterns are relatively stable over long periods of time.

Amongst the *advantages* of pretimed signal operation are the following.

(1) Usually, the cost of a pretimed installation is considerably less than that of a traffic-actuated one.
(2) The simplicity of the pretimed equipment results in relatively easy servicing and low maintenance costs.
(3) Controller timing is easily adjusted on site.
(4) Consistent starting times and durations of cycle length facilitate linking of adjacent traffic signals. This linking can permit progressive vehicle movement at a constant speed through many intersections along a given route, as well as providing positive speed control along the route.
(5) Pretimed controllers are not dependent for operation upon the movement of vehicles past detectors, so that a stopped vehicle or construction work does not interfere with its proper usage.

Pretimed traffic signals are generally recognized as having the following main *disadvantages*.

(1) Normally, pretimed traffic signals are geared to peak traffic requirements, with the result that excessive and frustrating delays to vehicles can occur during off-peak times.
(2) Pretimed signals cannot recognize or deal with short-period variations in traffic demand, with the result that one phase in a cycle may be under-utilized whilst another is over-saturated.

As may be gathered from the above, the timing of a pretimed traffic signal is of critical importance. If the signal setting is inadequate, excessive delays will occur, and motorists will be tempted to disobey the signals. There are many methods in use to estimate capacities, green times and consequent vehicle delays at signal-controlled intersections. For example, the Transport and Road Research Laboratory has devised a basic method[135, 136] of determining the optimum cycle length and phasing for fixed-time signals. This method is based on a combination of theory and computer simulation that was used to develop a mathematical model for average vehicular delay at a signalized junction; from this delay model, expressions for cycle length and timing splits for minimum overall intersection delay were derived.

It should be noted that this method of determining the signal timing is applicable to both pretimed signals and traffic-actuated signals which are being operated as fixed-time signals due to heavy traffic demands.

Traffic signal setting determination   It is convenient to explain the manner in which signal settings are determined by considering an example, as follows.

Assume that measurements of actual flows and saturation flows at a particular two-phase, four-way intersection gave the following data.

Arm	Actual flow, f_a (veh/h)	Saturation flow, f_s (veh/h)	Ratio $(f_a/f_s = y)$	Critical ratio
North	600	2400	0.250 ⎫	0.250
South	450	2000	0.225 ⎭	
East	900	3000	0.300 ⎫	0.300
West	750	3000	0.250 ⎭	

In addition, the lost time is composed as follows: starting delays = 2 s per phase (typical value), all-red periods = 3 s at each change of right-of-way, and red–amber periods = 3 s at each change of right-of-way.

With the TRRL method, the traffic controller settings which will give the minimum overall delay to vehicles are determined as follows.

The length of a fully-utilized green phase for an approach to an intersection can be considered to consist of an effective green period during which saturation flow occurs, and a lost time during which no flow takes place (see Fig. 6.7). The saturation flow can be defined as the flow which would be obtained if a continuous queue of vehicles was given 100 per cent green time. If the ratio of the actual flow to the saturation flow on the critical arm of each phase is denoted by y, then

$$C_0 = \frac{1.5L + 5}{1 - y_1 - y_2 - \ldots - y_n}$$

In the above equation, the symbols are as follows. C_0 is the optimum cycle length for minimum intersection delay (s). y_1, y_2, ..., y_n are the maximum ratios of actual flow to saturation flow for phases 1, 2, ..., n. L is the total lost time per cycle (s), which can be determined from the equation $L = nl + R$, where n = number of signal phases, l = average lost time per phase due to starting delays (s), and R = time during each cycle when all signals display red, including red plus amber, simultaneously (s). (It should be noted that if the lost time and saturation flows are both different for the different approaches of the same phase, then each arm of the phase should be considered in turn as the 'predominant' one and the longest cycle deduced is then the optimum one.)

Of considerable practical importance in relation to the above equation is that, when the cycle length is varied within the range $0.75C_0$ to $1.5C_0$, the minimum delay is never exceeded by more than 10–20 per cent. This helps in deducing a compromise cycle time for pretimed signals when the traffic flow changes during the day.

For the given data, the total time per cycle when red or red–amber aspects are shown to all phases is 12 s. The total lost time is therefore

$$L = nl + R = (2 \times 2) + 12 = 16 \, \text{s}$$

Then the optimum cycle length is obtained from

$$C_0 = \frac{1.5L + 5}{1 - y_1 - y_2} = \frac{(1.5 \times 16) + 5}{1 - 0.250 - 0.300} = 64 \, \text{s}$$

In one cycle, therefore, there will be $64 - 16 = 48$ s of total effective green time. Before the actual green times which will give the least overall delay can be obtained, it is necessary to get the optimum effective green times for each phase; this is done by setting the ratio of the effective green times equal to the ratio of the critical y-values. Thus, for the two-phase cycle,

$$g_1/g_2 = y_1/y_2 = 0.250/0.300$$

where g_1 and g_2 are the effective green times of phases 1 and 2, respectively. The effective green times for the N–S and E–W phases are therefore 20 s and 24 s, respectively. The controller settings then obtained by including the 2 s waste time in each phase are thus

N–S phase $= 21$ s green $+ 3$ s amber

E–W phase $= 25$ s green $+ 3$ s amber

Queue lengths When designing a signal system at a junction, it is important to know what queues are likely to develop—especially if there are other intersections nearby with which they might interfere—and what delays vehicles are likely to experience. The Transport and Road Research Laboratory method also provides formulae which can be used to answer these questions.

The queue at the beginning of the green period is usually the maximum in each cycle, and its length is given by

$$N = 0.5qr + qd \quad \text{or} \quad qr \qquad \text{(whichever is the larger)}$$

where $N =$ average number of vehicles in the queue, $r =$ red time (s), $q =$ flow (veh/s), and $d =$ average delay per vehicle (s).

The average delay per vehicle on any particular arm of the intersection can be expressed to a close approximation by the equation:

$$d = \frac{c(1 - \lambda)^2}{2(1 - \lambda x)} + \frac{x^2}{2q(1 - x)} - 0.65 x^{(2 + 5\lambda)} (c/q^2)^{1/3}$$

where d=average delay per vehicle (s), c=cycle time (s), λ=proportion of the cycle time that is effectively green for the arm under consideration ($=g/c$), q=flow (veh/s), and x=degree of saturation, i.e. the ratio of the actual flow to maximum possible flow under the given signal settings. Thus $x=qc/gs$, where s=saturation flow (veh/s) and g=effective green time (s).

The last term in the above equation amounts to 5–15 per cent of the average delay in most cases. As a result, the following equation is often used in practice to obtain the approximate delay:

$$d=0.9\left(\frac{c(1-\lambda)^2}{2(1-\lambda x)}+\frac{x^2}{2q(1-x)}\right)$$

Traffic-actuated signals   With traffic-actuated signals, a detector pad is placed in the carriageway at some distance back from each stop line so that every vehicle approaching the junction registers its presence to the controller by actuating the appropriate detector. The traffic-actuated controller differs from the pretimed controller in that the signal phases are not of fixed length, but vary according to the information being continuously input regarding the traffic flow, whilst the length of the cycle, and even the sequence of phases, may vary from cycle to cycle.

In contrast to pretimed signals, which generally can be only properly used at junctions in central areas and on main roads where traffic flows are consistent, stable and predictable, traffic-actuated signals can also be used at isolated intersections where traffic flows fluctuate considerably and where multiple phasing is required.

The main *advantages* of traffic-actuated signals are as follows.

(1)   Intersection capacity can be increased by the traffic-actuated controller reapportioning green time upon demand, so that the more heavily-travelled approaches are given favourable treatment.
(2)   Vehicle delay is minimized, especially during off-peak travel periods.
(3)   They provide maximum efficiency of movement at intersections where one or more of the traffic movements are subject to short-term fluctuations.
(4)   They are especially effective at intersections requiring the use of more than two signal phases.
(5)   When compared with pretimed signals, they tend to have less of the rear-end type of collisions associated with the arbitrary stopping of vehicles in the middle of a traffic stream.

The two major *disadvantages* of traffic-actuated signal systems are as follows.

(1)   The capital cost of a traffic-actuated system (including the detector) is significantly more than that of the equivalent pretimed signal installation.
(2)   Generally, servicing and maintenance costs are also higher due to the greater complexity of the equipment.

Normally, the advantages of traffic-actuated signals so far outweigh the disadvantages that they are favoured very much by traffic engineers; indeed, pretimed signals are now rarely, if ever, used in Britain.

With traffic-actuated signals, the most commonly used vehicle detector is the inductive loop[137, 138] that is inserted into the carriageway. Push-button detectors, mounted on traffic signal supports or on special posts, are used by pedestrians, whilst special push-button detectors may be placed on conveniently located posts for use by cyclists. Whatever the type of detector, the signal is automatically adjusted by the actuation in accordance with a previously decided action programme.

Most traffic-actuated signals in Britain are of the fully-actuated type. *Fully-traffic-actuated signals* utilize detectors on all approaches to the intersection; the detectors are used to assign the right-of-way to a particular route as a result of actuation on that route. In the event of the continued actuation of a given detector, the right-of-way (in the case of a two-phase, four-way junction) is retained with the route until a preselected maximum length of time has expired after which it is transferred to the cross-street. The right-of-way then remains with the cross-street for at least a (variable) minimum predetermined period of time which can be extended according to the traffic demand up to a maximum length.

An example of a fully-traffic-actuated, three-phase control arrangement, as it might appear on a drawing for a junction between a two-lane road and a dual carriageway, is illustrated in Fig. 7.27. This

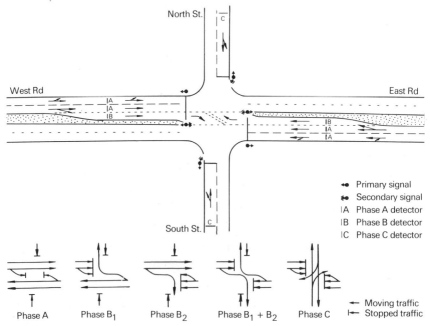

Fig. 7.27   Illustrative layout and phase diagram for a fully-traffic-actuated, three-phase signal installation on a dual carriageway

arrangement can be very useful when there are heavy right-turning movements from the major highway. Its operation is as follows.

(1) Vehicles travelling straight ahead and turning to the left on both East and West Roads are given the right-of-way during phase A.

(2) After the completion of a minimum length of green time, phase A can be interrupted by demands from vehicles on East Road waiting to turn to the north. This manoeuvre is carried out during the split-phase B_1.

(3) Similarly, phase A can be interrupted to allow vehicles to travel to the south during a separate split-phase B_2.

(4) If there are vehicles waiting in the right-turning lanes on both East and West Roads, then phase A can be interrupted to allow a separate phase $(B_1 + B_2)$ which permits only the right-turning movements.

(5) At the end of phase B_1 or B_2 or $(B_1 + B_2)$, phase C is initiated and the right-of-way is transferred to the vehicles on the cross-streets, North and South.

A *semi-traffic-actuated signal* is most effectively used at a junction between a major street carrying heavy volumes of traffic and a minor 'street', e.g. a factory entrance, with only sporadic traffic flow. With this type of installation, detectors are not placed in the main route carriageway; instead, preference is given to traffic movement on the main road and the green aspect is transferred to the minor road only upon actuation of the detectors placed in the minor approaches. The length of the minor street green time varies with the traffic demand but cannot be extended beyond a maximum limit. When the required or maximum minor street phase has expired, the green aspect is automatically returned to the major street where it remains for at least a predetermined interval.

Simple linked-signal systems

When dealing with the movement of heavy traffic volumes, it is obviously desirable that the actions of the individual signals should be coordinated where possible. Proper coordination of signals can lead to more efficient traffic movement, less delay to vehicles at intersections, and an increase in the capacity of the linked route due to the ease with which vehicles can move through the linked intersections.

There are a great many ways of coordinating fixed-time traffic signals. These may be grouped into the following three basic systems: (a) simultaneous, (b) alternate, and (c) flexible progressive.

Simultaneous system With a simultaneous coordinated system, all of the signals show the same indication to the same street at essentially the same time. At all intersections, the signal timing is essentially the same and when orders are received from a master controller all indications change simultaneously, so that all signals show green to the major road traffic and red to all cross-streets.

This was one of the earliest types of coordinated signal system, as it is very easily adapted to usage with pretimed signals. However, it has limited applications in modern traffic signal practice because of the following disadvantages.

(1)   The simultaneous stopping of all traffic along the roadway prevents continuous movement of vehicles.

(2)   It promotes high speeds, especially during off-peak hours, as drivers attempt to pass as many intersections as possible while the lights are green.

(3)   The cycle length and green time allocations are usually governed by the requirements at one or two major intersections within the system; this often results in serious inefficiencies at the remaining intersections.

(4)   The net effect is low journey speeds and possibly a reduction in route capacity.

The simultaneous system can be very useful if only two adjacent intersections are to be linked. In this instance, the green time given to the same major traffic flow during each cycle may have to be relatively long so that there is ample time for a major portion of the main road traffic to clear through both intersections.

Alternate system   Under this type of operation, all the signals are again operated by a master controller so that all the indications are changed simultaneously. It is different from the simultaneous system, however, in that adjacent signals or groups of signals show opposite aspects alternately along the major route. Typically, the cycle lengths are the same at all signals, and the green and red periods within a cycle are also the same length. Thus, if the intersections are equidistant, each block can be travelled in half the cycle time so that the driver meets a continuous green aspect along the route provided that travel takes place at the design speed.

With the irregular street systems prevalent in most British cities, the alternate signal system has limited possibilities for usage. Even in American cities, which tend to have gridiron road patterns, its use is limited to routes where the traffic on the cross-streets is approximately equal in volume to the traffic on the through road. If the flows are unbalanced, the result may be that the major route has too little green time and the minor roads have too much.

The alternate system does, however, limit speeding since motorists exceeding the design speed of the system have to stop at most of the intersections.

Flexible progressive system   This is the most satisfactory of the simple linking systems. The signals are so timed that the driver of a vehicle released by the green aspect at the first signal and proceeding along the major route at a predetermined speed will find that every signal changes from red to green upon approaching an intersection. If the design speed is maintained, the vehicle can then travel the entire length of the system without having to stop in front of one red aspect at any intersection.

Although the operation of the flexible progressive system requires that the cycle time for each intersection should be fixed, vehicle-actuated signals can also be used in the system. If vehicle-actuated signals are used, the local controllers at the individual intersections operate according to the overriding progressive plan while there are continuous demands from all detectors. When, however, the volume of traffic falls below a predetermined figure, each installation reverts to its original independent state and operates according to the individual requirements of each intersection.

The phasing of the signals in a flexible progressive system can be determined with the aid of a time–distance diagram, a basic example of which is shown in Fig. 7.28. On this diagram, the distances between intersections along the route to be controlled are plotted along the abscissa and the times along the ordinate axis. The usual problem is one of determining the best possible through-band speed and width for a given cycle length; in this simple example, however, it will be assumed that a design speed of 36.2 km/h has been decided upon as being suitable.

First a sloping line is drawn from zero time starting at a point on the distance scale opposite the first intersection; this indicates the movement of the first vehicle released at the beginning of the green period and travelling north at the chosen speed. Since the selected speed is 36.2 km/h (10.06 m/s), this line intersects a horizontal line drawn through the last intersection at 80 s from zero. The point where this line cuts the horizontal is taken as the beginning of a green period there also; the progress of the first vehicle travelling south at the same speed is then indicated by a second sloping line starting at this point and striking the ordinate axis at 160 s from zero. Under the progressive system, the point where this north-to-south line strikes the axis is again the beginning of a main road green period, and so on. It therefore follows that this 160 s period should contain exactly one or more complete signal cycles.

In practice, cycle lengths are rarely less than 25 s or more than 120 s. Short cycle lengths are undesirable because of possible insufficient running times as compared with stopping times. Often also, in urban areas, the minimum duration of the green period is governed by the length of time required by pedestrians to cross the road safely. A cycle time of more than 2 minutes is rarely used as it may cause an undue accumulation of vehicles in the intersection approaches with resulting undesirable delays. In the example in Fig. 7.28, a cycle length of 53.3 s is arbitrarily chosen for illustrative purposes. Further northbound and southbound sloping lines can now be drawn from the ordinate axis at intervals of 53.3 s. Each of these represents the progress in each direction of the first vehicles to be released during all the green periods at the first and last intersections.

The conditions at each intersection must now be studied to determine the lengths of the green periods. In this instance, let it be assumed that the conditions at the first and last intersections control the design, and these require that two-thirds of the cycle length should be in favour of the main road traffic and one-third in favour of the cross-road traffic. Green and amber periods in accordance with this division are drawn on the horizontals from the two intersections, beginning at the progress line for

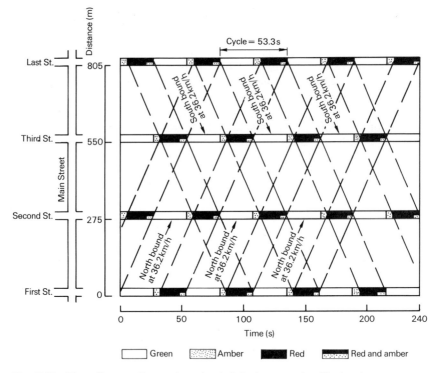

Fig. 7.28   Time–distance diagram for a simple linked system of traffic signals

the first vehicle in each direction. (If this is not exactly possible, the green periods should, anyway, include the moment at which the vehicles pass.) Having filled in the green periods, the progress lines for the last vehicles in each direction can now be drawn through the end of the green periods. These lines are parallel to those already drawn, and the band between each pair represents the progress of a platoon of vehicles in each direction along the main road. Green periods can now be filled in at the intermediate intersections so that, in every case, the combined lengths of the green plus amber periods contain the bandwidth already determined.

It should be appreciated that, unless the intersections are more or less equidistant, a perfect time–distance graph of this kind is not possible. For example, it will often be found that the bandwidth will be significantly narrower than the green plus amber times. In the case of a long route, it may be necessary to change the design speed and/or the bandwidth at some intermediate intersection. At some intersections, the green periods will be shorter than at more important intersections; indeed, some minor intersections may not require signals at all.

From a historical aspect, it might be noted that the coordination of adjacent signal installations was begun in Britain in the 1930s, with the local linking (in London and Glasgow) of simple vehicle-actuated controllers by underground cable. By the 1960s, flexible progressive systems had been developed that enabled the linking of a number of closely related intersections by means of underground cabling and a master controller.

Area-wide or urban traffic control (UCT) systems

The above three systems, on the whole, are most applicable to movement along major traffic routes; as such they have obvious drawbacks with reference to cross-movement traffic. In the early 1960s, interest was developed in the concept of extending linked-signal control by creating an area-wide system (perhaps city-wide) which would be under the control of a master computer. It was again in London and Glasgow that the first experiments in Britain into the use of area-wide computer control of traffic signal operation were attempted. These early experiments, which used fixed cycle plans, with green times being recalculated for particular traffic conditions or flow groups using a program called *TRANSYT* (see reference 139 for an updated version), were found to give considerable reductions in overall delays when compared with uncoordinated vehicle-actuated control), and encouraged further developments in this area.

The following introductory discussion is concerned only with the basics of area-wide or urban traffic control systems. For details, the reader is referred to a number of excellent publications that are readily available in the technical literature[128, 140, 141].

The implementation of area traffic control requires local traffic controllers to be situated at intersections. These local points of control, which are responsible for tactical signal control, are normally linked with a central master controller or computer which is responsible for the overall strategic control of the scheme. Two types of area-wide control have emerged so far, viz.: fixed-time control and traffic-responsive (or dynamic, or adaptive) control systems.

Fixed-time control, which is by far the more common, is, as the term implies, based on the use of traffic signals operating under fixed-time settings. This form of control relies upon historical data in the preparation of timing plans for a signalized area. Typically, signal plans are prepared off-line for at least three flow groups, e.g. morning and evening peak periods, and off-peak travel, and a particular strategy is then brought into play by the central control according to the time of the day. The strategy is normally based on such specific criteria as the maximization of journey speed on all routes within the controlled area, or the minimization of the number of stops for all vehicles in the network, etc.

Normally, the local controllers are linked with the master controller by means of a system of cables or radios. A fairly recent boon to fixed-time operation has been the development of 'cableless' link systems which make use of crystal clocks in local controllers. After synchronization with a central master clock, the local crystal clocks are left to run on their own timings; timing offsets between intersections therefore rely heavily upon the accuracy of these clocks. Timing plans are stored in programmable read-only memories installed in the local controllers, so that plans can be changed locally according to the time of day without the need for a master controller.

Fixed-time UTC systems are particularly suited to simple road networks with predictable flow patterns, and good historical traffic patterns, e.g. in the central areas of large cities. Their major *advantages* are as follows.

(1)   They are relatively simple in concept, implementation and operation.
(2)   The signal timings are predictable and motorists soon learn to adapt their routes and speeds to suit their own needs.
(3)   They are less costly, and are not dependent upon the operating reliability of vehicle detectors for their success.
(4)   They have proved their effectiveness.

The main *disadvantage* to fixed-time systems is that there is generally no automatic feedback to the master controller from the field, so that unexpected traffic incidents, e.g. accidents and illegal parking, cannot be catered for in the predetermined plans. To overcome this, the UTC control rooms are usually equipped with closed circuit television to facilitate prompt manual intervention as and when necessary.

Whilst fixed-time systems have given and continue to give excellent results, experience with them clearly indicated that greater benefits could be derived if fully-vehicle-responsive coordinated control systems could be utilized. Consequently, intensive research in the 1970s led to the development of a number of area-wide *dynamic traffic control systems*. This has been associated with the development (and reduction in price) of the computer as a traffic engineering tool, i.e. whilst, on the whole, a conventional master-controller-type computer only sends instructions based on a previously prepared plan to local controllers, the dynamic digital computer controllers have the capability of receiving information from detectors in the field, analysing and optimizing these data on-line at regular intervals (say every 5–10 minutes), and introducing a new control plan if the data indicate that a change is justified.

Traffic-responsive control systems, whilst more capable of responding to traffic fluctuations, generally have yet to justify the significant extra expense associated with their development and operation. In the long term, however, it is likely that these systems will predominate, particularly in dynamically developing urban areas, where variations in traffic patterns, changes in land use, and alterations to the highway network are regular happenings.

At the present time, there are well over two hundred area-wide traffic control schemes in use throughout the world (half of these are in the USA), most of which are of the fixed-time variety. In the USA, the advantages accruing from area traffic control would appear to have been more spectacular than in Europe; this is perhaps not unrelated to the fact that most traffic signals in the USA used to be operated under fixed-time control, whereas the more effective traffic-actuated signals have been the norm in Europe for many years. In Britain, there are well-established area-wide systems in Cardiff, Coventry, Glasgow, Leicester, Liverpool, London, Nottingham, Sheffield and Wolverhampton.

Governmental criteria used[142] to influence the decision as to whether or not to introduce an area-wide traffic control scheme are as follows: firstly, the number of signals in the proposed system must be not less than thirty and, secondly, *either* the signal density in the area is not less than four installations/km² *or* the average spacing of signals along the route(s) to be controlled is not less than four installations/km. Notwithstanding the above, satellite urban areas whose signal numbers do not justify a separate computer installation may be connected to main urban traffic control systems.

Reference 143 contains a most comprehensive bibliography of traffic signal research and development in Britain, and the reader is referred to this publication for further information regarding useful detailed reading in this area.

Priority for buses

Since the 1950s, the history of bus public transport usage has been one of continued decline (see Table 1.4). There are many reasons for this—all related to the phenomenal growth in the numbers and usage of the motor car—and three of the most important are given below.

(1)   The increasing availability of the motor car has meant less demand for travel by bus, i.e. a person can only travel in one vehicle at one time.
(2)   Land use changes made possible by the availability of the motor car have encouraged less centrally-oriented patterns of travel which are more difficult for public transport to cater for.
(3)   Traffic congestion associated with increased car usage has resulted in longer bus passenger journey times, less reliable services, less comfortable rides (more stops and starts), and greater operator running costs and higher passenger fares, all of which make travel by public transport less attractive.

Many and various land use, transport and traffic planning policies and strategies have been developed over the past decade in particular in order to help to reverse this trend (see Chapter 3). Public transport operators have implemented many innovations to improve their operational efficiency (e.g. see Table 7.30), and have streamlined their activities to an extent never previously envisaged. At the same time, the traffic engineer has implemented many traffic management and control measures in the effort to cope with the rising tide of car travel and traffic congestion; many of these have improved the overall flow of traffic but have tended to give greater benefits to traffic other than buses, whilst some have actually disadvantaged buses, e.g. by diverting bus routes and/or increasing travel distances.

In recent years, traffic engineers have paid particular attention to the development and application of traffic management measures which are

Table 7.30 Aspects of bus unreliability, their causes, possible remedial actions, and potential reductions in overall variation[144]

Source	Cause	Remedial action	Reduction
1 Bus fails to operate	Bus or crew shortage, vehicle breakdown, severe traffic congestion, other form of hold-up, tight schedules	Meet target percentages of operation every day, schedule buses to give low probability of late departure or failure	Substantial
2 Bus departure times from terminals and timing points	Tight schedules (actual running times commonly exceed scheduled times), traffic congestion, poor dispatch control, poor crew adherence to timetables	Reschedule on a more realistic basis, improve dispatch control, train and discipline crews, improve cqntrol strategies (e.g. radio control)	Substantial
3 'Penalty' time for stopping at bus stops (deceleration and acceleration times)	Variable probability of stopping at each stop (due to presence/absence of passengers boarding/alighting)	Reduce number of bus stops, increase number of compulsory stops	Slight
4 Passenger boarding times	Variable number/type of passengers, with driver-only operation=slow ticketing equipment, awkward fare-scale levels, passengers unprepared to tender fares	Improve ticketing equipment, adjust fare-scale levels, introduce limited change-giving policy, increase off-bus ticket sales, improve entrance layout, train drivers	Substantial
5 Passenger alighting times	Variable number/type of passengers, exit layout	Improve bus design	Slight
6 'Dead' time for doors to open/close and for buses to pull out into traffic	Speed of doors opening, driver's skill, variable gaps in traffic flow	Modify door interlock mechanism, move bus stops to enable buses to pull out into traffic with minimum delay	Slight
7 Delay time due to traffic controls or traffic congestion	Queues at junctions, pedestrian crossings, etc., variable traffic flow	Traffic management measures (e.g. bus lanes, bus priority at junctions, area-wide bus priority, parking controls, removal of conflicting traffic manoeuvres)	Substantial
8 Several services operate on one route	Buses scheduled independently and irregularly, uncoordinated dispatch control	Reschedule services together, introduce timing points, have one dispatcher	Substantial

deliberately aimed at reducing the delays to buses and increasing their reliability (see number 7 in Table 7.30). These schemes are generally known as bus priority schemes. The following discussion re bus priority schemes is based mainly on four excellent reviews of the literature[145-8] to which the reader is referred for further information.

Characteristics of bus priority measures

A bus priority scheme can be used tactically to solve local problems, or strategically to help to implement certain urban policies/strategies, e.g. traffic restraint (see Chapter 3). Whichever is the situation, the measures used in Britain fall into six main groups[148], as follows:

(1) bus lanes (with-flow and contraflow),
(2) bus-only streets and the tactical use of traffic regulations to reduce traffic flows on certain streets, thus enabling buses to move more freely in less congested conditions,
(3) purpose-built busways,
(4) priority at junctions (e.g. exemption from turning prohibitions and special optimum signal setting),
(5) setting signal progressions so as to favour buses,
(6) protection of bus stops.

Linked traffic signals, with-flow bus lanes, contraflow bus lanes, and priority turns tend to be most appropriately considered for use on general traffic streets in *central areas*. If a policy of car restraint is adopted in the central area, then it is essential that the bus system be given the capability of performing as an attractive alternative to the private car; in such instances, bus-only streets may also be developed both to help heavy concentrations of buses at particular locations and to give tangible evidence of the priority afforded to the bus services and the desirability of using them.

In the case of *radial routes* into town centres, the main concerns are with the journey speeds of the buses and their reliability. Both of these can be improved by the provision of with-flow bus lanes and/or priority at junctions.

Very often a *residential area* is deliberately designed so as to exclude through traffic and, as a result, distributor roads which would normally be considered suitable for use by buses have to follow long, indirect routes—which means that the buses are also discriminated against. In such instances, a separate purpose-built busway may be provided on its own right-of-way. In both new and old residential areas, bus-only streets or 'plugs' may be provided between adjacent distributors or culs-de-sac suitable for use by buses, so that buses can cross quickly between housing precincts.

Bus priority schemes may also be designed to help other classes of road user. Table 7.31 shows practice in this respect as it existed in 1978. What this table also suggests, however, is that whether or not other road users are allowed to have access to a bus priority scheme is often dependent upon practical policy considerations in the urban area in question.

With-flow bus lanes
By definition, with-flow bus lanes are traffic lanes reserved for bus usage where the buses continue to operate in the same direction as the normal

Table 7.31 Percentage of bus priority schemes in Britain available for usage by different road users[149]

Scheme	Bus services			Taxis	Delivery vehicles for access	Emergency vehicles	Cyclists	Disabled persons' vehicles
	Stage	Works/ contract	Express					
With-flow lanes	100	0	98	87	0	98	98	0
London								
provinces	100	48	58	8	10	58	56	0
Contraflow lanes	100	55	72	2	6	78	20	2
Pedestrian area								
access	100	19	28	16	75	81	12	25
Bus roads	100	40	46	10	26	65	19	5
Priority at								
junctions	100	24	31	4	6	33	4	0

traffic flow. The reserved lane is normally the kerb lane; however, on heavily-travelled arterial roads, the lane next to the central reservation may be used as it results in less interference to the buses on long runs.

In practice, most with-flow bus lanes are relatively short, e.g. over half of those in use in Britain in 1978 were less than 250 m long. They can be taken up and through intersections or they may stop short of the cross-street. Figure 7.29 shows the form of a commonly used pre-intersection bus lane layout. Note that the bus lane is actually terminated back from the stop line at the intersection; this arrangement allows other motor vehicles to use the full width of the junction, thereby maximizing its physical capacity, whilst allowing the buses to get sufficiently close to the junction that they will be able to travel through it within one signal phase.

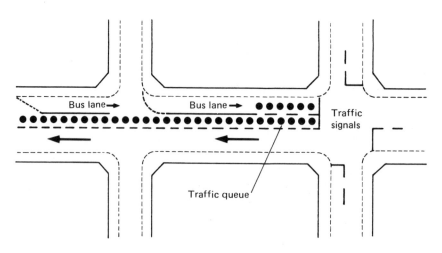

Fig. 7.29 General layout of a pre-intersection with-flow bus lane

The *advantages of with-flow bus lanes* are as follows.

(1)   They act as queue-jumping devices, particularly at intersections, and provide free-running conditions for buses which result in improved journey speeds (typically, by up to 5 km/h) and reduced travel times (typically, by up to 25 per cent through scheme areas).

(2)   Disruptions to normal traffic patterns are minimal.

(3)   Priority can be restricted to peak periods only, allowing the lanes to revert to mixed traffic usage during off-peak travel periods if desired.

(4)   The preferential treatment given to public transport improves its status amongst the general public and may attract new users (typically, 10 per cent increases).

(5)   The capital cost of implementing bus lane schemes is very low.

The *disadvantages for kerbside reserved lanes* are as follows.

(1)   They require continual policing since other non-priority vehicles will not only attempt to use the lanes because of their free-running conditions, but they may also attempt to stop/park in the lanes.

(2)   They cut across access to kerbside properties by commercial vehicles.

(3)   They may sometimes produce an accident problem initially, although this usually eases as drivers become more familiar with the lanes.

(4)   There may be an increase in traffic congestion (and queue lengths) for other traffic on bus lane streets, so that the journey times of non-priority vehicles are increased.

(5)   If non-priority traffic chooses to divert into adjacent alternative streets, there may be a spread of congestion with associated environmental detriment.

(6)   In rare instances, bus speeds may be reduced, e.g. if the kerbside lane is narrow and the carriageway is particularly rough due to manholes, gutters and camber changes at junctions.

The *disadvantages for central or median reserved lanes* are as follows.

(1)   It can be dangerous and inconvenient for bus passengers if they are required to cross lanes carrying moving traffic in order to reach the buses.

(2)   Streets may not be sufficiently wide to accommodate the mixed traffic lane, reserved lane, and median pedestrian refuge.

(3)   Buses are not normally adapted for right-side (in Britain) passenger loading.

(4)   There may be increased traffic congestion and journey times for non-priority vehicles.

General considerations   With-flow bus lanes are most justified on road sections where inadequate capacity manifests itself in repeated and long pre-intersection traffic queues which it is desirable for buses to avoid. There are no absolute rules as to when or where to introduce a bus lane, e.g. a reserved lane for only a small number of buses may be justified if there are considerable advantages for the buses and only minor disadvantages for the other traffic.

The two most important considerations in with-flow bus lane design are the beginning and end of the lane. The start of the bus lane must have a long, gradual taper to allow non-priority vehicles to merge into adjacent lanes safely and comfortably. If a bus lane starts immediately after a signal-controlled junction, care has to be taken to ensure that the merging queue does not block back into the intersection. The layout of the terminating end of a bus lane at a signal-controlled intersection depends upon whether it is to be brought right up to the stop line or whether a 'setback' is to be included. A procedure has been developed by the Transport and Road Research Laboratory whereby the setback length can be calculated from knowledge regarding the degree of saturation which existed prior to the bus lane, and the green time (giving due regard, of course, to such local conditions as siting of bus stops, location of side roads, etc.). This setback length can make an appreciable difference to the benefits derived from a bus lane, e.g. if it is too long, full capacity is maintained but the buses are further back in the queue and may not get through the intersection in the first green period, whilst if it is too short, capacity will be restricted with consequential effects resulting from additional delays to other traffic. Typically, the setback is between 50 and 80 m long.

The quality of the carriageway pavement can also be an important consideration in the decision whether or not to proceed with a bus lane proposal. Continued use of a particular lane by heavy buses can lead to rutting if the carriageway pavement is structurally inadequate, and this can create significant maintenance problems and costs to delayed traffic in the process of repairing the roadway.

Clear signing of bus lanes is essential both for efficient operation and for safety; to be fully effective, a combination of signing and carriageway markings is required. Publicity and enforcement are also critical; experience generally is that respect for the reserved lane decreases sharply unless it is properly and obviously policed. Even with periodic enforcement 'blitzes', violations will still happen, e.g. on Brixton Road in London, violations went from 1 per cent of all non-priority traffic after the first week, to 5 per cent after four months, to a stable figure of about 15 per cent after one year.

Reserved with-flow bus lanes are sometimes used to create bus-oriented one-way-street pairs. Bus lane schemes are often suited to one-way-street operation since turning conflicts are minimized and any loss in street capacity to other non-priority traffic is more capable of being handled.

Contraflow bus lanes
These priority lanes are normally installed in the kerb lane of one-way streets. In general, contraflow lanes shorten bus routes (as compared with one-way operation only) and provide buses with unhindered passage, thereby saving bus running time, increasing service reliability, and lowering operating costs. Furthermore, since lengthy diversions are avoided, bus passengers are normally able to alight closer to their

destinations than the one-way system would otherwise permit. British experience is that, where several contraflow lanes have been introduced within a central area network, average savings of up to 30 per cent in journey time have been achieved.

Contraflow bus lanes located in areas of important pedestrian activity encourage greater public transport usage, simply by being visible and requiring less passenger walking time. Since they enable both directions of a bus route to be kept in the one street, the service is also easier for the public to understand.

A major disadvantage of a contraflow lane is that it reintroduces traffic conflicts at intersections that the one-way system had eliminated. Thus the relatively high cost of these lanes (as compared to with-flow bus lanes) is mainly associated with the additional signalling, signing and channelization that may be required to obviate this problem. In addition, the capacity of the traffic system may be appreciably reduced, whilst linked-signal progression (which can be very efficient in a one-way-street system) may have to be compromised in order to give buses reasonable progression in the opposite direction.

A significant difference between contraflow and with-flow bus lanes is that the former are essentially self-policing, and have very few violations in practice. In addition, with-flow lanes very often operate during peak hours only, whereas contraflow lanes almost invariably operate for the full 24 hours.

The introduction of a contraflow bus lane into a one-way-street system can increase the potential accident risk to pedestrians; there are also dangers for traffic emerging from side streets and inadvertently turning into the bus lane. However, in most cases, the accident rate for a pair of one-way roads, one of which contains a contraflow bus lane, will be lower than that for the original two-way pair, although the road containing the bus lane can be expected to have a higher accident rate than the one without the bus lane. This increased accident potential means that clear, well-positioned signing is essential for safe contraflow operation.

Problems associated with loading/unloading by commercial vehicles are harder to solve for contraflow lanes than for with-flow ones. This is partly due to the 24-hour operation of these lanes, and partly due to the greater accident risk associated with allowing contraflow buses to move around loading/unloading vehicles.

Whereas with-flow bus lanes generally only give appreciable benefits when the degree of traffic saturation is very high, contraflow lanes are in general more easily justified when the degree of saturation of the traffic flow is small. Indeed, if the contraflow lane design is such that it does not cause appreciable delays to non-priority traffic, e.g. if the ends are controlled by roundabouts, the bus flows required to justify the lane are typically less than 40 buses per hour, and can be as low as 2–15 buses per hour[150].

Bus-only streets

A bus-only street in an existing urban area is normally created by banning all other traffic from the street or by admitting buses to a previously-pedestrianized street. The various aims of such a street are[145]:

(1)   to speed-up buses by removing other traffic from the way,
(2)   to create sufficient carriageway space so that buses can stop and wait as necessary without hindrance to other buses, e.g. bus termini are often sited in bus-only streets so that, in effect, the street becomes a mini bus station,
(3)   to assist pedestrians to cross the street more easily and carefully,
(4)   to improve the environment.

In many instances, taxis and vehicles may be allowed to enter a 'bus-only' street in order to gain access to premises which front it. In some cases, streets are operated as bus-only streets for less than the full 24 hours, e.g. Oxford Street (one of London's busiest shopping streets) is closed to all traffic except buses, taxis and access vehicles between 11 a.m. and 7 p.m. each weekday. In such cases, however, particular care has to be taken in relation to the design of an adequate signing system to ensure safe usage. Whilst careful design can produce a very pleasing environment, it is usually not appropriate to provide wall-to-wall pedestrian paving in such streets, and the danger of accidents is always present for pedestrians.

Bus-only streets are usually highly regarded by local authorities, public transport operators, pedestrians, shopkeepers and the public generally. Even motorists do not object too vociferously, particularly if the schemes are given considerable publicity before they are introduced.

Busways

As discussed previously (see Chapter 3), this term is used to describe segregated roadways that are specifically designed and built for buses only. For practical reasons, busways are relatively rare in existing urban areas, i.e. once the existing fabric of a town is built, it is usually very difficult to find a suitable continuous strip of land in the right place without demolishing properties. However, when a new town is being built, the opportunity is more easily taken to reserve rights-of-way for busways which are closer to houses, shops and workplaces than might otherwise be possible with conventional public transport services, and which can also provide for high-speed bus travel between stops.

If 2.50 m is taken as the width of a bus, and 0.75 m is allowed for manoeuvring on either side, the minimum carriageway width of a single lane busway is 4 m. On such a busway, however, a disabled bus could not be overtaken so it would be necessary for all buses to be fitted with special bumpers to allow for the pushing of a bus into the nearest layby. A one-way busway of 6.00 m would allow for overtaking, whilst a carriageway width of 6.75 m should allow comfortable two-way operation.

Ideally, busways should be grade-separated from other general-purpose roads. However, depending upon the traffic volumes to be catered for, there is no reason why traffic signals which give priority to the buses cannot be used to control at-grade intersections of busways and roads.

Generally, buses can tackle busway gradients of up to about 3 per cent without speeds being adversely affected.

Priority at junctions
A simple and widely used method of giving priority to buses at intersections is to exempt them from right- or left-turn prohibitions and, as appropriate, to provide special signal phases for buses. These are very effective and well-proven measures that are well accepted by motorists.

In recent years, considerable attention has been paid to the development of another form of bus priority at isolated signal-controlled intersections, viz.: the adaptation of the signal timing and phasing to decrease the delay to approaching buses. This can be done either by adjusting the signal settings to suit the arrival of buses or by having the approaching bus interact with the signal control.

The former type of signal priority is most appropriately applied to fixed-time or vehicle-actuated traffic signals with long cycles and three or more phases, *provided that buses are not contained in the traffic flow on more than one approach*[148]. At these intersections, it is possible to achieve a reduction in bus delay simply by dividing the green time which contains the bus flow into two periods, and then separating the two shortened green periods by the other traffic phases, e.g. a three-phase system A, B and C would become a four-phase system A_1, B, A_2 and C. As a result of this simple change, bus bunching will be significantly reduced, and the maximum delay to buses almost halved if the junction is not badly congested. Maximum effectiveness is achieved when the buses are able to depart during the first green period after their arrival. Where properly implemented, the cost is very small and the total capacity of the intersection is either unaffected or only slightly reduced.

The latter type of signal priority operates by detecting the bus on its own approach lane and/or by the use of selective bus detection equipment. The Department of Transport's selective bus detector system[151] comprises two parts: (a) a passive 'transponder' unit mounted on the underside of a bus, which is powered from either the bus electrical system or its own batteries, and (b) road equipment composed of a wire loop buried in the carriageway, which is connected via a feeder cable with a traffic signal controller at the roadside. When the bus-mounted unit passes over the road loop, a message is relayed to the controller to call the signal phase which favours the bus, or to extend the phase if the bus arrives towards the end of its green period; in other words, the bus gets a green period when it is needed rather than having to wait its turn with other traffic.

An 'inhibit' facility is built into the Department of Transport detector which does not permit more than one forced change per cycle; in this way, a steady bus flow cannot continually call the priority stage to the detriment of traffic on the other arms of the junction. A second 'compensating' facility allows side road traffic an extra time to clear if a phase is missed or curtailed as a result of a forced change which gives priority to buses.

As yet there would appear to be only relatively limited information published with regard to experiences with on-bus detectors, and decisions to use them should be taken with care and deliberation. Whilst substantial benefits can be obtained by buses, it has been found that the total delays caused to other vehicles can outweigh the savings to the buses which receive priority; it has also been found that improper usage can result in the overall delay to buses being increased because the extra delay to buses on the non-priority phases is greater than the savings to the priority buses.

Signal progressions to favour buses
As noted previously, it is usual for signals in area-wide traffic control schemes to operate according to a number of fixed-time programmes which vary according to the time of day or on the basis of traffic information provided to a central controller from a number of strategically-placed in-road detectors. Within these schemes, the coordinated traffic signal settings are developed by means of computer optimization methods which minimize total vehicle delay and/or the total amount of stopping and starting within the controlled network. In these calculations, a bus is treated in the same way as any other vehicle.

One of the best known optimization procedures is *TRANSYT*[139]. With this program, which was developed by the Transport and Road Research Laboratory, the basic traffic model is built up from a description of the average behaviour of vehicles in terms of hourly flows, journey times between intersections, platoon dispersion, the lengths of queues on the approaches to the signalized junctions, and the proportions of left- and right-turning vehicles at the junctions. The optimization procedure predicts the average queue size at each intersection in the model until a specified performance index (based on minimizing delay) is achieved.

The *TRANSYT* method for optimizing the traffic signal operation has been extended to produce settings which preferentially reduce the delays to buses. Thus *BUS TRANSYT*, which is the program used for this purpose, calculates the behaviour of the different classes of vehicles within the traffic stream and in the process takes into account the variation in bus journey times along a road as a function of their running speed and the average time spent at a bus stop. The bus delays are then weighted in proportion to the average bus occupancy as compared with that for other vehicles, and the optimization routine then adjusts the signal settings to minimize the total delay for *passengers* rather than vehicles.

The effectiveness of *BUS TRANSYT* has been examined in Glasgow, which has ninety-five signals in its urban traffic control network[152]. Compared with the basic *TRANSYT* method, *BUS TRANSYT* showed an 8 per cent average improvement in bus journey speeds throughout the working day; furthermore, the total delay to other traffic was not increased.

An advantage of this way of ensuring priority to buses is that no extra equipment is required to implement the method wherever the facilities already exist for linking traffic signals. Thus the only extra costs are the obvious ones of the personnel and computing times required to gather the extra survey data on bus flows and occupancies, routes, etc., and to compute the *BUS TRANSYT* signal timings.

Protection of bus stops
A major advantage of a bus bay is that it removes a stopped bus from the traffic stream whilst passengers are loading/unloading; this tends to improve road safety as well as obviously reducing the delays to other traffic. A disadvantage of a bus bay, however, which is particularly noticeable on roads that are well travelled, is that buses using the bay very often have difficulty in regaining positions in the traffic stream. Thus the provision of bus bays can sometimes act, in practice, against a policy of bus priority.

In certain circumstances, this problem can be obviated by providing a lengthened bus bay close to a junction so that it acts as an extra approach lane at the intersection. As well as decreasing the delays to buses, this also increases the capacity of the intersection. Furthermore, the positioning of the bus stop is likely to be very convenient for bus passengers. For safety reasons, the bus stop in this situation should be set back from the stop line at the junction as otherwise pedestrians waiting to cross the road may see a stationary bus and assume that all traffic is stopped—and walk into the paths of moving vehicles.

At non-junction locations, it may be appropriate to solve the bus bay problem by providing relatively shallow bays which are not intended to accommodate the full width of a bus but enable other traffic to pass it. This approach allows buses to re-enter the traffic stream more easily than with full-width bus bays[145].

It might be noted that in some countries, e.g. France and Belgium, buses are given priority under the law when moving away from bus stops, and other vehicles must give way to them. This requirement is often stated on the back of buses, where it can be seen by overtaking drivers.

Pricing control of traffic movement

Much of the present congestion in and about town centres is due to the inability of the inherited street system to handle the ever-increasing traffic demands, especially during the peak periods of the day. It is not possible to increase the road provision sufficiently in most towns to cater for all forms of vehicular movement—economic and environmental considerations render this 'solution' impracticable—so the general policy adopted (see Chapter 3) is to operate the road system as efficiently as possible using conventional traffic management techniques, and to maintain a high level of public transport provision whilst simultaneously seeking to restrain the numbers of private cars. At this time, parking control is the most widely accepted traffic restraint procedure.

Pricing control of traffic movement has been advocated[153] as a means by which such traffic might be restrained. With this approach, which is in the nature of a congestion tax, vehicle owners would have to pay higher charges in order to use congested roads and lower charges would be imposed for travel on other ones. Thus, in theory, only vehicles which have essential reasons for travelling on the high-priced roads would be willing to pay for the privilege of doing so, whilst other vehicles would be influenced not to make unnecessary trips on these facilities.

A practical example of a road pricing scheme developed as part of an overall traffic planning strategy for an urban area (Singapore) is described in Chapter 3. The Greater London Council did examine one such method of traffic restraint (see reference 154 for an excellent description of this proposal) which would have required all motorists to purchase and display a supplementary licence before entering Inner and/or Central London; this method was deemed to be unacceptable, however, apparently because of what were considered to be 'its inequitable and potentially regressive effects'[155].

Parking control is limited in its ability to restrain traffic, particularly in very large towns, i.e. it does not operate on through traffic and is only effective on terminating traffic if control can be imposed on the operation of *all* parking spaces. These limitations are illustrated in Fig. 7.30, from which it can be seen that, although parking control was able to reduce peak period traffic to public parking facilities (on- and off-street) in

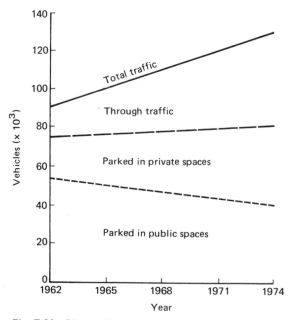

Fig. 7.30 Diagram illustrating the limited effect of parking control upon car movements in the Central London morning peak period, for the years 1962–74

Central London by 30 per cent between 1962 and 1974, traffic to private spaces and through traffic were both doubled.

Operational requirements

In order to be workable and to come close to attaining its objectives, any road pricing mechanism must satisfy the following basic requirements.

(1) The charges must be closely related to the use made of the roads. This is probably the most important consideration since it ensures that people using the congested roads a great deal will pay more than people who use them only rarely. It can be achieved by making the charges proportional either to the time spent on the congested roads or to the distance travelled on them.

(2) It must be possible to vary prices for different roads (or areas), at different times of the day, week or year, and for different classes of vehicles. This criterion is also a most important one. Thus the heavy commercial vehicles which contribute most significantly to congestion might be charged very highly for peak period use of the roads. Again, the price might be varied so that there would be higher charges for all vehicles, say, during the weeks immediately preceding Christmas, which might encourage people to shop earlier and thereby spread the congestion load.

(3) Prices must be stable and readily ascertainable by road users before they embark upon a journey. The object of road pricing is to influence the decision of the motorist *before* entering the congested area. Hence automatic systems which vary the price according to the existing congestion would be of little value.

(4) The method must be simple for road users to understand and the police (or traffic wardens) to enforce.

(5) The system must be accepted by the public as fair to all.

(6) Payment in advance must be possible, although credit facilities might also be permissible under certain conditions. In view of the large amounts of money which would be involved, payment in advance would be the only practical way of ensuring the applicability of the pricing mechanism.

(7) Any equipment used must possess a high degree of reliability. Any kind of equipment used to operate the pricing mechanism must be designed to last many years under conditions of tough usage. Any meters attached to vehicles must be robust and reasonably secure against fraud.

(8) The method must be amenable to gradual introduction commencing with an experimental phase.

(9) It must be capable of being applied to the whole country and to a future vehicle population of 30 to 40 million. It must also allow for temporary road users such as those entering from abroad.

(10) The pricing control mechanism should, if possible, indicate the strength of the demand for road space in different places so as to give guidance to the planning of new road improvements.

Indirect methods of charging

In addition to varying the parking charge(s), there are three main 'indirect' ways of charging for the use of roads in congested locations. These are by means of a differential fuel tax, a system of differential licensing, and a poll tax.

Differential fuel tax
It is technically possible to levy fuel tax at different rates in different areas, relating it to the amount of congestion in each area. The effectiveness of the differential would depend upon the opportunities and incentives for avoidance, and upon its ability to discriminate in detail against journeys on congested roads. Avoidance by the motorist could occur in two ways: firstly, by taking-on fuel while travelling on ordinary journeys through low-tax areas and, secondly, by making special fuel-fetching journeys to low-tax areas. Avoidance by the former could not be stopped and must be accepted as an inherent weakness of the method; avoidance by the latter could be minimized by the careful planning of area boundaries and differentials.

In urban-based countries such as Britain, the differential fuel tax scheme commends itself most obviously to conurbations, where a high tax could be levied throughout most of the built-up areas and then reduced gradually as one moved away from these areas, starting in the outer suburbs. However, unless the high-tax areas are so large that it is possible both to penalize regular users and to make special fuel-fetching journeys uneconomical, this method of control must be inefficient, as well as discriminatory to such businesses as garages.

Differential licences
Under the differential licensing scheme, the road user would require a specially-priced licence in order to drive within a particular congested area. Worldwide, there appears to be a general consensus that (next to control by varying parking charges) differential licensing has the greatest potential for being accepted as an effective method of imposing traffic restraint. Thus, for example, both the Singapore scheme and the London proposal represent practical variations of differential licensing.

In concept, differential licensing schemes can be divided according to whether they would be varied according to geographical area (with differentially-priced licences for different parts of the urban area) and/or by time (with different licences for different times of the day).

Area variations would require congested areas to be divided into zones of different classes such that, for example, areas of heavy congestion could be designated as 'red' zones, areas of moderate congestion 'blue', and less congested areas 'yellow'. Thus a relatively cheap yellow licence would provide access to yellow zones, a more expensive blue licence to blue and yellow zones, and a still more expensive red licence to all three classes of zone.

Time variations would distinguish between day and night usage, and possibly between peak and off-peak usage. A zone might be declared 'red' at the peaks, 'blue' during daytime off-peak hours, and 'yellow' at night. Only vehicles holding the required licence colour would then be able to enter these zones during these times.

To be effective, a system of *annual* differential licensing for a congested area would have to be expensive and transferable so that people could obtain licences for a short time. The high value of the licence could give rise to problems of theft, fraud and finance. A much greater difficulty is the technical problem of enabling the general public to hire the licences for limited periods of time.

There are no purely technical difficulties associated with the introduction of a system of *daily* differential licensing. The licences could be numbered in large figures or letters to denote the day and sold in two kinds of books. The first of these books might be termed a daily book since it would consist of a number of licences for a given day; this would be for retailers of licences such as garages. The second type of book would contain one licence for each day of the month or year; this book would be for the use of individual vehicle owners. In both cases, a rebate would be given for licences not detached from the book.

The main advantage of the daily licence is that it is simple and should not be difficult to implement. In addition, it has the advantage over the parking charge approach of being applicable to all traffic entering the congested area, and not just those parking. It has the disadvantages that not only does the short journey pay the same as the long journey, but through traffic with an expensive licence would be encouraged to use zones where congestion has been decreased, thereby diminishing the apparent fairness and reasonableness of the system.

Poll tax
It has been suggested that congestion would be reduced if a poll tax were to be levied on all employees in congested areas. The concept envisages that organizations with large numbers of employees either would move out of these areas and be replaced by organizations with fewer employees, or would initiate further mechanization of labour such as lift-operating, book-keeping, dictation and retail service.

There are so many complex variables of a social nature associated with the introduction of a poll tax that it is difficult to examine it objectively. On balance, however, the effect of a poll tax on the volume of road traffic would only be of limited value. This conclusion is based on the premise that the people most affected by the tax would be in the lower income groups and these groups have the lowest level of car ownership.

Direct methods of charging

Whereas the indirect methods of charging rely upon a tax on some product or service that is more or less closely correlated with road usage, the methods now to be discussed all make a direct charge to the motorist for usage of the congested roadway. The simplest example of direct charge control is the toll-gate which is used on bridges and tunnels and on some (foreign) motorways with few points of access. Although simple in concept, toll-gates cannot be considered for usage on roads in urban areas because of their cost and impedance to the flow of traffic. Direct methods that might be possible are off-vehicle recording systems and vehicle metering systems.

Off-vehicle recording systems
These systems are analogous to telephone charging methods in that every vehicle would be fitted with a piece of equipment which could be automatically identified and recorded by detectors placed in, over or beside the roadway at suitable pricing locations. The recording that a vehicle has passed a certain pricing location then, in theory, could be used in one of two ways. The first, termed 'point pricing', would involve setting up pricing points within congested areas so that vehicles could be debited with the appropriate charge when passing any pricing point. The alternative, 'continuous pricing', would involve setting up pricing points on the borders of congested areas and charging according to the time spent in the zone, as deduced from recordings at the zone entry and exit locations.

The utilization of any continuous pricing system cannot be considered practical for two reasons. Firstly, there is the considerable technical problem of monitoring private entrances and parking places within pricing zones so that charging would either cease or be reduced when vehicles enter parking places. Secondly, there is the intolerable error associated with a vehicle being recorded as entering a zone and its corresponding record of exit being missed or mislaid; this could result in a vehicle being charged for an indefinite period.

It appears, therefore, that any feasible off-vehicle system could work only on a point pricing basis. Such a system might comprise the following stages.

(1) *Identification of vehicles at the pricing points*   Vehicles would have to carry identification units which would enable their presence to be detected and recorded by roadside apparatus. In addition, the system would have to be capable of distinguishing between perhaps up to 40 million different identities. On technical grounds, these requirements could probably be met, but only at a relatively high cost.
(2) *Transmission of information to a central computing station*   The simplest method would be to accumulate the records at each pricing point on magnetic tape and then transport these to the central station at regular intervals.
(3) *Processing of data*   All charges would have to be sorted by vehicle identification number, and charges multiplied by an appropriate charging factor such as one for a private car, two for a commercial vehicle, etc. Bills could be issued at monthly or quarterly intervals.
(4) *Collection of payment*   The difficulties and costs of bill collection would need careful study. If there were ten million accounts and each

year one account in a thousand failed to pay after the usual reminders, there would be some 10 000 cases a year to follow up, many of them through the Courts. While some protection could be obtained by making renewal of vehicle licences conditional on all road charges being paid, many difficult cases would still remain; in addition, an important question which needs to be answered is whether or not it is sociologically acceptable to allow large numbers of people to become debtors for the considerable amounts which could be involved.

The advantages of a fully-automatic system of road pricing with ample price flexibility, in which the road user has no more to do than periodically send a payment cheque, are obvious. However, the disadvantages raised in the above discussion suggest its impracticability. In addition, the utilization of any such system is a definite threat to the privacy of road users in so far as it would be possible for public authorities to trace the vehicle movement of perhaps the entire populace. Notwithstanding the advantage of such a system in aiding, for instance, crime detection, this would appear to be such an invasion of privacy as to render its usage most undesirable in any democratic community.

Vehicle metering systems
The on-vehicle metering systems can be divided into those that are driver-operated and those that are automatically-operated.

Driver-operated meters Systems using these meters could only work on a continuous charging basis. Zones would have to be defined in congested areas and then allocated colours which would indicate the prices charged. The colours, which would be displayed electrically at all points of entry and exit, would have to be capable of being switched so that, for example, during off-peak hours or Sundays or public holidays, an expensive zone could be de-rated or perhaps de-zoned entirely.

Every vehicle entering a congested zone would carry a road meter displayed on the windshield and within comfortable reach of the driver's hand. When the meter was switched on and the timing mechanism working, it would show a coloured light, indicating the rate being charged. Thus a driver would switch to purple when entering a purple zone, and if the vehicle was driven later into a pink zone the meter would be switched accordingly. One or two lower rates for parking could be added, thereby providing each vehicle with what is in effect a personal parking meter and removing the need for footpath meters. Different types of vehicle would carry meters with different charging rates.

The method by which the motorist would pay for usage of the roadway could vary with the type of timing mechanism in the meter. If a clockwork mechanism were used, the motorist would purchase a complete meter which, as it was used, would run down at the predetermined rates. When the meter expired, it would be returned to an authorized station and a complete new one purchased. If an electrical timing mechanism were used, it would not be necessary to obtain a complete new meter whenever the meter time ran out. Instead, a cylindrical 'throw-away' electrolytic timer could be inserted into the meter by the driver, to form part of a circuit connecting the meter lamp with the vehicle battery. Also included would be a network of resistors so that the timer could run down at the correct rates. When the timer reached its designed end-point, the circuit would be broken and the meter lamp cut out. The timer could then be redrawn and a new one easily inserted.

The meter with the disposable electrolytic timer is to be preferred to the mechanical one—not only is it cheaper but it also considerably reduces difficulties of payment. Timers could be bought over the counter from garages or post offices in units of, say, £1, £2 or £5, thereby eliminating the need for the meter to be read or removed for resetting. The revenue due to the Government could be collected as an excise duty from the timer manufacturers.

Automatic meters   The principal drawback associated with the use of the driver-operated meter is that it adds to the responsibilities of both the motorist and the traffic authority. The automatic meter attempts to eliminate this by the placement of a control apparatus in the highway pavement which would activate and switch meters installed in passing vehicles. These meters could be used to operate within the point pricing and continuous pricing systems.

Point pricing systems   In this type of system, the meter carried by the vehicle would be basically an instrument capable of counting electrical impulses generated by electrical cables carrying very low currents and laid in the roadway at fixed pricing points. The vehicle meter could be a small unit, probably of the size and shape of a small book. Since it would have to be near the ground in order to pick up the signals, the meter could form a part of the vehicle number plate. In its simplest form, this meter would probably be in the form of a 'solid-state' counter, of the form used in computers. Although it could not be read in the usual way, the meter could be made to change colour to show how its economic life was expiring.

Two methods of payment suggest themselves for use under point pricing operation. Either the meter could contain a given economic capacity so that a new one would have to be bought when this was exhausted, or else a credit meter could be issued that would be capable of being fixed permanently to the vehicle and then the vehicle would have to be taken at intervals to registered stations so that the meter could be read and paid for.

The automatic meter point pricing system is essentially a sophisticated form of toll-gate control. Its principal advantage is that highly flexible pricing—and hence traffic control—is possible due to the feasibility of varying the price from point to point while the points themselves can be deployed as densely as required within the control area. A disadvantage is that, unless the pricing points are spread in very large numbers, they can to some extent be avoided by the irresponsible motorist. In addition, the fewer there are the less accurately do they relate to costs and the more arbitrary are their effects upon traffic.

Continuous pricing systems Under automatic meter-operated systems of continuous pricing, vehicles would be charged continuously while within pricing zones, the charges commencing when the vehicles enter a zone and ending when they leave. The main difference between this system and the driver-operated one is that, whereas in the latter the motorist sets the meter to the appropriate rate, under the automatic system this function would be performed by a switching circuit which would operate in response to signals received from road-sited transmitters installed at the entry and exit points and, if required, at intermediate points within the zone. A manual override control would also be provided to enable the motorist to select the parking rate when the vehicle is stopped.

Selected bibliography

(1) Sabey BE and Taylor H, *The Known Risks We Run: The Highway*, TRRL Report SR567, Crowthorne, Berks., The Transport and Road Research Laboratory, 1980.
(2) Department of Transport, *Road Accidents in Great Britain 1983*, London, HMSO, 1984.
(3) Smeed RJ and Jeffcoate GO, Effects of changes in motorisation in various countries on the number of road fatalities, *Traffic Engineering & Control*, 1970, **12**, No. 3, pp. 150-151.
(4) Hutchinson TP and Harris RA, Recent trends in traffic injury, *Injury*, 1978, **10**, No. 2, pp. 133-137.
(5) *Road Accidents 1977*, TRRL Leaflet LF694, Crowthorne, Berks., The Transport and Road Research Laboratory, 1978.
(6) Department of Transport, *Transport Statistics Great Britain 1972-1982*, London, HMSO, 1983.
(7) Williams MC, *Tabulations of 1977 Road Casualties Indicating Risks of Injury to Road Users in Relation to Vehicles Involved*, TRRL Report SR576, Crowthorne, Berks., The Transport and Road Research Laboratory, 1980.
(8) Cameron C and Macdonald WA, *A Review of Driver/Rider Training in Relation to Road Safety*, ARR Report No. 3, Melbourne, Victoria, The Australian Road Research Board, 1975.
(9) Whitaker J, *A Survey of Motorcycle Accidents*, TRRL Report LR913, Crowthorne, Berks., The Transport and Road Research Laboratory, 1980.
(10) Dalby D, *Space-sharing by Pedestrians and Vehicles*, TRRL Report LR743, Crowthorne, Berks., The Transport and Road Research Laboratory, 1976.
(11) Sabey BE, *Road Safety and Value for Money*, TRRL Report SR581, Crowthorne, Berks., The Transport and Road Research Laboratory, 1980.
(12) *Comparison of British Number-plate and Snellen Vision Tests for Car Drivers*, TRRL Leaflet LF76, Crowthorne, Berks., The Transport and Road Research Laboratory, 1977.
(13) Davison PA and Irving A, *Survey and Visual Acuity of Drivers*, TRRL Report LR945, Crowthorne, Berks., The Transport and Road Research Laboratory, 1980.
(14) Storie VJ, *Male and Female Car Drivers: Differences Observed in Accidents*,

TRRL Report LR761, Crowthorne, Berks., The Transport and Road Research Laboratory, 1977.

(15) Palmer PO, *The Effect of Biorythms on Accidents*, TRRL Report SR535, Crowthorne, Berks., The Transport and Road Research Laboratory, 1979.

(16) *Relation of Alcohol to Road Accidents*, London, The British Medical Association, 1960.

(17) Smeed RJ, Methods available to reduce the number of road casualties, *International Road Safety and Traffic Review*, 1964, **12**, No. 4.

(18) Sabey BE, *A Review of Drinking and Drug Taking in Road Accidents in Great Britain*, TRRL Report SR441, Crowthorne, Berks., The Transport and Road Research Laboratory, 1978.

(19) Clayton AB, Booth AC and McCarthy PE, *A Controlled Study of the Role of Alcohol in Fatal Adult Pedestrian Accidents*, TRRL Report SR332, Crowthorne, Berks., The Transport and Road Research Laboratory, 1977.

(20) Tillman WA, *The Psychiatric and Social Approach to the Detection of Accident-prone Drivers*, Unpublished MSc Thesis, University of Western Ontario, London, Ontario, 1948.

(21) Clayton AB, McCarthy PE and Breen JM, *The Male Drinking Driver; Characteristics of the Offender and His Offence*, TRRL Report SR600, Crowthorne, Berks., The Transport and Road Research Laboratory, 1980.

(22) Shaw L and Sichel HS, *Accident Proneness: Research in the Occurrence, Causation, and Prevention of Road Accidents*, Oxford, Pergamon Press, 1971.

(23) Linklater DR, Fatigue and long-distance truck drivers, *Proceedings of the Australian Road Research Board*, 1980, **10**, Part 4, pp. 193–201.

(24) Watts GR and Quimby AR, *Aspects of Road Layout That Affect Drivers' Perception and Risk Taking*, TRRL Report LR920, Crowthorne, Berks., The Transport and Road Research Laboratory, 1980.

(25) Charlesworth G, Design for safety, in *Proceedings of the Conference on Engineering for Traffic*, London, Printerhall, 1963.

(26) Tanner JC, Johnson HD and Scott JR, *Sample Survey of Roads and Traffic of Great Britain*, Road Research Technical Paper No. 62, London, HMSO, 1962.

(27) Tanner JC, Accidents at rural three-way intersections, *Journal of the Institution of Highway Engineers*, 1953, **2**, No. 11, pp. 56–57.

(28) Colgate MG and Tanner JC, *Accidents at Rural Three-way Junctions*, RRL Report LR87, Crowthorne, Berks., The Road Research Laboratory, 1967.

(29) *Design of Major/Minor Priority Intersections*, Technical Memorandum H11/76, London, The Department of the Environment, August 1976.

(30) Spicer AH, Wheeler AH and Older SJ, *Variation in Vehicle Conflicts at a T-junction and Comparison with Recorded Collisions*, TRRL Report SR545, Crowthorne, Berks., The Transport and Road Research Laboratory, 1980.

(31) Cameron JWM, *The Influence of the Layout of the Road Network on Road Safety: A Literature Review*, Technical Report RF/3/77, Pretoria, SA, The National Institute for Transport and Road Research, April 1977.

(32) Box PC and Associates, Chapter 4: Intersections, in Mayer PA (Ed.), *Traffic Control and Roadway Elements—Their Relationship to Highway Safety*, Washington DC, The Automotive Safety Foundation, 1970.

(33) Webster FV and Newby RF, Research into relative merits of roundabouts and traffic signal controlled intersections, *Proceedings of the Institution of Civil Engineers*, 1964, **27**, pp. 47–75.

(34) *Access to Highways—Safety Implications*, Advice Note TA4/80, London, The Department of Transport, February 1980.

(35) OECD Research Group TS2, *Traffic Safety in Residential Areas*, Working Paper RR/TS2/79.1, Paris, OECD, April 1979.

(36) Codling PJ, *Thick Fog and Its Effect on Traffic Flow and Accidents*, RRL Report LR397, Crowthorne, Berks., The Road Research Laboratory, 1971.

(37) Kockmond WC and Perchonok K, *Highway Fog*, NCHRP Report 95, Washington DC, The Highway Research Board, 1970.

(38) Heiss WH, *Highway Fog: Visibility Measures and Guidance Systems*, NCHRP Report 171, Washington DC, The Highway Research Board, 1976.

(39) Sabey BE and Storie VJ, *Skidding in Personal-injury Accidents in Great Britain in 1965 and 1966*, RRL Report LR173, Crowthorne, Berks., The Road Research Laboratory, 1968.

(40) Ludema KC and Gujrati BD, *An Analysis of the Literature on Tire-Road Skid Resistance*, ASTM Special Technical Publication 541, Philadelphia, Pa., The American Society for Testing and Materials, July 1973.

(41) OECD Road Research Group, *Road Safety at Night*, Paris, OECD, December 1979.

(42) Sabey BE, *Road Accidents in Darkness*, TRRL Report LR536, Crowthorne, Berks., The Transport and Road Research Laboratory, 1973.

(43) Sabey BE and Johnson HD, *Road Lighting and Accidents: Before and After Studies on Trunk Road Sites*, TRRL Report LR586, Crowthorne, Berks., The Transport and Road Research Laboratory, 1973.

(44) Scott PP, *The Relationship Between Road Lighting Quality and Accident Frequency*, TRRL Report LR929, Crowthorne, Berks., The Transport and Road Research Laboratory, 1980.

(45) Beaumont K and Newby RF, *Traffic Law and Road Safety Research in the United Kingdom—British Counter Measures*, TRRL paper presented at the National Road Safety Symposium held at Canberra, Australia, on 14-16 March 1972.

(46) Department of the Environment, *Drinking and Driving*, Report of the Department (Blennerhessett) Committee, London, HMSO, 1976.

(47) Grime G, *The Protection Afforded by Seat Belts*, TRRL Report SR449, Crowthorne, Berks., The Transport and Road Research Laboratory, 1979.

(48) Grattan E and Hobbs JA, *Injuries to Occupants of Heavy Goods Vehicles*, TRRL Report LR854, Crowthorne, Berks., The Transport and Road Research Laboratory, 1978.

(49) Russam K, *Road Safety for Children in the United Kingdom*, TRRL Report LR678, Crowthorne, Berks., The Transport and Road Research Laboratory, 1975.

(50) Wells P, Downing CS and Bennett M, *Comparison of On-road and Off-road Cycle Training for Children*, TRRL Report LR902, Crowthorne, Berks., The Transport and Road Research Laboratory, 1979.

(51) Bennett M, Sanders BA and Downing CS, *Evaluation of a Cycling Proficiency Training Course Using Two Behaviour Recording Methods*, TRRL Report LR890, Crowthorne, Berks., The Transport and Road Research Laboratory, 1979.

(52) Colbourne HV and Sargent KJ, *A Survey of Road Safety in Schools: Education and Other Factors*, RRL Report LR388, Crowthorne, Berks., The Road Research Laboratory, 1971.

(53) Fazakerley JA, Davies RF, Henderson R and Sheppard D, *Evaluation of 'Better Driving' Courses Run by Police Forces for the Public*, TRRL Report LR949, Crowthorne, Berks., The Transport and Road Research Laboratory, 1980.

(54) *Accident Investigation and Prevention Manual*, London, The Department of Transport, March 1974 (amended 1981).

(55) Cleveland DE, *Manual of Traffic Engineering Studies*, Washington DC, The Institute of Traffic Engineers, 1964.

(56) Bennett GT, Vehicle speeds in residential areas: some preliminary results, *The Highway Engineer*, 1978, **25,** No. 6, pp. 2–5.

(57) Leake GR, Fuel conservation—is there a case for stricter motorway speed limits?, *Traffic Engineering & Control*, 1980, **21,** No. 11, pp. 551–558.

(58) Duff JT, Road accidents in urban areas, *Journal of the Institution of Highway Engineers*, 1968, **15,** No. 5, pp. 61–69.

(59) Committee on Traffic and Highway Safety, The 55-mph speed limit: a review, *Transportation Engineering Journal*, 1980, **106,** No. TE3, pp. 299–308.

(60) *Local Speed Limits*, Circular Roads No. 1/80, London, The Department of Transport, February 1980.

(61) Mostyn BJ and Sheppard D, *A National Survey of Drivers' Attitudes and Knowledge About Speed Limits*, TRRL Report SR548, Crowthorne, Berks., The Transport and Road Research Laboratory, 1980.

(62) Persuading drivers to reduce speed: some recent work from the TRRL, *Traffic Engineering & Control*, 1974, **15,** No. 16/17, p. 776.

(63) *Transverse Yellow Bar Markings at Roundabouts*, Circular Roads No. 17/78, London, The Department of Transport, 1978.

(64) Watts GR, *Road Humps for the Control of Vehicle Speeds*, TRRL Report LR597, Crowthorne, Berks., The Transport and Road Research Laboratory, 1973.

(65) Duffell JR and Hopper R, 'Sleeping policemen'—their effectiveness in regulating vehicle speeds, *Chartered Municipal Engineer*, 1975, **102,** No. 8, pp. 151–159.

(66) Sumner R and Baguley PC, *Speed Control Humps on Residential Roads*, TRRL Report LR878, Crowthorne, Berks., The Transport and Road Research Laboratory, 1979.

(67) Rutley KS, *Advisory Speed Signs for Bends*, TRRL Report LR461, Crowthorne, Berks., The Transport and Road Research Laboratory, 1972.

(68) O'Flaherty CA and Coombe RD, Speeds on level rural roads—a multivariate approach, Part 3, *Traffic Engineering & Control*, 1971, **13,** pp. 108–111.

(69) Webb PJ, *The Effect of an Advisory Speed Signal on Motorway Traffic Speeds*, TRRL Report SR615, Crowthorne, Berks., The Transport and Road Research Laboratory, 1980.

(70) Goodwin PB and Hutchinson TP, The risk of walking, *Transportation*, 1977, **6,** pp. 217–230.

(71) Sheppard D and Valentine SD, *The Provision of Road Safety Instruction for the Elderly*, TRRL Report SR533, Crowthorne, Berks., The Transport and Road Research Laboratory, 1979.

(72) Department of the Environment, *Pedestrian Safety*, London, HMSO, 1973.

(73) Ministry of Transport, *Roads in Urban Areas*, London, HMSO, 1966, and as corrected in Technical Memorandum H12/73, October 1973.

(74) *Pedestrian Crossings—Revised Criteria*, Circular Roads No. 19/74, London, The Department of the Environment, 1974.

(75) Jacobs GD and Wilson DG, *A Study of Pedestrian Risk in Crossing Busy Roads in Four Towns*, RRL Report LR106, Crowthorne, Berks., The Road Research Laboratory, 1967.

(76) *Pelican Crossings: Pelican Crossing Operation*, Departmental Standard TD4/79, London, The Department of Transport, May 1979.

(77) Transport and Road Research Laboratory, *Transport and Road Research 1975*, London, HMSO, 1976.

(78) Garwood F and Moore RL, Pedestrian accidents, *Traffic Engineering & Control*, 1962, **4,** No. 5, pp. 274–276 and 279.

(79) Institute of Traffic Engineers, Pedestrian overcrossings—criteria and priorities, *Traffic Engineering*, October 1972, pp. 34–39 and 68.

(80) Road Research Laboratory, *Research on Road Traffic*, London, HMSO, 1963.

(81) Goldschmidt, J, *Pedestrian Delay and Traffic Management*, TRRL Report SR356, Crowthorne, Berks., The Transport and Road Research Laboratory, 1977.

(82) *Pedestrian Subways: Layout and Dimensions*, Departmental Standard TD2/78, London, The Department of Transport, July 1978.

(83) *Combined Pedestrian and Cycle Subways: Layout and Dimensions*, Departmental Standard TD3/79, London, The Department of Transport, May 1979.

(84) Nye C, Pedestrianisation, *The Highway Engineer*, 1978, **25**, No. 6, pp. 6 and 8.

(85) Tuohey GJ, *Traffic Aspects of Pedestrian Malls*, RRU Bulletin No. 36, Wellington, NZ, The National Roads Board, 1978.

(86) *Woonerf: A New Approach to Environmental Management in Residential Areas and the Related Traffic Legislation*, The Hague, Royal Dutch Touring Club, 1980.

(87) Department of the Environment, *Residential Roads and Footpaths; Layout Considerations*, London, HMSO, 1978.

(88) *Roads, Bicycles and Bikeways*, Brickfield Hill, New South Wales, The National Association of Australia State Road Authorities, 1978.

(89) Stores A, *Cycle Ownership and Use in Great Britain*, TRRL Report LR843, Crowthorne, Berks., The Transport and Road Research Laboratory, 1978.

(90) Downing CS, Cycle safety, in *Cycling as a Mode of Transport*, TRRL Report SR540, pp. 19–41, Crowthorne, Berks., The Transport and Road Research Laboratory, 1980.

(91) *Guide for Bicycle Routes,* Washington DC, The American Association of State Highway and Transportation Officials, 1974.

(92) Robinson K, Cycle routes in Peterborough, in the TRRL report as reference 90, pp. 97–106.

(93) Trevelyan P, The design of cycle facilities, in the TRRL report as reference 90, pp. 47–60.

(94) Ratcliffe JT and Turver MG, Middlesbrough cycleways, *Chartered Municipal Engineer*, 1979, **106,** No. 5, pp. 146–150.

(95) *Ways of Helping Cyclists in Built-up Areas*, Local Transportation Note 1/78, London, The Department of Transport, 1978.

(96) Hawley L, *Cycle Ways: A Review of the Literature on Cycle Ways*, Adelaide, Faculty of Architecture and Planning, The University of Adelaide, 1975.

(97) Hudson M, *The Bicycle Planning Book*, London, Open Books/Friends of the Earth, 1978.

(98) Maryland Department of Transportation, *The ABCDs of Bikeways*, Report FHWA-TS-77-201, Washington DC, US Government Printing Office, 1977.

(99) *Manual of Uniform Traffic Control Devices for Streets and Highways*, Washington DC, US Department of Transportation, 1978.

(100) Walton KN, Signs of the times: a commentary on the international influence on United Kingdom traffic signs, *Proceedings of Seminar K on Traffic and Environmental Management*, PTRC Report P139, pp. 245–250, London,

Planning and Transport Research and Computation (International) Company Ltd, 1976.

(101) Department of Transport, *Traffic Signs Manual*, London, HMSO (as amended from time to time).

(102) Worboys W et al., *Traffic Signs*, Report of the Committee on Traffic Signs for All-purpose Roads, London, HMSO, 1963.

(103) Huddart KW, Communicating with the driver and the pedestrian, *The Highway Engineer*, 1974, **21**, No. 5, pp. 25–32.

(104) *Lettering Styles for Traffic Signs*, Transport and Road Digest No. 19, Pretoria, SA, The National Institute for Transport and Road Research, February 1980.

(105) Christie AW and Rutley KS, Relative effectiveness of some letter types designed for use on road traffic signs, *Roads and Road Construction*, 1961, **39**, No. 464, pp. 239–244.

(106) Hoffman ER and Macdonald WA, A comparison of stack and diagrammatic advance direction signs, *Australian Road Research*, 1977, **7**, No. 4, pp. 21–26.

(107) Odescalchi P, Conspicuity of signs in rural surroundings, *Traffic Engineering & Control*, 1960, **2**, No. 7, pp. 390–393 and 397.

(108) Trevelyan P, Highway directional signing, *Proceedings of Seminar H on Traffic and Environmental Management*, PTRC Report P165, London, Planning and Transport Research and Computation (International) Company Ltd, 1978.

(109) *Design of Rural Motorway to Motorway Interchanges: General Guidelines*, Technical Memorandum H6/75, London, The Department of Transport, April 1975.

(110) Lozano RD, The visibility, colour and measuring requirements of road signs, *Lighting Research and Technology*, 1980, **12**, No. 4, pp. 206–210.

(111) Sessions GM, *Traffic Devices: Historical Aspects Thereof*, Washington DC, The Institute of Traffic Engineers, 1971.

(112) Biggs NL, Directional guidance of motor vehicles—a preliminary survey and analysis, *Ergonomics*, 1966, **9**, No. 3.

(113) Bali SG, McGee HW and Taylor JI, *State-of-the-Art on Roadway Delineation Systems*, Report FHWA-RD-76-73, Washington DC, US Government Printing Office, May 1976.

(114) O'Flaherty CA, *Delineating the Edge of the Carriageway in Rural Areas*, London, Printerhall, 1972 (42 pp.), also (in abridged form) in *Research and Development of Roads and Road Transport 1971*, Washington DC, The International Road Federation, 1971, pp. 403–437.

(115) Jobson AJ, *Delineation: Road Studs and Post Delineators*, Internal Report RF/1/76, Pretoria, SA, The National Institute for Transport and Road Research, May 1976.

(116) BS 873: *The Construction of Road Traffic Signs and Internally Illuminated Bollards*, Part 4—*Road Studs*, London, The British Standards Institution, 1973.

(117) Reid JA and Tyler JW, Reflective devices as aids to night driving, *Highways and Traffic Engineering*, July 1969, **37**, No. 1715.

(118) Walker AE and Chapman RG, *Assessment of Anti-dazzle Screen on M6*, TRRL Report LR955, Crowthorne, Berks., The Transport and Road Research Laboratory, 1980.

(119) Carlson RD, Allison JR and Bryden JE, *Performance of Highway Safety Devices*, Research Report 57, Albany, NY, The New York State Department of Transportation, December 1971.

(120) Fox JC, Good MC and Joubert PN, Development of breakaway utility poles, *Proceedings of the Australian Road Research Board*, 1980, **10,** Part 4, pp. 202–219.

(121) Jobson AJ, *Crash Cushions: Review of Vehicle-impact Attenuation Systems*, Internal Report RF/5/75, Pretoria, SA, The National Institute for Road Research, 1975.

(122) *Guide to Selecting, Locating, and Designing Traffic Barriers*, Washington DC, The American Association of State Highway Officials, 1977.

(123) *Safety Fences*, Technical Memorandum H9/73, London, The Department of the Environment, 1973.

(124) Lokken EC, Concrete safety barrier design, *Transportation Engineering Journal of ASCE*, 1974, **100,** No. TE1, pp. 151–169.

(125) Newby RF and Johnson HD, London–Birmingham motorway accidents, *Traffic Engineering & Control*, 1963, **4,** No. 10, pp. 550–555.

(126) Johnson HD, *Cross-over Accidents on All-purpose Dual Carriageways*, TRRL Report SR617, Crowthorne, Berks., The Transport and Road Research Laboratory, 1980.

(127) Ferguson JA and Jenkins IA, Incidents on high-speed roads, *Traffic Engineering & Control*, 1977, **18,** No. 5, pp. 240–245.

(128) US Department of Transportation, *Traffic Control Systems Handbook*, Washington DC, US Government Printing Office, June 1976.

(129) Bruce JA, One-way major arterial streets, in *Improved Street Utilization Through Traffic Engineering*, Special Report 93, Washington DC, The Highway Research Board, 1967, pp. 24–36.

(130) Duff JT, Traffic management, in *Proceedings of the Conference on Engineering for Traffic*, London, Printerhall, 1963.

(130) Duff JT, Traffic management, in *Proceedings of the Conference on Engineering for Traffic*, London, Printerhall, 1963.

(131) Duff JT, One-way streets, *Traffic Engineering & Control*, 1963, **4,** No. 9, pp. 518–520.
TRRL Report LR841, Crowthorne, Berks., The Transport and Road Research Laboratory, 1978.

(133) Focus on signs and markings; movable traffic devices in Australia, *Traffic Engineering & Control*, 1975, **16,** No. 2, pp. 91–93.

(134) BS 505: *Road Traffic Signals*, London, The British Standards Institution, 1971.

(135) Webster FV, *Traffic Signal Settings*, Road Research Technical Paper No. 39, London, HMSO, 1958.

(136) Webster FV and Cobbe BM, *Traffic Signals*, Road Research Technical Paper No. 56, London, HMSO, 1966.

(137) Anderson RL, Electromagnetic loop vehicle detectors, *IEEE Transactions on Vehicular Technology*, 1970, **VT-19,** No. 1, pp. 23–30.

(138) Morris DJ, Hulsher FR, Dean KG and Macdonald DE, Loop configurations for vehicle detectors, *Proceedings of the Australian Road Research Board*, 1978, **9,** Part 5, pp. 3–11.

(139) *TRANSYT Users Guide*, TRRL Report LR888, Crowthorne, Berks., The Transport and Road Research Laboratory, 1980.

(140) Middleton G and Luk JYK, Area traffic control systems: some aspects of planning and design, *Australian Road Research*, 1979, **9,** No. 2, pp. 25–34.

(141) Tomecki AB, *Area Traffic Control Overseas*, Technical Report TR/21/77, Pretoria, SA, The National Institute for Transport and Road Research, 1977.

(142) *Development of Area Traffic Control Systems*, Circular Roads No. 26/75, London, The Department of the Environment, June 1975.

(143) Traffic management: state of the art report, *Municipal Engineer*, 1984, **1,** No. 3, pp. 253–274.

(144) Skinner RJ, Bus planning methods: (3) service reliability, *Traffic Engineering & Control*, 1980, **21,** No. 11, pp. 554–558.

(145) NATO Committee on the Challenges of Modern Society, *Bus Priority Systems*, CCMS Report No. 45, Crowthorne, Berks., The Transport and Road Research Laboratory, 1976.

(146) Rapson G (Ed.), *Bus Priority Schemes*, London, Planning and Transport Research and Computation (International) Company Ltd, 1978.

(147) OECD Road Research Group, *Bus Lanes and Busway Systems*, Paris, OECD, December 1976.

(148) *Implementation of Bus Priorities*, Technical Memorandum H6/76, London, The Department of Transport, June 1976.

(149) National Bus Company, *Bus Priority Schemes*, Research Report No. 19, Peterborough, The National Bus Company, April 1978.

(150) Bly PH, Webster FV and Oldfield RH, Justification for bus lanes in urban areas, *Traffic Engineering & Control*, 1978, **19,** No. 2, pp. 56–59 and 63.

(151) Dow IM, The Department of Transport bus detector, *Traffic Engineering & Control*, 1977, **18,** No. 1, pp. 10–11 and 18, and No. 2, pp. 63–67 and 74.

(152) Robertson DI and Vincent RA, *Bus Priority in a Network of Fixed-time Signals*, TRRL Report LR666, Crowthorne, Berks., The Transport and Road Research Laboratory, 1975.

(153) Smeed RJ et al., *Road Pricing: The Economic and Technical Possibilities*, London, HMSO, 1964.

(154) May AD, Supplementary licensing: an evaluation, *Traffic Engineering & Control*, 1975, **16,** No. 4, pp. 162–167.

(155) Lane R and Hodgkinson DH, A permit system for traffic restraint, *Traffic Engineering & Control*, 1976, **17,** No. 3, pp. 94–97 and 100.

8
Highway lighting

Since ancient times, man has striven to overcome his inability to see in the dark and has made attempts, albeit with varying degrees of success prior to the 20th century, to light the streets of urban areas. It is reported[1] that the Arabs in Spain lit many kilometres of street in Cordova in the 10th century, whilst in England, in 1405, citizens of London were required to hang 'lanthorns' outside their homes by order of the City Council; in 1736, regularly-spaced oil lamps in city streets were paid for out of local rates. Better standards were achieved in the 19th century with gas lighting, and the introduction of electric lighting in 1913 began a new era in respect of street lighting.

Whereas originally lighting was installed as an amenity, the development of the motor vehicle and the consequent increases in traffic using the highways resulted in demands for better street lighting in order to reduce accidents.

As noted previously (see Chapter 7), the road accident rate at night is currently about one-third higher than the rate during the day; were it not for street lighting, the ratio of night accidents to day accidents would be considerably higher. Numerous before-and-after studies (see, for example, reference 2) have shown that where good road lighting is properly installed accident rates on most roadways are reduced, particularly at intersections and on road sections where the night-to-day accident ratio is high and standards of design are low. Furthermore, the greatest beneficiary of good road lighting in urban areas is very often the pedestrian, i.e. the percentage reduction in pedestrian accidents is normally greater than the reduction in other types of injury accidents. Various analyses of accident studies have suggested that nightly road deaths and accidents would be most significantly reduced if adequate lighting were to be provided on roadways at the following critical areas of driver decision:

(1) entrances and exits,
(2) interchanges and intersections,
(3) bridges, overpasses and viaducts,
(4) tunnels and underpasses,
(5) guide sign locations,
(6) dangerous hills and curves,
(7) well-travelled roads in urban areas,
(8) rest areas and connecting roads,

(9)   railway grade crossings,

(10)   elevated and depressed roadways.

The installation of a proper highway lighting system provides other fringe benefits which are not always appreciated. For example, traffic flow during evening peak periods and at night is considerably improved since drivers are more confident in their movements and can more easily observe traffic management intentions. The development of late evening shopping is partly due to the commercial centres being attractively and well lit. In addition, there is no doubt but that good street lighting is a powerful weapon in the fight against crime in urban areas.

In the limited space available here, it is not possible to attempt to discuss the many and varied lighting designs at the critical roadway areas listed above. The approach therefore taken in this chapter is concentrated on some of the basic features which underlie the rational design of lighting installations on traffic routes in Britain.

Terminology and practice

Before examining the basic concepts and considerations underlying the design of highway lighting, it is helpful to define some of the terms in common usage. These may be divided into two main groups: photometric terms which relate to light and its measurement units, and lighting installation terms which relate to the physical equipment and its layout on the roadway.

Firstly, however, it is useful to comment briefly upon the overall approach used in Britain in respect of lighting design.

British practice

As with many other aspects of traffic engineering, the design of a modern road lighting installation depends upon the careful amalgamation of theoretical and empirical knowledge. British practice in this respect is summed up in a British Standard[3]—this is amended at irregular intervals—which lays down standardized design criteria which assure that lighting installations are always reasonably satisfactory. The Standard divides the lighting requirements for streets and highways into the following eight main groups.

Lighting for traffic routes   This group, which is concerned with the lighting of all-purpose traffic routes up to 15 m wide (single carriageways) and up to 2 m × 11 m wide (dual carriageways) and bearing traffic volumes of up to 60 000 vehicles per day, is subdivided into two sub-classes, viz. Group A10 and Group A12.

Group A10 has a mounting height of 10 m, and is used to light narrow traffic routes, traffic routes on which a mounting height of 12 m is aesthetically undesirable, and wide, heavily-travelled routes where a large number of intersections, bends, etc., lead to short spacings which make the use of a 12 m mounting height uneconomical. Group A12 has a mounting height of 12 m and is used to light wide, heavily-travelled routes with comparatively few intersections, bends, hills, etc., where proper advantage can be taken of a longer spacing between lanterns.

Lighting of subsidiary routes Roads falling into this group can be functionally described as being either distributor or access routes. The importance of environmental considerations in relation to the design of lighting for minor roads is given a particular emphasis in this grouping.

Lighting for single-level road junctions, including roundabouts Into this category of lighting fall single-level T-junctions (square and oblique), cross-roads, symmetrical Y-junctions, intersections on road bends, channelized intersections, and roundabouts.

Lighting for grade-separated junctions Interchanges and flyovers are dealt with in this group.

Lighting for bridges and elevated roads Lighting systems for highway bridges, footbridges, bridges of historical interest, and elevated viaducts and roads fall into this grouping.

Lighting for underpasses and bridged roads This category includes lighting for all types of covered street and highway, including underpasses, short tunnels, and roads below overbridges. It excludes long tunnels.

Lighting for roads with special requirements Lighting systems in the vicinity of aerodromes, railways, docks and navigable waterways fall into this group.

Lighting for town and city centres and areas of civic importance This group deals with urban areas carrying vehicular traffic, pedestrian precincts, public car parks, and pedestrian subways and stairways.

As can be seen, the definitions of a number of these groups are rather vague, and this is deliberately so. The placing of a roadway into any group is left to the discretion of the highway authority concerned, as it is generally considered that an outside body could not make the correct decision without having an intimate knowledge of particular and local circumstances.

It might be noted here also that the Standard does not consider it practicable to lay down levels of luminance, uniformity, visibility, etc., because the lighting designer usually has neither the data on which to compute them nor the means of ensuring their permanence. Instead, the designer has to select a lamp and lantern and then decide on the geometry of the installation. The recommendations in the Standard are therefore in terms of the desired amount of distribution of light and the geometry of the installation.

Photometric terms

The following definitions have been made as simple as possible, rather than absolutely precise, and the units and abbreviations are those used in Britain.

Luminous flux   This is the light given by a light source or received by a surface irrespective of the directions in which it is distributed. The unit of luminous flux is the lumen (lm); this is the flux emitted through a unit solid angle (a steradian) from a uniform point source of one candela. A steradian is equal to the solid angle subtended at the centre of a sphere by a unit area of its surface.

Luminous intensity   This describes the light-giving power (candlepower) of a lantern in any given direction. The unit of luminous intensity is the candela (cd); this is an international standard and is related to the luminous intensity of a 'black body' (one which absorbs all radiation incident on it) at the temperature of freezing of platinum.

Lower hemispherical flux   By this is meant the luminous flux emitted by a light source in all directions below the horizontal.

Illumination   This is the luminous flux *incident* on a surface per unit area. The unit of illumination is the lumen per square metre (lm/m^2) or lux.

Luminance   The term luminance is now used instead of 'brightness' to describe the rate at which light is reflected from a unit projected area of an illuminated surface in a given direction; it is the luminous intensity per unit projected area of the surface. (Whereas illumination is a measure of the amount of light flux falling on a surface, luminance is a measure of the amount of light which the area reflects towards the eye of the observer. In other words, the luminance varies not only with the amount of light reaching the surface but also in the manner in which it is reflected.) The usual unit is the candela per square metre (cd/m^2). Thus, if a very small portion of an illuminated surface has an intensity I cd in a particular direction, and if the projection of the surface on a plane perpendicular to the given direction has an area $D\,m^2$, then the luminance in this direction is $I/D\,cd/m^2$.

Luminosity   Sometimes loosely called 'brightness', this is the visual *sensation* indicating that an area appears to emit more or less light; it correlates approximately with the term luminance. It is not measurable.

Mean hemispherical intensity   This is the downward luminous flux divided by 2π; it is the average intensity in the lower hemisphere.

Peak intensity ratio   The ratio of the maximum luminous intensity to the mean hemispherical intensity is called the peak intensity ratio.

Beam This is the portion of the luminous flux emitted by a lantern which is contained by the solid angle subtended at the effective light centre of the lantern containing the maximum intensity but no intensity which is less than 90 per cent of the maximum intensity.

Installation terms

Lighting installation This term refers to the entire equipment provided for lighting a highway section. It comprises the lamps, lanterns, means of support and the electrical and other auxiliaries.

Lighting system This is an array of lanterns having a characteristic light distribution. Systems are commonly designated by the name of the light distribution, i.e. cut-off, semi-cut-off, and non-cut-off.

Lamp The bulb or light source is called the lamp.

Lantern The lantern consists of the lamp together with its housing and such features as refractors, reflectors and diffusers which are integral with the lamp and housing.

Outreach This is the horizontal distance measured between the centre of a lantern mounted on a bracket and the centre of the supporting column or wall face.

Overhang The overhang is the horizontal distance between the centre of a lantern and the adjacent edge of the carriageway.

Mounting height This is the vertical distance between the surface of the carriageway and the centre of the lantern.

Spacing By this is meant the distance measured parallel to the centreline of the carriageway between successive lanterns in an installation. The successive lanterns may or may not be arranged on the same side of the carriageway.

Arrangement The pattern according to which lanterns are sited on plan is termed the arrangement. In a *staggered* arrangement, the lanterns are located alternately on either side of the carriageway. When the lanterns are placed on either side but opposite each other, the pattern is called an *opposite* arrangement. With a *central* arrangement, the lanterns are sited in an axial line close to the centre of the carriageway. A *single-side* arrangement is one in which the lanterns are placed on one side only of a carriageway (see Fig. 8.1).

Width of carriageway This is the distance between kerblines as measured in a direction at right-angles to the length of the carriageway.

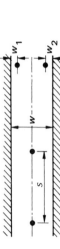
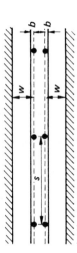
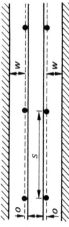
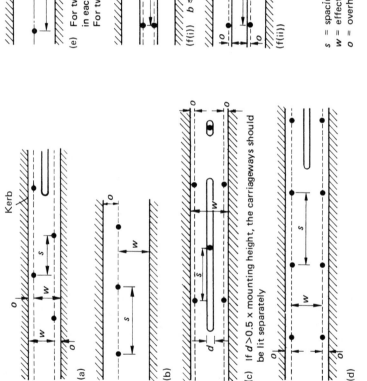

(a)

Kerb

(b)

(c) If $d > 0.5 \times$ mounting height, the carriageways should be lit separately

(d)

(e) For twin central lanterns, the distance between lanterns in each pair should not exceed 0.5 × mounting height. For twin central lanterns overhung, $w = w_1 + w_2$

(f(i)) $b \leqslant 0.25 \times$ mounting height

(f(ii))

s = spacing
w = effective width
o = overhang

Fig. 8.1 Lantern mounting arrangements: (a) staggered, (b) single-side, (c) opposite, (d) opposite plus central, (e) central on single carriageway, and (f) twin central on dual carriageway: (i) lanterns mounted behind the kerb of the central reservation and (ii) lanterns mounted beyond the kerb of the central reservation

Effective width of carriageway   This is the width of carriageway which it is intended to make bright. The relation between effective width and carriageway width for a single carriageway is given in Table 8.1.

Table 8.1 Effective widths of carriageway associated with various lighting arrangements

Arrangement	Effective width
Central	Carriageway width
Opposite plus central	Carriageway width
Staggered	Carriageway width minus overhang
Single-side	Carriageway width minus overhang
Opposite	Carriageway width minus twice overhang

Span   The part of the highway which lies between successive lanterns is called the span.

Geometry   The geometry of a lighting system refers to the interrelated linear dimensions and characteristics of the system, i.e. the spacing, mounting height, effective width, overhang and arrangement.

Basic concepts and considerations

Objectives

Virtually all of the rural highway system, and a significant proportion of the urban/suburban road system, depends upon vehicle headlights for illumination. On many of these roadways, this form of illumination is quite adequate; on others, especially on urban streets where there is a need to promote the safe and efficient movement of pedestrians as well as vehicles, the road user must be provided with the means by which additional visual information can be gathered. The objectives in providing fixed roadway lighting can therefore be summarized as follows[4]:

(1) to supplement vehicle headlights, extending the visibility range beyond their limits both laterally and longitudinally,
(2) to improve the visibility of roadway features and objects on or near the roadway,
(3) to delineate the roadway ahead,
(4) to provide visibility of the environment,
(5) to reduce the apprehension of those using the roadway.

Means of discernment

From the above, it can be seen that the basic objective of road lighting is different from that of interior lighting. With interior lighting, the aim is to reproduce daylight as closely as possible so that the forms and textures of objects are clearly seen. Thus interior lighting requires not only high but also even illumination. In road lighting, these detailed qualities are unimportant; what is important is that the motorist should be able to discern clearly the presence and movements of any object on or adjacent to the roadway which may be a potential hazard. This is achieved not by having an even illumination on the road and its surroundings but by appearing to have an even luminance on the road surface *as it is seen by the motorist.*

An object is visible to a driver if there is sufficient contrast of luminosity or colour between the object and its background or between different parts of the object. This implies that discernment is by either silhouette, reverse silhouette or surface detail.

When an unilluminated object on the carriageway is discerned by a driver because its luminosity is less than its background, then it is said to be seen by silhouette. It might perhaps be expected that road lanterns would make dark objects bright—as do vehicle headlights—but in fact the usual effect is to make the road surface more luminous and the object is seen in dark silhouette. It is only occasionally that the luminosity of the object is greater than that of the surface, and then visibility is by means of reverse silhouette. This occurs, for example, when a pedestrian in white clothing stands just behind a lantern and receives a very high amount of illumination on the side which faces the oncoming driver. When a high amount of illumination is directed on the side of an object facing the motorist, but discernment is by means of variations in luminosity within the object itself rather than in contrast with its background, then discernment is said to be by surface detail. This is the principle underlying the design of illuminated roadside traffic signs[5].

Light distribution on the carriageway

When designing a lighting system on inter-junction sections of a major traffic route, the visibility criterion is that of discernment by silhouette. In Britain, the general aim with respect to lighting main roads in urban areas is to provide a sufficient contrast between the object and the carriageway so that the results in most situations are at least adequate for safe driving at about 50 km/h without headlights. On residential and other 'non-traffic' routes, the motorist is expected to use headlights to help to achieve the desirable level of visibility; on these roads, the lighting is intended to suit the needs of the pedestrian rather than the needs of the motorist.

If the desired result is to be obtained on the major traffic routes, then the power and geometry of a lighting system must be such that it appears to the driver that the carriageway has high luminosity.

The bright patch

Each lantern in a lighting installation contributes a single 'bright patch' on the carriageway and, in the ideal case, the lanterns are sited so that the patches link up to cover the entire road surface. In general, the shape and luminance of this bright area depend upon the following six main factors:

(1) the reflection properties of the surface,
(2) the distribution of light from the lantern,
(3) the power of the lantern,
(4) the height of the light source,
(5) the distance of the light source from the observer,
(6) the height of the observer.

It is easiest to compare the reflection characteristics of carriageways by considering two extreme types of 'surfacings'. Let these be a near-perfect diffuser and a mirror. Assuming that the light source has a uniform luminous intensity in all directions (which in practice it does not have), then the contours of equal illumination on the road are concentric circles. If the carriageway is very diffuse, then the ratio of luminance to illumination is a constant, the contours of equal luminance coincide with those of equal illumination, and the bright patch produced on the road is nearly a perfect circle. To the motorist, this circle appears as the narrow 'ellipse' on the road surface just beneath the lantern. On the mirror surface, however, the bright patch would appear simply as the image of the light source itself just in front of the vehicle.

In practice, no dry road surface gives either a mirror image or a perfect ellipse. Instead, as is indicated in Fig. 8.2, a bright area or 'patch' is obtained which is roughly T-shaped, with the head of the T stretching across the road but reaching only a short distance behind the lantern, and the tail extending towards the observer. This T-shape has a number of very interesting characteristics, as follows.

(1) The tail of the T always extends towards the observer, wherever he or she may be. This is due to the preferential reflection properties of the carriageway which only enable it to reflect the light back to the observer which reaches the surface in the direction *towards* him or her.
(2) If the carriageway adjacent to a light source is viewed from two different locations, then two different light patches will be seen.
(3) The point of maximum luminous intensity occurs neither in the head of the T nor at the end of the tail, but somewhere in between.
(4) The shape of the luminous area depends very much upon the reflection characteristics of the road surface. A rough-textured surface gives a patch in which the head predominates and the tail is very small; as the road surface becomes polished, or is more fine-textured, the head becomes less pronounced and the tail longer and more luminous. A smooth wet surface produces hardly any head but, instead, the tail becomes very luminous and is long and thin, e.g. the 'streaks' commonly noted on a flooded road surface.
(5) Since the tail is formed by light rays which leave the lantern near to the horizontal, then cutting off these light rays will shorten the length of the tail of the T.
(6) The lower the lantern-mounting height, the shorter will be the tail of the T and the greater will be the luminance of the central part of the patch. This also causes a reduction in the luminance of the darker regions of the road, since they are normally 'lit' by the edges of the T-shaped

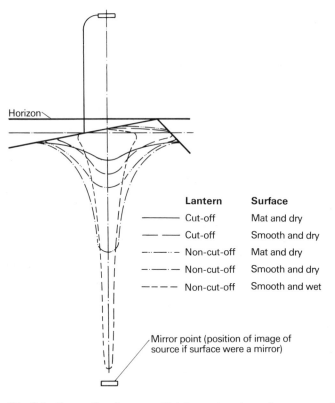

Lantern	Surface
—————— Cut-off	Mat and dry
— — — Cut-off	Smooth and dry
—··——··— Non-cut-off	Mat and dry
—·——·— Non-cut-off	Smooth and dry
— — — — Non-cut-off	Smooth and wet

Mirror point (position of image of source if surface were a mirror)

Fig. 8.2 Perspective diagrams of bright patches shown in contours of equal luminance formed on the carriageway by a single lantern, according to the type of its light distribution curve and the nature of the road surface (after reference 6)

patches; if the lanterns are very low, the centre of the carriageway may be so dark that silhouette discernment will be practically impossible.

(7) Light-coloured surfacings, e.g. concrete roads or bituminous ones containing light-coloured aggregates, produce higher carriageway luminances at lower levels of illumination than do dark-coloured ones. Indeed, luminance levels up to 30 per cent greater can be achieved from a given layout simply by selecting an appropriate light-coloured surfacing[7]. The lightness of colour affects mainly the size and brightness of the heads of the patches and has little effect upon the tails. Thus, in areas where good-quality, light-coloured aggregates are available, serious consideration should be given to their use either as the main aggregate in a surfacing or as a surface dressing, even though their usage might be slightly more expensive.

(8) The further the observer is from the light source, the more clearly defined is the shape of the bright patch.

(9) Within the limits of observers' heights which are generally met with in practice, the shape and luminosity of the bright patch are little affected by the height of the observer.

The glare problem

As can be gathered from the above discussion, the problem of lighting a major traffic road essentially reduces to selecting and locating lanterns so that luminous patches of suitable shape are formed on the roadway which link up in such a way that the complete surface appears well lit to the driver. A major factor influencing the design therefore is the manner in which the light is distributed from the lantern. This brings into prominence the problem of glare.

If the total output of light from a point source is allowed to radiate with uniform luminous intensity in all directions, then not only will much of the flux be wasted, e.g. that which goes upwards, but the amount which falls on the carriageway at any point will vary inversely with the square of the distance of that point from the light source. To overcome these disadvantages, the designers of road lanterns limit the directions in which the flux may be distributed and regulate the directional intensities in the vertical plane so as to obtain the desired luminosity on the roadway. To avoid flux wastage, lanterns are furnished with redirectional apparatus which receives the luminous flux emitted by the lamp and redistributes it in a desirable fashion. A bright patch having a particular luminosity is then obtained by varying the luminous intensity distribution so that the higher intensities are emitted at greater angles from the vertical.

The vertical angle at which the maximum luminous intensity is emitted may affect the safety of road users because of the glare effects to which the drivers of vehicles may be subjected. It is customary to describe two main types or effects of glare; these are *disability* glare and *discomfort* glare. Disability glare has been likened to the production of a veiling luminance or luminous fog over the whole visual field which effectively reduces the contrast between object and background, and therefore produces a loss of visual efficiency. Discomfort glare (also described as psychological glare) is the term used to describe the sensations of distraction and annoyance which are experienced when glare sources are present in the field of view.

The effects of glare sources decrease the further they are removed from the line of sight and, for this reason, lanterns are normally placed above the driver's line of vision. Since a driver cannot usually see more than about 20 degrees above the horizontal because of the cut-off effect of the roof of a vehicle, it means that when a vehicle is close to a lighting column the luminous intensities emitted at angles of less than about 70 degrees from the downward vertical cannot reach the driver's eyes and cause glare; there is therefore no necessity to limit the intensities below this angle. Glare is, however, caused by light leaving the lanterns within about 20 degrees of the horizontal and, in practice, the more lanterns that are visible to the driver, the greater is the glare effect. To minimize this problem, lighting engineers have designed lanterns that emit relatively low luminous intensities within this 20-degree glare-zone.

When the lanterns are designed so that there is a rapid reduction of

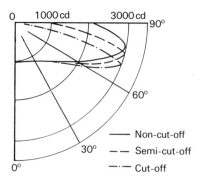

—— Non-cut-off

– – Semi-cut-off

—·— Cut-off

Fig. 8.3   Typical lighting distributions with equal maximum intensities; the polar curves of intensity are in a vertical plane approximately parallel to the axis of the road

luminous intensity between about 20 degrees below the horizontal and the horizontal, then they are said to belong to the *cut-off* system of lighting. If, however, there is a less severe reduction in intensity in the same region, the lanterns are characterized as belonging to the *semi-cut-off* system. If there is no reduction in intensity, then the lanterns are said to be *non-cut-off*. Figure 8.3 illustrates two typical examples of cut-off and semi-cut-off lighting distributions. For comparison purposes, an example of a non-cut-off distribution is also shown.

The relative advantages of the cut-off and semi-cut-off systems are as follows[3, 8]. The cut-off system produces less glare and gives a better performance on rough, harsh and light-coloured surfaces. Cut-off lanterns are better suited to central or opposite lighting arrangements because these lanterns require a small and nearly uniform spacing-to-height ratio. There are advantages in the use of cut-off lanterns at grade-separated intersections to reduce glare on adjacent carriageways. The semi-cut-off system allows a longer spacing between lanterns, a greater flexibility of siting of columns, and lower overall costs owing to the greater spacing-to-height ratios. It is also better suited to staggered arrangements; in addition, it gives a better performance with smoother surfaces and a better appearance to buildings.

Conditions which favour the cut-off system are:

(1)   rough and harsh carriageway surfaces,
(2)   an absence of buildings,
(3)   the presence of large trees,
(4)   long, straight sections of roadway,
(5)   slight humps and bridges,
(6)   few intersections and obstructions,
(7)   the proximity of railways, docks or airports.

Conditions which favour the semi-cut-off system are:

(1)   smooth carriageway surfaces,
(2)   the presence of buildings close to the carriageway, especially those of architectural interest,
(3)   many intersections and obstructions.

Lantern arrangements

Many types of lantern arrangement are used to light carriageways between intersections. Of those shown in Fig. 8.1, the ones in most common usage on single carriageways are single-side, staggered, and central plus opposite, with the choice in any given situation being primarily influenced by the width of the road. Overall, however, a staggered arrangement is generally considered to give the 'best' light distribution on most single carriageway roads. Central arrangements are generally considered most appropriate for dual carriageways.

Single carriageways

The cases for and against a central mounting arrangement that is suspended over the centre of the carriageway involve considerations of appearance, visibility and safety, as well as of erection and maintenance.

In favour of a central arrangement as compared with the side-mounted one is the fact that, for a given number of lanterns per kilometre, the central arrangement has half the spacing per row of lanterns. This means that the light patches overlap very well to produce a uniformly-illuminated roadway centre, the luminosity of which falls off smoothly to the sides. This results in an attractive appearance which makes the roadway very inviting to the motorist. It is, however, these very features which tend to cause the lighting engineer to hesitate about installing a central lighting arrangement. For instance, in wet weather, the driver of a vehicle is faced with a brilliant array of lamps in a near-direct line, which distracts his or her attention from more important regions of the roadway and makes them appear less bright. The luminous centre invites the driver to increase speed and to keep to the crown of the road. Most important of all, the sides of the carriageway are in comparative darkness, with the result that pedestrians or vehicles entering from side roads may be well on to the carriageway before the major road driver is aware of their presence. From the point of view of maintenance, central mounting has the additional disadvantage that the suspended lanterns can be difficult to service. On busy roads, the maintenance crew may be exposed to some danger, especially if they have to work at night; in daylight hours, they are a source of obstruction to traffic while servicing the lanterns.

On balance, therefore, suspended central arrangements are to be discouraged and it is only when special conditions justify them that they should be used, e.g. on a carriageway flanked with trees on either side or on a narrow road. When a central arrangement is utilized, it should be in conjunction with a semi-cut-off or cut-off lighting system, since both of these cause a reduction in the completeness of overlap of the light patches and hence the difference in luminance between the centre and the sides of the roadway is reduced.

Single-side mounting is also to be discouraged on wide single carriageways since it results in one side of the roadway being well lit, whilst the other side is in comparative darkness; with this arrangement, there is a tendency for drivers to veer towards the lit side, and the safety problems noted above again arise. In general, single-side mounting is most applicable to narrow one- and two-way streets where the carriageway width is less than 1.0–1.5 times the mounting height of the lantern[4].

The staggered arrangement, as a general rule, is used on streets of medium width, i.e. 1.5–2.0 mounting heights.

The opposite system, which is essentially two independent single-side systems, tends to be used where single carriageway streets are very wide (and where central reservations on dual carriageway roads are too wide to allow for the effective accommodation of a central arrangement). When an opposite arrangement is utilized, a very uniform carriageway luminosity can be obtained, albeit at the cost of increased expenditure for extra lanterns, etc.

Dual carriageways

From the viewpoints of economics and service, central lighting is the preferred arrangement for dual carriageways with an acceptable central reservation. Locating the lantern supports in the central reservation reduces the number of supports by one-half from that required for single-side arrangements; also, since there is only one row of supports, there is need for only one run of electrical conductor, and savings in material and construction costs are significant.

Situations where the use of a central lighting arrangement would not be appropriate include the following: (a) where the central reservation is to be used for public transport vehicles, (b) where there is no central crash barrier within which the lantern supports can be encased to ensure that the relative danger of fixed-object collisions is not increased, (c) where the two carriageways are on separate alignments or where the central reservation is too wide to allow both carriageways to be illuminated from one line of supports, and (d) on dual carriageways in urban areas where the highway is depressed and wall mounting may be more appropriate.

Intersections

The large variety of intersections which exist makes it difficult to prescribe standard arrangements which will fit all site and traffic conditions. However, some principles have been formulated[7] to guide the designer in the achievement of good design practice at junctions, and these are given below.

(1) Seeing by silhouette vision is unlikely to occur at intersections and hence the lamp arrangement should aim to illuminate conflict areas and 'objects' in and around them (e.g. pedestrians, cars, kerbed islands, and carriageway markings and signs) so that they are seen by direct vision.
(2) The level of illumination and its uniformity should be such that the layouts of islands and of the various carriageways and turning roadways are clearly discernible by drivers approaching the junction and negotiating the required movements within the junction.

(3) Points where traffic streams merge and diverge should be well illuminated.

(4) Lanterns should be placed to provide the best possible illumination of pedestrian crossing areas.

(5) The lamp arrangement as seen in perspective should provide route guidance to lead traffic through the junction.

(6) The number of lighting poles near the conflict area should be minimized. Where traffic signals are installed, joint sharing of the posts should be achieved wherever possible. Where large channelizing islands exist, consideration should be given to using high-mast floodlighting techniques to reduce the number of poles about the junction.

(7) Lighting poles should not be located: (a) close to the approach ends of narrow residual medians and median islands, (b) in the nose area of islands where traffic streams diverge, (c) in areas where the poles might obstruct the sight lines of drivers waiting to enter or cross another traffic stream, (d) in the vulnerable areas along the outside of curved 'turning roadways', and (e) at roundabouts in the small approach 'spitter' islands, on the central island opposite entry roadways, or on the left-hand side immediately downstream of an entry point to a roundabout.

Spacings

Most lanterns emit two main beams, one up and one down the roadway; thus the maximum intensities are emitted in vertical planes which are nearly parallel to the direction of the road. The spacing of the lanterns then depends upon the length of the bright patches produced by these beams, and the extent to which it is desired that they should overlap. For example, a typical cut-off lantern mounted 7.62 m above the road surface produces a patch which is only about 6 m long; at the other end of the scale, a non-cut-off lantern with a peak intensity at about 80 degrees, an intensity exceeding half the peak value at 86 degrees, and an appreciable intensity right up to the horizontal may produce light patches of nearly 46 m in length.

The basic dimension used in the design of highway lighting is the mounting height of the lantern. Provided that the spacing, overhang and effective width dimensions do not exceed certain values which are proportional to the mounting height, then the light output required from each lantern in the lower hemisphere in order to achieve a given level of illumination can be assumed to vary approximately as the square of the mounting height.

Table 8.2 gives empirical design-spacing relationships, derived on the basis of the above assumptions, for the arrangements shown in Fig. 8.1 for both cut-off and semi-cut-off lanterns. The mounting height, h, normally used in these relationships is either 10 or 12 m on major traffic routes. The relationships apply to lanterns with the following minimum light flux in the lower hemisphere: 12 000 lm at 10 m mounting height or 20 000 lm at 12 m mounting height. The spacings obtained from Table 8.2 are normally such as to give an installation of high quality, and good

Table 8.2 British practice in respect of design lantern spacings for cut-off and semi-cut-off systems (reproduced by permission of the British Standards Institution)[3]

Arrangement	Type*	Spacing		Maximum effective width (w_{max})
		s	s_{max}	
Staggered	CO	$2.2h^2/w$	$3.0h$	$1.5h$
	SCO	$3.6h^2/w$	$4.0h$	$1.4h$
Opposite	CO	$4.4h^2/w$	$3.3h$	$2.0h$
	SCO	$7.2h^2/w$	$4.4h$	$2.0h$
Single-side	CO	$1.9h^2/w$	$3.3h$	$0.7h$
	SCO	$2.5h^2/w$	$4.4h$	$0.6h$
Central on single carriageway	CO	$3.8h^2/w$	$3.3h$	$1.4h$
	SCO	$5.0h^2/w$	$4.4h$	$1.2h$
Opposite plus central	CO	$4.4h^2/w$	$3.0h$	$3.0h$
	SCO	$7.2h^2/w$	$4.0h$	$2.8h$
Twin central on dual carriageway	CO	$2.8h^2/w$	$3.3h$	$0.9h$
	SCO	$3.7h^2/w$	$4.4h$	$0.8h$

·*CO = cut-off; SCO = semi-cut-off

uniformity of brightness, on straight sections of road of moderate skid-resistance. On sharp bends and at intersections, the spacings will have to be closer (and the illumination greater) than on straight roads.

Outreach and overhang

Lanterns are very often located so as to overhang the traffic lanes. This presents a rather pleasing roadway appearance to the motorist, which is somewhat similar to that associated with a central arrangement of lanterns. On wide roadways with side lighting, overhanging the lanterns may be necessary in order to light the centre of the carriageway, which would otherwise appear unduly dark. It is not normally necessary to overhang carriageways less than about 9 m wide. The overhang distance should preferably not exceed about 1.8 m on columns up to 9 m high, especially on heavily-travelled roads in built-up areas, as otherwise the footpaths and kerbs may be in undesirable shadow. Another practical consideration limiting the amount of overhang is that the lantern should be easily accessible for maintenance purposes. As a general practice, therefore, the overhang should be as small as possible; indeed, there are advantages in locating the lanterns over or even behind the kerbs, provided that the effective width is not excessive.

The amount of outreach is governed by the extent of the overhang and by safety considerations affecting the location of the lighting columns. A lighting column close to the edge of the carriageway is a potential cause of accidents to vehicles which leave the roadway. Where there is a footpath close to the carriageway, the lighting columns should be located behind the footway. In some instances, it·may be possible to support the lanterns on wall brackets. Whatever the method of support, the normal practice in Britain is for the outreach not to exceed 2 m.

Lamps

The lamps considered suitable for highway lighting have varied considerably over the years, so that, in practice, street lighting systems in most urban areas tend to be formed of a mixture of light sources, each of which has its advantages and disadvantages. The light sources with which the traffic engineer usually becomes familiar include the tungsten filament lamp and the sodium vapour, fluorescent and mercury gaseous discharge lamps.

The *incandescent or tungsten lamp*, in which the illumination is provided by passing an electric current through a filament of tungsten in order to raise it to incandescence, is probably the light source that is most widely used in homes today. Whilst this was at one time the most commonly used lamp in highway lighting practice, its use is now generally confined to minor streets of a residential nature. Its major advantages are low capital cost, good colour and 'instant-on' characteristic. The major disadvantages associated with the tungsten lamp are its low light-producing efficiency (i.e. luminous efficacy measured in lumens per watt) and short life.

A *sodium vapour lamp* is, in principle, one in which an electrical discharge takes place in a vapour of this metal. It consists of a discharge tube filled with a mixture of neon and argon gases (in which the discharge takes place in order to start the lamps), and contains small drops of sodium distributed along the length of the tube. When heated by the discharge, the sodium forms a vapour and the end result is visible electromagnetic radiation or light. Two types of sodium vapour lamps are now in use, viz. low-pressure and high-pressure versions.

The low-pressure sodium lamp produces a very characteristic monochromatic light which makes all objects it illuminates appear more or less yellow, so that exact colour recognition is impossible; as such its usage has tended to be frowned upon in environmentally-sensitive areas. With the exception of the tubular fluorescent lamp, this sodium lamp is the least dazzling of all the types used for road lighting[9]. Visual acuity increases in monochromatic light owing to the elimination of the phenomenon of chromatic aberration, whilst the advantages of the yellow light with respect to vision in foggy weather are very obvious to the motorist. Of all the street light sources, the low-pressure sodium vapour lamp has the highest luminous efficacy (see Fig. 8.4). It also has a long rated life, this being the survival life of 50 per cent of a large group of lamps.

The City of London pioneered the widespread use of high-pressure sodium lamps as recently as 1967[10]. Whilst having a longer life than its low-pressure counterpart, the high-pressure sodium lamp produces a 'white light' that is much more acceptable in environmentally-sensitive areas. As well as having good lumen maintenance, this lamp also has an excellent luminous efficacy. Its capital cost is, however, relatively high.

The *fluorescent lamp* consists of a long tube, the inside of which is coated with fluorescent powder commonly called phosphor. When an

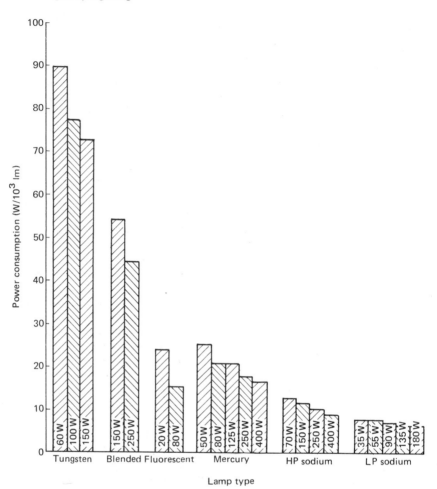

Fig. 8.4 Power consumption of various types of lamp used for street lighting

electrical discharge is caused in an inert gas in the tube, the fluorescent powder is excited and it produces a visible white light that is very pleasing in appearance. This light is particularly suited to locations where good colour appearance and rendering are important. Apart from its high cost, an important disadvantage is the bulk and weight of the lantern, which often adds to the structural requirements of the pole and mast arm, i.e. because of the relatively low output per lamp, at least two are normally used in each lantern. Also, the light output is noticeably affected by temperature changes, so that its usage is restricted in areas where wide variations are encountered.

The principle of the *mercury vapour lamp* (now becoming obsolete for street lighting) is similar to that of the sodium vapour lamp except that mercury takes the place of sodium. The most common type of mercury lamp uses a clear bulb, and the light emitted from this lamp has a characteristic bluish-white colour; although this light makes the surroundings more pleasing to the eye than does the yellow of the low-pressure sodium lamp, it tends to emphasize blue, green and yellow colours, whilst orange and red colours appear brownish. The desire to improve the colour rendition led to the development of the *high-pressure mercury fluorescent lamp* with a phosphor coating on an internal wall of each bulb. This coating fluoresces when excited by the mercury discharge, and the net result is an off-white light which gives a much more pleasing effect than does the bluish-white of the clear lamp. Its luminous efficacy is normally moderate.

The choice of lamps for street lighting involves many considerations including such factors as[11]: (a) life, (b) lumen maintenance, (c) lumen efficacy, (d) capital costs, (e) annual operating costs, (f) colour appearance, (g) colour rendering qualities, (h) reliability, and (i) available lamp ratings. However, in recent years, particular attention has been paid to annual operating costs, as energy costs have risen considerably (e.g. between April 1974 and June 1975, electricity supply charges for street lighting in the County of Humberside rose by 79 per cent[12], whereas the rise had been only 12.5 per cent during the previous twenty years). As a result, the low-efficacy tungsten, tungsten–mercury blended, and fluorescent lamps are now rarely installed on new sections of roadway, whilst high-pressure mercury (MBF) lamps are also less favoured than previously. Low-pressure sodium vapour (SOX) lighting is the most efficient source of public lighting available at the present time, yielding some 200 lumens per watt of energy consumed; it tends to be used most frequently on motorways, trunk roads and high-speed traffic routes in rural areas, i.e. on highways on which high efficacies and luminance levels are required and good colour discrimination is not so important. However, where good colour rendering is important, e.g. on prestige routes, in tourist areas, and in historical towns, the high-pressure sodium vapour (SON) lamp—which has an efficacy currently in excess of 120 lm/W, and its technology is improving—is more suitable.

Before-and-after studies have compared the effects of different types of road lighting upon accidents[13]. These investigations have shown that there are no significant differences between the accident reductions resulting from the installation of good fluorescent, mercury, and sodium lighting systems.

Selected bibliography

(1) Zuman N, The case for improved street lighting, *The Highway Engineer*, 1980, **27,** No. 2, pp. 8–11.

(2) Jobson AJ, *Freeway Lighting Literature Survey*, Internal Report RE/6/73, Pretoria, SA, The National Institute for Road Research, 1973.

(3) BS 5489, London, The British Standards Institution, 1973. (Complete copies can be obtained from BSI at Linford Wood, Milton Keynes, MK14 6LE.)

(4) Walton NE (Ed.), *Roadway Lighting Handbook*, FHA Implementation

Package 78-15, Washington DC, US Government Printing Office, December 1978.

(5) Reid JA, The lighting of traffic signs and associated traffic control devices, *Public Lighting*, 1964, **29,** No. 127, pp. 252-264.

(6) Waldram JM, International recommendations for public thoroughfares, *Traffic Engineering & Control*, 1966, **7,** No. 12, pp. 753 and 755.

(7) Barton EV, Road and traffic engineering considerations, in Hall RR (Ed.), *The Design of Fixed Lighting for Arterial Roads and Freeways*, ARR Research Report No. 106, pp. 63-75, Melbourne, Victoria, The Australian Road Research Board, February 1980.

(8) Road lighting: state of the art report, *Municipal Engineer*, 1984, **1,** No. 2, pp. 171-183.

(9) Cohu M, Public lighting by sodium vapour lamps, *Traffic Engineering & Control*, 1962, **4,** No. 3, pp. 169-177.

(10) Austin BR, Lamp life performance of high-pressure sodium lamps in the City of London, *Traffic Engineering & Control*, 1976, **17,** No. 8/9, pp. 364-365.

(11) Hawkins MR, Street lighting—some county problems, *Chartered Municipal Engineer*, 1979, **106,** No. 4, pp. 130-132.

(12) Roberts G, A policy for more efficient street lighting, *Chartered Municipal Engineer*, 1979, **106,** No. 5, pp. 151-153 and 156-158.

(13) Cleveland DE, Chapter 3: Illumination, in Mayer PA (Ed.), *Traffic Control and Roadway Elements—Their Relationship to Highway Safety*, Washington DC, The Automotive Safety Foundation, 1969.

Index